FLOW THROUGH POROUS MEDIA

CONTRIBUTORS

JACOB BEAR
STANLEY N. DAVIS
G. DE JOSSELIN DE JONG
ROGER J. M. DE WIEST
DANIEL DICKER
H. J. MOREL-SEYTOUX
RALPH R. RUMER, JR.
E. S. SIMPSON
DALE SWARTZENDRUBER
ARNOLD VERRUIJT

FLOW THROUGH POROUS MEDIA

Edited by

Roger J. M. De Wiest

PROFESSOR
PRINCETON UNIVERSITY
PRINCETON, NEW JERSEY

ACADEMIC PRESS New York and London 1969

ACADEMIC PRESS, INC.
111 Fifth Avenue, New York, New York 10003

United Kingdom Edition published by
ACADEMIC PRESS, INC. (LONDON) LTD.
Berkeley Square House, London W1X6BA

LIBRARY OF CONGRESS CATALOG CARD NUMBER: 68-59161

PRINTED IN THE UNITED STATES OF AMERICA

MORRIS MUSKAT

Morris Muskat, a member of the Advisory Council of Civil and Geological Engineering at Princeton University, has been admirable in many respects. Having had his share of criticism and misunderstanding, which beset all scientific writers, he was an example of tolerance. Furthermore, having made his contributions while very young, he retired at an early age, leaving his opponents to be presented with the wooden dagger, symbolic of the compulsory retirement of a Roman gladiator.

For these reasons, on the sixty-fifth anniversary of his birth, we are pleased to dedicate this volume to him.

CONTENTS

Chapter 8. **Elastic Storage of Aquifers**

Arnold Verruijt

Chapter 9. **Generating Functions in the Theory of Flow through Porous Media**

G. de Josselin de Jong

Chapter 10. **Green's Functions in the Flow through Porous Media**

Roger J. M. De Wiest

Chapter 11. **Introduction to Flow of Immiscible Liquids in Porous Media**

H. J. Morel-Seytoux

LIST OF CONTRIBUTORS

Numbers in parentheses indicate the pages on which the authors' contributions begin.

JACOB BEAR (109), Department of Civil Engineering, Technion, Israel Institute of Technology, Haifa, Israel

STANLEY N. DAVIS (53), Department of Geology, University of Missouri, Columbia, Missouri

G. DE JOSSELIN DE JONG (377), Civil Engineering Department, Delft University of Technology, Delft, Netherlands

ROGER J. M. DE WIEST (1, 401), Department of Civil and Geological Engineering, Princeton University, Princeton, New Jersey

DANIEL DICKER (293), College of Engineering, State University of New York at Stony Brook, Stony Brook, New York

H. J. MOREL-SEYTOUX (455), Department of Civil Engineering, Colorado State University, Fort Collins, Colorado

RALPH R. RUMER, JR. (91), Faculty of Engineering and Applied Sciences, State University of New York at Buffalo, Buffalo, New York

E. S. SIMPSON (201), Hydrology and Water Resources Office, University of Arizona, Tucson, Arizona

DALE SWARTZENDRUBER (215), Department of Agronomy, Purdue University, Lafayette, Indiana

ARNOLD VERRUIJT (331), Department of Civil Engineering, Delft University of Technology, Delft, Netherlands

PREFACE

The National Science Foundation–sponsored Institute in Hydrology for College Teachers, which was held at Princeton University in the summer of 1965 under my direction, provided an excellent opportunity to bring together many people actively interested in the flow of fluids in the ground. Although most were college teachers, there were also participants from the petroleum industry and from federal agencies engaged in research related to groundwater.

Lectures covering topics in groundwater and surface water hydrology and in hydrogeology and open channel flow presented ample opportunities for lively discussions. It was only natural, therefore, that the participants of the institute expressed the desire that a book be published which would make available the material covered and updated by the various contributors.

Since it was clear that the material on subterranean fluid flow would far outweigh the notes on surface water hydrology and open channel flow, the latter material was dropped from the volume for uniformity. This book lends itself to use by graduate students, either as a textbook or as a supplement to assigned reading. I have successfully tested the various chapters at Princeton over a number of years and I find that the student who masters the material of this book—and it is not impossible for a first-year or second-year graduate student to do so—is well equipped for independent work.

I have dedicated this volume to Morris Muskat, who is considered by many to be the "father of the flow through porous media." Through his numerous research papers, and his monumental books on flow through porous media (1937) and oil production (1949), which have been translated into Russian, he has secured a place for his favorite research subject in the general domain of mathematical physics.

Roger J. M. De Wiest

March 1969
Princeton, New Jersey

1 Fundamental Principles of Ground-Water Flow

Roger J. M. De Wiest
Department of Civil and Geological Engineering
Princeton University
Princeton, New Jersey

1. Darcy's Law

Darcy's law of laminar flow of homogeneous fluids in homogeneous porous media may be derived by considering the equilibrium of the forces acting on the fluid flowing within a macroscopic elementary volume, as demonstrated in Chapter 3 (see also Collins, 1961).

Ground water moves from levels of higher energy to levels of lower energy, whereby its energy is essentially the result of elevation and pressure. Kinetic energy, proportional to the square of the velocity, is neglected because ground-water velocities are very small, at least in laminar flow. While flowing, ground water experiences a loss in energy due to friction against the walls of the granular medium along its seepage path. This loss per unit length of distance traveled, or hydraulic gradient, is simply proportional to the velocity of ground water for laminar flow in aquifers or seepage through earth embankments. When the proportionality of hydraulic gradient and ground-water velocity is expressed by a mathematical equation, a linear law of flow, called *Darcy's law*, arises.

The similarity between ground-water flow and laminar pipe flow was recognized by Darcy (1856) and Dupuit (1863). Experiments with pipe flow were first conducted by Hagen in 1839 and almost simultaneously by Poiseuille in 1841, after whom the law for laminar flow in pipes was named. Darcy also was oriented toward experimental work, and in particular he investigated the friction factor of the pipe flow formulas. In 1856 Darcy conducted an experiment on a vertical pipe of cross-sectional area A filled with sand, under conditions simulated by Fig. 1 (Hubbert, 1953). From his investigations of the flow through horizontally stratified beds of sand, Darcy concluded that the flow rate Q was proportional to the energy loss, inversely

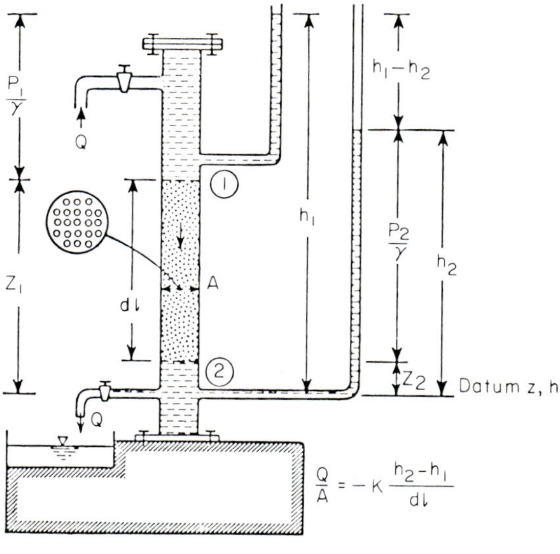

FIG. 1. Apparatus to demonstrate Darcy's law (after Hubbert, 1953).

proportional to the length of the flow path, and proportional to a coefficient K, depending on the nature of the sand. (Both Darcy and Dupuit failed to recognize the fact that K depends on properties of the fluid as well as on characteristics of the medium.) Darcy's law may be expressed as

$$Q = KA[(h_1 - h_2)/dl] = -KA\, dh/dl \tag{1}$$

in which

$$h = z + (p/\gamma) + \text{arbitrary constant} \tag{2}$$

and

$$K = k(\gamma/\mu) \tag{3}$$

where K is the hydraulic conductivity. In these equations h is the energy per unit weight of fluid or hydraulic head in the case of water, z is the elevation above an arbitrary datum plane, p is the pressure sustained by the fluid in the pores of the medium, γ is the specific weight of the fluid, μ is the dynamic viscosity of the fluid, and k is the intrinsic permeability of the medium. When the fluid is water, the hydraulic gradient $\mathfrak{I}$ is defined as

$$\mathfrak{I} = dh/dl \tag{4}$$

The subscripts 1 and 2 in Eq. (1) refer to the value of h at elevations z_1 and z_2.

Ground-water flow could be treated microscopically by the laws of hydrodynamics if the granular skeleton of the porous medium were a simple geometrical assembly of prismatic, unconnected tubes. The seepage path, far from being a prismatic channel, however, is tortuous, branching into a multitude of tributaries. In its original form, Darcy's law avoids the insurmountable difficulties of the hydrodynamic microscopic picture by introducing a twofold averaging macroscopic concept.

(1) It considers a fictitious flow velocity, the Darcy velocity or specific discharge q, through a given crosssection of a porous medium rather than the true velocity between the grains, as is evident from Eq. (1):

$$q = Q/A = -K\, dh/dl \qquad \text{or} \qquad q = -K\mathfrak{I} \tag{5}$$

(2) It considers average hydraulic values rather than local hydrodynamic values of this velocity.

These simplifying concepts were introduced by the nature of Darcy's experiment, which only permitted the measurement of average hydraulic values in the sand-filled cylindrical pipe.

In 1857 Dupuit's (De Wiest, 1965c) familiarity with pipe flow and open-channel flow led to his derivation of Eq. (5) in the following way. Dupuit used the equation for uniform flow in open channel

$$\mathfrak{J} = (P/A)(\alpha \underset{\sim}{Q} + \beta \underset{\sim}{Q}^2) \tag{6}$$

in which $\mathfrak{J}$ is the hydraulic gradient, $\underset{\sim}{Q}$ is the average velocity, P is the wetted perimeter of the cross-sectional area A, and α and β are coefficients. He then assumed that the channel was filled with sand and, therefore, that the water would flow much slower than in an ordinary open channel. In fact, he pointed out, water would flow through the grain interstices as through an infinite number of tiny parallel pipes, and if the sand were homogeneous, Eq. (6) would be applicable with $P\alpha = -1/K$, characteristic constant of the medium. Furthermore the term in $\underset{\sim}{Q}^2$ would be neglected because of the small velocities of ground-water flow. Thus Dupuit arrived at Eq. (5) and then described Darcy's experiment of the previous year (1856) in support of his formula. Dupuit, 1863 stated: "The fundamental equation $q = -K\mathfrak{J}$ may therefore be considered as an experimental result. If we have derived it first from the equations of uniform flow, it is only to indicate the similarity which exists between the two kinds of flow."

Darcy's law in vectorial form may be written as

$$\mathbf{q} = -K \operatorname{grad} h \tag{7}$$

If K is a constant, i.e., if the medium is homogeneous and isotropic, and if the fluid is of constant density and viscosity, Eq. (7) becomes

$$\mathbf{q} = -\operatorname{grad} \Phi \tag{8}$$

in which

$$\Phi = Kh = K(z + p/\gamma) + \text{constant} \tag{9}$$

is called the velocity potential, and has the dimensions $[L]^2/[T]$. It is well known that Φ may be combined with a streamfunction Ψ to form the complex potential function $W = \Phi + i\Psi$. In two-dimensional problems, aquifers and porous media of unit width (measured perpendicularly into the plane of the paper) are commonly treated. For these problems Ψ may be expressed in discharge units per unit width.

The velocity potential concept is very useful for the solutions of problems in which K, as a first approximation, is treated as a constant, as is the case in most simplified civil engineering applications. Its continued use may be defended for this reason alone. In general, however, the theory of ground-water flow is more complete when the specific discharge is derived via the gradient of a force potential, as will be shown later. Again it is emphasized that Eq. (8) is valid only for constant K and therefore is not a correct expression of Darcy's law in general.

Darcy's law states that macroscopically, the Darcy velocity or specific discharge $\mathbf{q}$ is equal to the negative gradient of Kh when K is constant. The true velocity in the pores of the medium is $\mathbf{q}/n$, in which n is the porosity of the medium. Several researchers (Hubbert, 1956) have derived it from the general Navier–Stokes equations for viscous flow. Such derivations invariably stem from statistical considerations and simplifications of the complicated microscopic flow picture. Although they do not contribute to the formulation of a new law, they confirm the early belief that Darcy's law is of the nature of a statistical result giving the empirical equivalent of the Navier–Stokes equations. A simple proof (De Wiest, 1965a), more heuristic than rigorous, follows.

Consider the Navier–Stokes equations in two-dimensional form:

$$(1/n)(\partial u/\partial t) + (u/n^2)(\partial u/\partial x) + (v/n^2)(\partial u/\partial z) = -(1/\rho)(\partial p/\partial x) + (\mu/\rho n)\,\nabla^2 u \tag{10}$$

and

$$(1/n)(\partial v/\partial t) + (u/n^2)(\partial v/\partial x) + (v/n^2)(\partial v/\partial z) = -(1/\rho)(\partial p/\partial z) + (\mu/\rho n)\,\nabla^2 v - g \tag{11}$$

In Eqs. (10) and (11), u/n and v/n are the components of the true velocity $\mathbf{q}/n$. The statistical averages may essentially be summarized by the following assumptions, dimensionally correct,

$$\nabla^2 u = -(1/c)(u/L^2), \qquad \nabla^2 v = -(1/c)(v/L^2) \tag{12}$$

where L may be thought of as being a characteristic length (for example, the pore size d of the medium), and where c is a dimensionless parameter. If it is further assumed that

$$u = -\partial\Phi/\partial x, \qquad v = -\partial\Phi/\partial z \tag{13}$$

the Navier–Stokes equations become

$$\begin{aligned} -(1/n)\,\partial/\partial x\,(\partial\Phi/\partial t) + (1/n^2)\,\partial/\partial x\,[\tfrac{1}{2}(\partial\Phi/\partial x)^2 + \tfrac{1}{2}(\partial\Phi/\partial z)^2] &= -(1/\rho)(\partial p/\partial x) + (\mu/c\rho d^2 n)(\partial\Phi/\partial x) \\ -(1/n)\,\partial/\partial z\,(\partial\Phi/\partial t) + (1/n^2)\,\partial/\partial z[\tfrac{1}{2}(\partial\Phi/\partial x)^2 + \tfrac{1}{2}(\partial\Phi/\partial z)^2] &= -(1/\rho)(\partial p/\partial z) + (\mu/c\rho d^2 n)(\partial\Phi/\partial z) - g \end{aligned} \tag{14}$$

If μ and ρ are constant, both equations may be integrated directly and their integrals should be the same, i.e.,

$$-(1/n)(\partial\Phi/\partial t) + n^2\tfrac{1}{2}[(\partial\Phi/\partial x)^2 + (\partial\Phi/\partial z)^2] \\ + p/\rho - \mu\Phi/c\rho n d^2 + gz = F(t) \tag{15}$$

For steady flow, provided the inertia terms here represented by the kinetic energy of the liquid may be neglected, Eq. (15) reduces to

$$\Phi = (k\gamma/\mu)(p/\gamma + z) + \text{constant} \tag{16}$$

in which

$$k = cnd^2 \tag{17}$$

Equation (16) is identical with Eq. (9) and therefore Darcy's law for constant K has been derived. Equation (17) shows that the intrinsic permeability k is only a function of the properties of the medium, that it is proportional to $[L]^2$, and that it depends also upon a dimensionless shape factor, here represented by the product cn.

2. Hubbert's Force Potential and Darcy's Law

More physical insight into fluid flow in porous media, especially as related to petroleum engineering, may be gained by the introduction of a force potential Φ^* (Hubbert, 1940) as follows:

$$\Phi^* = g\int_{z_0}^{z} dz + \int_{p_0}^{p} dp/\rho(p) \tag{18}$$

where z_0 and p_0 are elevation and pressure at a point of a given datum plane in the porous medium, z and p are the elevation and fluid pressure at the point where the potential is to be determined, and ρ is the density of the fluid, restricted to be a function of p only. The restriction on ρ is actually required to make the integral

$$\int_{s_0}^{s} (\nabla p/\rho) \cdot d\mathbf{s}' \tag{19}$$

independent of the path of integration. Indeed, this integral is independent of the path when $(1/\rho)\ \nabla p = \nabla A$, where A is a scalar function. Since $\nabla \times (\nabla A) = 0$, the independence of the path requires $\nabla\rho \times \nabla p = 0$, which in turn requires $\rho = \text{constant}$ or $\rho = \rho(p)$, a function of p only.

If the gradient of Eq. (18) is taken, the result is

$$-\nabla\Phi^* = \mathbf{g} - (1/\rho)\,\nabla p \tag{20}$$

Equation (20) is a force equation per unit mass of fluid. The left side of this equation states that the unit mass of fluid is subject to a force oriented from surfaces of higher Φ^* to surfaces of lower Φ^* and perpendicular to these surfaces. The right side of the equation states that this force has a gravity component and a component caused by the gradient of the fluid pressure. In Eq. (20), only the direction of $\mathbf{g}$ is fixed, but the vectors $(1/\rho)\,\nabla p$ and grad Φ^* may have arbitrary directions.

The magnitude of $-\nabla\Phi^*$ in most cases of ground-water flow is only a small fraction of $\mathbf{g}$ so that $-(1/\rho)\,\nabla p$ becomes nearly vertical, which means that the surfaces of constant pressure are nearly horizontal. Only in the special case $(1/\rho)\,\nabla p = \mathbf{g}$ do we find that $\nabla\Phi^* = 0$, and this means that there is hydrostatic equilibrium. The vectors $-\nabla p$ generate a vector field everywhere normal to the isobaric surfaces $p =$ constant. The vectors $-\nabla\Phi^*$ generate a family of field lines or a vector field everywhere normal to the equipotential surfaces $\Phi^* =$ constant. How may the latter vector field be related to the streamlines? Let it be recalled that a streamline is a line tangent in each of its points to the velocity vector of a particle inserted at these points. The relationship asked for is Darcy's law, which may be written (Hubbert, 1953) as

$$\mathbf{q} = -(K/g)\,\nabla\Phi^* \tag{21}$$

If K is constant, as for a homogeneous and isotropic medium and for water, then $\mathbf{q}$ will have the same direction as $-\nabla\Phi^*$, i.e., streamlines are normal to surfaces or lines of $\Phi^* =$ constant. But for anisotropic media, where K has preferential properties in given directions, flow lines will in general be oblique to the direction of $-\nabla\Phi^*$ and hence, not normal to equipotential surfaces (lines).

In the past, attempts have been made to relax the restrictions put on Hubbert's potential. In particular, the existence of a generalized potential Φ^G as an extension of Hubbert's force potential Φ^* was advanced.

It was proposed to extend the concept of potential as defined above and to apply it to the flow of saline groundwater, with variable salt concentration $c(x, y, z, t)$ and variable temperature $T(x, y, z, t)$, where x, y, z are coordinates of space and t is the time variable. Therefore the density would vary as

$$\rho = \rho[p, c(x, y, z, t), T(x, y, z, t)] \tag{22}$$

and the proposed Φ^G takes the form

$$\Phi^G = g \int_{z_0}^{z} d_z + \int_{p_0}^{p(x,y,z,t)} \frac{dp}{\rho[p, c(x, y, z, t), T(x, y, z, t)]} \tag{23}$$

The second term of Eq. (23) may be rewritten as

$$\Phi^{G'} = \int_{p_0}^{p} \frac{dk}{\rho(k, x, y, z, t)} \tag{24}$$

where k is a dummy variable of integration. The question arises whether Φ^G, which has the dimensions of mechanical energy per unit mass, represents a true potential. It cannot be argued that $\Phi^{G'}$ is indeterminate, but it can be stated that Eq. (24), which must be the transformation of a line integral (work of a force along a certain path, in this case), does not reflect the variation of ρ as a function of x, y, z along the path of integration. Equation (24) is therefore, in my opinion, an incomplete expression of

$$\int_{s_0}^{s} \left[\frac{\nabla p}{\rho(p, x, y, z, t)}\right] \cdot d\mathbf{s}' \tag{25}$$

in which [Eq. (24)] the variation of ρ along the path of integration has been neglected. To examine whether $\Phi^{G'}$ can be a potential, one has to look at the line integral of Eq. (25). Because in this case $\nabla\rho \times \nabla p \neq 0$, there is no potential.

The above argument is based on the fact that in flow through porous media, when inertia forces are neglected, the only acting forces on a unit mass of fluid are $\mathbf{g}$ and $(1/\rho)\,\nabla p$, whether ρ is constant, a function of p only, or a function of x, y, z as well as of p. These are the only forces that occur when Darcy's law is derived either from the Navier–Stokes equations or from a generalized approach of dispersion in porous mediums based on considerations of transport phenomena (Bird *et al.*, 1960).

It has sometimes been speculated that there would be another force present, apart from those mentioned above, due to variations in density. This speculation arose from the application of Leibnitz's rule for differentiation of an integral to Eq. (24) as follows:

$$\nabla\Phi^{G'} = \frac{\nabla^* p}{\rho(p, x, y, z, t)} + \int_{p_0}^{p} \nabla\left(\frac{1}{\rho}\right) dk \tag{26}$$

However, the term $(1/\rho)\,\nabla^* p$ should not be taken for $(1/\rho)\,\nabla p$. The terms of the right side of Eq. (26) have the dimensions of forces per unit mass, but they are only mathematical components without physical meaning. Equation (26) can also be derived by writing Eq. (24) as

$$\Phi^{G'} = \Phi(p, x, y, z, t) \tag{27}$$

and by taking the directional derivatives to build up the gradient. It follows, for example, that

$$\begin{aligned} \partial\Phi^{G'}/\partial x &= (\partial\Phi/\partial p)(\partial p/\partial x) + (\partial\Phi/\partial x) \\ \nabla\Phi^{G'} &= (\partial\Phi/\partial p)\, \nabla^* p + \nabla^{**}\Phi \\ &= (\nabla^* p)/\rho + \int_{p_0}^{p} \nabla(1/\rho)\, dk \qquad [26] \end{aligned}$$

in which ∇^* and ∇^{**} are partial gradients.

Equation (26) is therefore of no practical use. From these considerations, it follows that a correct expression for Darcy's law, and also for $\rho(p, x, y, z)$, is

$$\mathbf{q} = -(k/\mu)[\rho g\, \nabla z + \nabla p] \tag{28}$$

where k is the intrinsic permeability, dimensions L^2, and μ is the dynamic viscosity.

The density ρ may vary with x, y, z through independent variations of p, c and T. In general, therefore,

$$d\rho = (\partial\rho/\partial p)\, dp + (\partial\rho/\partial c)\, dc + (\partial\rho/\partial T)\, dT \tag{29}$$

As is shown hereafter, the transformation of the equation of conservation of mass into the equation for ground-water flow in a homogeneous, isotropic medium becomes simple only if the second and third terms of Eq. (29) may be neglected. The significance of the third term was appraised by Hubbert (1940, p. 841) as follows:

"For liquids under a temperature gradient, there actually may be a slight rotational component due to changes of density with temperature, but in ground water, this is ordinarily of a negligible magnitude compared with the potential forces acting."

The second term of the second member of Eq. (29) usually cannot be neglected when strong concentration gradients are present. In this case, transformation of the equation of conservation of mass leads to a differential equation with two unknowns, p and c, or h and c, in which h becomes a point water head as defined by Lusczynski (1960). An additional equation in c can then be obtained by expressing conservation of mass of the dispersing salt. The solution of the system of equations with unknown p and c in general will not be easy.

Therefore, in the present derivations, only variations of ρ due to variations in pressure are considered.

3. Equations of Ground-Water Flow and Storage Coefficient

The fact that artesian aquifers are apparently all more or less compressible and elastic was first recognized by Meinzer (1928) and further accepted by Theis (1935, 1938), who introduced the concept of coefficient of storage, defined as the volume of water, measured in cubic feet, released from storage in each column of an aquifer having a base 1 ft^2 and a height equal to the thickness of the aquifer, when the water table or other piezometric surface is lowered 1 ft. Theis postulated that in unconfined flow the coefficient of storage for long periods of pumping is approximately equal to the specific yield of the material through which the water table falls and therefore is independent of the thickness of the aquifer. In artesian aquifers, according to Theis, the coefficient of storage is presumably directly proportional to the thickness of the aquifer. Jacob (1940) made the first attempt to arrive at an analytical expression for the storage coefficient in his study of the flow of water in an elastic artesian aquifer. His thinking on this subject over a number of years was finalized in an article on the flow of ground-water (Jacob, 1950), and his analytic expression for the storage coefficient was never questioned. It appeared as such in the most recent textbooks by Hantush (1964) and De Wiest (1965a) and in the most recent research articles. At the 1965 Hydrology Institute, sponsored by the National Science Foundation at Princeton University, some participants became critical of Jacob's derivation of the equation of conservation of mass. The criticism concerned the fact that in one side of the equation the net inward mass flux was calculated for a volume element without deformation while in the other side of the equation, to compute the rate of change of the mass inside the volume element, the element itself was deformed.

I have arrived in a more rigorous manner at the same final expression of storage coefficient as was obtained by Jacob. In the process, however, I redefine the specific storage whereby the relation between specific storage and storage coefficient is no longer the thickness of the aquifer but the result of the integration of the equation of flow over the thickness of the aquifer. The equation of flow itself is derived by means of Darcy's law as expressed by Hubbert (1940), or by the use of $\mathbf{q} = -K\nabla h$, whereby the dependence of K on the compressibility of water is not neglected.

3.1. Equation of Conservation of Mass†

Consider an elemental volume δU of a porous medium completely saturated with fluid of density ρ centered around $P(x, y, z)$, where the specific discharge is $\mathbf{q}(x, y, z, t)$ with components u, v, w. The principle of conservation of mass (Fig. 2) for the volume element requires

mass inflow rate − mass outflow rate = change of mass storage in time

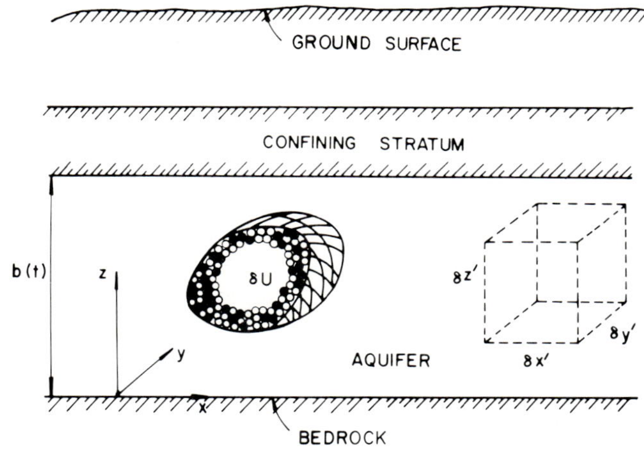

Fig. 2. Conservation of mass (after De Wiest, 1966a).

or, considering that the mass of fluid stored in the volume element is equal to $\delta M = n\rho\, \delta U$, where n is the porosity,

$$-\nabla \cdot \rho\mathbf{q} = \partial(n\rho)/\partial t \tag{30}$$

In this derivation the control volume δU is kept constant. Changes of n within the volume δU are essentially due to vertical compression of the porous medium, as variations of the lateral dimensions of aquifers are negligible because of the constraints of these aquifers by their surroundings. To find the change of n with time as a function of the elastic properties of the aquifer, it is convenient to consider an elemental parallelepiped $\delta x'\, \delta y'\, \delta z'$ with porosity n and to subject it to deformation in the vertical direction. This elemental volume $\delta x'\, \delta y'\, \delta z'$ should be kept separate from the control volume δU, however, which is not deformed.

† De Wiest (1966a).

Let α be the one-dimensional vertical compressibility of the granular skeleton of the medium, treated as a continuum and let σ_z be the stress on the intergranular skeleton in the vertical direction. By definition

$$\frac{1}{\alpha} = -\frac{d\sigma_z}{[d(\delta z')]/\delta z'}$$

so that

$$d(\delta z') = -\alpha(\delta z')\, d\sigma_z$$

and

$$\partial(\delta z')/\partial t = -\alpha(\delta z')(\partial\sigma_z/\partial t) \tag{31}$$

The volume of solid grains $\Delta\forall_s = (1 - n)\,\delta x'\,\delta y'\,\delta z'$ may be considered as constant because the compressibility of the individual grains is considerably smaller than that of their skeleton as considered above, and is also smaller than the compressibility of water. The total derivative of this quantity is zero, or

$$d(\Delta\forall_s) = d[(1 - n)\,\delta x'\,\delta y'\,\delta z'] = 0 \tag{32}$$

As said before, variations in $\delta x'$ and $\delta y'$, the lateral dimensions of the volume element, are negligible compared to the change in $\delta z'$. Therefore, $\delta x'\,\delta y'$ may be treated as a constant in Eq. (32), so that

$$dn = [(1 - n)/\delta z']\, d(\delta z')$$

and

$$\partial n/\partial t = [(1 - n)/\partial z']\, \partial(\delta z')/\partial t \tag{33}$$

Equation (33) may be reduced further by insertion of Eq. (31) as follows

$$\rho\, \partial n/\partial t = -\rho(1 - n)\alpha\ \partial\sigma_z/\partial t \tag{34}$$

Since there is complete saturation, the volume of fluid in the deformed elemental volume $\delta U'$ is $n\,\delta U'$ and the compressibility β of the fluid may be defined as

$$\beta = -[d(n\,\delta U')/n\,\delta U']/dp \tag{35}$$

in which p is the pore pressure. Conservation of mass of the fluid stored in $\delta U'$ further requires that

$$d(\rho n\,\delta U') = 0 \tag{36}$$

A combination of Eqs. (35) and (36) yields

$$d\rho = \rho\beta\, dp \tag{37}$$

and therefore,

$$n\,\partial\rho/\partial t = n\rho\beta\,\partial p/\partial t \tag{38}$$

At any depth, intergranular pressure σ_z and pore pressure p add to render the total or combined pressure. The combined pressure is numerically equal to the weight of all the matter that rests above per unit area if arching effects of the overlying strata are neglected. It follows that

$$\sigma_z + p = \text{constant}$$

and

$$\partial\sigma_z/\partial t = -\partial p/\partial t \tag{39}$$

From Eqs. (34), (38), and (39), the final form of the equation of conservation of mass becomes

$$-\left[\frac{\partial(\rho u)}{\partial x} + \frac{\partial(\rho v)}{\partial y} + \frac{\partial(\rho w)}{\partial z}\right] = \rho[(1-n)\,\alpha + n\beta]\,\frac{\partial p}{\partial t} \tag{40}$$

3.2. Transformation of $-\nabla\cdot(\rho\mathbf{q})$ by Means of the Exact Formulation of Darcy's Law

Let Darcy's law be expressed by Eq. (28)

$$\mathbf{q} = -(k/\mu)[\rho g\,\nabla z + \nabla p]$$

or

$$u = -\frac{k}{\mu}\frac{\partial p}{\partial x}; \qquad v = -\frac{k}{\mu}\frac{\partial p}{\partial y}; \qquad w = -\frac{k\rho}{\mu}\left[g + \frac{1}{\rho}\frac{\partial p}{\partial z}\right] \tag{41}$$

$$-\nabla\cdot(\rho\mathbf{q}) = -\rho\left(\frac{\partial u}{\partial x} + \frac{\partial v}{\partial y} + \frac{\partial w}{\partial z}\right) - \left(u\frac{\partial \rho}{\partial x} + v\frac{\partial \rho}{\partial y} + w\frac{\partial \rho}{\partial z}\right) \tag{42}$$

By differentiation of Eq. (41) we obtain

$$\frac{\partial u}{\partial x} = -\frac{k}{\mu}\frac{\partial^2 p}{\partial x^2}; \qquad \frac{\partial v}{\partial y} = -\frac{k}{\mu}\frac{\partial^2 p}{\partial y^2}; \qquad \frac{\partial w}{\partial z} = -\frac{k}{\mu}\frac{\partial^2 p}{\partial z^2} - \frac{k}{\mu}g\frac{\partial \rho}{\partial z} \tag{43}$$

From Eq. (37) it follows that

$$\partial\rho/\partial x = \rho\beta\,\partial p/\partial x; \qquad \partial\rho/\partial y = \rho\beta\,\partial p/\partial y; \qquad \partial\rho/\partial z = \rho\beta\,\partial p/\partial z \tag{44}$$

Substitution of Eqs. (44) and (43) into Eq. (42) gives

$$\frac{\rho k}{\mu}\nabla^2 p + \frac{kg\rho^2\beta}{\mu}\frac{\partial p}{\partial z} + \frac{k}{\mu}\left[\rho\beta\left(\frac{\partial p}{\partial x}\right)^2 + \rho\beta\left(\frac{\partial p}{\partial y}\right)^2 + \rho^2\beta\frac{\partial p}{\partial z}\left(g + \frac{1}{\rho}\frac{\partial p}{\partial z}\right)\right] = -\nabla\cdot(\rho\mathbf{q})$$

or

$$-\nabla \cdot (\rho \mathbf{q}) = \rho \frac{k}{\mu} \nabla^2 p + 2\rho^2 g\beta \frac{k}{\mu} \frac{\partial p}{\partial z} + \frac{k}{\mu} \rho\beta \left[\left(\frac{\partial p}{\partial x}\right)^2 + \left(\frac{\partial p}{\partial y}\right)^2 + \left(\frac{\partial p}{\partial z}\right)^2 \right] \tag{45}$$

In general, especially when observation wells are used, it is convenient to work with the piezometric head

$$h = z + g^{-1} \int_{p_0}^{p} dp/\rho(p) \tag{46}$$

which is proportional to Hubbert's force potential Φ^* defined by Eq. (18). The partial derivatives are derived from Eq. (46):

$$\frac{\partial p}{\partial x} = \rho g \frac{\partial h}{\partial x}; \qquad \frac{\partial p}{\partial y} = \rho g \frac{\partial h}{\partial y}; \qquad \frac{\partial p}{\partial z} = \rho g \left(\frac{\partial h}{\partial z} - 1\right) \tag{47}$$

These partial derivatives may be compared with Eq. (44) to give

$$\frac{\partial \rho}{\partial x} = \rho^2 \beta g \frac{\partial h}{\partial x}; \qquad \frac{\partial \rho}{\partial y} = \rho^2 \beta g \frac{\partial h}{\partial y}; \qquad \frac{\partial \rho}{\partial z} = \rho^2 \beta g \left(\frac{\partial h}{\partial z} - 1\right) \tag{48}$$

Partial differentiation of Eq. (47) and insertion of the results into Eq. (45) lead to

$$-\nabla \cdot (\rho \mathbf{q}) = (\rho^2 kg/\mu)(\nabla^2 h - 2\rho\beta g \, \partial h/\partial z) \tag{49}$$

when the sum of the quadratic terms in the derivatives of h is neglected.

Finally, from Eq. (46) it also follows that

$$\partial p/\partial t = \rho g \, \partial h/\partial t \tag{50}$$

By means of Eqs. (49) and (50), Eq. (40) may be rewritten as

$$K(\nabla^2 h - 2\rho\beta g \, \partial h/\partial z) = \rho g[(1 - n)\, \alpha + n\beta] \, \partial h/\partial t \tag{51}$$

where

$$K = k\rho g/\mu \tag{52}$$

is the fluid (hydraulic) conductivity. Equation (51) may also be obtained by the use of Darcy's law in its form $\mathbf{q} = -K \nabla h$, if the variation of $\rho = \rho(p)$ in K is taken into consideration. If this variation is neglected (see Jacob, 1950), the multiplier of $\partial h/\partial z$ in Eq. (51) becomes $\rho\beta g$ rather than $2\rho\beta g$.

3.3. Equation for Ground-Water Flow in a Homogeneous, Isotropic Medium

Equation (51) may be rewritten as

$$\nabla^2 h - 2\rho\beta g \, \partial h/\partial z = (S_s{}^*/K) \, \partial h/\partial t \tag{53}$$

where

$$S_s{}^* = \rho g[(1 - n)\, \alpha + n\beta] \tag{54}$$

is the specific storage with respect to an undeformed unit volume of aquifer, inside which the porosity n changes because of compaction of the medium, and also inside which ρ changes with p. This specific storage is therefore different from that defined by Jacob, who considered deformation of the volume element in the continuity equation.

To adapt Eq. (53) to the case of a confined aquifer of uniform thickness b, which in fact changes with time, i.e., $b = b(t)$, it is no longer possible to multiply and divide Eq. (53) by b and to introduce $S = S_s b$ as Jacob did. Instead, the following procedure is adopted. The dimensionality of Eq. (53) is reduced by 1 through integration of Eq. (53) in the z direction over the thickness $b(t)$ of the aquifer as follows.

The average hydraulic head is introduced as

$$\bar{h}(x, y, t) = [1/b(t)] \int_0^{b(t)} h(x, y, z, t) \, dz$$

(this averaging process can be tolerated for mildly sloping aquifers), and Eq. (53) is integrated along z over b. Application of Leibnitz's rule for differentiation of an integral gives

$$\frac{\partial^2}{\partial x^2} [b(t)\, \bar{h}] + \frac{\partial^2}{\partial y^2} [b(t)\, \bar{h}] + \left.\frac{\partial h}{\partial z}\right|_0^{b(t)} = \frac{S_s{}^*}{K} \int_0^b \frac{\partial h}{\partial t} \, dz \tag{55}$$

whereby use is made of the fact that $h(b) = h(0) = \bar{h}$.

From Darcy's law,

$$w = -K \, \partial h/\partial z \tag{56}$$

Furthermore, for $z = b(t)$,

$$w/n = \partial b/\partial t \tag{57}$$

and for $z = 0$,

$$w = 0 \tag{58}$$

Also, in analogy with Eq. (31),

$$\partial b/\partial t = \alpha b \, \partial \bar{p}/\partial t \tag{59}$$

where

$$\bar{p}(x, y, t) = (1/b) \int_0^b p(x, y, z, t) \, dz$$

and, in analogy with Eq. (50),

$$\partial \bar{p}/\partial t = \rho g \, \partial \bar{h}/\partial t \tag{60}$$

Therefore,

$$\begin{aligned} \partial h/\partial z &= 0 && \text{at} \quad z = 0 \\ \partial h/\partial z &= -(n/K)\, \alpha b \rho g \, \partial \bar{h}/\partial t && \text{at} \quad z = b(t) \end{aligned} \tag{61}$$

and

$$\begin{aligned} \int_0^b \partial h/\partial t \, dz &= \partial/\partial t \int_0^b h \, dz - h(x, y, b, t) \, \partial b/\partial t \\ &= \partial/\partial t \, [b(t) \, \bar{h}] - \bar{h} \, \partial b/\partial t = b \, \partial \bar{h}/\partial t \end{aligned}$$

so that

$$\int_0^b \partial h/\partial t \, dz = b \, \partial \bar{h}/\partial t \tag{62}$$

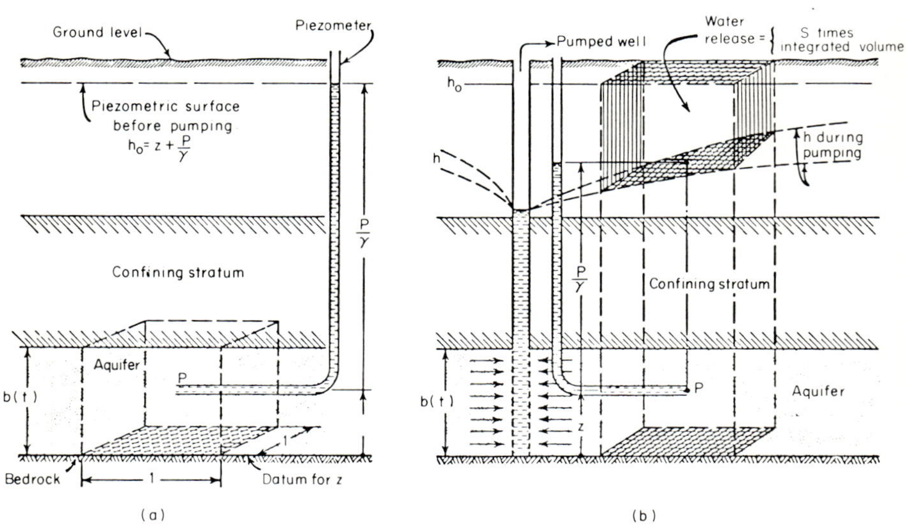

FIG. 3. Physical interpretation of storage coefficient S. Piezometric head: (a) before pumping; (b) during pumping (after De Wiest, 1965a).

The insertion of Eqs. (61) and (62) into Eq. (55) leads to

$$b \nabla^2 \bar{h}(x, y, t) = (\rho g b/K)[(1 - n) \alpha + n\beta] \partial \bar{h}/\partial t + (\rho g b/K) n\alpha \partial \bar{h}/\partial t \qquad (63)$$

or

$$\nabla^2 \bar{h}(x, y, t) = (S/T) \partial \bar{h}/\partial t \qquad (64)$$

where the storage coefficient S (see Fig. 3)

$$S = \rho g b(\alpha + n\beta) \qquad (65)$$

and the transmissivity T

$$T = Kb \qquad (66)$$

are introduced.

It is evident that S and T are both variable in time. Equation (64), however, is often integrated with the assumption that S/T is independent of time, whereas in the determination of S and T from pumping tests, average values of S and T in time are obtained.

3.4. Discussion

A better understanding of the previous derivations, in particular of Eqs. (57) and (61), may be achieved by considering Fig. 4, in which water is withdrawn via the sink in P. This withdrawal creates flow

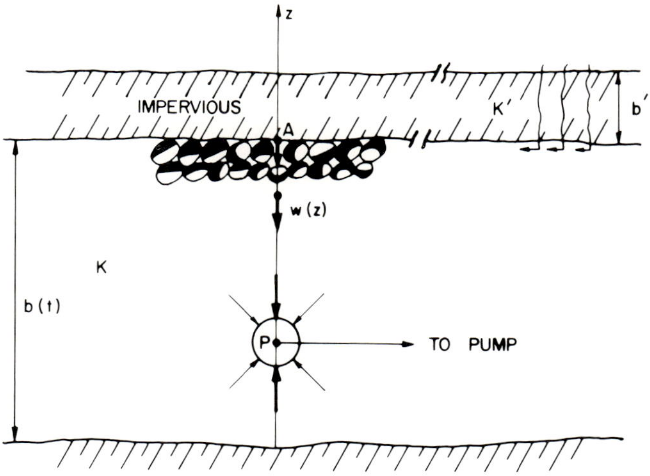

Fig. 4. Boundary condition $\lim_{z \to b}[(-K/n)(\partial h/\partial z)] = \partial b/\partial t$ (after De Wiest, 1966a).

toward the sink at velocities which increase in magnitude when the sink is approached. Of particular interest here is the vertical component of the Darcian velocity $w(z)$, which varies from the top of the aquifer to the sink. It is due to the fact that there is a gradient of the head in the z direction, and therefore the product $K(\partial h/\partial z)$ is not meaningless but represents a Darcian velocity. At the top of the aquifer the actual velocity is matched by the subsidence velocity of the confining stratum in such a way that

$$\lim_{z\to b}[-(K/n)\,\partial h/\partial z] = \partial b/\partial t$$

This equation holds only for the top of the aquifer, and at a point an infinitesimally small distance below the top of the aquifer the velocity would be different. In other words, in Eq. (55), $\partial h/\partial z$ remains a function of z, as is evident from the integration, and it is evaluated for $z = b$ and $z = 0$.

$$\int_0^b \frac{\partial^2}{\partial z^2}\, h(x, y, z, t)\, dz = \int_0^b \frac{\partial}{\partial z}\left[\frac{\partial h}{\partial z}(x, y, z, t)\right] dz$$

Another form of interpretation is to say that, although the introduction of $\bar{h}$ means that the net average component of the Darcian velocity in the z direction is zero, some local velocities are not zero in the z direction.

The withdrawal of water as sketched in Fig. 4 is quite common in nature, as exemplified by the flow to a well screened over part of the thickness of the aquifer. Equation (64) has been derived independently of assumptions of the way in which water is withdrawn. Often this equation is applied to the flow in a confined aquifer completely penetrated by a well, and for practical purposes the results have been good, in spite of the underlying assumptions.

In any case, water must be withdrawn in order to justify the use of Eq. (56). Should this not be the case, and should a deformation of the pervious stratum have been caused by an extraneous force, say the load of a building on top of the confining layer, we would have the so-called undrained case of soil mechanics. The load would cause a uniform pore pressure rise and therefore a uniform increase of head in the aquifer. Hence there would be no gradient in the head and no Darcian velocity. In hydrology we are not interested in this case.

There would be no basic changes in the derivations if the confining stratum of Fig. 4 were replaced by a semipervious one of thickness b' and hydraulic conductivity K', overlain by an unconfined aquifer under constant head H. The boundary condition expressed by Eqs. (56) and (57) would become

$$-\frac{K}{n}\frac{\partial h}{\partial z} = \frac{\partial b}{\partial t} + \frac{K'}{b'}(\bar{h} - H) \tag{67}$$

and Eq. (64) would be replaced by

$$\nabla^2 \bar{h}(x, y, t) - \frac{K'}{bb'K}(\bar{h} - H) = \frac{S}{T}\frac{\partial \bar{h}}{\partial t} \tag{68}$$

which has generally been accepted for the flow in leaky aquifer conditions.

Finally, I have not used the relative velocity between fluid particles and solid matrix, because this concept is of little value to the hydrologist and to the petroleum engineer. It may, however, be significant in applications of soil mechanics, and therefore the topic of elastic storage of aquifers has been treated separately in Chapter 8 of this volume by A. Verruyt. The concept of a generalized Darcy law, using the relative velocity between fluid and grain skeleton, is by no means novel. It was used first by the Russians Gersevanov, Florin, Krylov, and Barenblatt and was introduced in the Western literature by Biot (1955).

4. Potential Flow

In many applications of civil engineering and hydrogeology, K is assumed to be constant, so that the concept of a velocity potential $\Phi = Kh$ as defined by Eq. (9) is valid. In this case, the powerful tools of hydrodynamics are available, and are especially useful in the solution of two-dimensional problems.

For steady state flow, it follows from Eq. (64) that $\nabla^2\Phi = 0$. By means of the Cauchy–Riemann equations, the conjugate harmonic function Ψ (stream function) can be determined such that the complex potential function $W(z) = \Phi(x, y) + i\Psi(x, y)$ is an analytic function of the complex variable $z = x + iy$.

4.1. Uniform Flow

Uniform flow is the ground-water flow analogous to parallel flow in hydrodynamics. The slope of the hydraulic grade line or hydraulic gradient $\mathfrak{I}$ is constant. Therefore,

$$\begin{aligned} \partial\Phi/\partial x &= -K\mathfrak{I} \quad \text{or} \quad \Phi = -K\mathfrak{I}x + c \\ \partial\Psi/\partial y &= -K\mathfrak{I} \quad \text{or} \quad \Psi = -K\mathfrak{I}y + c \end{aligned} \tag{69}$$

4.2. Well Flow

The equations for a line sink of strength $-Q/b$ per unit length may be borrowed from hydrodynamics to render Φ and Ψ for the steady radial flow to a well in a confined aquifer.

$$\Phi = (Q/2\pi b) \ln r + c, \qquad \Psi = (Q/2\pi b)\, \theta + c \tag{70}$$

Here b is the thickness of the aquifer.

Example: A Well in a Uniform Flow. Consider a confined aquifer with $b = 100$ ft and having a uniform flow with

$$K\mathfrak{I} = 6.9 \text{ ft/day} = 8 \times 10^{-5} \quad \text{ft/sec}$$

A completely penetrating well is pumped at $Q = 107.5$ gal/min., i.e., $Q/b = 24 \times 10^{-3}$ ft²/sec, and it is necessary to determine the groundwater divide for the water entering the well.

The flowlines of the composite flow $\Psi =$ constant are determined by adding the values of Ψ_u (uniform flow) and Ψ_w (well flow) at points of intersection, as indicated on Fig. 5

$$\Psi_u = -K\mathfrak{I}y + C = -8 \times 10^{-5}\, y \quad \text{ft}^2/\text{sec}$$

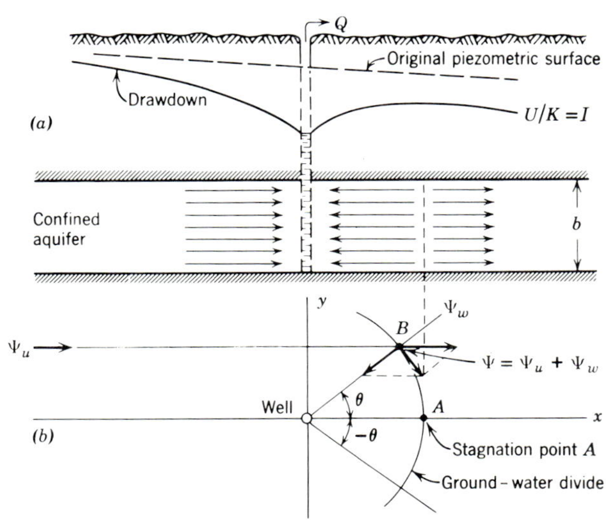

Fig. 5. Well in uniform flow. (a) Vertical cross section. (b) Plan view for superposition of stream functions (after De Wiest, 1965a).

if $C = 0$ by assumption. The lines $\Psi_u =$ constant are represented in Fig. 6 as parallel lines for $y = 0, \pm 25, \pm 50, ..., \pm 200$ ft.

$$\Psi_w = (Q/2\pi b)\, \theta + C = 24(\theta/2\pi) \times 10^{-3} \quad \text{ft}^2/\text{sec}$$

if $C = 0$ by assumption.

In Fig. 6, θ varies from 0 to π counterclockwise and from 0 to $-\pi$ clockwise measured from the x-axis. The well is placed at the origin of the coordinate axis. The rays $\Psi_w =$ constant are labeled for $C = 0$; for example, for $\theta = \pi/2$,

$$\Psi_w = (Q/2\pi b)(\pi/2) = 6 \times 10^{-3} \quad \text{ft}^2/\text{sec}$$

To construct a streamline, points where the sums of the added stream functions assume the same value are joined by a smooth line. With the particular choice $C = 0$ in the expressions of Ψ_u and Ψ_w, the ground-water divide happens to be the streamline $\Psi = 0$ up to the stagnation point A. For another choice of the constant C a different value of Ψ along the ground-water divide would be found, emphasizing once more that Ψ is determined only up to an arbitrary constant. It would also be possible to construct in any point B (Fig. 5) of the aquifer two velocity vectors: one horizontal and of constant length, the other inversely proportional to r according to $\partial\Phi/\partial r$ as taken from Eq. (70), and pointed towards the origin. Their resultant would be tangent to the streamline passing through B. In point A the vectors are in the opposite direction. If these vectors are equal, A is a stagnation point. Its location is determined by the equality

$$Q/2\pi b x_s = K\mathfrak{J}$$

i.e., in the present case $x_s = 47.7$ ft. The ground-water divide defines the boundary of the aquifer region contributing to the well discharge. The width $2y$ of the contributing region infinitely far to the left follows from

$$\Psi = (Q/2\pi b)\, \theta - K\mathfrak{J}y$$

in which $\Psi = 0$ and $\theta = \pi$. In the present case $2y = 300$ ft.

From the foregoing it is evident that the relative strengths of uniform flow and well flow determine the location of the ground-water divide. The ratio of these characteristics is very important in the study of the recirculation between a recharge well and a discharge well when uniform ground-water flow is present. For each of these ratios there is a critical distance between the well and the recharge well above which recirculation between the wells ceases.

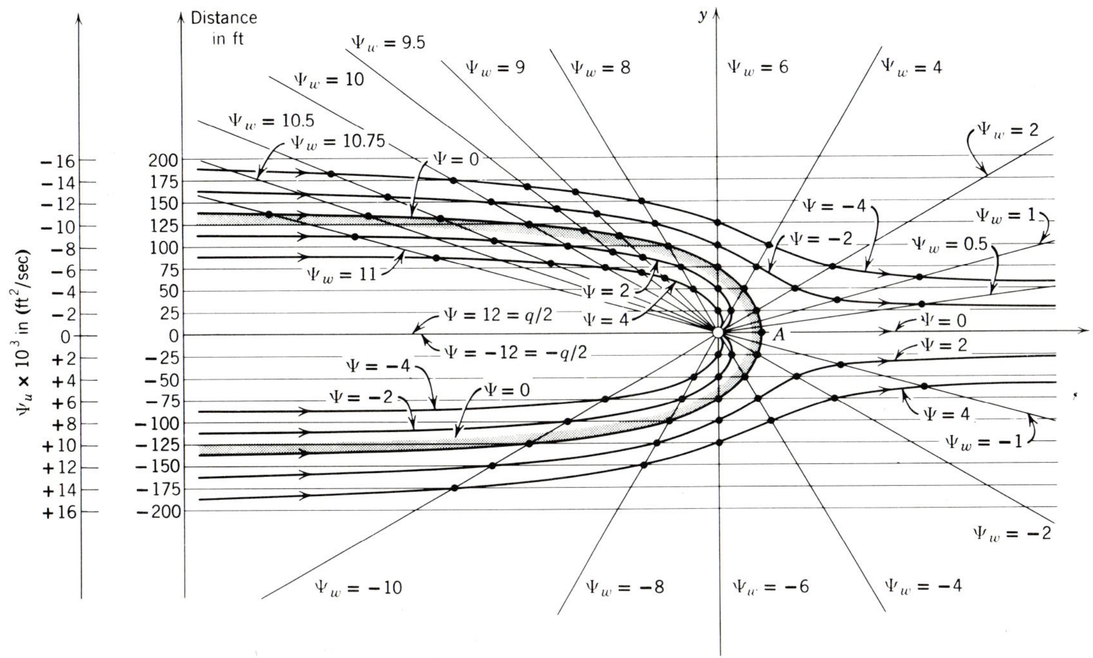

Fig. 6. A well in uniform flow. Graphical construction of flow lines. All Ψ_w and Ψ are multiplied by 10^3 (after De Wiest, 1965a).

4.3. Several Wells

Equations (70) are borrowed from hydrodynamics and govern the flow to a single singularity (the well in ground-water flow) in an infinite domain. These equations may be extended without difficulty to include any number of wells ($i = 1, 2, \ldots, n$) located at (x_i, y_i). This follows from the linearity of Laplace's equation, allowing the superposition of solutions. In the case of confined flow, for example, Φ_i may be considered the potential required in any point (x, y) to sustain a discharge Q_i of the ith well located in the point (x_i, y_i). Hence

$$\Phi_i = (Q_i/4\pi b) \ln r_i^2 + \text{constant}$$

To sustain simultaneous discharges $\sum_{i=1}^{n} Q_i$ of all the wells, a potential $\Phi(x, y) = \sum_{i=1}^{n} \Phi_i$ will be needed. Therefore,

$$\Phi(x, y) = (4\pi b)^{-1} \sum_{i=1}^{n} Q_i \ln r_i^2 + \text{constant} \tag{71}$$

in which the arbitrary constant may be conveniently determined in each problem.

4.4. Method of Images

The principle of superposition is also applied when the domain is not infinite but semi-infinite or bounded. In the case of a single well near a stream for example, a recharge well may be considered as the negative image of the discharging well reflected in the line $x = 0$ which is coincident with the axis of the stream. The system "well–stream" is replaced by the system "well–image well with negative discharge" and both systems observe the same boundary conditions along the line $x = 0$ and at infinity. This artificial device called "method of images" makes it possible to combine solutions of the type given by Eq. (70) in the case of noncircular aquifers.

The method of images is also valid when the flow is unsteady. In particular, it may be used to prove that there is a hydraulic connection between an aquifer and its intersecting stream. Such a hydraulic connection tends to offset the unsteady nature of the flow, when a well is pumped near the stream.

5. Mathematical Techniques in Potential Flow

The problems of ground-water flow often consist in finding the potential Φ in any point of the medium. From the knowledge of Φ, the ground-water velocities are readily determined, as is the pore pressure p. Mathematically, the construction of the function $\Phi(x, y)$ is equivalent to finding the complex potential function $W(z)$.

In many problems, the domain $W = \Phi + i\Psi$ is represented by a simple geometrical configuration and can be mapped onto the physical z-plane. Often this is accomplished by mapping both the physical plane and the W-plane onto a common intermediate t-plane, and by subsequent elimination of t between W and z. This technique is therefore almost entirely based on familiarity with conformal mapping.

5.1. Conformal Mapping

5.1.1. *Flow underneath a Dam on a Finite Stratum*†

The physical z-plane and the intermediate t-plane are represented in Fig. 7.

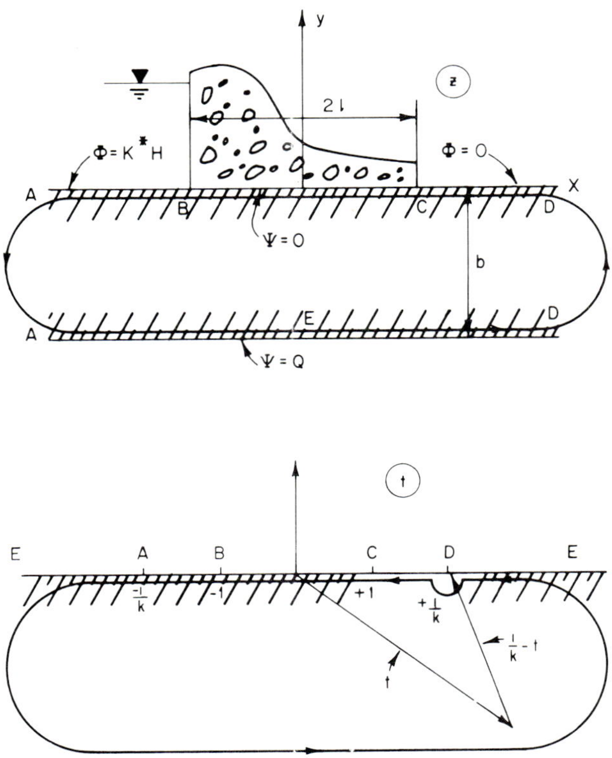

Fig. 7. Flow underneath dam on finite stratum. Physical plane and t-plane.

† Polubarinova-Kochina (1952).

By means of the Schwarz–Christoffel mapping, the infinite strip of the z-plane is mapped onto the lower half of the t-plane such that

$$\frac{dz}{dt} = A\left(t + \frac{1}{k}\right)^{-1}\left(t - \frac{1}{k}\right)^{-1}$$

and therefore,

$$z = A'\frac{k}{2}\ln\frac{1+kt}{1-kt} + B \tag{72}$$

in which A and A' remain arbitrary. B must be zero because $t = 0$ corresponds to $z = 0$. A' and k are determined by the boundary conditions:

(1) For point C, $z = l$, $t = 1$. Therefore,

$$l = A'\frac{k}{2}\ln\frac{1+k}{1-k} \tag{73}$$

(2) When the boundary is contoured in the z-plane, in D there is a transition from $z = -bi + \infty$ to $z = +\infty$, and therefore, a jump $\Delta z = bi$ (end value—begin value). Correspondingly there is a change in the argument of $z = -A'(k/2)\ln(1 - k)$. The argument of the vector $1 - kt$ varies from π to 0, or its change is $-\pi$. The change in the natural logarithm is $-i\pi$ and the change in z is $\Delta z = -(A'k/2)(-i\pi) = (A'k/2)\,i\pi$. From the two values of Δz, it follows that $b = (A'k/2)\,\pi$ and therefore Eq. (72) becomes

$$z = \frac{b}{\pi}\ln\frac{1+kt}{1-kt} \tag{74}$$

and from Eq. (73)

$$k = \tanh(\pi l/2b) \tag{75}$$

The boundary conditions in the physical plane are

$$\Phi = K^*H \quad \text{along } AB; \qquad \Psi = 0 \quad \text{along } BC$$

$$\Phi = 0 \quad \text{along } CD; \qquad \Psi = +Q \quad \text{along } AED$$

(Note: K^* is used instead of K for the hydraulic conductivity.) In the complex potential plane $W = \Phi + i\Psi$, the seepage domain is represented by the rectangle $ABCD$. This rectangle (Fig. 8) is mapped onto the lower half of the t-plane in such a way that the points A, B, C, D, E are mapped onto the same points as was the case in the t-plane of Fig. 7, and the boundaries are contoured such that the interior of the rectangle corresponds to the lower half of the t-plane. The Schwarz–Christoffel mapping renders

$$dW/dt = A[(t^2 - 1)(t^2 - 1/k^2)]^{-1/2}$$

and therefore,

$$W = A' \int_0^t \frac{dt}{[(1 - t^2)(1 - k^2t^2)]^{1/2}} + B \tag{76}$$

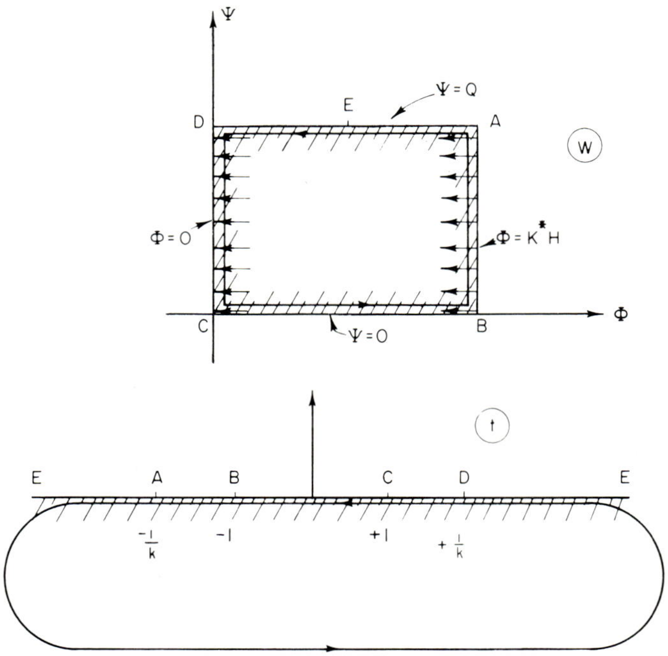

FIG. 8. Flow underneath dam on finite stratum. Seepage rectangle and t-plane.

A' and B are determined from the boundary conditions:

(1) $W(0) = B = K^*H/2$.

(2) For $t = 1$, $W = 0$.

or

$$0 = A' \int_0^1 \frac{dt}{[(1 - t^2)(1 - k^2t^2)]^{1/2}} + \frac{K^*H}{2}$$

$$0 = A'K + \frac{K^*H}{2} \qquad \text{and} \qquad A' = -\frac{K^*H}{2K}$$

in which K is the complete elliptic integral of the first kind. The final expression for W may be written as

$$W = \frac{K^*H}{2K} [K - F(\arcsin t, k)] \tag{77}$$

in which

$$u = \int_0^t \frac{dt}{[(1 - t^2)(1 - k^2t^2)]^{1/2}} = F(\arcsin t, k) \tag{78}$$

is the elliptic integral of the first kind in canonical form. The inverse of the elliptic integral is the elliptic sine

$$t = \operatorname{sn} u$$

Elimination of t between Eqs. (74) and (77) leads to the final solution. From Eq. (77) it follows that $u = K - (2KW/K^*H)$ and therefore

$$t = \operatorname{sn}[K - (2KW/K^*H)] \tag{79}$$

Equation (79) may be inserted in Eq. (74) to give the final solution

$$z = \frac{b}{\pi} \ln \frac{1 + k \operatorname{sn}[K - (2KW/K^*H)]}{1 - k \operatorname{sn}[K - (2KW/K^*H)]} \tag{80}$$

The discharge $\Delta\Psi = -Q$ in particular follows from the condition that for $t = 1/k$, $W = iQ$. Indeed

$$\begin{aligned} iQ &= -\frac{K^*H}{2K} \int_0^{1/k} \frac{dt}{[(1 - t^2)(1 - k^2t^2)]^{1/2}} + \frac{K^*H}{2} \\ &= -\frac{K^*H}{2K} (K + iK') + \frac{K^*H}{2} \end{aligned}$$

or $Q = -K^*HK'/2K$ and therefore,

$$\Delta\Psi = (K^*H/2)(K'/K) \tag{81}$$

in which K' is the complete elliptic integral of the first kind with the complementary modulus k'. The ratio K'/K is tabulated as a function of k, which is itself a function of the length of the dam and the thickness of the stratum (Aravin and Numerov, 1955; Harr, 1962).

5.1.2. *Flow underneath a Dam on an Infinite Stratum*

In this example, use is made of the hodograph method whereby the conjugate velocity-plane $\bar{w}$ acts as an intermediate mapping plane between the physical plane and the complex potential plane. The hodograph is a picture of the velocity variations along the boundaries of the physical plane. Complex potential W, conjugate velocity $\bar{w}$ and physical plane z are related through the equation

$$dW/dz = -\bar{w} \tag{82}$$

The physical plane is represented in Fig. 9 and the W, $\bar{w}$ and t-planes are represented in Fig. 10. The $\bar{w}$-plane is mapped onto the upper half of the t-plane by the mapping

$$t = (1/a^2)\,\bar{w}^2 \tag{83}$$

in which a remains to be determined.

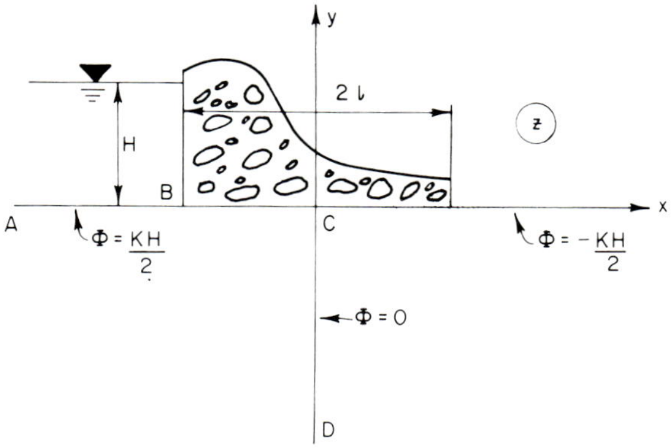

FIG. 9. Flow underneath dam on infinite stratum. Physical plane.

The arbitrary constant in $\Phi = K\,(h + \text{constant})$ is chosen to be equal to $-H/2$. Therefore $\Phi = K[(p/\gamma) + y - (H/2)]$. Because of the symmetry of the flow, only one quadrant has been analyzed.

The mapping of the semi-infinite strip of the W-plane onto the same t-plane (homologous points) is accomplished by

$$t - 1 = \tan^2 W\pi/KH \tag{84}$$

The elimination of t between Eqs. (83) and (84) is straightforward. It follows that

$$\bar{w} = a[\cos\,(W\pi/KH)]^{-1} \tag{85}$$

Equation (82) is now used to find the relationship between W and z.

$$dz = -dW/\bar{w} = -(1/a)\cos(W\pi/KH)\,dW$$

or

$$z = -(1/a)(KH/\pi)\sin(\pi W/KH) + b$$

The arbitrary constants a and b are determined by the boundary conditions in B ($z = -l$, $W = KH/2$) and in C ($z = 0$, $W = 0$). It follows that

$$z = -l\sin(\pi W/KH) \tag{86}$$

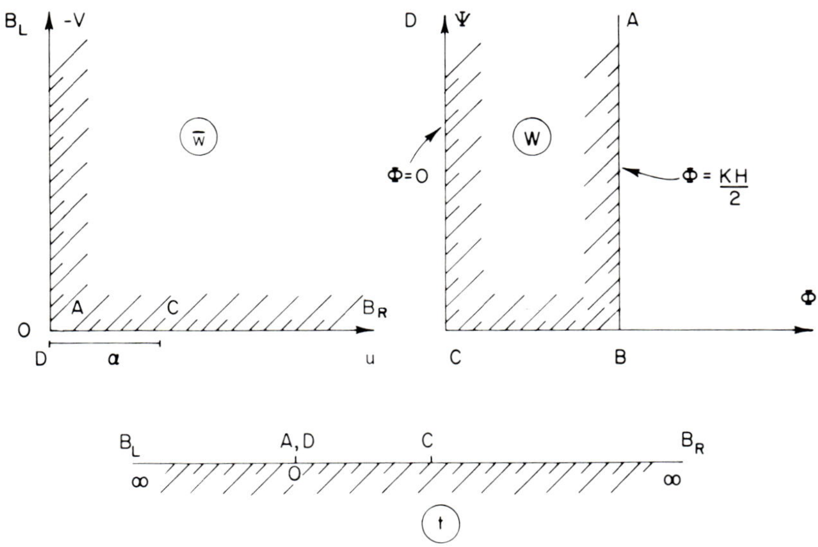

FIG. 10. Flow underneath dam on infinite stratum. Conformal mappings.

The conjugate velocity $\bar{w}$ is given by

$$\bar{w} = (KH/\pi)(l^2 - z^2)^{-1/2} \tag{87}$$

The streamlines $\Psi =$ constant are given by

$$\frac{x^2}{l^2\cosh^2(\pi\Psi/KH)} + \frac{y^2}{l^2\sinh^2(\pi\Psi/KH)} = 1 \tag{88}$$

The lines of equal potential $\Phi =$ constant are given by

$$\frac{x^2}{l^2\sin^2(\pi\Phi/KH)} - \frac{y^2}{l^2\cos^2(\pi\Phi/KH)} = 1 \tag{89}$$

Note that in Eqs. (88) and (89), x and y must be negative, $0 < \Phi < KH/2$ and $0 < \Psi < \infty$ (Polubarinova–Kochina, 1952).

5.1.3. *The Inverse Function*

The idea of using the technique of conformal mapping to take fully into consideration the infinite extent of some domains has been attributed (de Jong, 1962) to Boothroyd *et al.* (1949). It was subsequently used by de Jong (1962) to investigate by means of an electric analog the potential distribution resulting from pumping a well near a rectilinear coast of infinite extent. de Jong showed that significant deviations in the potential distribution exist between the electric analog with a finite domain and the analog that makes up for the infinite extent of the domain by means of the inverse function.

The same principle was also applied by the author (De Wiest 1966b) in an experimental investigation of two overlying aquifers of infinite extent, penerated by a common well.

Figure 11 shows the plan view of the horizontal Hele–Shaw apparatus of one aquifer, Fig. 12 shows a close-up photograph of the entire apparatus. Each aquifer, as indicated in Figs. 11 and 12, consists of an interior domain in the form of a square and an exterior domain in the form of a clover leaf. As shown further, the clover-leaf domain arises from the conformal mapping of the exterior of the square domain by means of the inverse function. The boundaries of the square domain and the clover-leaf domain are interconnected by means of tubes in

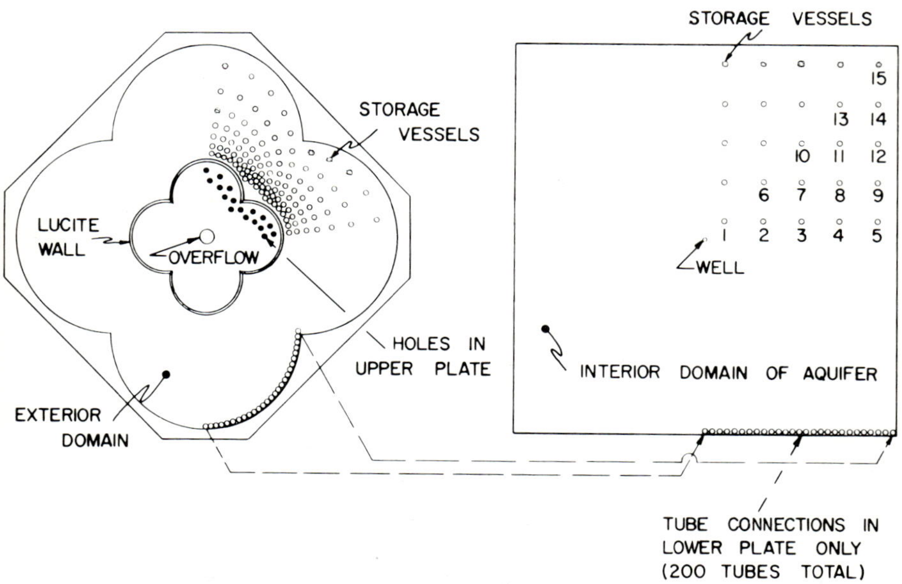

FIG. 11. Plan view of horizontal Hele–Shaw apparatus (after De Wiest, 1966b).

such a way that points of the same potential are joined together. The storage vessels for the exterior domains were designed by means of the theory of analytic functions as follows.

Consider the function

$$w = \mathscr{L}^2/z \tag{90}$$

in which $w = u + iv$ is an analytic function of $z = x + iy$ and $\mathscr{L}$ is a

FIG. 12. Close-up photograph of Hele–Shaw apparatus (after De Wiest, 1966b).

characteristic length, say half the width of the square in Fig. 13. The boundaries of the square domain in the z-plane map onto the clover leaf of the w-plane, each side of the square in the z-plane mapping onto a circle in the w-plane. The square element A maps onto the curvilinear element A'.

The diameter of the storage vessels inside the clover leaf of the w-plane can now be determined as follows.

Let the equation for the head h in the z-plane be

$$\nabla^2 h(x, y) = (S/T)\,\partial h/\partial t \tag{91}$$

The corresponding equation for h in the w-plane is

$$\nabla^2 h(u, v) = (S'/T')\, \partial h/\partial t \tag{92}$$

On the other hand, we have

$$\nabla^2 h(x, y) = \nabla^2 h(u, v) |\, dw/dz\, |^2 \tag{93}$$

and

$$|\, dw/dz\, |^2 = \mathscr{L}^4/z^4 \tag{94}$$

Substitution of Eqs. (93) and (94) into Eq. (92) leads to

$$\nabla^2 h(x, y) = (S'/T')[\mathscr{L}^4/(x^2 + y^2)^2](\partial h/\partial t) \tag{95}$$

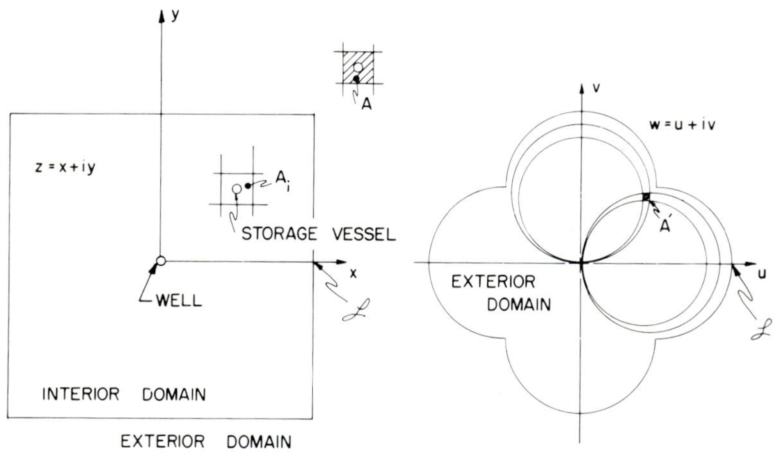

FIG. 13. Conformal mapping of exterior domain of aquifers (after De Wiest, 1966b).

By construction $T' = T$, and therefore compatibility of Eqs. (91) and (95) is ensured if

$$S' = S(x^2 + y^2)^2/\mathscr{L}^4 \tag{96}$$

From Eq. (96) it is evident that there is a practical limit to the construction of S', which is reached when the storage vessel occupies exactly its total assigned area A'. Indeed, from Eq. (96) it also follows that the diameter of the storage vessel is not affected by the conformal mapping and remains the same in the w-plane as it was in the z-plane. To see this, let (Santing, 1957)

$$S = \pi r^2/A \tag{97}$$

$$S' = \pi r'^2/A' \tag{98}$$

Substitution of Eqs. (97) and (98) into Eq. (96) shows that

$$r'^2/r^2 = (A'/A)(dz/dw)^2 = 1 \tag{99}$$

since $| dw/dz | = R_0$; the coefficient of magnification R_0 (Churchill, 1948) is variable from point to point and its square is equal to A'/A.

It appears, therefore, that there is as "saturation" value of the storage coefficient S' for the exterior domains beyond which practical modeling becomes impossible. When this level is reached, all available space inside the exterior domain is used for storage and a single central vessel may be built. In Figs. 11 and 12 these central storage vessels have the shape of a clover leaf. The upper plate of the apparatus is perforated with holes at the interior of these larger vessels.

5.2. The Hodograph Method

The hodograph method, as already seen, is a special case of conformal mapping, especially useful in the solution of steady two-dimensional ground-water problems involving free-surface boundaries (De Wiest, 1960, 1961; Bear and Dagan, 1962, 1964) and interface problems (de Josselin de Jong 1965). The advantage of using the hodograph-plane lies in the fact that only the boundaries in the $\bar{w}$-plane are known, although the location of the corresponding boundaries in the physical z-plane is not known, as is the case for a free surface or the interface between salt and fresh water, treated as immiscible.

The relationship between W, $\bar{w}$, and z is given by Eq. (82) and $\bar{w}$ plays the role of an intermediate plane, while the ultimate goal remains to establish W as a function of z.

The construction of the hodograph in the $\bar{w}$-plane is reduced to finding the representation of the various boundaries, as indicated in Fig. 14.

5.2.1. *Seepage Surface*

Along a seepage surface

$$\Phi = K\left(y + \frac{p_a}{\gamma}\right) + \text{constant}$$

in which p_a is the atmospheric pressure. By differentiation of both sides of the equation, it follows that

$$\frac{d\Phi}{ds} = K\frac{dy}{ds} \quad \text{or} \quad \frac{\partial\Phi}{\partial x}\frac{dx}{ds} + \frac{\partial\Phi}{\partial y}\frac{dy}{ds} = K\frac{dy}{ds}$$

or if α is the angle made by the seepage surface with the horizontal

$$-u\cos\alpha - v\sin\alpha = K\sin\alpha$$

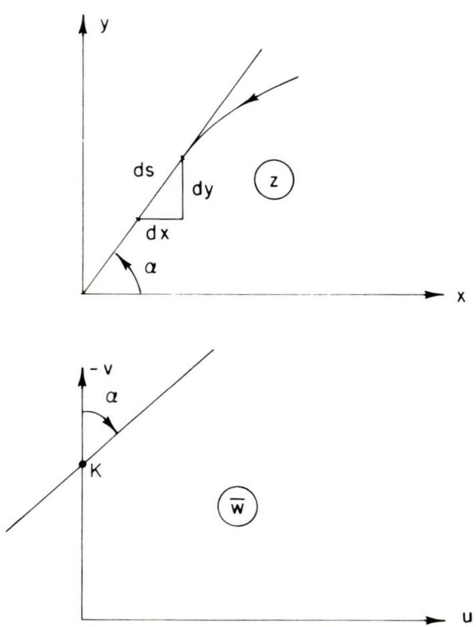

Fig. 14(a)

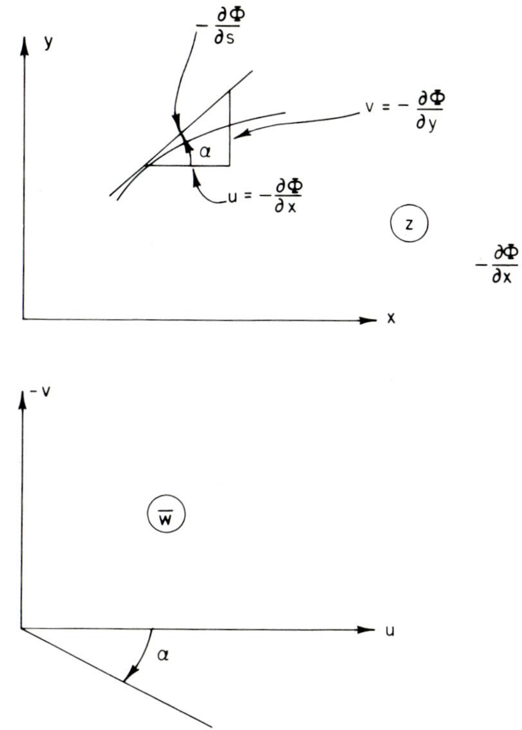

Fig. 14(b)

Fig. 14. Hodograph representations: (a) Seepage surface; (b) streamline; (c) equipotential line; (d) free surface; (e) salt water–fresh water interface.

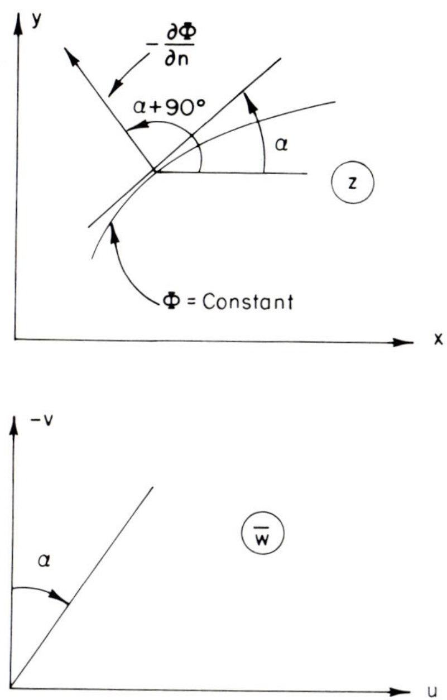

FIG. 14(c)

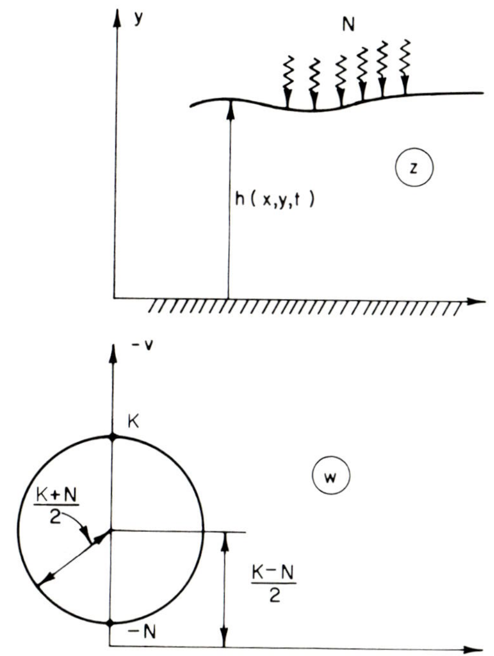

FIG. 14(d)

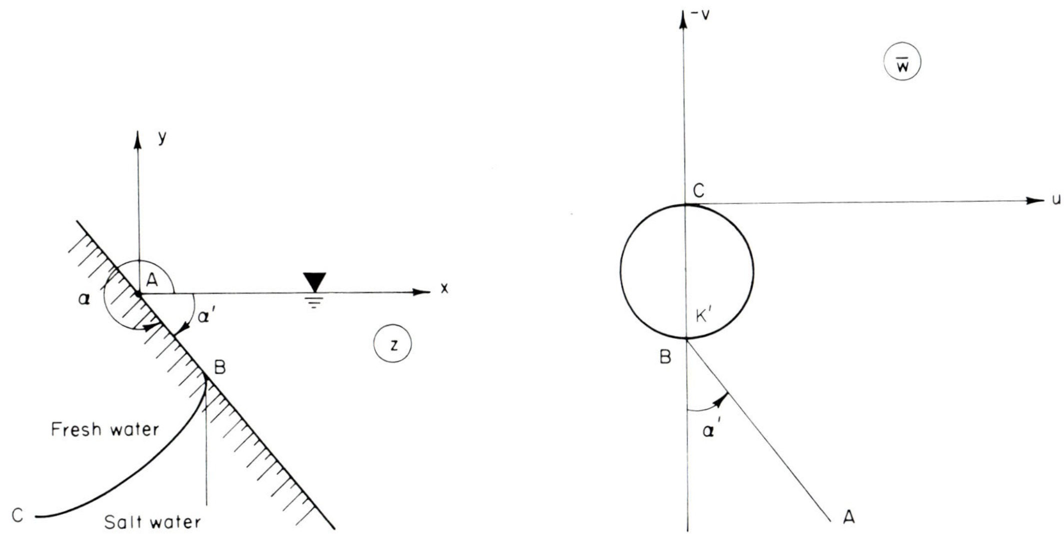

FIG. 14(e)

Therefore, the hodograph of a seepage surface is the straight line

$$-v = K + u \tan(90^\circ - \alpha) \tag{100}$$

5.2.2. *Streamline, Impermeable Boundary*

At each point of a streamline, where the tangent makes the angle α with the horizontal

$$u = -\frac{\partial \Phi}{\partial x} = -\frac{\partial \Phi}{\partial s} \cos \alpha; \qquad v = -\frac{\partial \Phi}{\partial y} = -\frac{\partial \Phi}{\partial s} \sin \alpha$$

Therefore the hodograph of a streamline (locally) or of an impermeable boundary is

$$-v = -(\tan \alpha)\, u \tag{101}$$

5.2.3. *Equipotential Line*

The velocity components normal to the tangent in a point of the equipotential line making an angle α with the horizontal are

$$u = -\frac{\partial \Phi}{\partial x} = -\frac{\partial \Phi}{\partial n} \cos(\alpha + 90^\circ) = \frac{\partial \Phi}{\partial n} \sin \alpha$$

$$v = -\frac{\partial \Phi}{\partial y} = -\frac{\partial \Phi}{\partial n} \sin(\alpha + 90^\circ) = -\frac{\partial \Phi}{\partial n} \cos \alpha$$

Therefore, the hodograph of an equipotential line where its tangent makes an angle α with the horizontal is

$$-v = u \tan(90^\circ - \alpha) \tag{102}$$

5.2.4. *Free Surface*

In general, the head in an unconfined aquifer may be expressed as

$$h(x, y, t) = y + \frac{p(x, y, t)}{\gamma} + \text{constant} \tag{103}$$

At the free surface, p/γ becomes a constant and y becomes a function of x, t, say $y = y_f$. The total derivative may be taken of Eq. (103) to render, at the free surface,

$$\frac{\partial h}{\partial x}\frac{dx}{dt} + \frac{\partial h}{\partial y}\frac{dy}{dt} + \frac{\partial h}{\partial t} = \frac{dy}{dt} \tag{104}$$

At the free surface, the following equations hold.

$$\frac{dx}{dt} = -\frac{1}{S_y} K \frac{\partial h}{\partial x}; \qquad \frac{dy}{dt} = -\frac{1}{S_y} K \frac{\partial h}{\partial y} - \frac{N}{S_y} \tag{105}$$

in which S_y is the specific yield and N is the rate of replenishment (negative) or evaporation (positive).

Insertion of Eqs. (105) into Eq. (104) renders

$$K^2 \left(\frac{\partial h}{\partial x}\right)^2 + K^2 \left(\frac{\partial h}{\partial y}\right)^2 + NK \frac{\partial h}{\partial y} - K^2 \frac{\partial h}{\partial y} - NK = S_y K \frac{\partial h}{\partial t}$$

and for steady state flow this equation becomes

$$u^2 + v^2 + v(K - N) - NK = 0$$

or

$$u^2 + \left(-v - \frac{K - N}{2}\right)^2 = \left(\frac{K + N}{2}\right)^2 \tag{106}$$

Equation (106) is the equation of a circle and represents the hodograph of the free surface as indicated in Fig. 14. Equation (106) may also be derived by adding to the property $-v = K + u \tan(90° - \alpha)$ which the free surface has in common with the seepage surface, the condition $d\Psi = N dx$ or $\Psi = Nx +$ constant, implying that for $N = 0$ the free surface is a streamline.

5.2.5. *Interface between Fluids of Different Densities*

The location of the interface between fluids of different densities is the subject of studies in salt water intrusion (Hubbert, 1940; Charmonman, 1965). Let Φ_f be the fresh water potential, Φ_s be the salt water potential. Pressure p and elevation y, which are unique for any point in the two regions, may be expressed in either Φ_f or Φ_s :

$$\Phi_f = K_f(y + p/\gamma_f); \qquad \Phi_s = K_s(y + p/\gamma_s) \tag{107}$$

Elimination of p between Eqs. (107) leads to

$$\Phi_f = \frac{\gamma_s}{\gamma_f} \frac{K_f}{K_s} \Phi_s - \frac{\gamma_s - \gamma_f}{\gamma_f} K_f y \tag{108}$$

In the study of the movement of fresh water with respect to stagnant salt water, whereby immiscibility of the fluids is assumed, the potential of fresh water in the salt water zone is constant, and therefore Eq. (108) may be rewritten as

$$\Phi = \text{constant} - K'y \tag{109}$$

in which the subscript f has been dropped and in which

$$K' = K \frac{\gamma_s - \gamma_f}{\gamma_f} \tag{110}$$

Equation (109) resembles the equation $\Phi =$ constant $+ Ky$ which holds along the free surface. If it is further assumed that the interface between the two fluids is a streamline, as was the case of the free boundary for $N = 0$, then the hodograph of the interface is a circle centered at $(0; -K'/2)$ with radius $K'/2$.

5.2.6. *Outflow Surface of Fresh Water into Ocean*

Equation (109) may be used to express the variation of Φ along the outflow surface AB of Fig. 14.

$$\frac{d\Phi}{ds} = -K' \frac{dy}{ds} \quad \text{or} \quad \frac{\partial \Phi}{\partial x}\frac{dx}{ds} + \frac{\partial \Phi}{\partial y}\frac{dy}{ds} = -K' \frac{dy}{ds}$$

or if α is the angle made by the outflow surface with the horizontal

$$-u \cos \alpha - v \sin \alpha = -K' \sin \alpha$$

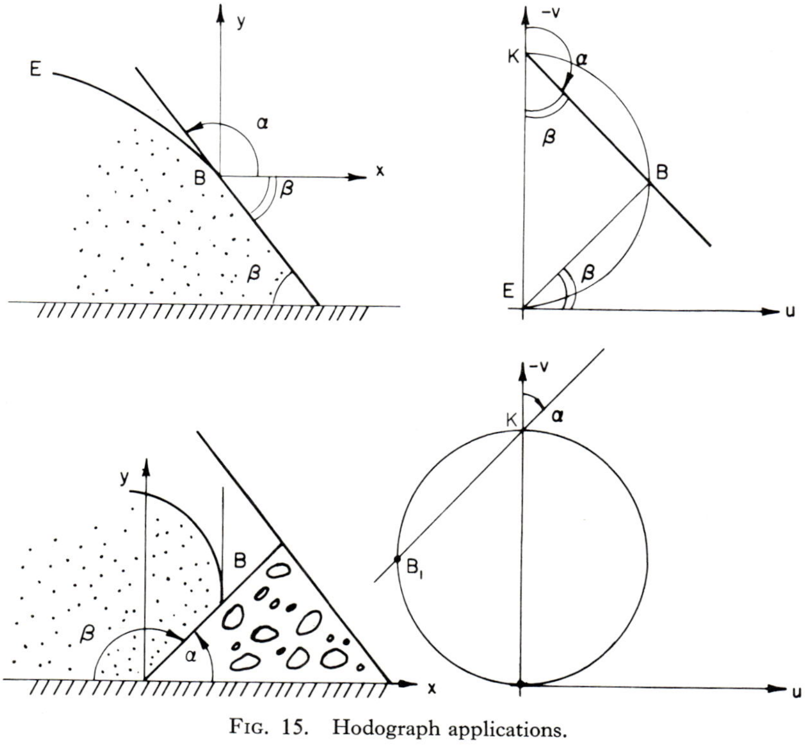

FIG. 15. Hodograph applications.

Therefore the hodograph of the outflow surface is the straight line

$$-v = -K' + u \tan(90^\circ - \alpha) \tag{111}$$

5.2.7. *Applications of Hodograph Method*

(1) Prove that the free surface (Fig. 15) is tangent to the seepage surface when $\beta < 90^\circ$. Draw hodograph of free surface and seepage surface. B must be at intersection of both lines. From the equality of the angles $(180^\circ - \alpha)$, it follows that the free surface is tangent to the seepage surface in B.

(2) Prove that the free surface (Fig. 15) is tangent to the vertical when $\beta > 90^\circ$. Draw hodograph of free surface and seepage surface. B_1 cannot be the image of B, because then u would be negative there, which is not compatible with the physics of the flow. Therefore, B must map into $u = 0$, $-v = K$. QED

6. Leaky Aquifers

Aquifers in nature are not always perfectly confined between completely impervious strata. This became more evident when the recharge conditions were studied for aquifers that apparently had larger yields than were available from the replenishment in the outcrop area. It was found that those aquifers were overlain by poorly pervious yet water-transmitting strata which, over large contact areas, contributed significantly to the recharge of the aquifer. The semiconfining beds in turn were overlain by ponded water, as in the case of the Dutch polder areas, or by other more pervious strata which again were confined or semiconfined. The percolation of water into and away from the main aquifer through the semiconfining strata is called leakage. The phenomenon was subjected to mathematical analysis first by Dutch hydrologists and engineers (De Glee, 1930; Steggewentz and Van Nes, 1939) and later, in this country, by Jacob and Hantush (see Hantush, 1964). From 1954 to 1964, Hantush developed the theory of the leaky aquifer. The first important Russian contribution on this subject seems to come from Mjatiev, who analyzed the interaction of pervious strata separated by semiconfining beds (see Polubarinova–Kochina, 1952).

The equation of the leaky aquifer has been derived earlier as [Eq. (68)],

$$\nabla^2 \bar{h} - (K'/Kbb')(\bar{h} - H) = S/T\,(\partial \bar{h}/\partial t)$$

In this equation, the drawdown $s = H - \bar{h}$ and the leakage factor $B = (Kbb'/K')^{1/2}$ are introduced so that

$$\nabla^2 s - (s/B^2) = S/T\,(\partial s/\partial t) \tag{112}$$

The leakage factor has been introduced only because of convenience in the computations. A large leakage factor means that little leakage takes place and vice versa. To characterize the amount of leakage, values of another coefficient, the leakance or leakage coefficient K'/b', are usually given numerically. This coefficient may be defined as the quantity of water that flows across a unit area of the boundary between the main aquifer and its semiconfining bed, if the difference between the head in the main aquifer and that of the ponded water supplying leakage is unity.

7. Dispersion†

Hydrodynamic dispersion (Bachmat and Bear, 1964) in porous media describes the mechanical or convective process by which one fluid displaces another fluid while mixing with it in a zone of transition, the dispersion zone. This phenomenon is studied by itself, as opposed to that of immiscible fluid displacement where a sharp interface separates regions occupied by fluids of different nature. Dispersion results in a variation of concentration of the displacing fluid in the dispersion zone, essentially because individual fluid particles travel at variable velocities through the irregularly shaped pore channels of the medium. The random distribution and orientation of these channels impose a tortuous pattern upon the microscopic streamlines although these streamlines do not intersect with each other.

Dispersion may be visualized by the introduction of a slug of dye (of initial volume V_0 and concentration c_0) in a fluid flowing through a porous medium. The center of the slug will travel along a conventional streamline with the average velocity of the entire fluid. As time goes on (Day, 1958), the slug will increase in volume, displace the surrounding fluid, and mix with it in such manner that its concentration c per unit volume of porous medium may be expressed as:

$$c/c_0 = (4\pi Dt)^{-3/2} \exp(-r^2/4Dt) \tag{113}$$

where r indicates any ray emanating from the center of the slug,

† De Wiest (1965b).

t represents the time, c_0 is the initial concentration, D is the coefficient of disperson, introduced by Scheidegger (1954). D has the dimensions L^2T^{-1}. Since concentration may be thought of as the relative density of particles of a given substance in its surroundings, the right side of Eq. (113) may also be interpreted as the probability that a given particle after a time t will be found in a unit volume located a distance r away from the expected position of the center of the slug. This spatial distribution (Day, 1958) of the particles is Gaussian with standard deviation σ:

$$\sigma = (2Dt)^{1/2} \tag{114}$$

If instead of dispersion of a slug in three dimensions, unidirectional flow is considered and the dispersion of a thin layer (say of unit thickness) of infinite extent moving with average velocity u in the direction x perpendicular to the plane of the layer, then Eq. (113) is replaced by:

$$c/c_0 = (4\pi Dt)^{-1/2} \exp[-(x - ut)^2/4Dt] \tag{115}$$

This equation shows that, at least in this simplified framework of average unidirectional flow, lateral components of dispersion cancel out or neutralize one another, while the individual particles are normally distributed in the x-direction around the plane of maximum concentration that moves with average velocity $u = x/t$. Equation (115) is a solution of the partial differential equation (so-called diffusion equation, by analogy with heat conduction):

$$\partial c/\partial t = D\, \partial^2 c/\partial X^2 \tag{116}$$

where $X = x - ut$, as may be certified by direct insertion of Eq. (115) into Eq. (116). Here the origin of the coordinate axis $X = 0$ moves with the average velocity u; in other words, it is attached to the dispersing layer.

Unidirectional flow with longitudinal dispersion has been tested in the laboratory by means of columns (lucite tubes filled with a water-saturated porous medium), at the entrance of which a tracer solution of given concentration is introduced. The dispersion equation in a coordinate system fixed to the column becomes:

$$\partial c/\partial t = D(\partial^2 c/\partial x^2) - u(\partial c/\partial x) \tag{117}$$

where u is again the average velocity of flow and where the other symbols are as defined before.

As shown by Rifai *et al.* (1956), and Harleman and Rumer (1962), the solution of Eq. (117) for the initial condition

$$c(x, 0) = 0, \qquad x > 0$$

and boundary conditions

$$c(0, t) = c_0, \qquad t \geqslant 0$$

$$c(\infty, t) = 0, \qquad t \geqslant 0$$

is given, as a first approximation, by (see Fig. 16)

$$\frac{c}{c_0} = \frac{1}{2} \operatorname{erfc}\left[\frac{x - ut}{2(Dt)^{1/2}}\right] \tag{118}$$

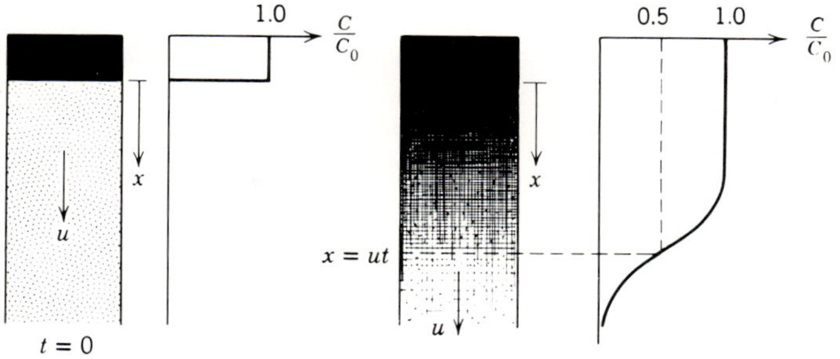

FIG. 16. Graphic representations of longitudinal dispersion (after Harleman and Rumer, 1962).

where erfc(x) is the complementary error function:

$$1 - \operatorname{erfc}(x) = 1 - \frac{2}{(\pi)^{1/2}} \int_0^x e^{-\alpha^2}\, d\alpha$$

Actually the solution of Eq. (117) for the given conditions contains a second term which may be neglected only for $D < ux$. Although the above variation of the concentration of the tracer in the displacing front is derived for a tube of infinite length, its typical sigmoid shape has been obtained in a very similar way as a so-called "break through" curve for a column of finite length. Here the ratio of the concentration c at the column outlet over the constant concentration c_0 at the inlet is plotted against the ratio of effluent volume V_{ef} over the water volume V_w

in the voids of the column. (The latter ratio is commonly called pore-volume.)

Figure 17 represents an experimental breakthrough curve and a curve for piston flow, a fictitious flow in which the velocities would all be the same along parallel streamlines. If the effect of molecular diffusion were neglected, then the inflection point of the sigmoid curve would have the coordinates $c/c_0 = 0.5$ and $V_{\mathrm{ef}}/V_{\mathrm{w}} = 1.0$. This would essentially mean that the breakthrough curve has a mean concentration that moves with the average flow velocity. Biggar and Nielsen (1960), in particular, investigated the effects of the molecular or ionic diffusion as part of the general dispersion phenomenon. They proved that molecular diffusion becomes important at small flow velocities, and also when the medium consists of a natural soil skeleton instead of an assembly of washed sands or glass beads of uniform diameter size,

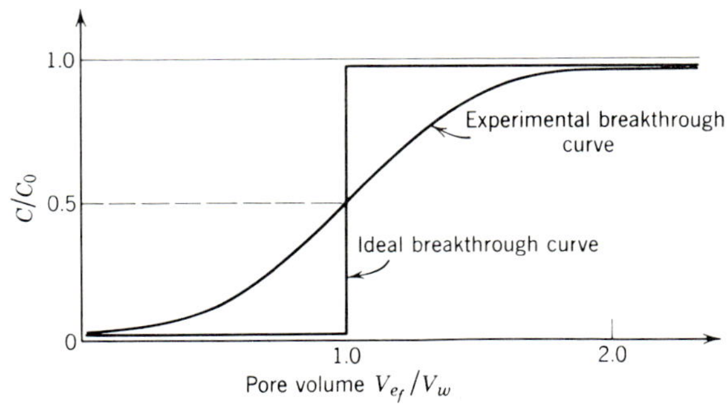

FIG. 17. Breakthrough curves (after Rifai *et al.*, 1956).

and when the medium is not completely saturated. Due to the presence of "dead-end pores," a significant fraction of the total pore volume of a soil does not contribute to the volume of effluent measured, and this results in "holdback," or a translation of the breakthrough curve to the left. This translation is a relative measure of the volume of water not displaced but remaining within the sample, and its magnitude depends on the amount of nonmoving water (a characteristic of the soil), so that clays would have breakthrough curves more displaced than sands. Ionic diffusion also becomes important at low flow velocities when the medium is not completely saturated and this is manifested by a translation of the breakthrough curve to the left, with an early arrival of tracer particles in the effluent, and a slower asymptotic approach of the curve to the value $c/c_0 = 1.0$. Biggar and Nielsen's work clearly showed that intrinsic differences among porous materials such as clays and sands became evident through the shapes and positions of breakthrough curves obtained from saturated and unsaturated media.

Saffman (1959) studied a statistical model of porous medium of a more general nature than that proposed by Scheidegger (1954) and also suitable for the investigation of lateral dispersion. He objected to Scheidegger's assumption that a particle carries out a random walk consisting of statistically independent straight steps in equal small intervals of time, because it would be expected that a particle stays longer in a region where the velocity is small than in one where it is large. Instead he proposed that the path followed by each particle would be a random walk in which length, direction and duration of each step are random variables. The probability distribution function of the displacement of a fluid particle after a given time may then be calculated and a value for the dispersion obtained. Again the assumption was made that molecular diffusion would be small as compared to mechanical dispersion.

As to the expression for the coefficient D of longitudinal dispersion, there is general agreement among researchers (Harleman and Rumer, 1962) that it is of the form:

$$D = D_{\mathrm{m}} + \alpha u^n \tag{119}$$

where D_{m} is the effective molecular diffusion coefficient, u is the average flow velocity and the value of the exponent n is close to one. For $n = 1.0$, α has the dimension of a length, characteristic of the medium only. In a similar way, an expression for the lateral dispersion coefficient may be derived. Both dispersion coefficients have been experimentally determined for various porous materials.

The study of miscible displacement is of particular interest to the civil engineer in charge of water supply from coastal aquifers, where saline water from the sea intrudes into the aquifer and mixes with fresh water. The study of the static and dynamic equilibrium between fresh water and salt water has first been treated as that of the position of an interface between two immiscible fluids. Later on (Harleman and Rumer, 1962), effects of dispersion were incorporated in the study, using the results available from the above analysis in which the dispersion equation was solved for certain idealized boundary conditions. Considerable progress has been made in the late 1950's and early 1960's in the experimental and analytical investigation of transfers of water in porous media in multiple phase conditions (de Josselin de Jong, 1958, 1960; Ogata, 1961; Ogata and Banks, 1961; Simpson, 1962).

8. Flow of a Nonhomogeneous Fluid in a Porous Medium

Perhaps the most significant contribution to the theory of flow of chemically inhomogeneous fluids has been made by Lusczynski (1960). Contingent upon the reading of waterlevels in some observation wells, it allows for the computation of the velocity vector field in a medium where the salt content of the water gradually varies. The value of this theory for engineering purposes cannot be overemphasized because the velocity vector field may be constructed regardless of the often complicated boundary conditions created by the geological nature of the water bearing strata, which preclude a complete analytical solution of the problem.

Lusczynski introduced the concepts of point-water head, fresh-water head and environmental-water head to compute the velocity distribution in the zone of dispersion and to determine the front of the dispersed water in the fresh water. In the sequel we will use his concepts of point-water head and fresh-water head but will replace his definition of environmental-water head by one of "true environmental-water head." This leads to a simpler derivation of the velocity field but it does no longer allow for the determination of the fresh-water dispersed water interface. For the computation of the latter, the reader is referred to the paper by Lusczynski.

8.1. Point-Water Head

Figure 18(a) shows a real well cased over its entire depth and screened at the point where one wants to measure the head in function of the density of the water at the point. Here the well is screened in salt water and filled only with salt water of uniform density as found at the bottom of the well. The overlying fresh or dispersed water that may have entered the well during its construction has been removed. Similarly, point-water head may be determined in the zones of fresh water and dispersed water. Point-water head H_{ip} at a point i in ground water of variable density ρ_i is defined as the water level, measured here from mean sea level as datum plane, in a well filled with an amount of water of the type found at the point and sufficient to balance the existing pressure p at the point. By definition:

$$H_{ip} = z + (p/\rho_i g) \tag{120}$$

where the first subscript in H_{ip} refers to the point at which the head is measured, and the second subscript indicates that the water in the well is that at the point in question.

8.2. Fresh-Water Head

If one end of a standpipe were introduced at the point i of Fig. 18(b) and if the pipe were filled with fresh water of density ρ_f only to such an extent that it would balance the existing pressure at point i, then the level of the fresh water in the standpipe would measure the fresh-water head H_{if} defined as:

$$H_{if} = z + (p/\rho_f g) \tag{121}$$

where again mean sea level is chosen as datum plane for elevation.

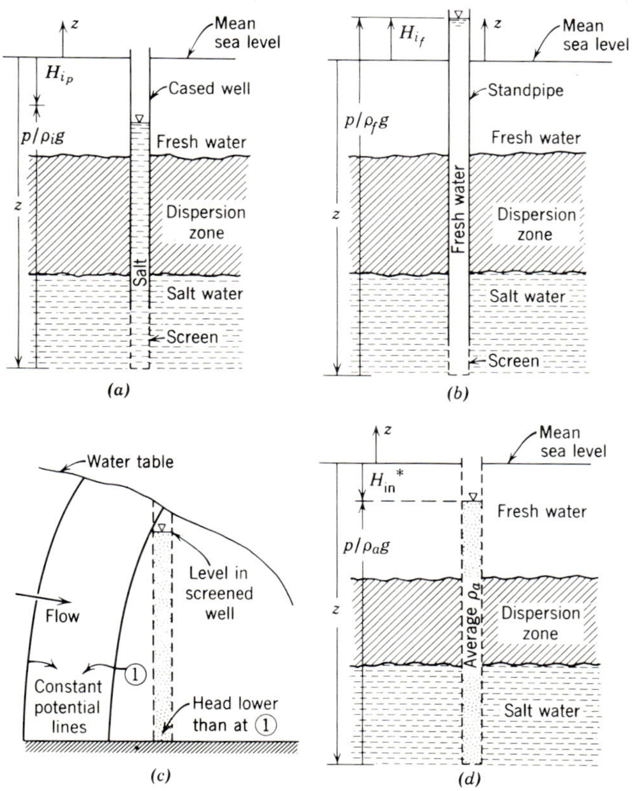

Fig. 18. Fresh-water, salt-water, and environmental-water heads (after De Wiest, 1965a).

Fresh-water heads could only be measured in fresh water with such a device, but not in salt nor in diffused water, whereas point-water heads can always be measured. Therefore, it is useful to express fresh-water heads in terms of measurable point-water heads. This is done by stating that the pressure p is unique and therefore, the same in Eqs. (120) and (121). Elimination of p/g from these equations leads to:

$$\rho_f H_{if} = \rho_i H_{ip} - z(\rho_i - \rho_f) \tag{122}$$

As is indicated in Fig. 18(a) and (b) the point-water head for a point in the salt water zone may be negative whereas the fresh-water of the same point may be positive.

8.3. True Environmental-Water Head

This type of head may be visualized by the water level in a real well screened over its entire depth. Water of variable density would penetrate in such a well and essentially occupy the same position in depth as under natural conditions.

If, for example, a borehole is drilled in an unconfined aquifer of water of constant density and screened over its entire depth, then it will fill with water up to a level slightly below the water table [Fig. 18(c)]. Indeed this level is indicative of the average head over the vertical through a point of the watertable, because the equipotential lines are curved under dynamic flow conditions. If the water had a variable density, with the denser water below, as is the case of Fig. 18(a) and (b), then the head at the bottom of the well would even be smaller and hence the true environmental-water head would always be indicated in the well by a water level below the top of the zone of saturation. Define now

$$H_{in}^* = z + (p/\rho_a g) \tag{123}$$

where H_{in}^* is the environmental-water head, and ρ_a the average density of water between point i and the free surface as indicated in Fig. 18(d):

$$\rho_a = (H_{in}^* - z)^{-1} \int_z^{H_{in}^*} \rho(x, y, l)\, dl \tag{124}$$

where x, y, z are the coordinates of the point i and where l is a dummy variable of integration. Elimination of the pressure p from Eqs. (121) and (123) leads to

$$\rho_a H_{in}^* = \rho_f H_{if} - z(\rho_f - \rho_a) \tag{125}$$

The generalized form of Darcy's law for flow of a chemically inhomogeneous fluid may be expressed as:

$$\mathbf{q} = \frac{k_i}{\mu_i} (\rho_i \mathbf{g} - \operatorname{grad} p) = -\frac{k_i g}{\mu_i} \left(\rho_i \operatorname{grad} z + \frac{1}{g} \operatorname{grad} p\right) \tag{126}$$

where $\mathbf{q}$ is the specific discharge or bulk velocity in point i; k the intrinsic permeability of the medium; μ, ρ, p, respectively, the dynamic viscosity, and pressure of the fluid. [See Eq. (28).]

In order to derive an equation for $\mathbf{q}$ in terms of H_{if}, H_{in}^*, ρ_f, and ρ_a, equivalent expressions of the right side of Eq. (126) are developed. First, the gradient of Eq. (125) is taken to render:

$$\rho_a \operatorname{grad} H_{in}^* = \rho_f \operatorname{grad} H_{if} + (\rho_a - \rho_f) \operatorname{grad} z + (z - H_{in}^*) \operatorname{grad} \rho_a$$

or (127)

$$\rho_a \operatorname{grad} H_{in} - (z - H_{in}^*) \left[\frac{\partial \rho_a}{\partial x} \mathbf{u}_x + \frac{\partial \rho_a}{\partial y} \mathbf{u}_y\right] = \rho_f \operatorname{grad} H_{if} + (z - H_{in}^*) \frac{\partial \rho_a}{\partial z} \mathbf{u}_z + (\rho_a - \rho_f) \mathbf{u}_z$$

where $\mathbf{u}_x$, $\mathbf{u}_y$, $\mathbf{u}_z$ are unit vectors along the coordinate axes. Furthermore, differentiation of ρ_a with respect to z renders:

$$\frac{\partial \rho_a}{\partial z} = \frac{\rho_a - \rho_i}{H_{in}^* - z} + \frac{\partial H_{in}^*}{\partial z} \frac{\rho_f - \rho_a}{H_{in}^* - z} \tag{128}$$

and Eq. (127) becomes, after insertion of Eq. (128)

$$\rho_a \left(\frac{\partial H_{in}^*}{\partial x} \mathbf{u}_x + \frac{\partial H_{in}^*}{\partial y} \mathbf{u}_y + \frac{\partial H_{in}^*}{\partial z} \mathbf{u}_z\right) + (\rho_f - \rho_a) \frac{\partial H_{in}^*}{\partial z} \mathbf{u}_z - (z - H_{in}^*)\left[\frac{\partial \rho_a}{\partial x} \mathbf{u}_x + \frac{\partial \rho_a}{\partial y} \mathbf{u}_y\right] = \rho_f \operatorname{grad} H_{if} + (\rho_i - \rho_f) \mathbf{u}_z$$

and finally,

$$\rho_a \left[\frac{\partial H_{in}^*}{\partial x} \mathbf{u}_x + \frac{\partial H_{in}^*}{\partial y} \mathbf{u}_y\right] + \rho_f \frac{\partial H_{in}^*}{\partial z} \mathbf{u}_z - (z - H_{in}^*) \left[\frac{\partial \rho_a}{\partial x} \mathbf{u}_x + \frac{\partial \rho_a}{\partial y} \mathbf{u}_y\right] = \rho_f \operatorname{grad} H_{if} + (\rho_i - \rho_f) \mathbf{u}_z \tag{129}$$

Also, the gradient of Eq. (121) renders, after addition of ρ_i grad z to both members of the equation:

$$\rho_i \operatorname{grad} z + (1/g) \operatorname{grad} p = \rho_f \operatorname{grad} H_{if} + (\rho_i - \rho_f) \mathbf{u}_z \tag{130}$$

In the case of an anisotropic medium, if k_x, k_y, k_z are the intrinsic permeabilities in the principal directions, coinciding with the coordinate axes, the components of Eq. (126) may be written as:

$$q_x = -\frac{k_x g}{\mu_i}\left[\rho_i \operatorname{grad} z + \frac{1}{g} \operatorname{grad} p\right]_x, \qquad q_y = -\frac{k_y g}{\mu_i}\left[\rho_i \operatorname{grad} z + \frac{1}{g} \operatorname{grad} p\right]_y$$

$$q_z = -\frac{k_z g}{\mu_i}\left[\rho_i \operatorname{grad} z + \frac{1}{g} \operatorname{grad} p\right]_z$$

Because the right sides of Eqs. (129) and (130) are the same, both sides of Eq. (129) may be used as equivalents of the left side of Eq. (130) and hence be inserted in Eq. (126) to render the velocity components of the specific discharge. The interesting feature of Eq. (129) is that its left side, when projected on a vertical, reduces to $\rho_f(\partial H_{in}^*/\partial z)$, which is simply a product of fresh-water density and a gradient defined by environmental-water heads. Also, along any horizontal the right side of Eq. (129) reduces to $\rho_f(\partial H_{if}/\partial l)$, which is simply a product of fresh-water density and a directional derivative defined by fresh-water heads. Therefore,

$$q_x = -\frac{k_x g}{\mu_i}\rho_f\frac{\partial H_{if}}{\partial x}, \qquad q_y = -\frac{k_y g}{\mu_i}\rho_f\frac{\partial H_{if}}{\partial y}, \qquad q_z = -\frac{k_z g}{\mu_i}\rho_f\frac{\partial H_{in}}{\partial z} \qquad (131)$$

When the velocity components are computed by means of Eq. (131) and when the resultant velocity is constructed, it will follow that the direction of flow is usually not parallel to the resultant hydraulic gradient.

The strength of Lusczynski's theory lies in the fact that Hubbert's equation of the interface of immiscible liquids can be derived as a special case of Lusczynski's more general concepts. Even the elementary Ghyben–Herzberg equation for the depth of salt water underlying fresh water may be derived as an extreme limit case of Lusczynski's theory. The restrictions on the Ghyben–Herzberg equation, as seen from Lusczynski's theory, may be summarized in three conditions, respectively requiring that there be no zone of dispersion, that there be no vertical velocity components due to differences in environmental-water head between points along the vertical in salt-water and fresh-water zones, and that the point-water head in the salt-water zone be negligible.

Hubbert's equation, to be valid from the Lusczynski point of view, requires only two of the three aforementioned conditions. Hubbert's equation indeed has not neglected the point-water head in the salt-water zone.

REFERENCES

Aravin, V. I., and Numerov, S. (1955). "Seepage Computations for Hydraulic Structures." Moscow State Press.

Bachmat, Y., and Bear, J. (1964). The general equations of hydrodynamic dispersion in homogeneous, isotropic porous media, *J. Geophys. Res.* **69**, 2561–2567.

Bear, J., and Dagan, G. (1962). The Use of the Hodograph Method for Ground Water Investigations, Publ. No. 24, Faculty of Civil Eng., Technion, Haifa, Israel.

Bear, J., and Dagan, G. (1964). Some exact solutions of interface problems by means of the hodograph method, *J. Geophys. Res.* **69**, 1563–1572.

Biggar, J. W., and Nielsen, D. R. (1960). Diffusion effects in porous materials, *J. Geophys. Res.* **65**, 2887–2895.

Biot, M. A. (1955). Theory of elasticity and consolidation for a porous anisotropic solid, *J. Appl. Phys.* **26**, 182–185.

Bird, R. B., Stewart, W. E., and Lightfoot, E. N. (1960). "Transport Phenomena." Wiley, New York.

Boothroyd, A. R., Cherry, E. C., and Makar, R. (1949). An electrolytic tank for the measurement of steady state response, transient response and allied properties of networks, *Proc. Inst. Elec. Engrs. (London)* **96**, 163–177.

Charmonman, S. (1965). A Solution of the pattern of fresh-water flow in an unconfined coastal aquifer, *J. Geophys. Res.* **70**, 2813–2819.

Churchill, R. V. (1948). "Introduction to Complex Variables and Applications." McGraw-Hill, New York.

Collins, R. E. (1961). "Flow of Fluids through Porous Materials." Reinhold, New York.

Darcy, H. (1856). "Les fontaines publiques de la ville de Dijon." Dalmont, Paris.

Day, P. R. (1958). Dispersion of a moving salt-water boundary advancing through saturated sand, *Trans. Am. Geophys. Union* **39**, 67–74.

De Glee, G. J. (1930). "Over Grondwaterstromingen by Wateronttrekking by middel van Putten." T. Waltman, Jr., Delft.

de Jong, J. (1962). Een eenvoudige methode voor het nabootsen van een oneindig potentiaalveld, *Water* **46**, 185–186.

de Josselin de Jong, G. (1958). Longitudinal and transverse diffusion in granular deposits, *Trans. Am. Geophys. Union* **39**, 67–74.

de Josselin de Jong, G. (1960). Singularity distributions for the analysis of multiple-fluid flow through porous media. *J. Geophys. Res.* **65**, 3739–3758.

de Josselin de Jong, G. (1965). A many valued hodograph in an interface problem, *Water Resources Res.* **1**, 543–555.

De Wiest, R. J. M. (1960). Unsteady flow through an underdrained earth dam, *J. Fluid Mech.* **8**, 1–9.

De Wiest, R. J. M. (1961). Free surface flow in homogeneous porous medium, *J. Hydraulics Div.*, Am. Soc. Civil. Engrs. 181–220. Also with discussion: (1962). *Trans. Am. Soc. Civil Engrs.* **127**, 1045–1089.

De Wiest, R. J. M. (1965a). "Geohydrology." Wiley, New York.

De Wiest, R. J. M. (1965b). Dispersion and diffusion phenomena in porous media. *Bull. RILEM, Paris* **29**, 17–22.

De Wiest, R. J. M. (1965c). History of the Dupuit–Forchheimer assumptions in groundwater hydraulics. *Trans. Am. Soc. Agricultural Engrs.* **8**, 508–509.

De Wiest, R. J. M. (1966a). On the storage coefficient and the equations of ground water flow, *J. Geophys. Res.* **71**, 1117–1122, 5016–5017.

De Wiest, R. J. M., (1966b). Hydraulic model study of nonsteady flow to multiaquifer wells, *J. Geophys. Res.* **71**, 4799–4810.

Dupuit, J. (1863). "Études théoriques et pratiques sur le mouvement des eaux dans les canaux découverts et à travers les terrains perméables," 2nd ed. Dunod, Paris.

Hantush, M. S. (1964). Hydraulics of wells. *Advan. Hydrosci.* **1**, 281–432.

Harleman, D. R. F., and Rumer, R. R. (1962). The dynamics of salt-water intrusion in porous media, Civ. Eng. Dept. Rept. No. 55, M.I.T. Press, Cambridge, Massachusetts.

Harr, M. E. (1962). "Ground Water and Seepage." McGraw-Hill, New York.

Hubbert, M. K. (1940). The theory of ground-water motion, *J. Geol.* **48**, 785–944.

Hubbert, M. K. (1953). Entrapment of petroleum under hydrodynamic conditions, *Bull. Am. Assoc. Petrol. Geologists.* **37**, 1954–2026.

Hubbert, M. K. (1956). Darcy's law and the field equations of the flow of underground fluids, *Trans. Am. Inst. Mining Met. Engrs.* **207**, 222–239.

Jacob, C. E. (1940). The flow of water in an elastic artesian aquifer, *Trans. Am. Geophys. Union*, **21**, 574–586.

Jacob, C. E. (1950). *In* "Engineering Hydraulics," (H. Rouse, ed.), Chap. 5, pp. 321–386. Wiley, New York.

Lusczynski, N. J. (1960). Head and flow of ground water of variable density, *J. Geophys. Res.* **66**, 4247–4256.

Meinzer, O. E. (1928). Compressibility and elasticity of artesian aquifers, *Econ. Geol.* **23**, 263–291.

Ogata, A. (1961). Transverse diffusion in saturated isotropic granular media, USGS, Professional Paper 411-B. Washington, D.C.

Ogata, A., and Banks, R. B. (1961). A solution of the differential equations of longitudinal dispersion in porous media, USGS, Professional Paper 411-A, Washington, D.C.

Polubarinova-Kochina, P. Ya. (1952). "Theory of Ground Water Movement." Princeton Univ. Press, Princeton, New Jersey.

Rifai, M. N. E., *et al.* (1956). Dispersion phenomena in laminar flow through porous media, "Sanit. Eng. Res. Lab. Rept.," No. 3, Univ. of California Press, Berkeley, California.

Saffman, P. G. (1959). A theory of dispersion in a porous medium, *J. Fluid Mech.* **6**, 321–349.

Santing, G. (1957). A horizontal scale model, based on the viscous flow analogy, for studying ground water flow in an aquifer having storage, Intern. Assoc. Sci. Hydrol. 105–114, General Assembly, Toronto.

Scheidegger, A. E. (1954). Statistical hydrodynamics in porous media, *J. Appl. Phys.* **25**, 997–1001.

Simpson, E. S. (1962). Transverse dispersion in liquid flow through porous media, USGS Professional Paper 411-C, Washington, D.C.

Steggewentz, J. H., and Van Nes, B. A. (1939). Calculating the yield of a well taking account of replenishment of the ground water from above. *Water Water Eng.* **41**, 561–563.

Theis, C. V. (1935). The relation between the lowering of the piezometric surface and the rate and duration of discharge of a well using ground water storage. *Trans. Am. Geophys. Union*, **16**, 519–524.

Theis, C. V. (1938). The significance and nature of the cone of depression in ground-water bodies, *Econ. Geol.* **33**, 889–920.

2 Porosity and Permeability of Natural Materials

Stanley N. Davis
Department of Geology
University of Missouri
Columbia, Missouri

1. Introduction

Few properties of natural materials have received more attention from hydrogeologists than porosity and permeability. Despite the vast amount of work expended in measuring these two properties in countless samples, reliable values can rarely be estimated on the basis of general geologic knowledge alone. The purpose of this chapter is to review the extent of our general knowledge concerning porosity and permeability and to relate this knowledge to the more analytical discussions which follow in subsequent chapters.

1.1. Porosity

Porosity, as used in this chapter, is defined as the ratio of pore volume to the total volume of a given sample of material. This definition is adequate for many purposes. Microscopically and submicroscopically, nevertheless, a complete gradation exists from large pores easily accessible to fluids to very small openings in minerals that are caused by minor lattice imperfections. The smallest openings contain fluids that are rendered essentially static by the close proximity of atomic force fields at the surfaces of the solids (Kemper *et al.*, 1964). Whether or not these submicroscopic openings are measured in the laboratory as part of the void space depends on the methods of the measurement employed. Techniques using heat, centrifuging, liquid saturation under a vacuum, and other conditions abnormal to near-surface rocks will certainly displace static fluid that might be best considered part of the solid rock.

A large number of terms have been introduced into the literature to describe various physical measures related to porosity. Moisture equivalent, effective porosity, specific retention, drainage coefficient, coefficient of storage, and specific yield are some of the terms used to qualify variables such as degree of saturation, forces applied to the sample, length of test, degree of interconnection of pores, and fluid chemistry. Besides simple porosity, the only other closely related term used in this chapter will be specific yield, which will be defined as the volume of water drained by gravity from an initially saturated sample divided by the total volume of that sample. Ideally, specific yield should be measured using samples having column lengths of more than 3 m to avoid excessive capillary effects. Drainage time should be measured in many years for clay-size material. As a result of these inconvenient requirements, few reliable measurements of specific yield of fine-grained materials have been made. Modern laboratories tend to rely on centrifuging to speed drainage and to overcome capillary effects. Strain induced by centrifuging, however, makes the resultant specific yield values of fine grained materials open to question. This is particularly true of nonindurated silts and clays.

1.2. Permeability

Permeability is a measure of the ease with which fluids pass through a porous material. Permeability, or more properly, "intrinsic permeability," is a property of the solid material and is independent of the density and viscosity of the fluid. A common expression for permeability k is

$$k = -\frac{Q}{A}\frac{\mu}{\rho g}\left(\frac{\partial h}{\partial s}\right)^{-1}$$

in which Q is the volume of fluid discharged per unit time through a cross section having an area of A, μ is the viscosity of the fluid, ρ is the density of the fluid, g is the acceleration of gravity, and $\partial h/\partial s$ is the hydraulic gradient in the direction of flow s.

Permeability k has the dimensions of L^2. The darcy, which is almost universally used in the petroleum industry, will be used as the unit of permeability in this chapter. One darcy has the value of 0.987×10^{-8} cm^2. Units of hydraulic conductivity, which superimpose fluid properties on permeability and have the dimensions of LT^{-1} are most common in ground-water work. The "meinzer" is one of the popular units of hydraulic conductivity in the United States. It is related to a fixed viscosity and density of water (pure water at 60° F or 15.6° C). Media having a permeability of 1 darcy have a hydraulic conductivity of 18.2 meinzers.

Analytical work with single-phase flow of fluids almost universally assumes that permeability is constant with respect to time. Several important exceptions should always be kept in mind. Largest changes take place with changes of state or dissolution of minerals, as for example the slow formation of limestone caverns, the more rapid dissolution of gypsum, or the extremely rapid melting of granular ice as warm rain water percolates through a snow bank. Significant changes also take place in response to stresses associated with declining artesian pressures or with external compaction of sediments. If fluid chemistry changes, sedimentary materials will respond dynamically. Effects of partial or complete hydration which will cause clays, colloids, and some organic material to expand and close pore spaces are the most important of the responses to changing fluid chemistry. Permeabilities are largest with air and other nonpolar fluids. Partial hydration by brines will cause values to lower. Full hydration using distilled water as the measuring fluid will cause a drastic reduction of permeability (Johnston and Beeson, 1945). These effects are greatest in fine-grained sediments and could account for permeability differences of several orders of magnitude in the sample.

Effects of measuring techniques on the permeability of fine-grained sediments and rocks is particularly striking. Young *et al.* (1964) have shown that increasing confining pressures of samples of argillaceous rocks in order to duplicate subsurface conditions produces a decrease in permeability of more than tenfold over results of unconfined tests. Permeability contrasts in some nonindurated materials may be even larger.

Natural material is commonly assumed to be homogeneous and isotropic with respect to permeability. In general, computational problems associated with heterogeneous aquifers are easier to deal with than those of nonisotropic aquifers. The measurement and/or estimation of the actual extent of heterogeneity is also much easier geologically than is the measurement of anisotropy. Further attention will be given to these matters in later sections of this chapter as well as in other parts of this volume.

2. Dense Rocks

2.1. Introduction

A large number of rocks have porosities of less than 2% and permeabilities of less than 10^{-2} darcy. Granite, dolerite, quartzite, slate, and gabbro are but a few of the common dense rocks which are abundant in the earth's crust. Of the major rock groups, most metamorphic, all plutonic igneous rocks, most volcanic rocks in near-surface intrusives, and many dense sedimentary rocks are grouped together as far as their gross hydrogeologic properties are concerned. Such rocks in their unmodified state will not yield fluids to wells nor will they store significant amounts of fluids in their limited pore spaces. Typical values of porosity and permeability for dense rocks are given in Table 1.

2.2. Weathering and Fracturing

Weathering and fracturing of dense rocks will increase overall permeabilities by two to four orders of magnitude (Table 2). Weathering will increase rock porosity from initial values generally less than 5% to final values of between 30 and 60% (Stewart, 1964). Fracturing alone,

TABLE 1

Rocks of Low Permeabilities and Porosities

Rock name	Porosity (%)	Permeability (darcys)	Reference
Chert	—	1.9×10^{-7}	Stuart *et al.* (1954)
Diabase	0.1	—	Brace (1965)
Dolomite	6.3	1.0×10^{-3}	Murray (1960)
	0.4	—	Brace (1965)
Granite	0.3	—	Brace (1965)
Graywacke	—	3.0×10^{-6} [a]	Stuart *et al.* (1954)
Limestone	8.4	1.0×10^{-3}	Murray (1960)
Marble	0.3	—	Brace (1965)
	0.6[b]	—	Manger (1963)
Mica schist	—	2.1×10^{-6}	Stuart *et al.* (1954)
Quartzite	—	1.9×10^{-6}	Stuart *et al.* (1954)
	0.6	—	Brace (1965)
Rock salt	0.6	7.3×10^{-6} [c]	Gloyna and Reynolds (1961)
Slate	—	1.3×10^{-6} [d]	Stuart *et al.* (1954)
	3.4[e]	—	Manger (1963)

[a] Median of 5 samples.

[b] Mean of 100 samples.

[c] Representative value taken from several different determinations. Permeability varied widely with confining pressure on specimen and with the type of fluid used.

[d] Median of 9 samples.

[e] Mean of 19 samples.

however, does not generally increase overall rock porosities by more than 2–5%. The size of fractures may range from large dislocations having open cracks of more than 1.0 mm to microfractures that develop throughout the rock prior to failure. Recent laboratory experiments (Brace *et al.*, 1966) have shown that initially dense rocks will gain from 0.004 to 3.6% in porosity after being compressed under moderately high confining pressures. Brittle rocks such as aplite will gain the least porosity, and slightly ductile rocks such as limestone will gain the most porosity. The hydrogeologic importance of this rock dilatancy under stress is yet to be investigated. It is possible that large volumes of otherwise dense rock will undergo small but significant increases in porosity near major faults and possibly even in zones of regional stress accumulation that are not broken by major faults.

Both weathering and fracturing are most evident within 20 m of the surface. In tropical countries weathering may extend to depths of more than 100 m, but most temperate regions have not experienced significant weathering of dense rocks below about 50 m.

TABLE 2

Porosity and Permeability of Originally Dense Rocks Modified by Fracturing and Weathering

Rock name	Type of test	Porosity (%)	Specific yield (%)	Permeability (darcys)	Reference
Quartz-mica schist, weathered	Laboratory	48[a]	20.6[a]	3.3×10^{-2} [a]	Stewart (1964)
	Field	—	0.61	0.97[b]	Stewart (1964)
Graywacke, fractured	Field	—	—	4.5×10^{-2} [c]	Lewis *et al.* (1966)
Metasediments fractured	Field	2.4 (estimated)	—	3.1×10^{-2} [d]	Lewis *et al.* (1966)
Iron formation	Field	—	—	5.5×10^{-4} (minimum) 5.8×10^{-2} (maximum)	Stuart *et al.* (1954)
Metabasalt	Field	—	2.0	1.9[e]	Meyer and Beall (1958)
Schist	Field	—	3.0	1.4[e]	Meyer and Beall (1958)
Marble	Field	—	0.4	18[f]	Meyer and Beall (1958)

[a] Median values for 21 samples.
[b] Assumed aquifer thickness of 70 ft.
[c] Median of 11 tests.
[d] Median of 6 tests.
[e] Assumed aquifer thickness of 200 ft.
[f] Assumed aquifer thickness of 160 ft.

Fracturing of rocks may extend to depths of several thousands of meters, and water has been encountered in some mines and tunnels at depths of more than 1000 m. Large excavations far below the surface, nevertheless, generally encounter little water (Samuelson, 1965).

Fractures of several types exist in dense rocks. Faults are extensive fractures that mark zones of differential movement within the earth's crust. They generally extend to an unknown depth below the surface. Some faults most certainly exist at depths of more than 10 km below the surface and hence penetrate all fluid-bearing zones of current economic interest. Fractures within the fault zones will tend to close with depth owing to the weight of overlying rocks. Joints are single fractures, commonly with planar surfaces along which macroscopic movement parallel with the surface is lacking. Joints arise from tensile stresses associated with the shrinkage of rocks during the drying, cooling, or chemical transformations; shearing stresses associated with the expansion of rocks during warming, hydration of minerals, or unloading; and compressive forces most commonly associated with large-scale earth movements. Although, theoretically, some joints can propagate to great depths, the frequency of many joints decreases rapidly with depth (Jahns, 1943). Besides joints and faults, fractures of many origins are encountered near the surface. Landslides, impact effects, explosions, subsidence, and various human activities can disrupt rocks to depths of a few hundred meters.

2.3. Relation of Permeability and Porosity to Depth

The permeability and porosity of dense rocks decrease with depth. This decrease is not only caused by a smaller amount of rock weathering with depth but is also caused by the fact that many forces causing fracturing are localized near the surface. The rate of decrease varies widely from one locality to the next. As a rough generalization, however, both the porosity and permeability of originally dense rocks should be about 10 times greater at a depth of only 10 m than at a depth of 100 m.

Numerous ground-water studies have noted a decrease of yields with depth, which is related to the decrease of permeability and porosity with depth (Frommurze, 1937; Johnston, 1962; Le Grand, 1954; and Dingman *et al.*, 1956). Several possible sources of bias, nevertheless, are present in the water-well data. First, testing methods used are not uniform. Second, climatic conditions vary widely in regions used for comparison. Third, and probably most serious, wells started in areas of permeable rocks will be successful and therefore will be terminated at a shallow depth. On the other hand, wells drilled in areas of relatively impermeable rocks will tend to be drilled much deeper in unsuccessful attempts to reach water-bearing horizons. A comparison of wells within the entire region will show high water yields per foot of depth in shallow zones and a low yields per foot of depth in deeper wells. This may suggest falsely that permeability decreases with depth, whereas the variations of permeability are actually lateral rather than vertical.

Data from tests made on holes drilled for dam and tunnel site exploration have been studied in an attempt to evalute the possibility of bias in the water-well data (Davis and Turk, 1964). The depth and spacing of these holes are commonly determined before drilling, and many are positioned so as to intercept potentially permeable zones near the bottoms of the holes. Thus, the data are biased, if at all, in the direction of an increase in the permeability of rocks with depth. The data from hydrogeologic testing of the holes are also gathered in a more uniform manner, so that differences related to testing methods are not as important as in water wells. Results of the study showed essential agreement between test-hole and water-well data (Fig. 1); namely, a rapid decrease with depth of the water that can be injected into the drillholes. The general conclusion is, therefore, that whatever biases exist in the water-well data, they are not important enough to mask the gross relationships of water yield and depth.

The relation between well depths and the yield of wells has been studied mostly for igneous and metamorphic rocks, and only to a minor degree for sedimentary rocks. Rather limited data suggest, however, that the same general type of depth-yield relationships holds for all dense rocks.

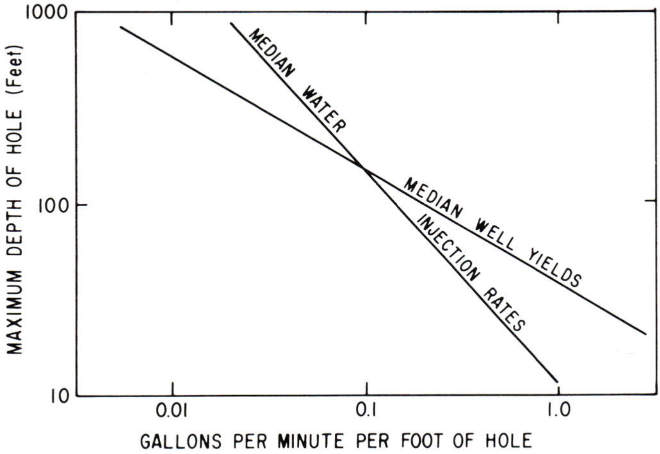

FIG. 1. The relation between depth, and yield and injection rates for granitic rocks of the Sierra Nevada (Davis and Turk, 1964).

Lack of detailed information on the actual configuration of flow in the vicinity of wells in dense rocks makes it difficult to relate well yields to gross permeabilities of the rocks. Tracer dilution tests require a prior knowledge of the local hydraulic gradient in the vicinity of the well. Conventional aquifer tests, if valid for dense rocks, require that the effective thickness of the aquifer is known. The common assumption, which is also made in compiling Table 2, is that the thickness of saturated rock in the well is equal to the effective thickness of the aquifer, so that the hydraulic conductivity can be obtained by simply dividing the transmissivity of the aquifer by the thickness of the aquifer. Pressure or packer tests are even more difficult to interpret because the shorter length of hole involved in the test will be influenced much more by the local fracture patterns developed in the rock. As a consequence of the foregoing uncertainties, most studies have given the relationship between depth and well yield or injection rates rather than attempting to estimate the actual hydraulic properties of the rocks as a function of depth.

The relation of well yields to well depths would suggest that a close relation must exist between the local topography and well yields. This is borne out by a number of studies (see a review in Davis and De Wiest, 1966). In general, recharge from streams and production of water directly from alluvium within valleys strongly favor large well yields in valleys. Valleys also tend to form along lines of structural weakness that may contain fractured and permeable rocks. After valleys, the next best water yielding areas are found on flat uplands. Here, erosion is slow, so that weathered rocks that are permeable and porous tend to accumulate. Inasmuch as the underlying rocks have a low permeability, water will tend to accumulate within the basal part of the weathered material. Valley walls and sharp-crested hills and ridges are the least favorable localities for water-well development. These areas have maximum rates of erosion with bare, dense rock at or near the surface. In addition, opportunities for recharge are minimal. Relative yields of wells suggest that rocks 5–50 m below the surface average from two to three times more permeable under broad upland areas than at or near tops of steep hills.

2.4. Importance of Rock Type

The influence of rock type on gross permeability of rocks is not as large as one might expect (Davis and De Wiest, 1966). Maximum yields of domestic wells ranging in depth from 10 to 100 m will generally be between about 40 and 80 liters/min. This is true of granite, diorite, dense basalt, gneiss, schist, and many other rocks of diverse origin. Softer metamorphic rocks such as serpentine and some slates and phyllites will tend to have lower well yields, generally from 10 to 40 liters/min. Marble and dolomite will have somewhat greater maximum yields, generally from 50 to 150 liters/min. Truly large differences in well yields between areas having different rock types are usually due to differences in the histories of weathering and/or of fracturing of the rock rather than to lithologic differences. As an example, rocks in a thrust sheet may be more than 100 times more permeable than similar rocks in an adjacent autochthonous mass, yet in another region the two rock types may have nearly identical permeabilities.

2.5. Effects of Solution on the Permeability and Porosity of Dense Rocks

All minerals are soluble to some extent. Water circulating through minor cracks can, if given enough time, open the cracks and increase greatly the permeability. Abundant evidence of this type of permeability increase is so well known from limestone, dolomite, and marble that it would hardly seem worth more than passing mention. A number of fascinating problems concerning solution permeability, however, remain to be investigated. Factors controlling the initiation and selective development of solution openings are poorly understood (Davis, 1966). Structural control of solution is almost always present but details are puzzling. For example, a cavern may follow a joint for several meters and suddenly come to a dead end or turn abruptly along another intersecting joint. The apparent randomness is such that even if joint and fault patterns are mapped in detail at the surface, the exact position of subsurface solution openings cannot be predicted with confidence. Another interesting question concerning solution openings is the extent to which they develop in rocks that are ordinarily thought of as being rather insoluble. Tolman (1937) suggested several years ago that solution of siliceous rocks was important, but little work has been done to test his hypothesis.

Water in equilibrium with the minerals in basaltic rocks will contain about 60 ppm (parts per million) silica, calculated as SiO_2. Water in equilibrium with the minerals in granite will have from 15 to 30 ppm, and water in equilibrium with quartz will have only about 10 ppm silica (Davis, 1964). Rain water has less than 0.5 ppm silica, so that if it entered directly into cracks, a significant amount of silica should be removed in solution. A hypothetical example is now given.

Assumptions:

Water passes through quartzite (nearly pure SiO_2).

A single set of parallel joints receives per year 10 liters of water for every meter of joint. This would represent 1.0% of a 1000 mm/year rain in joints spaced 1.0 m apart.

Recharge water increases uniformly in silica content from 0.0 ppm at the surface to 10.0 ppm at a depth of 10 m.

Calculations:

Area of joint being considered is 10.0 m^2.

Annual removal of silica is 100 mg.

Silica removed in 10^5 years is 10^4; this would represent a rock volume of about 3.8×10^3 cm^3.

Under the assumptions made, the crack marking the joint would be widened by 0.38 mm in 10^5 years.

The widening of a crack by 0.38 mm is very significant from the standpoint of fluid flow. Normal cracks along joints are mostly less than 0.1 mm wide (Davis, 1966), and the discharge of ground water along the cracks is a function of the width of the crack cubed. This assumes that the flow through cracks can be approximated by the equations for the flow between parallel plates. Thus, even if the increase in an exceptionally large cracks were only from 1.0 to 1.3 mm (representing slightly less than the 10^5 years solution in the previous example), the discharge through the rock would be more than doubled, providing the hydraulic gradient remains constant.

Several factors, however, argue against the rapid opening of cracks in crystalline rocks by solution. First, considerable silica is taken into solution as the water passes through the soil horizon (Davis, 1964; Laney, 1965), so that water moving downward into the underlying rock will be much nearer saturation than rain water. Second, surface and subsurface runoff will tend to bypass the rather tight joints without significant circulation of ground water through the joints. Third, unlike most carbonate rocks, the average silica-rich rock will have large amounts of insoluble residue such as iron, aluminum, and titanium oxides that will tend to clog the small cracks in the rock.

Large initial permeability of the rock and a high water temperature both would favor rapid solution enlargement of primary openings in silica-rich rocks. Hot springs in crystalline rocks may have discharges of more than 100 liters/min (Waring, 1965). Many of these springs are probably localized along fault zones. Such large discharges would suggest that solution openings have developed in the rocks. This is logical inasmuch as silica is almost as soluble in boiling water as limestone is in the average near-surface ground water.

2.6. Anisotropic Characteristics

Reliable measurements in the field of the extent of anisotropy in dense rocks is virtually lacking. Stuart *et al.* (1954) made aquifer tests in low grade metamorphic rocks of the iron mining district of Michigan and found the transmissivity of the rocks to be roughly twice as large along the strike as perpendicular to the strike. Similar results were obtained by Stewart (1964) in tests of higher grade metamorphic rocks of Georgia in which the permeability was greater in the direction of the schistocity than perpendicular to the schistocity. On the other hand, Yokota (1963) compared theoretical flow nets under a dam with the actual head measurements made after the construction of the dam and concluded that for the rock studied, which was granite, isotropy could be assumed with very little error.

2.7. Log-Normal Distribution of Well Yields

Wells drilled at essentially random locations will have yields that will have a log-normal frequency distribution (Fig. 2). Well yields are probably related closely to average rock permeabilities inasmuch as most wells range through rather limited depths. Thus, it is likely that the overall permeabilities, if they could be measured, would also follow a similar log-normal frequency distribution.

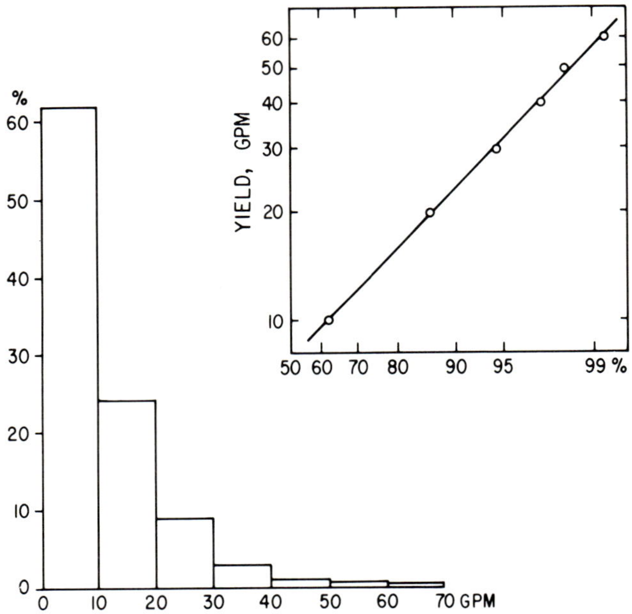

Fig. 2. Frequency distribution of yields of 1675 wells in crystalline rocks of Maryland. The inset graph shows the same data as the histogram but plotted as a cumulative curve on log-normal probability paper (data from Meyer and Beall, 1958; Dingman *et al.*, 1956).

3. Volcanic Rocks

3.1. Introduction

Rocks that have originated from the solidification of a magma at or near the surface of the earth are classified as volcanic rocks. This definition not only includes the solidified material in ancient lava flows, but also rocks within sills, dikes, and other near-surface intrusions as well as a great variety of material that was originally fragmental ejecta from volcanic eruptions. Hydraulic properties of these rocks vary from those of dense fractured rocks to nonindurated fragmental material which is almost identical to alluvium. Typical values of permeability and porosity are given in Table 3.

TABLE 3

Porosity and Permeability of Volcanic Rocks

Rock name	Porosity (%)	Permeability (darcys)	Reference
Basalt			
very dense	0.80[a]	—	Schoeller (1962)
porous	11.4	—	Schoeller (1962)
moderately dense	7.7[b]	1.4×10^{-5}	
Obsidian	0.52	—	Schoeller (1962)
Phonolite	1.98[c]	—	Schoeller (1962)
Pumice	87.3	—	Schoeller (1962)
Tuff	31.0[a]	—	Schoeller (1962)
zeolitized	39[d]	4×10^{-5}	Keller (1960)
pumiceous	40[d]	1.15×10^{-2}	Keller (1960)
friable	36[d]	1.4×10^{-3}	Keller (1960)
welded	14[d]	3.3×10^{-4}	Keller (1960)

[a] Average of 8 samples.
[b] Unpublished test results, K. Shizaki (1966), Univ. of Hawaii, Honolulu.
[c] Average of 5 samples.
[d] Porosity and permeability are averages of several samples.

Although, when viewed on a microscopic scale, volcanic rocks may be almost entirely made of interlocking crystals, they possess several distinctive hydrologic characteristics that set them apart from the normal crystalline rocks which have just been discussed. The solid matrix of volcanic rock is probably less permeable than the average granite, gabbro, or slate, but unlike these rocks, a number of primary features give rise to permeable zones within the otherwise solid rock. These features are related to the final history of the rocks and only indirectly to the chemical characteristics of the rocks. For this reason, considerable attention will be given to details of genesis.

3.2. Origin of Porosity and Permeability

When molten rocks cool in dikes and sills a few hundred to a few thousand feet below the surface, the resulting rocks are dense and almost nonpermeable. The same materials reaching the surface as lava flows can form some of the most permeable aquifers known. The physical differences in surrounding pressures, rates of cooling, and amount of associated fractured debris will account for the contrasts in hydrogeologic properties. In the subsurface, the slow rate of cooling will inhibit the development of large shrinkage cracks, will allow a gradual emission of gases that are in solution, and will favor a homogeneous rock. In contrast, at the surface the rapid cooling and degassing produces a rock with well developed vesicular texture, particularly in the upper parts, and closely spaced cooling joints. The movement of the lava also has a bearing on the ultimate permeability of the rocks. While the lava is in motion, the top tends to form a crust due to cooling. Continued movement of the molten lava will drag the upper crust along, while at the same time fracturing it and producing masses of blocky rubble. The rubble will be pulled under the leading edge of the flow in a manner similar to the tread on a tracked vehicle. The final result in many places is a solidified flow with coarse rubble both above and below the more dense flow rock. Even in flows of very low viscosity which lack thick rubble zones, some space will remain unfilled between individual flows. The low viscosity will favor thin flows that tend to cool rapidly along the margins so that small depressions in the older underlying flows will not be filled by the overlying lava that tends to cool and bridge the depressions.

Interflow stream gravels together with the interflow rubble discussed previously will produce a high overall permeability in most recent basalts. Other less common causes of high permeability are lava tubes, gas vents, and tree molds. Some lava tubes are more than 5 m in diameter and 2 km in length. For practical purposes, the "permeability" of the lava tube is infinite and its presence will mean that the average permeability of the lava tube together with all the adjacent rocks is very high even though most of the rocks are locally nonpermeable.

Compaction, washing in of surface sediments, and the formation of secondary minerals will gradually reduce the permeability of basalts so that most pre-Cenozoic basalts have rather low permeabilities. Specific capacities of most older basalts are less than 100 liters/min m of drawdown. In contrast, wells in geologically young basalts commonly have at least a specific capacity of 1000 liters/min m of drawdown. Values of ten times this amount are not uncommon.

Volcanic rocks such as dacite, andesite, and latite that tend to be lighter in color and contain larger amounts of silica will originate as lava flows of higher viscosity than the average basalt. These lava flows will tend to be thicker and contain fewer permeable zones. The most viscous lava is generally rhyolite. Lavas with a high viscosity will tend to clog volcanic vents, which in turn produces explosions that eject pyroclastic material.

The porosity of pyroclastic material most commonly ranges from about 25 to 50%. The higher values are of recent volcanic ash and the lower values are of mud flow debris. The only pyroclastic materials with lower porosities are ancient deposits that have undergone compaction and some recrystallization, and welded tuffs that are formed from the welding together of volcanic debris which is still semimolten when it settles to earth. Porosities of welded tuffs may be from 5 to 10% and those of metavolcanics may be less than 1%.

Permeabilities of pyroclastic material are generally below 10^{-2} darcy. Exceptions are in very coarse volcanic breccia and in pyroclastic material that is reworked by stream water (Davis and Hall, 1959). Low permeabilities are caused by the very fine size of much of the material making up ash and tuff deposits, and by the poor sorting of the coarser agglomerates and volcanic mud flows.

Weathering will tend to increase the porosity of the volcanic rocks but may drastically reduce the permeability. Thick, tropical soils may commonly exceed 60% porosity but permeabilities of the water-saturated soils are mostly below 1.0 darcy.

3.3. Aquifer Tests of Permeable Basalt

Numerous attempts have been made to complete meaningful tests of highly permeable basalt aquifers. Most of the tests have been less than satisfactory inasmuch as the drawdowns produced by pumping are very small and are developed too rapidly to obtain useful data. In addition, extensive turbulent flow probably exists in the vicinity of the wells which eliminates the use of equations based on Darcian flow.

Pumping tests of aquifers near Pearl Harbor on the island of Oahu, Hawaii suggest transmissivity values of roughly 10^7 gal/day ft (1.24×10^5 m^2/day) (S. Lau, personal communication, 1966). The total thickness of the aquifers tested is not known, but may range from 300 to 1000 m. If so, the average permeability of the rocks must be between 100 and 300 darcys. Tests of the Snake River basalt in Idaho (Walton and Stewart, 1961) suggest transmissivity values similar to those of Oahu. Of a total of 11 determinations, extreme values were 1.6×10^5 and 1.8×10^7 gal/day ft, with a median of 1.7×10^6 gal/day ft. Interbedded gravels are more important in the Idaho basalts and the total thickness of the aquifer is probably less than in Oahu, so that a direct comparison is not possible. Nevertheless, an average permeability of 100 darcys might be a reasonable guess for young basalts in this area.

Walton and Stewart (1961) gave 5 determinations of the coefficient of storage of the Snake River basalt, which ranged from 0.02 to 0.06 and averaged 0.048. These are reasonable values for water-table conditions in basalt.

3.4. Anisotropy of Basalt

Permeability of basalt and related rocks is commonly anisotropic when considered in large volumes, although this anisotropy may not be too evident in small laboratory samples. The centers of thick lava flows will be relatively impermeable. This fact together with the fact that buried soils will be developed on the tops of flows means that most of the permeability will be parallel with the flows themselves. Distinct head differences encountered in drilling through permeable basalt support these general conclusions concerning anisotropy. Also, within the plane of the flows, the permeability should be greatest in the direction of the steepest original dip of the flows. Lava tubes as well as minor

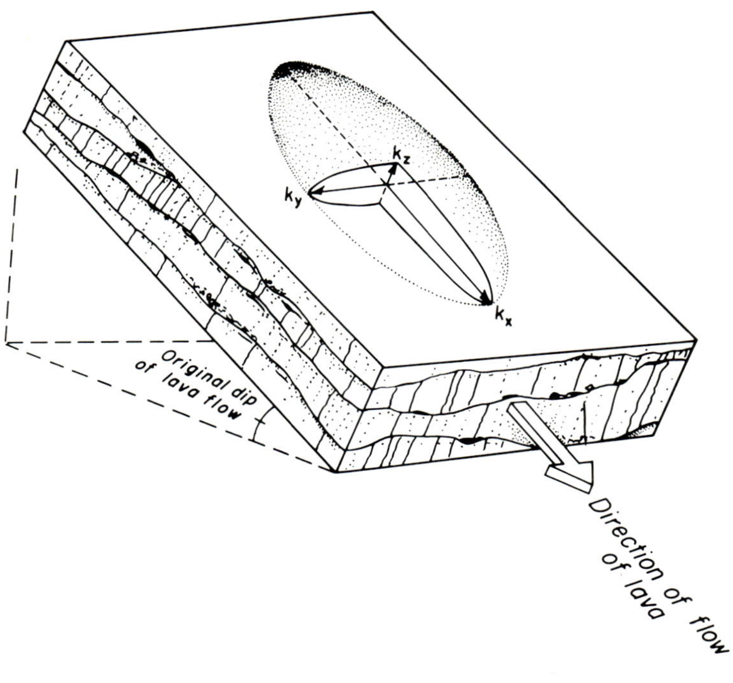

Fig. 3. Diagram illustrating the probable orientation and relative magnitude of the overall permeability of young basaltic rocks.

rills on the top of solidified flows, and later, stream gravel filling post-volcanic valleys, will all tend to be aligned with the steep slopes. Qualitatively, then, the anisotropy might be represented by the ellipsoid of Fig. 3.

4. Indurated Sedimentary Rocks

4.1. Introduction

More is probably known about the hydraulic characteristics of permeable sedimentary rocks than of any other major group of rocks. This is due to the economic interest in these rocks by the petroleum industry. Indeed, today, the bulk of data available in company files is so large that it will probably never be possible to analyze and publish more than a small fraction of the total. An extensive summary of porosity and permeability determinations is given in Muskat's classic work (Muskat, 1937). Some of his data along with those of other researchers are given in Table 4.

Most sedimentary rocks can be classified broadly as sandstones, shales, or carbonate rocks. The total amount of salt, gypsum, conglomerate, and other minor sedimentary types is probably less than 5% of the overall volume of sedimentary rocks. Inasmuch as each of the major types is hydrogeologically more or less distinct, they will be discussed separately.

4.2. Shale

The term *shale* is used in this section to describe all fine-grained rocks that may range in hardness from rather soft but compact clay to rock resembling slate but lacking extensive recrystallization of constituent minerals. In a strict petrographic sense, the term shale should be reserved for only those fine-grined sedimentary rocks which display a marked tendency to split into platy fragments when broken. Terms such as mudstone, stilstone, and claystone are normally used for various other rocks which are included here with the shales.

Shales, as used in the broad sense, constitute roughly 50% of all sedimentary rocks. Despite their abundance, however, they have received little attention by past researchers. Prior to 1950, most interest in shale was centered around geologic problems of structures produced by differential compaction of shale and the long-term expulsion of hydro-

TABLE 4

Permeability and Porosity of Sedimentary Rocks

Rock name	Porosity (%)	Permeability (darcys) Vertical	Orientation not given	Horizontal	Reference
Arkose					
coarse-grained	10.9	3.8×10^{-4}		5.5×10^{-4}	Rima *et al.* (1962)
fine-grained	14.4	1.6×10^{-4}		1.6×10^{-3}	Rima *et al.* (1962)
medium-grained	25.6	5.5×10^{-4}		1.1×10^{-3}	Rima *et al.* (1962)
Chalk, Cretaceous	29.2[a]	—		—	Schoeller (1962)
Chert, Mississippian	3.8	—		—	Manger (1963)
Conglomerate, coarse grained	17.3	3.8×10^{-4}		4.9×10^{-4}	Rima *et al.* (1962)
Dolomite					
Mississippian	27.8		2.9×10^{-1}		Murray (1960)
Ordovician	0.4[b]	—		—	Manger (1963)
	11.9		1.6×10^{-2}		Murray (1960)
Limestone					
Cretaceous	4.6[c]	—		—	Schoeller (1962)
oolitic	21.6		3.4×10^{-1}		Archie (1952)
Pennsylvanian	6.3[d]	—		—	Manger (1963)
Permian	10.1		7.7×10^{-3}		Archie (1952)
Sandstone					
Cambrian	11.2[e]	—		—	Manger (1963)
Cretaceous	—	4.8×10^{-2}		1.5×10^{-4}	Young *et al.* (1964)
Cromwell	16.6	1.7×10^{-1}		4.1×10^{-1}	Muskat (1937)
Gilcrest	27.4	5.6×10^{-1}		8.0×10^{-1}	Muskat (1937)
Pennsylvanian	17.4[f]	—		—	Manger (1963)
Prue	11.4	4.7×10^{-4}		3.4×10^{-3}	Muskat (1937)
Wilcox	15.6	3.6×10^{-1}		7.6×10^{-2}	Muskat (1937)
	12.4	8.5×10^{-2}		8.8×10^{-2}	Muskat (1937)
Shale					
Cretaceous	—		4×10^{-6}		Gondouin and Scala (1958)
Pennsylvanian	—		9×10^{-8}		Gondouin and Scala (1958)
Oligocene and Miocene	21.1[g]	—		—	Manger (1963)
Silurian	5.2[h]	—		—	Manger (1963)
Siltstone	9.7	1.6×10^{-4}		1.2×10^{-4}	Rima *et al.* (1962)
	—	1.5×10^{-8}		—	Young *et al* (1964)

[a] Mean of 16 samples.
[b] Mean of 56 samples.
[c] Mean of 24 samples.
[d] Mean of 2109 samples.
[e] Mean of 24 samples.
[f] Mean of 587 samples.
[g] Mean of 9 samples.
[h] Mean of 5 samples.

carbons and connate water. The fluid-bearing properties of the shales were of secondary importance. Recently, questions related to water chemistry and unusual formation pressures have generated considerable interest in the permeability of shale. Unfortunately, little of the modern data has been published.

Determinations of shale permeabilities found in the older literature should be considered as only a rough approximation to actual subsurface conditions. Sampling and testing will disturb the natural state of the shale. Original fractures will either be opened or closed by the sampling process. The sample as a whole will expand as it is brought to the surface from great depths. Moreover, the high pressures necessary to force fluids through the sample will physically alter the sample. Effective hydrated volumes of minerals in shale are also particularly sensitive to variations of the chemistry of fluids used in permeability tests. Descriptions of testing procedures are needed before the full significance of reported permeabilities can be determined.

Even with the limitations given above, some useful generalizations can be made concerning the permeability of shale. Laboratory work as well as aquifer tests suggest that permeabilities of unfractured shale are always less than 10^{-3} darcy and most commonly less than 10^{-5} darcy. Such low permeabilities further suggest that water movement through shale can only be significant if the length of time allowed for flow is large or if the area through which the water moves is very large. For example, on the basis of aquifer tests, Walton (1962) has concluded that in northern Illinois the average vertical permeability of the Maquoketa Formation, which is largely shale, is only 2.7×10^{-6} darcy, yet the total vertical leakage over a single square mile amounts to 21,000 gal during one year. This amount is highly significant in terms of the recharge of a regional aquifer.

Porosities of shales are much easier to determine than permeabilities. Porosity of most shale is a direct function of the original environment of deposition and present or past depths of burial. Shales near the surface that have never been deeply buried will have porosities of from 10 to 25%. Older shales that have been compacted by burial and have also undergone some cementation may have porosities of from 2 to 10% (Table 4).

All gradations exist between soft clays and brittle chert or carbonate rocks. The more brittle the shale is, the more likely the shale will contain secondary cracks, particularly within 100 m of the surface of the land. Some shales are fractured to such an extent that they constitute locally important aquifers. Yields of wells suggest that the effective permeability of the fractured shale may range from 1 to 10 darcys.

4.3. Sandstones

Sandstones make up roughly 25% of the sedimentary rocks of the world. Although compositions vary widely, quartz is the predominant mineral in most sandstones. In general, sandstones are considerably more permeable than shales. Some sandstones, nevertheless, are so firmly cemented by quartz or other mineral matter that they have permeabilities that are as low as the average dense plutonic rock.

Porosities of sandstone are almost entirely a function of the amount of cementing material between the sand grains and the extent to which the larger grains interdigitate. Nonindurated sands will have porosities of between 30 and 45%. Maximum porosities of from 10 to 25% are more common. From the standpoint of geological processes, two main reasons account for loss of porosity. The first is cementation by calcite, silica, and other minerals introduced into the rock by circulating ground water. Cementation may take place at any depth. The second is compaction that squeezes fine-grained matrix material into open pores and interdigitates larger grains by grain-boundary solution at points of highest stress. Compaction is most important at considerable depths below the surface where temperatures and pressures are both high. Grain-boundary solution is without doubt most effective when temperatures are above 90°C, when silica becomes quite mobile in an aqueous environment.

Recent studies by Maxwell (1964) and Atwater (1966) both confirm the fact that the porosity of sandstone tends to diminish systematically with depth so that sandstone below about 10 km has less than 5% porosity. Atwater suggested that the decrease averages about 1.3% for every 300 m of burial in the Louisiana gas and oil reservoirs which were studied. This conclusion has profound effects on the economics of petroleum exploration, because not only is it unreasonable to expect significant permeability at depths of more than 10 km, but the total amount of fluids in storage decreases rather rapidly with depth. Thus, even if permeability is relatively high, both the added cost of drilling and the decrease in potential yield suggest marginal economic returns at great depths.

The permeability of sandstone is a direct function of the size of the interconnected pores. For sandstones with only a moderate amount of cementing material, the pore size is related directly to the size of the sand grains making up the rock. Hence, permeability can be correlated directly with grain size (Johnson and Greenkorn, 1963). On the other hand, if a formation has about the same grain size throughout but varying amounts of cement, then a direct relation exists between porosity and permeability (Levorsen, 1954). For example, samples of the Woodbine sandstone which have a mean porosity of 28% will have a mean permeability of about 1.5 darcy, but samples having a mean porosity of 21% will only have a mean permeability of about 10^{-2} darcy (Collins, 1961). In general, however, the number of variables affecting permeability are so large that just a simple measure of either grain size or porosity will not prove to be a reliable guide to the permeability of sandstones.

The presence of horizontal stratification in most sandstone means that the permeability of very large samples, say more than 100 m^3, should be considered to be uniformly anisotropic. Actual samples taken in a direction perpendicular to the bedding will show large variations of permeability (Muskat, 1937). The gross effect of this permeability stratification is that the effective vertical permeability of large masses of sandstone is very low even though the horizontal permeability may be quite high. Many practical problems, nevertheless, are concerned only with fluid flow that is essentially parallel with the bedding, so that the flow can be considered to be within a two-dimensional isotropic medium, with the larger horizontal permeability taken as the effective or "average" permeability.

As cementation and compaction take place, more of the permeability will be due to the fractures within the rock that cut through the original layers of low permeability. The overall effect of complete lithification will be to reduce the permeability parallel with bedding and to increase the vertical permeability by fracturing. Rocks should become more isotropic if the fracturing is regularly spaced and in more than one direction. Published data are lacking, however, to support the foregoing generalizations, which are based only on general geologic reasoning.

Our knowledge concerning anisotropy on a small scale within sandstone is rather incomplete, but is still much better founded than our understanding of the gross anisotropy of large volumes. Several studies have shown that the average permeability of even small samples will tend to be much greater parallel with the bedding than perpendicular to the bedding. Piersol *et al.* (1940) measured vertical as well as horizontal permeabilities of a large number of samples. The median ratio of the horizontal to the vertical permeability was 1.5, and about 12% of the samples had a ratio of more than 3.0. Slightly less than 6% of the sample had vertical permeabilities larger than the horizontal permeabilities. In a study by Greenkorn *et al.* (1964), 81 out of a total of 142 small samples of sandstone had significant anisotropy. Only 63 of the samples had recognizable stratification, but of these, 43 had their direction of maximum permeability parallel with the stratification.

Of the 23 samples in which vertical permeability was the greatest, 17 either had no visible stratification or the bedding made an angle of more than 45° with the horizontal plane.

Cardwell and Parsons (1945) have considered the problem of defining the effective or "average" permeability in heterogeneous sandstone. They concluded that the permeability should be taken as a value between the arithmetic mean and the harmonic mean of the various permeability measurements made in a given stratum. They further state that the "average" is closer to the arithmetic mean than the smaller harmonic mean. Several researchers have concluded that the geometric mean should be taken as the "average" permeability of a heterogeneous stratum (Greenkorn *et al.*, 1964; Warren and Price, 1961; Bennion and Griffiths, 1966). Inasmuch as the geometric mean has a value between the harmonic and arithmetic means of a given set of values, the two separate conclusions are not in conflict.

4.4. Carbonate Rocks

Carbonate rocks consist mostly of calcite and dolomite, with clay and quartz as common secondary minerals. If mineralogical determinations are made, the rock terms limestone and dolostone (more commonly called dolomite) can be used for rocks primarily composed of the minerals calcite and dolomite respectively. Almost all dolomite is secondary in origin, having been formed from pre-existing calcite. If the formation of dolomite takes place before the rock is lithified, the resulting dolostone can be dense and almost nonpermeable. On the other hand, dolostone formed after general lithification of limestone will result in considerable increase of porosity and permeability because the transition from calcite to dolomite results in a reduction of space occupied by the minerals of about 13%.

General changes in the character of carbonate rocks as a function of time are rather striking. Geologically young carbonate rocks mostly have porosities that may range from about 20% for coarse, blocky limestone to more than 50% for poorly indurated chalk. With time, the soft carbonate minerals will be deformed by overburden pressures, recrystallization will take place in zones of lower stress, and considerable transfer of calcite will take place by virtue of circulating ground waters. Mobilization and recrystallization of calcite will probably be most rapid in coarse-grained fragmental rocks and slowest in lime muds where the movement of water is slow due to the initially low permeability.

Permeabilities of young reef and fragmental limestone are about as high as those encountered in young basalts. For example, the Biscayne aquifer of the Miami area in Florida is reported to have an average permeability of about 1600 darcys (Parker, 1951). Buried coral reefs in the Waimanalo area of Oahu, Hawaii, are probably about as permeable as the Biscayne aquifer (D. Lum, personal communication, 1966). These and other highly permeable aquifers have considerable primary porosity which has been greatly augmented by secondary solution.

Permeabilities of unfractured portions of older carbonate rocks are generally less than 1.0 darcy and commonly less than 0.01 darcy. These permeabilities are of importance to the production of petroleum but are not very significant in terms of yields of water wells because the wells are usually too shallow to develop the high hydraulic gradients needed to produce the water.

High permeabilities are developed in older carbonate rocks only through fracturing and solution of the carbonate minerals. Local permeabilities are almost infinite in the vicinity of large solution openings, but even in areas famous for their caverns, most water wells have small yields. For example, wells in the vicinity of Mammoth Cave, Kentucky, have a maximum yield of only 36 gal/min (Cushman *et al.*, 1965). This is because the chances of striking a cavern with a well are small even with the benefit of technical guidance. Most of the more permeable zones encountered by wells in older limestone are probably small solution openings along irregularities in bedding planes. Because of their orientation, vertical solution openings along joints and faults are not encountered as often. The near proximity of vertical openings, nevertheless, favors the development of larger openings along adjacent bedding planes.

Regional studies of carbonate rocks most commonly show average porosities of from 4 to 15% for moderately dense rocks and from 20 to 45% for chalk. The total amount of water that these rocks will yield on gravity drainage is, nevertheless, very small and may range from about 0.2 to 1.0% (Mégnien, 1964; Walker, 1956). Such small specific yields suggest that the total volume occupied by interconnected solution openings is very small even in regions of numerous limestone caverns.

4.5. Statistical Distribution of Permeabilities and Porosities

Histograms of a large number of permeability determinations from single geologic units suggest that the normal distributions are skewed in a strongly positive direction. The same data, however, will generally fall along a straight line if plotted as a cumulative curve on a log-normal probability scale (Law, 1944), Fig. 4. Bennion and Griffiths (1966) have compiled data from several thousand samples of conglomerate, sandstone, and limestone units, and have found log-normal distributions that were skewed slightly to the right. Extensive studies of the specific capacities of several thousand water wells in various geologic units, including sedimentary rocks, have shown that their values also have log-normal distributions (Csallany and Walton, 1963; Seaber and Hollyday, 1966).

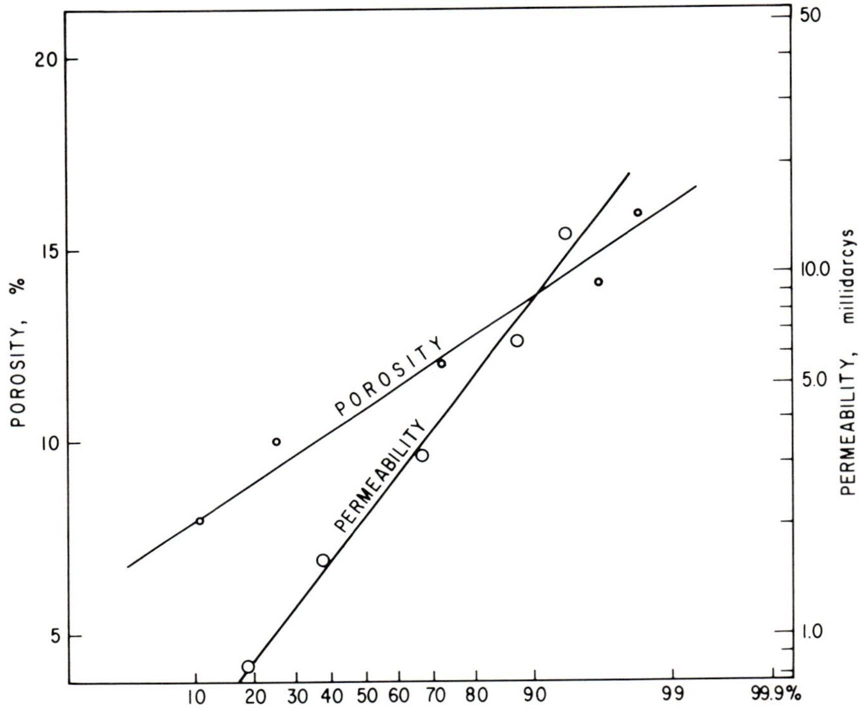

FIG. 4. Cumulative curves of porosity and permeability of "Bradford sand." Porosity is plotted on a normal probability scale and permeability is plotted on a log-normal probability scale (data from Muskat, 1937).

Inasmuch as specific capacities are an indirect function of aquifer permeabilities, it appears to be a general rule that the statistical distributions of permeabilities within given geologic units are log-normal.

Porosity values from a single formation are reported to have normal distributions rather than log-normal distributions (Law, 1944; Bennion and Griffiths, 1966). This is reasonable inasmuch as permeability is an exponential function of porosity for a given average geometric configuration of the pore space. The few sets of data that I have plotted, however, suggest that porosity values do not fit as well to a normal distribution as permeability data fit a log-normal distribution. Bennion and Griffiths (1966), however, found the opposite to be true.

Some very useful observations have been made by Seaber and Hollyday (1966) concerning the standard deviation or "statistical spread" of the data from various types of rocks. Carbonate rocks and metamorphic rocks have large standard deviations in specific capacity data whereas sandstone and shales have relatively small standard deviations. Glacial deposits have much larger means but also have large standard deviations. Data by Bulnes (1946) also suggest that within carbonate rocks, the larger standard deviations are associated with rocks of lower permeability (Fig. 5).

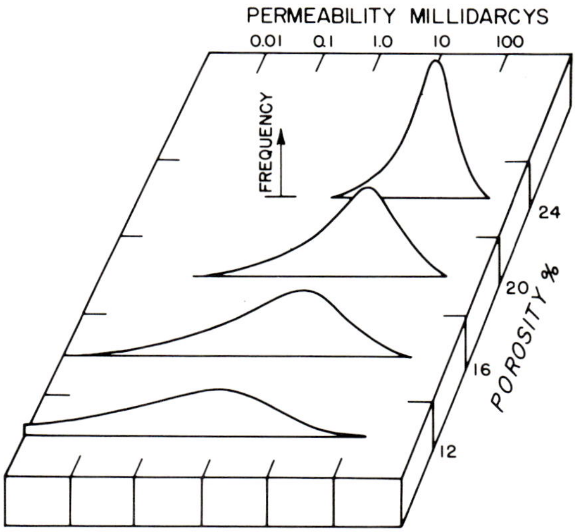

FIG. 5. Frequency curves of permeability of San Andreas dolomite, Wasson Field, Texas (Bulnes, 1946).

5. Nonindurated Sediments

5.1. INTRODUCTION

The hydraulic characteristics of nonindurated sediments have been of the greatest interest to soil scientists, civil engineers, and hydrogeologists who have made countless measurements of permeability and porosity of materials from diverse geologic origins. The measurements include both field tests that give average values for large volumes of material and laboratory measurements of samples having volumes of a few hundred cubic centimeters or less. Laboratory tests have the advantage of giving more reliable information on porosity and on the extent of heterogeneity and anisotropy than can be gained from field tests. Unfortunately, laboratory studies have two serious disadvantages. First, the number of samples needed to define average conditions in nature is commonly excessive. Second, obtaining samples that are effectively undisturbed and hence representative of natural conditions is a most difficult task.

In reviewing the literature I have found surprisingly few publications that give full information on the geologic origin, grain-size distribution, orientation, porosity, and permeability of the samples. Although data given in Table 5 are rather sparce, an attempt has been made to select values that appear to be representative of various geologic units. General ranges of permeability found in nonindurated sediments are shown in Fig. 6. Details of particle-size distributions for samples of different permeabilities are shown in Fig. 7.

5.2. Alluvium

By far the largest number of field and laboratory tests have been made of alluvium. Recovery of core material has been one of the major problems in the study of subsurface samples. For example, Tanaka and Hollowell (1966) found that hydraulic conductivities of cores of water-bearing sediments along the Arkansas River varied from 79 to 500 meinzers, but that aquifer tests in the same region indicated average hydraulic conductivities of water-bearing sediments of 2500–4100 meinzers. The large difference was attributed to poor recovery of the most permeable sections of the cores. On the other hand, improved methods of core recovery can yield laboratory data that correspond very closely to results of aquifer tests in the field (Johnson and Sniegocki, 1967).

Broadly speaking, coarse alluvial deposits have similar grain-shape and grain-size distributions. Considerable success has been obtained in correlating median grain size with permeability in sediments from similar geologic environments (Bedinger, 1961; Norris and Fidler, 1965). Major differences in permeabilities of similar sized material are most commonly related to differences in packing of the particles or differences in the amount of clay binding between larger particles.

Some success has also been obtained in using particle size analyses as an indication of permeability of sediments from diverse geologic origins. Masch and Denny (1966) have presented a review of past work

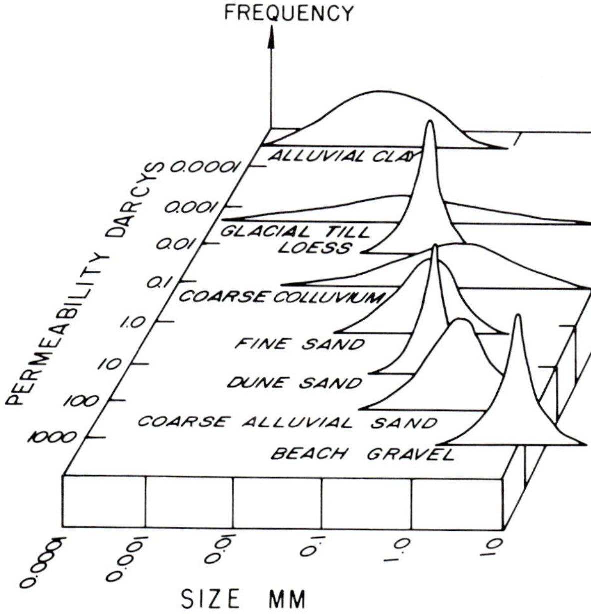

FIG. 6. Particle-size distribution related to permeability of various nonindurated sediments.

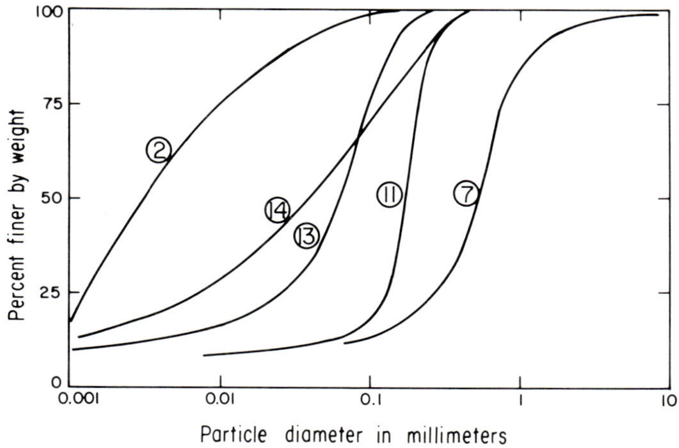

FIG. 7. Particle-size distribution of some of the samples listed in Table 5. Numbers in the diagram correspond with those in the table (Johnson and Morris, 1962).

TABLE 5

Porosity and Permeability of Nonindurated Material

Description of sample	Sample number	Porosity (%)	Permeability (darcys)		Median diameter (mm)	Reference
			Horizontal	Vertical		
Clay						
marine	1	48.5	—	1.6×10^{-5}	0.0005	MacGary and Lambert (1962)
silty, depth 1558 ft	2	41.1	—	1.1×10^{-4}	0.003	Johnson and Morris (1962)
946 ft, lacustrine?	3	40.5	2.8×10^{-4}	1.1×10^{-5}	0.0021	Johnson and Morris (1962)
838 ft, lacustrine?	4	44.1	1.1×10^{-5}	1.1×10^{-5}	0.0036	Johnson and Morris (1962)
Sand						
Ohio River alluvium	5	—	25.2	26.4	—	Gallaher (1964)
marine	6	41.0	55.0	38.5	0.5	MacGary and Lambert (1962)
coarse, depth 1174 ft, alluvium	7	41.0	—	2.2	0.5	Johnson and Morris (1962)
medium, depth 767 ft, alluvium	8	42.9	18.2	0.6	0.32	Johnson and Morris (1962)
fine, alluvium	9	51.3	25.3	26.4	0.2	MacGary and Lambert (1962)
	10	46.3	13.2	16.5	0.2	MacGary and Lambert (1962)
depth 1284 ft	11	40.1	—	1.48	0.17	Johnson and Morris (1962)
depth 911 ft	12	42.4	1.05	0.27	0.14	Johnson and Morris (1962)
Silt						
sandy, depth 472 ft	13	39.4	—	3.8×10^{-2}	0.059	Johnson and Morris (1962)
clayey, depth 1527 ft	14	34.1	—	5.5×10^{-3}	0.04	Johnson and Morris (1962)
loess	15	50.0	0.22	0.33	0.02	MacGary and Lambert (1962)

Orientation of remaining samples was not specified in the literature cited

TABLE 5 (*continued*)

Description of sample	Sample number	Porosity (%)	Permeability (darcys)	Median diameter (mm)	Reference
Clay					
kaolinite under confining load of $\frac{1}{2}$ kg/cm	16	58.8	2.38×10^{-3}	—	Olsen (1966)
Oligocene subsurface, Mississippi	17	33.3	1.1×10^{-5}	—	Unpublished data, U.S. Geol. Survey, Denver, Colorado
Gravel					
sand, silt; end moraine (median 7 samples)	18	—	2.2	—	Poth (1963)
silt, clay; colluvium	19	63.4	5.5×10^{-5}	0.13	Cohen (1965)
Sand					
aquifer Arkansas River alluvium	20	33.8	19.8	—	Tanaka and Hollowell (1966)
	21	37.4	27.5	—	Tanaka and Hollowell (1966)
point bar, Humboldt River, Nevada	22	38.3	77	0.82	Cohen (1965)
Patuxent formation, Cretaceous (median of 23 field tests)	23	—	36.8	—	Bennett and Meyer (1952)
from sand dune	24	35.8	28.0	—	Brown and Newcomb (1963)
Silt					
Miocene subsurface, Mississippi	25	33.7	3.85×10^{-5}	0.045	Unpublished data, U.S. Geol. Survey, Denver, Colorado
sandy overbank deposit, Humboldt River, Nevada	26	49.2	2.2×10^{-4}	0.02	Cohen (1965)
overbank deposit, Arkansas River	27	47.5	5.5×10^{-2}	—	Tanaka and Hollowell (1966)
	28	49.8	1.6×10^{-2}	—	Tanaka and Hollowell (1966)

as well as new data on artificially mixed samples of sand-size material. They concluded that (1) permeability increases with an increase in median diameter, (2) for a given median diameter, the permeability decreases with an increase in the standard deviation of particle sizes, (3) permeability is most sensitive to changes in standard deviation in samples with larger median grain sizes, (4) other factors being equal, unimodel distributions have higher permeabilities than bimodal distributions, and (5) permeabilities can be estimated by a correlation system developed in their study provided a size analysis of the sediment sample has been made.

The bedded character of alluvium imparts a strong anisotropy to the deposits if the average properties of large volumes are considered. Anisotropy is not so marked but is still present in small samples tested in the laboratory. Johnson and Morris (1962) report both the horizontal and vertical hydraulic conductivities of 61 laboratory samples of alluvium and lacustrine sediments from the San Joaquin Valley, California. Only 4 of the samples had a vertical conductivity greater than the horizontal conductivity; 11 of the samples indicated the same conductivity in both directions; and the remaining 46 samples had a horizontal conductivity greater than the vertical conductivity. In general, the horizontal conductivities were from 2 to 10 times as large as the vertical conductivities.

Alluvial deposits are characterized by their general heterogeneity. In detail, however, the materials can be quite homogeneous. Overbank deposits in large stream valleys may have uniform properties for many miles parallel with the stream and may have widths of several thousands of feet. Overbank deposits on broad floodplains are mostly silts and coarse clays which have permeabilities from 10^{-3} to 10^{-1} darcy (Tanaka and Hollowell, 1966).

Channel deposits tend to highly variable in detail but much more permeable than the overbank floodplain deposits. In valleys of large rivers, the coarser channel deposits generally have permeabilities of from 1 to 200 darcys, with a median value close to 50 darcys. Channel deposits of ephemeral streams will generally have permeabilities of from 0.1 to 100 darcys with a median permeability of perhaps near 5 darcys.

Porosities of alluvium are relatively uniform despite wide differences in permeability (Fig. 8). Some systematic differences of porosity, however, exist. In general, clay and silt-size materials are more porous than gravel and coarse sand. Also, within the coarser sediments a rather uniform decrease in porosity with an increase in particle size has been measured by various individuals. Typical data are shown in Fig. 9.

5.3. Eolian Deposits

Eolian deposits in a given area are surprisingly uniform in hydrogeologic properties. The low density and viscosity of air promote efficient size sorting. As a result of this, deposits have nearly the same size for a certain wind condition and distance of transportation. The traction load of the wind consists of sand having diameters of 0.1–1.0 mm that may accumulate in nearly homogeneous dunes, which are probably

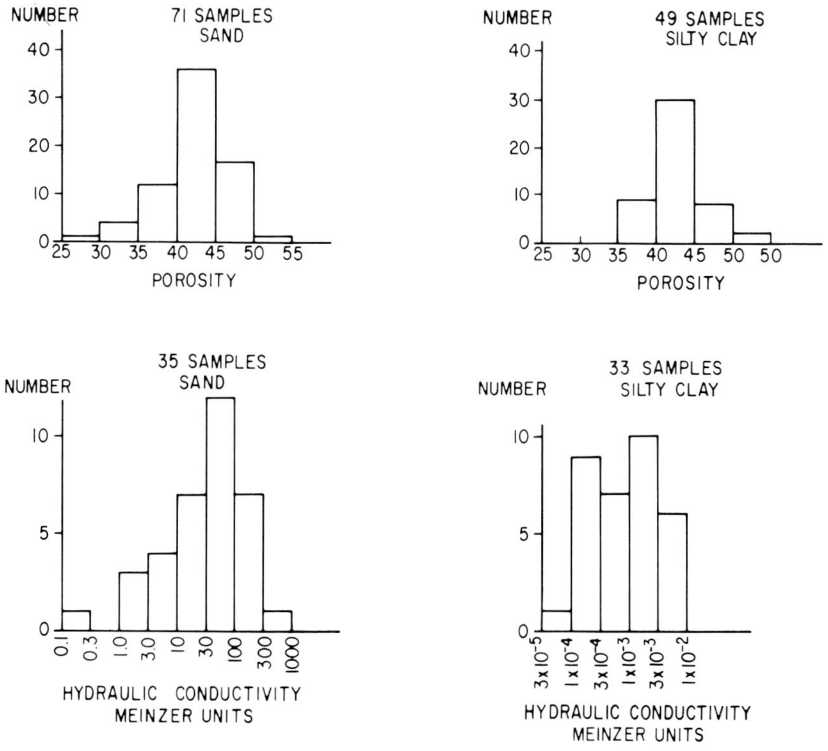

FIG. 8. Porosity and hydraulic conductivity of sediments from the San Joaquin Valley, California (data from Johnson and Morris, 1962).

about as isotropic with regard to permeability as any materials in nature. Porosities will be between 35 and 40% and permeabilities mostly between 5 and 50 darcys, with a median of about 25 darcys.

The suspended load of the wind consists mostly of particles having diameters between 0.005 and 0.1 mm. Coarser grained particles will form loess deposits near broad floodplains and areas of deflation in arid regions. The grain size diminishes slowly away from the source area until coarse clay sizes may predominate at distances of more than 100 km. Permeability will vary with the size. Minimum values of about 10^{-4} darcy will be found in fine grained material, and maximum values of about 1.0 darcy will be found in the coarse grained deposits. The median value of 193 tests from 14 localities is slightly more than 10^{-2} darcy (Gibbs and Holland, 1960). Porosities of most loess will range from 40 to 55%.

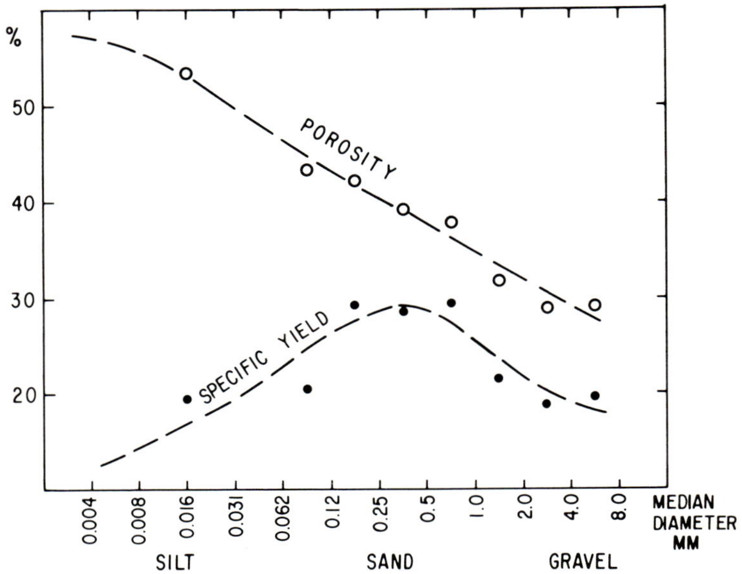

FIG. 9. Laboratory determinations of porosity and specific yield of 322 samples of valley alluvium from Humboldt River Valley, Nevada (data from Cohen, 1965).

5.4. Glacial Deposits

Sediments associated with glaciers vary in grain size from fine-grained tills and local lacustrine deposits to extremely coarse outwash gravels. Only the hydrogeologic properties of the coarser materials have been investigated widely. In general, outwash materials are identical in their hydrogeologic properties to normal alluvium. A nearby glacier, however, provides a local source of coarse materials. If meltwater runoff is vigorous and persists for several decades, if not centuries, the coarse outwash may be well sorted and have an exceptionally high permeability. Some of the most productive wells in the United States are in outwash near the cities of Spokane and Tacoma (Walters, 1963) in the state of Washington. Judging by specific capacities of more than 500 gal/min ft.

of drawdown for moderately shallow wells, permeabilities of at least 300 darcys are indicated.

Till from continental glaciers most commonly has a low permeability except where stringers of water-sorted material are present. Norris (1963) has tabulated available data on the permeability of clay-rich till. The permeability values of 37 determinations ranged from about 1.7×10^{-5} to 5×10^{-2} darcy, with a median of 2.2×10^{-3} darcy.

5.5. Fine-Grained Materials of Diverse Origins

Abundant fine-grained materials are found in lake, marine, and colluvial deposits. Soil development at and near the ground surface will also produce fine-grained materials which accumulate largely in the *B* horizon, or "subsoil." Although the actual size of individual particles controls to some extent the permeability and porosity of fine-grained materials, the degree and nature of the aggregation of the smaller particles into larger more-or-less stable masses is much more significant. Water with a high percentage of sodium ions will tend to disperse the aggregates and reduce permeability. On the other hand, some tropical soils have tightly packed soil nodules that are almost as stable as rock fragments. Such soils have exceptionally high permeabilities. Many zonal soils have a relatively rapid decrease in permeability and porosity with depth (Fig. 10). The decrease is owing to the fact that stable soil aggregates are found in the surficial, organically

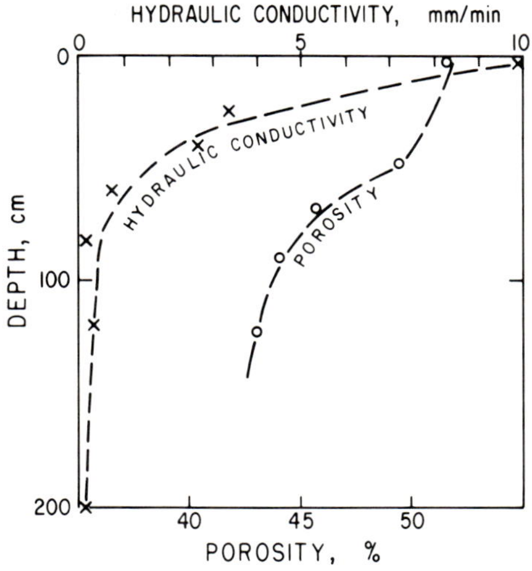

FIG. 10. Porosity and hydraulic conductivity variations with depth in a chernozem soil (data from Solov'yev, 1961).

rich *A* horizon, or "topsoil," whereas less permeable inorganic clays and colloids tend to lodge in the lower *B* horizon.

Most hydrogeologists would generally classify lake clays as being almost nonpermeable. Such is not always the case, however. For example, fine calcarious muds of the Bonneville Salt Flat of northwestern Utah have secondary shrinkage cracks that render them as permeable as coarse sands. Aquifer tests suggest average permeabilities of as much as 100 darcys in some of the subareas within the playa deposits. Although calcareous sands composed of fecal pellets of brine shrimp are interbedded with the muds, no apparent correlation exists between yields of test wells and percentage of sand encountered in the wells (C. P. Bingham, personal communication, 1966).

The fine-grained sediments at Bonnevill have shrinkage cracks to depths of at least 4 m. The cracks can be seen clearly in fresh excavations below the local water table. Cracks are a few millimeters wide and spaced from 5 to 20 cm apart. Individual beds are from 2 to 30 cm thick and have lateral continuities, measured in hundreds of meters, to more than 4 km.

Inasmuch as the playa deposits are in a low topographic position, they have not been subjected to desiccation during historical times except following construction of recent brine-collection ditches. The cracks, however, do not appear to open or close with wetting or desiccation along the ditch walls. A tentative hypothesis is that the fine-grained sediments were deposited originally in fresh to brackish water, and that brine almost saturated with salts has been introduced later. The increase in ionic strength in the fluids was then responsible for the shrinkage of the sediments.

A practical aspect of the foregoing discussion is the fact that disaggregated sediments as observed in drill cuttings or in laboratory size analyses may be highly misleading as far as conclusions concerning sediment permeabilities are concerned.

REFERENCES

Archie, G. E. (1952). Classification of carbonate reservoir rocks and petrophysical considerations, *Am. Assoc. Petroleum Geologists Bull.* **36**, 278–298.

Atwater, G. I. (1966). The effect of decrease in porosity with depth on oil and gas reserves in sandstone reservoirs, Unpublished paper presented to the School of Earth Sci., Stanford Univ., California, Jan. 10, 1966.

Bedinger, M. S. (1961). Relations between median grain size and permeability in the Arkansas River Valley, Arkansas, U.S. Geol. Surv. Prof. Paper 424-C, pp. 31–32.

Bennett, R. R., and Meyer, R. R. (1952). Geology and ground-water resources of the Baltimore area, *Maryland Dept. of Geol., Mines, and Water Resources Bull.* **4**.

Bennion, D. W., and Griffiths, J. C. (1966). A stochastic model for predicting variations in reservoir rock properties, *Trans. Am. Inst. Mining Met. Engrs.* **237**, 9–16.

Brace, W. F. (1965). Some new measurements of linear compressibility of rocks, *J. Geophys. Res.* **70**, 398–398.

Brace, W. F., Paulding, Jr., B. W., and Scholz, C. (1966). Dilatancy in the fracture of crystalline rocks, *J. Geophys. Res.* **71**, 3939–3953.

Brown, S. G., and Newcomb, R. C. (1963). Ground-water resources of the coastal sand-dune area north of Coos Bay, Oregon, U.S. Geol. Surv. Water-Supply Paper 1619-D.

Bulnes, A. C. (1946). An application of statistical methods to core analysis data of dolomitic limestone, *Trans. Am. Inst. Mining Met. Engrs.* **165**, 223–240.

Cardwell, W. T., Jr., and Parsons, R. L. (1945). Average permeabilities of heterogeneous oil sands, *Trans. Am. Inst. Mining Met. Engrs.* **169**, 34–42.

Cohen, P. (1965). Water resources of the Humboldt River Valley near Winnemucca, Nevada, U.S. Geol. Survey Water-Supply Paper 1795.

Collins, R. E., (1961). "Flow of Fluids Through Porous Materials." Reinhold, New York.

Csallany, S., and Walton, W. C. (1963). Yields of shallow dolomite wells in northern Illinois, Illinois State Water Surv. Rept. Invest. 46.

Cushman, R. V., Krieger, R. A., and McCabe, J. A. (1965). Present and future water supply for Mammoth Cave National Park, Kentucky, U.S. Geol. Surv. Water-Supply Paper 1475-Q, pp. 601–647.

Davis, S. N. (1964). Silica in streams and ground water, *Am. J. Sci.* **262**, 870–891.

Davis, S. N. (1966). Initiation of ground-water flow in jointed limestone, *Natl. Speleological Soc. Bull.* **28**, 111–118.

Davis, S. N., and De Wiest, R. J. M. (1966). "Hydrogeology." Wiley, New York.

Davis, S. N., and Hall, F. R. (1959). Water quality of eastern Stanislaus and northern Merced Counties, California. Stanford Univ. *Geol. Sci.* **6**, 112.

Davis, S. N., and Turk, L. J. (1964). Optimum depth of wells in crystalline rocks, *Ground Water* **2**, 6–11.

Dingman, R. J., Fergusun, H. F., and Martin, R. O. R. (1956). The water resources of Baltimore and Harford Counties, *Maryland Dept. of Geol., Mines, and Water Resources Bull.* **17**.

Frommurze, H. F. (1937). The water-bearing properties of the more important geological formations in the Union of South Africa, Union of South Africa Geol. Surv. Memoir No. 34.

Gallaher, J. T. (1964). Geology and hydrology of alluvial deposits along the Ohio River between the Uniontown Area and Wickliffe, Kentucky, U.S. Geol. Surv. Hydrologic Atlas 129.

Gibbs, H. J., and Holland, W. Y. (1960). Petrographic and engineering properties of loess, U.S. Bur. of Reclamation Eng. Monographs No. 28.

Gloyna, E. F., and Reynolds, T. D. (1961). Permeability measurements of rock salt, *J. Geophys. Res.* **66**, 3913–3921.

Gondouin, M., and Scala, C. (1958). Streaming potential and the SP log, *Trans. Am. Inst. of Mining and Met. Engrs. Paper* 8023.

Greenkorn, R. A., Johnson, C. R., and Shallenberger, L. K. (1964). Directional permeability of heterogeneous anisotropic porous media, *Trans. Soc. of Petrol. Engrs.* **231**, 124–132.

Jahns, R. H. (1943). Sheet structure in granite, its origin and use as a measure of glacial erosion in New England, *J. Geol.* **51**, 71–98.

Johnson, A. I., and Morris, D. A. (1962). Physical and hydrologic properties of water-bearing deposits from core holes in the Los Banos-Kettleman City area, California, U.S. Geol. Surv. Open-File Rept. Denver, Colorado.

Johnson, A. I., and Sniegocki, R. T. (1967). Comparison of laboratory and field analyses of aquifer characteristics at an artificial recharge well site, Paper presented at *Intern. Assoc. Scientific Hydrology Symp. on Artificial Recharge of Aquifers, Haifa, Israel.*

Johnson, C. R., and Greenkorn, R. A. (1963). Correlation of lithology and permeability for a Virgilian sandstone reservoir in central Oklahoma, *Geol. Soc. Am. Proc.* **73**, 180–181.

Johnston, N., and Beeson, C. M. (1945). Water permeability of reservoir sands, *Trans. Am. Inst. Min. Met. Engrs.* **160**, 43–55.

Johnston, P. M. (1962). Geology and ground-water resources of the Fairfax Quadrangle, Virginia, U.S. Geol. Surv. Water-Supply Paper 1539-L.

Keller, G. V. (1960). Physical properties of tuffs of the Oak Spring Formation, Nevada, U.S. Geol. Surv. Prof. Paper 400-B.

Kemper, W. D., Maasland, D. E. L., and Porter, L. K. (1964). Mobility of water adjacent to mineral surfaces, *Soil Sci. Soc. Am. Proc.* **28**, 164–167.

Laney, R. L. (1965). A comparison of the chemical composition of rainwater and ground water in western North Carolina, p. 187–189, U.S. Geol. Surv. Prof. Paper 525-C.

Law, J. (1944). A statistical approach to the interstitial heterogeneity of sand reservoirs, *Trans. Am. Inst. Mining Met. Engrs.* **155**, 202–222.

LeGrand, H. E. (1954). Geology and ground water in the Statesville area, North Carolina, North Carolina Dept. of Conservation and Development, Div. of Mineral Resources Bull. 68.

Levorsen, A. I. (1954). "Geology of Petroleum." Freeman, San Francisco, California.

Lewis, D. C., Kriz, G. J., and Burgy, R. H. (1966). Tracer dilution sampling technique to determine hydraulic conductivity of fractured rock, *Water Resources Res.* **2**, 533–542.

MacGary, L. M., and Lambert, T. W. (1962). Reconnaissance of ground-water resources of the Jackson Purchase region, Kentucky, U.S. Geol. Surv. Hydrologic Atlas 13.

Manger, G. E. (1963). Porosity and bulk density of sedimentary rocks, U.S. Geol. Surv. Bull. 1144-E.

Masch, F. D., and Denny, K. J. (1966). Grain size distribution and its effect on the permeability of unconsolidated sands, *Water Resources Res.* **2**, 665–677.

Maxwell, J. C. (1964). Influence of depth, temperature, and geologic age on porosity of quartzose sandstone, *Bull. Am. Assoc. Petrol. Geologists* **48**, 697–709.

Mégnien, Claude. (1964). Observations hydrogéologiques sur le sud-est du bassin de Paris, Mém. Bur. de Rech. Géol. et Minières No. 25.

Meyer, G., and Beall, R. M. (1958). Water resources of Carroll and Frederick Counties, *Maryland Dept. Geol., Mines, and Water Resources Bull.* 22.

Murray, R. C. (1960). Origin of porosity in carbonate rocks, *J. Sediment. Petrol.* **30**, 59–84.

Muskat, M. (1937). "The Flow of Homogeneous Fluids Through Porous Media." McGraw-Hill, New York.

Norris, S. E. (1963). Permeability of glacial till, U.S. Geol. Surv. Prof. Paper 450-E, pp. 150–151.

Norris, S. E., and Fidler, R. E. (1965). Relation of permeability to particle size in a glacial-outwash aquifer at Piketon, Ohio, U.S. Geol. Surv. Prof. Paper 525-D, pp. 203–206.

Olsen, H. W. (1966). Darcy's law in saturated kaolinite, *Water Resources Res.* **2**, 287–295.

Parker, G. G. (1951). Geologic and hydrologic factors in the perennial yield of the Biscayne aquifer, *Am. Water Works Assoc. J.* **43**, 817–835.

Piersol, R. J., Workman, L. E., and Watson, M. C. (1940). Porosity, total liquid saturation, and permeability of Illinois oil sands, Ill. Geol. Surv. Rept. Invest. No. 67.

Poth, C. W. (1963). Geology and hydrology of the Mercer Quadrangle, Mercer, Lawrence, and Butler Counties, Pennsylvania, Pennsylvania Topographic and Geol. Surv. Ground-Water Rept. W-16.

Rima, D. R., Meisler, H., Longwill, S. (1962). Geology and hydrology of the Stockton Formation in southeastern Pennsylvania, Pennsylvania Topographic and Geol. Surv. Ground-Water Rept. W-14.

Samuelson, W. J. (1965). Engineering geology of the Norad Combat Operations Center, Colorado Springs, Colorado, *Engineering Geol.* **2**, 20–30.

Schoeller, H. (1962). "Les eaux souterraines." Masson, Paris.

Seaber, P. R., and Hollyday, E. F. (1966). Statistical analysis of regional aquifers, A paper presented to the 1966 Ann. Meeting, Geol. Soc. of Am., San Francisco, California.

Solov'yev, P. Ye. (1961). Change in the physical properties of ordinary chernozems under the effect of forest vegetation, *Dokl. Soil Sci.* Section No. 2, pp. 29–33. (English transl. by Am. Inst. of Biol. Sciences).

Stewart, J. W. (1964). Infiltration and permeability of weathered crystalline rocks Georgia Nuclear Laboratory, Dawson County, Georgia, *U.S. Geol. Surv. Bull. 1133-D.*

Stuart, W. T., Brown, E. A., and Rhodehamel, E. C. (1954). Groundwater investigations of the Marquette Iron-Mining District: Michigan Geol. Surv. Div. Tech. Rept. 3.

Tanaka, H. H., and Hollowell, J. R. (1966). Hydrology of the alluvium of the Arkansas River, Muskogee, Oklahoma to Fort Smith, Arkansas, U.S. Geol. Surv. Water-Supply Paper 1809-T.

Tolman, C. F. (1937). "Ground Water." McGraw-Hill, New York.

Walker, E. H. (1956). Ground-water resources of the Hopkinsville Quadrangle, Kentucky, U.S. Geol. Surv. Water-Supply Paper 1328.

Walters, K. L. (1963). Highly productive aquifers in the Tacoma area, Washington, U.S. Geol. Surv. Prof. Paper 450-E, pp. 157–158.

Walton, W. C. (1962). Selected analytical methods for well and aquifer evaluation, Illinois State Water Surv. Bull. 49.

Walton, W. C., and Stewart, J. W. (1961). Aquifer tests in the Snake River Basalt, *Trans. Am. Soc. Civil Engrs.* **126**, 612–632.

Waring, G. A. (1965). Thermal springs of the United States and other countries of the world; a summary, U. S. Geol. Surv. Prof. Paper 492.

Warren, J. E., and Price, H. S. (1961). Flow in heterogeneous porous media, Trans. Soc. Petrol. Engrs, *Am. Inst. Mining, Met. Petrol. Engrs.* **222**, 153–169.

Yokota, J. (1963). Experimental studies on the design of grouting curtain and drainage for the Kurobe No. 4 Dam, *Rock Mech. Engineering Geol.* **1**, 104–119.

Young, A., Low, P. F., and McLatchie, A. S. (1964). Permeability studies of argillaceous rocks, *J. Geophys. Res.* **69**, 4237–4245.

3 Resistance to Flow through Porous Media

Ralph R. Rumer, Jr.
Faculty of Engineering and Applied Sciences
State University of New York at Buffalo
Buffalo, New York

1. Introduction

1.1. Nature of Resistance to Flow

The resistance offered by any stationary object immersed in a flowing fluid is composed of both the shear stress acting tangentially to the surface of the object and the normal stress (or pressure) acting perpendicularly to the surface of the object. The total resistance offered by the object in any given direction becomes the sum total of the corresponding components of the shear stress and pressure acting over the entire surface of the object.

This relatively simple definition of resistance has proven valid whenever the actual shear stress and pressure distributions are known over the surface of the object. This definition assumes that the surface of the object is impervious and hence no flow occurs through this surface.

Thus, it is seen that the task of predicting the resistance of any given object requires intimate knowledge of the fluid stress distribution over the surface of the object. This knowledge can be obtained when the velocity and pressure are known at every point throughout the flowing fluid. If we consider this fluid to be incompressible and Newtonian, the velocity and pressure at every point can be obtained from an exact solution to the Navier–Stokes equations of motion taking into account the boundary conditions for the particular flow situation.

Exact solutions to the Navier–Stokes equations have only been found for the regime of flow characterized as *laminar*, and then only for relatively simple flow configurations. Considering the complex geometry of a pore system, an exact solution to the Navier–Stokes equations for this problem appears a remote, if not impossible, task.

It is only natural that some sort of averaging process be employed in seeking a suitable resistance equation (or equation of motion) for flow through a porous medium. The equation referred to is one that relates the velocity of flow to the loss of energy in the flow. Any resistance to flow implies dissipation of energy and hence the task is to relate the appropriate velocity to the fluid pressure incorporating the characteristics of the flow boundary. After a more detailed consideration of the basic assumptions to be employed in the succeeding treatment, a method for determining the resistance equation will be presented and discussed.

Historically, the task of determining a suitable equation of motion (or resistance equation) has been one of seeking a theoretical justification for the empirical observation of Darcy. The so-called "Darcy law" has been repeatedly verified experimentally for certain flows and hence has provided a sound goal for the theorists to seek.

1.2. Microscopic versus Macroscopic View

The pore system of a medium represents the fluid conduits and is formed by the packing of many discrete particles. Normally the size of the particles is considered small. The pattern of flow through this conduit system is complex and at the microscopic scale, not amenable to a rigorous solution of the Navier–Stokes equations.

A simplified yet extremely useful picture is obtained by considering the fluid motion on a macroscopic scale while still retaining the artifice of a continuum. The macroscopic fluid properties and flow characteristics such as density, velocity, and pressure, can still be thought of as point quantities and continuous. The macroscopic volume must contain a sufficient number of particles such that its porosity is representative of the medium as a whole, but yet remain much smaller than the region of interest in the porous medium. This macroscopic approach is commonly used and forms the basis for most analyses of flow in porous media. The relative difficulty and impracticality of measuring anything except macroscopic quantities either in the laboratory or the field has only served to make this approach seem all the more rational.

2. General Theory

2.1. Assumptions and Restrictions

The scope of this treatment has been confined to porous media composed of individual discrete particles. The pore system geometry that results from the packing (either naturally or artificially) of these particles is considered to be fixed; i.e., transient phenomena such as consolidation of the soil matrix, and shrinking or swelling of the individual particles will not be considered. These restrictions are not intended to minimize their importance. Instead, the purpose of this treatment is to provide a basic understanding of the resistance phenomenon by concentrating on the hydraulic processes occurring within this constant pore system geometry.

The fluid flow considered will be assumed to be incompressible. Sorption effects at the fluid–solid interfaces will not be considered. Finally, the pore system will be assumed to be completely filled by the fluid, i.e., *saturated.*

If the flow through a granular bed is viewed in a manner that considers the resisting force of each individual particle, an equation can be derived describing the macroscopic or seepage velocity (Rumer and Drinker, 1966). By convention, the seepage velocity is considered to be the average fluid velocity in the neighborhood of a microscopic point obtained by averaging spatially the actual microscopic point fluid velocities through the pore system of the macroscopic volume. In equation form, the macroscopic or seepage velocity components become (Rumer and Harleman, 1963):

$$\bar{u} = (n\,\Delta\forall)^{-1} \iiint_{\Delta\forall} u_a \, d\forall, \qquad \bar{v} = (n\,\Delta\forall)^{-1} \iiint_{\Delta\forall} v_a \, d\forall$$
$$\bar{w} = (n\,\Delta\forall)^{-1} \iiint_{\Delta\forall} w_a \, d\forall \tag{1}$$

where $\bar{u}$, $\bar{v}$, $\bar{w}$ are the components of the seepage velocity; u_a, v_a, w_a are the components of the actual fluid velocity vector at a microscopic point; n is the porosity; and $\Delta\forall$ is the macroscopic volume. The actual fluid velocity vector is defined at all points within the pore system of the elemental volume shown in Fig. 1. It should be noted that the actual point fluid velocity is identically zero at every point occupied by the solid particles. For this reason, the volume integral includes the entire volume occupied by both solid particles and fluid. The use of volume integrals to obtain the average magnitude of scalar quantities is widely used (e.g., average salt concentration or average temperature). Its use here is approriate so long as the direction is specified.

For the basic development to follow, the porous medium will be assumed to consist of uniform-size spheres, although it will become evident that this assumption is not a necessary restriction.

2.2. Force-Balance Equation†

Consider the steady flow of a fluid through a cylindrical element that contains a sufficient number of particles such that the porosity of the element is representative of the medium as a whole (see Fig. 1). As already stated, this element is macroscopic with respect to the particles and pore spaces, yet much smaller than the region of interest within the porous medium. The element is centered on a streamline representing the direction of the mean motion of the fluid, at an angle θ

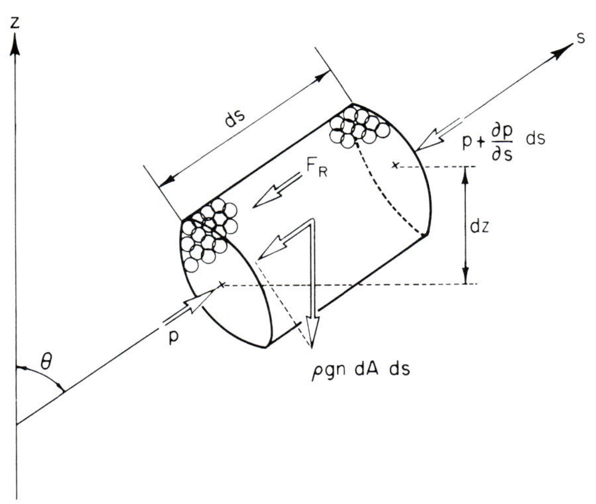

Fig. 1. Diagram of forces acting on the fluid within a cylindrical element of the porous field.

† See Collins (1961).

to the z-axis. For the spherical particles in the assumed model, the volume porosity and areal porosity are equal. The fluid volume within the element is given by $n\,dA\,ds$, in which dA is the total cross-sectional area of the element and ds is an incremental macroscopic length measured along the streamline. The solid particles each have a cross-sectional area given by αd^2 and a volume given by βd^3. For spheres, the shape factors α and β become equal to $\pi/4$ and $\pi/6$, respectively. For the uniform-size spheres, d is the particle diameter; for irregular-shape nonuniform particles, d would be the characteristic length of the average-size particle and would have to be determined by some appropriate technique.

A summation of forces acting in the s-direction on the fluid in the element results in a zero net force for this steady flow case. Thus,

$$pn\,dA - \left(p + \frac{\partial p}{\partial s}\,ds\right) n\,dA - (\rho g n\,dA\,ds)\cos\theta - F_{\mathrm{R}} = 0 \qquad (2)$$

in which the first two terms are the pressure forces acting at each end of the element, the third term is the weight component due to gravity, and F_{R} denotes the resisting force offered by all the solid particles contained within the element. ρ represents the average density of the fluid in the macroscopic pore volume and g is the gravitational acceleration. Replacing $\cos\theta$ by $\partial z/\partial s$ and simplifying, Eq. (2) becomes

$$-\frac{\partial p}{\partial s} - \rho g\,\frac{\partial z}{\partial s} - \frac{F_{\mathrm{R}}}{n\,dA\,ds} = 0 \qquad (3)$$

For those cases where the density ρ can be considered constant throughout the total pore region of the medium, Eq. (3) can be further simplified to

$$-\frac{\partial h}{\partial s} - \frac{F_{\mathrm{R}}}{\rho g n\,dA\,ds} = 0 \qquad (4)$$

in which h equals the piezometric head given by $h = (p/\rho g) + z$.

Further reduction of Eqs. (3) or (4) will depend on suitable representations of the resisting force F_{R}.

3. Resistance Functions

3.1. Linear Laminar Regime of Flow

The resisting force offered by a single sphere for the special case of slow viscous flow of a Newtonian fluid is given by Stokes' law

(Goldstein, 1938). This law can be expressed in a generalized form as

$$f_p = \lambda\mu \, d \, \bar{u}_s \tag{5}$$

in which f_p is the resistance or drag of a single particle, μ represents the dynamic viscosity of the fluid, $\bar{u}_s$ denotes the local average velocity of flow around the particle (i.e., the seepage velocity) in the s-direction, and λ represents a coefficient that takes into account the effects of neighboring particles. The coefficient λ, which reaches a limiting minimum value of 3π for a single sphere in an infinite fluid, will depend on the local streamline configuration around the particle and, hence, must be some function of the geometry of the pore system. More specifically, λ will depend upon the porosity, the shape of the particles, and the distribution of the sizes of the particles.

The total resistance F_R offered by all of the particles in the element will thus be

$$F_R = N f_p \tag{6}$$

in which N represents the number of particles in the element. The number of particles N can be given as

$$N = \frac{(1 - n) \, dA \, ds}{\beta d^3} \tag{7}$$

The numerator in Eq. (7) is the total volume of solids in the macroscopic element and the denominator represents the solid volume of a single particle.

With the aid of Eqs. (5), (6), and (7), Eq. (4) can be written as

$$-\frac{\partial h}{\partial s} - \frac{\lambda(1 - n)\,\mu}{\beta d^2 \, n\rho g} \bar{u}_s = 0 \tag{8}$$

Solving Eq. (8) for $\bar{u}_s$ and replacing $\bar{u}_s$ by the Darcian or superficial velocity (specific discharge), $q_s = n\bar{u}_s$, there results

$$q_s = -\frac{\beta n^2}{\lambda(1 - n)} d^2 \frac{\rho g}{\mu} \frac{\partial h}{\partial s} \tag{9}$$

The quantity $\beta n^2/\lambda(1 - n)$ is dependent only on the pore system geometry and can be replaced by a single coefficient C. Thus,

$$q_s = -Cd^2 \frac{\rho g}{\mu} \frac{\partial h}{\partial s} \tag{10}$$

Here the product Cd^2 is recognized as the intrinsic permeability k, and has the units of length squared. Finally, Darcy's law can be written in the familiar form

$$q_s = -k \frac{\rho g}{\mu} \frac{\partial h}{\partial s} \tag{11}$$

For the more general case given by Eq. (3) in which the density ρ may vary in space in a macroscopic sense, the resulting expression for q_s becomes

$$q_s = -\frac{k}{\mu}\left(\frac{\partial p}{\partial s} + \rho g \frac{\partial z}{\partial s}\right) \tag{12}$$

Equation (12) indicates the validity of using Darcy's law for flows of variable density while, at the same time, pointing out the macroscopic nature of ρ as used in this equation. As developed here, the actual fluid density at a microscopic point and the macroscopic fluid density ρ are identical only for constant density homogeneous fluids.

3.1.1. *Experimental Results for Linear Laminar Flow*

The intrinsic permeability coefficient k and its relationship to the measurable properties of a porous medium have received much study. The limiting factor in all experimental approaches is, of course, the inability to measure the actual pore geometry of a medium in place. Those quantities that are measurable and that have been included in most attempts to formulate this relationship include the mean particle size, porosity, and particle size distribution.

The relationship between the intrinsic permeability and media parameters derived from the preceding analysis can be obtained from Eqs. (9) and (11). This relationship is

$$k = \frac{\beta n^2}{\lambda(1-n)} d^2 \tag{13}$$

For media composed of spheres, the coefficient β is equal to $\pi/6$. Experimental results from investigations on spheres provide the data necessary to determinine λ by utilizing Eq. (13) (see Table 1).

Using an average value of λ obtained from the data of Harleman *et al.* (1963), and a porosity value of $n = 0.37$, Eq. (13) becomes

$$k = (6.54 \times 10^{-4})\, d^2 \tag{14}$$

A plot of the intrinsic permeability k versus particle diameter reveals fairly good agreement between the values of k obtained from Eq. (14) and experimentally determined values. Results for angular particles are also included (see Fig. 2).

TABLE 1

Experimental Results on Nearly Uniform Porous Media

Diameter of mean particle size d (cm)	Uniformity coefficient	Porosity n	Intrinsic permeability k (cm^2)	Calculated coefficient from Eq. (13) λ	Investigator
			Spherical shape		
0.53	1.0[a]	—	151×10^{-6}	—	Brownell *et al.* (1956)
0.30	1.0	—	64.5×10^{-6}	—	Sunada (1965)
0.2	1.0	0.40	34.6×10^{-6}	163	Harleman *et al.* (1963)
0.2	1.0	0.37	22.0×10^{-6}	207	Harleman *et al.* (1963)
0.2	1.0	0.38	24.5×10^{-6}	199	Harleman *et al.* (1963)
0.14	1.15	0.38	15.7×10^{-6}	152	Harleman *et al.* (1963)
0.092	1.08	0.37	5.7×10^{-6}	169	Harleman *et al.* (1963)
0.0545	1.0	—	2.62×10^{-6}	—	Ward (1964)
0.0458	1.0	—	1.87×10^{-6}	—	Ward (1964)
0.039	1.13	0.36	1.03×10^{-6}	157	Harleman *et al.* (1963)
0.0383	1.0	—	1.15×10^{-6}	—	Ward (1964)
0.0322	1.0	—	0.901×10^{-6}	—	Ward (1964)
0.0273	1.0	—	0.577×10^{-6}	—	Ward (1964)
			Angular shape		
0.546	—	0.33	2.65×10^{-4}	—	Banks *et al.* (1962)
0.14	1.15	0.39	9.45×10^{-6}	—	Harleman *et al.* (1963)
0.092	1.08	0.40	4.83×10^{-6}	—	Harleman *et al.* (1963)
0.045	1.12	0.39	1.10×10^{-6}	—	Harleman *et al.* (1963)
0.0115	1.0	0.33	8.1×10^{-8}	—	Raimondi *et al.* (1959)

[a] The uniformity coefficient is the ratio of the 60 percentile size to the 10 percentile size taken from a *gravimetrically* determined cumulative frequency distribution plot.

Apparently, Eq. (14) gives good approximations for the intrinsic permeability of media composed of relatively uniform-size particles, whether they be spherical or angular in shape. Extension of Eq. (14) to nonuniform media proves to be discouraging in that λ appears to be highly dependent upon the degree of nonuniformity. The actual magnitude of the volume coefficient β seems less critical, based on the results obtained using Eq. (14) for both spherical and angular particles (see Fig. 2).

3.1.2. *Limitations to the Resistance Equation for Linear Laminar Flow*

Equations (11) and (12) must cease to be valid when the resistance function F_R can no longer be given by a generalization of Stokes' law.

This occurs when the inertial forces neglected in the derivation of Stokes' law are no longer vanishingly small. These inertia terms or water particle accelerations are derived from the convective acceleration resulting from the constantly varying cross-sectional area exposed to the flow, and also from the centrifugal accelerations resulting from the curvature of the flow paths. A limiting value for the Reynolds number,

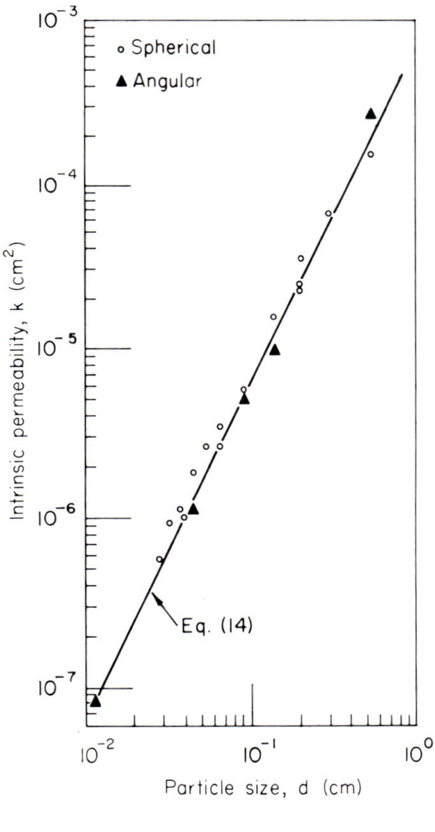

FIG. 2. Relation of intrinsic permeability to particle size for nearly uniform porous media.

as predicted by Eq. (11) or (12), gives excellent correspondence with observed values. Utilizing the limiting Reynolds number for Stokian flow around a sphere, a departure from Darcy's law [Eq. (11) or (12)] would be expected for a Reynolds number range 0.1–1.0. The limiting value of the Reynolds number for Stokes' solution for a single sphere has been experimentally determined to be 0.5 (Goldstein, 1938). Some variation in the actual magnitude of the Reynolds number results from differences in the definition of the characteristic particle length in the case of nonuniform or angular media and in the definition of the characteristic velocity; i.e., whether it is the superficial or seepage velocity. In either case, the comparison between observed values and predicted values is good.

3.2. Nonlinear Laminar Regime of Flow

A more general expression for the resistance function F_R can be obtained by making use of the so-called "drag" equation. For this, the expression for the resistance of a single particle can be written as

$$f_P = C_D \alpha d^2 (\rho \bar{u}_s^2/2) \tag{15}$$

in which C_D is the "drag coefficient" and for incompressible flow is generally considered to be a function of the Reynolds number of the flow and also the shape of the particle. For flow in the Stokes' range, $C_D = (2n\lambda)/(\alpha\mathscr{R})$, and Eqs. (5) and (15) become identical, leading again to Eq. (11) for linear laminar flow. In this case, $\mathscr{R}$ refers to the bed-Reynolds number, which by convention is defined as $\mathscr{R} = (q_s\, d\rho)/\mu$.

Making use of the more general expression for particle resistance given by Eq. (15), the general force-balance equation, Eq. (4), becomes

$$-\frac{\partial h}{\partial s} - \frac{\alpha(1-n)}{2\beta n^3 d}\frac{C_D}{g} q_s^2 = 0 \tag{16}$$

By using Eq. (13), Eq. (16) can be written as

$$-\frac{\partial h}{\partial s} - \frac{\alpha}{2n^2}\left(\frac{1-n}{\lambda\beta}\right)^{1/2} \frac{1}{g\sqrt{k}} C_D q_s^2 = 0 \tag{17}$$

The empirical nature of C_D, except for the case of Stokian flow around a sphere [in which case Eq. (17) becomes equal to Eq. (11)], seems at first to provide little help towards the problem of obtaining a resistance equation for flow beyond the linear laminar range. By introducing a dimensionless friction factor (Ward, 1964) defined as

$$f_k = -\frac{\partial h}{\partial s}\frac{g\sqrt{k}}{q_s^2} \tag{18}$$

Eq. (17) becomes

$$f_k = \frac{\alpha}{2n^2}\left(\frac{1-n}{\lambda\beta}\right)^{1/2} C_D \tag{19}$$

It will be noted in Eq. (18) that the minus sign is retained in the definition of f_k. f_k is always positive when evaluated for actual flow cases since the piezometric head gradient $\partial h/\partial s$ is always negative in the direction of flow. The friction factor f_k is dependent upon both the pore system geometry and the Reynolds number, as seen in Eq. (19).

For the case of linear laminar flow where $C_D = (2n\lambda)/(\alpha\mathscr{R})$, Eq. (19) results in the following simple expression for f_k,

$$f_k = 1/\mathscr{R}_k \tag{20}$$

where $\mathscr{R}_k$ is called the permeability Reynolds number and is given by $\mathscr{R}_k = (q_s\rho\sqrt{k})/\mu$ (Harleman *et al.*, 1963). It can be seen that Darcy's law is again obtained by combining Eqs. (20) and (18).

Upon examination of Eq. (19), it is seen that since the friction factor f_k is dependent upon both the pore system geometry and the Reynolds number of the flow, all experimental data on porous media flow will not fall precisely on a single curve in a plot of f_k versus $\mathscr{R}$. Although the data will appear to converge to a single curve, the scatter will be evidence of this dependence of f_k on the geometry of the pore system as well as the flow Reynolds number.

Even for a single sphere in an infinite fluid, there is no general solution for the drag coefficient in terms of the flow Reynolds number. Such factors as the characteristics of the turbulence in the flow approaching the sphere and even the surface texture (or roughness) of the sphere itself become important in the actual determination of the resistance offered by the stationary sphere to the flow. The complexity of the flow pattern around a single sphere with increasing velocity testifies to the analytical difficulties to be encountered in attempting to formulate an expression for the drag coefficient. Only for the orderly motion associated with Stokian flow has success been achieved. Beyond this range, the flow undergoes phases of development from stationary eddies behind the sphere, to a shedding of these eddies into the main flow, and finally to a turbulent wake composed of very disorderly motion. If the flow through and around a group of spheres is considered, the analytical difficulties are compounded and the problem seems almost insurmountable.

Some measure of success beyond the limits of Stokian motion has been achieved (Goldstein, 1938) and these results can be put to use here. For Reynolds numbers less than 2, Goldstein has obtained an approximate solution for the drag coefficient for laminar flow past a sphere taking partial account of the inertia forces neglected in Stokes' solution.

Again, in a generalized form, the drag coefficient as obtained by Goldstein (1938) can be written as

$$C_D = (\xi/\mathscr{R})\left[1 + \sum_{1}^{j} (-1)^{j+1} C_j \mathscr{R}^j\right] \tag{21}$$

in which ξ is a constant and equal to $(2n\lambda)/\alpha$ to be consistent with the preceding development, the C_j are coefficients that are dependent only upon the geometry of the object (in this case the pore geometry), and the index j takes on values of $j = 1, 2, 3, \ldots$. Inserting Eq. (21) into Eq. (19) results in an expression for the resistance coefficient f_k that is applicable to the first part of the nonlinear laminar regime of flow. This is

$$f_k = (d/\sqrt{k}\,\mathscr{R})\left[1 + \sum_1^j (-1)^{j+1} C_j \mathscr{R}^j\right] \tag{22}$$

or in terms of the permeability Reynolds number,

$$f_k = \mathscr{R}_k^{-1}\left[1 + \sum_1^j (-1)^{j+1} C_j \mathscr{R}^j\right] \tag{23}$$

As has already been noted, Eq. (23) reduces to Darcy's law for very small Reynolds numbers [i.e., when the series expression in Eq. (23) can be neglected]. If just the first term in the series is retained (similar to Oseen's approximation for nonlinear flow past a sphere), an expression for f_k results which is comparable to that proposed by Forchheimer (Muskat, 1946)

$$f_k = \mathscr{R}_k^{-1} + C_1(\mathscr{R}/\mathscr{R}_k) \tag{24}$$

More recently (Sunada, 1965), this equation was again presented as a general equation to cover all velocities of flow. Writing Eq. (24) in complete notation, one has

$$-\frac{\partial h}{\partial s} = \frac{\mu}{\rho g k} q_s + \frac{C_1 d}{g\sqrt{k}} q_s^2 \tag{25}$$

In this equation, C_1 is the first coefficient C_j in the series expansion.

It has not been possible to predict analytically the value for C_1 for porous media flow. Considering that the coefficients C_j are dependent upon shape (Goldstein's equation was for spheres), one would expect C_1 also to be shape dependent, or more generally, dependent upon the geometry of the pore system. By fitting the Eq. (25) to experimental data, values of C_1 can be determined. Unfortunately, this coefficient takes on different values when obtained in this manner, depending on the range of Reynolds numbers one tries to cover. For a range of Reynolds numbers of 0.3–420, Sunada's (1965) results give $C_1 = 0.2$ for spheres. According to this development, Eqs. (24) and (25) are only a partial representation for the resistance equation and higher order terms are probably necessary to describe more adequately the beginning part of the nonlinear laminar regime of flow through porous media.

3.3. Turbulent Regime of Flow

In accordance with the force balance development presented here, the resistance equation for turbulent flow through a porous medium must remain as given in Eq. (19). It is recognized that this formulation provides very little satisfaction to the analyst and various approximations will undoubtedly be made. Such approximations may be in the form of a truncated power series in terms of the velocity such as that proposed by Forchheimer (Goldstein, 1938) and again more recently by Sunada (1965) and Ward (1964), see Eq. (25).

If one considers the well-known plot of C_{D} versus $\mathscr{R}$ for a single sphere, it is apparent that over a certain range of Reynolds numbers, C_{D} remains virtually constant. This would indicate that over some comparable range of Reynolds numbers, Eq. (19) could be given approximately as

$$f_k = \text{constant} \tag{26}$$

The trend of experimental results for high Reynolds numbers indicates that Eq. (26) may have some validity if applied within the appropriate range of Reynolds numbers.

Certain statements regarding turbulent flow through porous media can be made with some degree of assurance. First, the inception of turbulent flow undoubtedly takes place in a somewhat random pattern at various locations throughout the medium. It can be expected that, at these locations, conditions are conducive to the generation of the disorderly fluctuating motion called "turbulent flow". As the mean velocity of flow through the pore system of the medium increases, the spatial density of these "turbulent spots" probably increases until finally the flow becomes turbulent everywhere. Based on our knowledge of flow around submerged objects (e.g., spheres or bluff bodies) one can expect these "turbulent spots" to first occur where the pore geometry exhibits sharp projections or abrupt changes in cross-sectional area. This concept of the inception of turbulence in flow through porous media is premised on the existence of those types of flow discontinuities that occur whenever fluid streams of different velocities come together. This concept is compatible with the "drag" model developed in this chapter. The establishment of wakes and points of separation is also consistent with this concept.

It is worth pointing out that the capillary tube model for porous media flow fails both qualitatively and quantitatively to predict the limit of validity of Darcy's law and it also fails to predict the existence of the smooth transitional flow from linear laminar flow to turbulent flow. This transitional region includes the nonlinear laminar flow regime. The capillary tube model does not allow for curvature of the streamlines, and hence, the inertial terms (both convective and centrifugal) are absent. In addition, in a tube, turbulence is generated by a velocity gradient without any abrupt discontinuities. This type of flow is called "shear flow." The experiment of Reynolds with a dyestreak in a tube is a good example of a shear flow becoming turbulent. The limit of validity (for the flow equation derived from the capillary tube model) as measured by the Reynolds number should be $\mathscr{R} \simeq 2000$, which does not agree with observed values for departure from Darcy's law. In this case, the Reynolds number is defined using the tube diameter as the characteristic length.

The scale of the eddies associated with turbulent flow through porous media is, by necessity, limited by the dimensions of the pore openings. A similar restriction does not exist for the intensity of the turbulence as measured by the root mean square of the fluctuating components of the actual fluid velocity at a point in the flow.

4. The Special Case of Upward Flow: Fluidization

For most situations regarding flow through porous media, the drag force exerted by the fluid on the particles is transmitted through the solid particle matrix by point contact and ultimately to some constraining boundaries. Occasions do arise when this is not the case. Most notable of these is the case of vertical upward flow. For this case the drag force on the individual particles is opposed by only the submerged weight of the particles. When the drag force becomes equal to the submerged weight of the particles, the point contact between adjacent particles vanishes and the particles become suspended in the flow. This condition is frequently termed "fluidization of the bed."

For this special case it can be seen that a slight expansion of the bed has the effect of reducing the seepage velocity due to the increase in the porosity and the resulting increase in the volume of voids contained in the macroscopic volume $\Delta\forall$. For a constant volume rate of flow (or discharge), a suspended bed of uniform-size particles will exhibit a fairly constant particle concentration per unit volume. Any increase in the discharge will cause a further expansion of the bed with a resulting decrease in the concentration of particles per unit volume. For all values of discharge, the seepage velocity remains essentially constant; thus, a balance exists between the submerged weight of the particle and the drag force exerted by the fluid on the particle. There is a limit to this condition of quasi-equilibrium. This limit occurs when the discharge increases in magnitude such that the particles cannot expand sufficiently so as to maintain the condition of constant seepage velocity. Hence, the drag force exceeds the submerged weight of the particle. For this case, the particles are then simply swept away by the flow.

The suspended or fluidized bed behaves quite differently from the corresponding "packed" bed for the same discharge or flow Reynolds number. In the case of the suspended particle, the interference of one particle on the flow approaching the next is demonstrated by a fluctuating motion of the particle. For this reason, the fluidized bed system becomes turbulent at a lower value of flow Reynolds number than does its counterpart in the packed bed system.

The process of fluidization has proven useful in many practical applications. Hydraulic sorting of nonuniform-size sand particles can be accomplished in this way. Hydraulic filters in water treatment plants are cleaned of accumulated deposits by this process.

For this special case of vertical upward flow, a porous bed of discrete particles becomes suspended when the piezometric head gradient is given by

$$-\partial h/\partial z = (1 - n)(S_s - 1) \tag{27}$$

in which the gradient of h is with respect to z, since the s-direction coincides with the z-coordinate axis. S_s represents the specific gravity of the solid. From Eqs. (27) and (16) the following expression is obtained for the drag coefficient when the condition of fluidization exists

$$C_D = \frac{2\beta n^3 d}{\alpha} \frac{g}{q_s^2} (S_s - 1) \tag{28}$$

For the special case of uniform-size spheres, $\beta = \pi/6$ and $\alpha = \pi/4$ and Eq. (28) becomes

$$C_D = \tfrac{4}{3}(S_s - 1)(gdn^3/q_s^2) \tag{29}$$

The results of experiments (Rumer and Drinker, 1966) performed in a vertical column using uniform-size spheres of diameter $d = \frac{1}{8}$ inch, and having a specific gravity $S_s = 1.28$, are plotted in Fig. 3. Notice that the extension of the experimental curve for the fluidized condition intersects the curve of C_D versus $\mathscr{R}$ for a single sphere at a Reynolds number using the experimentally determined free fall velocity of a single sphere.

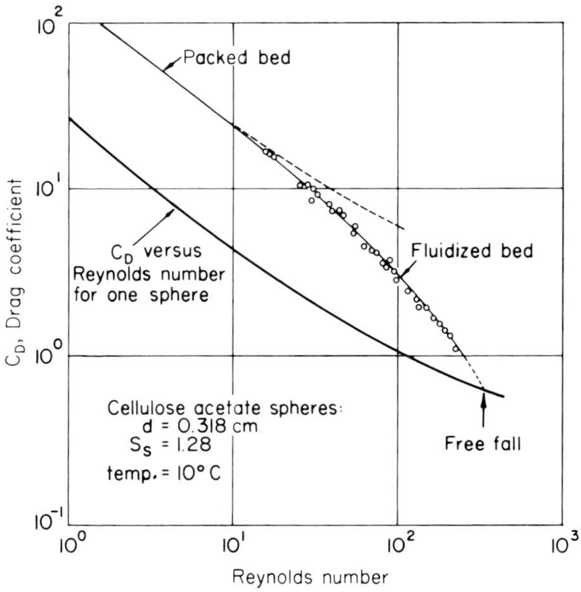

FIG. 3. Drag coefficients versus Reynolds number for flow through porous media.

5. Summary

A macroscopic force-balance equation has been derived for steady saturated flow through a porous medium composed of discrete particles. This equation states simply that the piezometric head gradient along a macroscopic streamline in flow through porous media is equal to the resisting force offered by the particles per unit weight of fluid.

Based on existing knowledge pertaining to hydrodynamic resisting forces on immersed bodies, it was shown that this macroscopic force-balance equation transformed into the familiar Darcy equation for very slow creeping motion. For higher velocity flows beyond the limit of validity of Darcy's law, the macroscopic force-balance equation transformed into a permeability Reynolds number power series expression for the resistance coefficient f_k. It was shown that a truncated form of this series expression led to the so-called "Forchheimer equation" for flow through porous media. The approach was further extended to the special case of upward flow and fluidization. For each case, a resistance equation was developed relating the resistance coefficient to the permeability Reynolds number and other parameters dependent either upon the geometry of the pore system or the physical properties of the particles and the fluid. These equations are:

linear laminar regime of flow,

$$f_k = \frac{1}{\mathscr{R}_k} \qquad [20]$$

nonlinear laminar regime of flow,

$$f_k = \frac{1}{\mathscr{R}_k}\left[1 + \sum_{1}^{j} (-1)^{j+1} C_j \mathscr{R}^j\right] \qquad [23]$$

fluidized bed or suspension,

$$f_k = \frac{1}{\mathscr{R}_k^{\,2}}\left[g(S_s - 1)\,\frac{[nd(\beta/\lambda)^{1/2}]^3}{(\mu/\rho)^2\,(1 - n)^{1/2}}\right] \qquad (30)$$

Equation (30) can be obtained from Eq. (28) utilizing Eqs. (19) and (13).

A meaningful presentation of the resistance of a single sphere to flow around it has been the traditional plot of the drag coefficient versus the flow Reynolds number. An analogous plot for flow through porous media (Fig. 3) indicates striking similarities to the single-sphere case, while at the same time permitting the transition from a packed bed to a fluidized bed to be portrayed as one curve. This should prove instructive to the reader who is seeking a unifying thread for these two cases that have so often been treated separately.

REFERENCES

Banks, R. B., Jerasate, S., and Owston, R. (1962). Studies on dispersion in porous media flow. Tech. Rept., Northwestern Univ. and SEATO School of Eng., Bangkok.

Brownell, L., Dombrowski, H., and Dickey, C. (1950). Pressure drop through porous media. *Chem. Eng. Progr.* **46**, 415.

Collins, R. E. (1961). "Flow of Fluids through Porous Materials." Reinhold, New York.

Goldstein, S. (1938). "Modern Developments in Fluid Dynamics," Vol. II. Oxford Univ. Press (Clarendon), London and New York.

Harleman, D. R. F., Mehlhorn, P. F., and Rumer, R. R. (1963). Dispersion-permeability correlation in porous media. *J. Hydraulics Div., Am. Soc. Civil Engrs.* **89**(2), 67.

Muskat, M. (1946). "The Flow of Homogeneous Fluids through Porous Media." Edwards, Ann Arbor, Michigan.

Raimondi, P., Gardner, G. H. F., and Petrick, C. (1959). Effect of pore structure and molecular diffusion on the mixing of miscible liquids flowing in porous media. *A.I.Ch.E.–S.P.E., Joint Symp. Fundamental Concepts Miscible Fluid Displacement, Part II, San Francisco, California*, Preprint 43.

Rumer, R. R., and Harleman, D. R. F. (1963). Intruded salt-water wedge in porous media, *J. Hydraulics Div., Am. Soc. Civil Engrs.* **89**(6), 193.

Rumer, R. R., and Drinker, P., (1966). Resistance to laminar flow through porous media, *J. Hydraulics Div., Am. Soc. Civil Engrs.* **92**(5), 155.

Sunada, D. (1965). Turbulent flow through porous media. Water Resources Center Contrib. No. 103, Univ. of Calif., Berkeley.

Ward, J. (1964). Turbulent flow in porous media. *J. Hydraulics Div., Am. Soc. Civil Engrs.* **90**(5), 1.

4 Hydrodynamic Dispersion

Jacob Bear

Department of Civil Engineering
Technion, Israel Institute of Technology
Haifa, Israel

1. Introduction

1.1. Definition of Hydrodynamic Dispersion

Consider saturated flow through a porous medium and let a portion of the flow domain contain a given mass of solute. This solute will be refrred to as a *tracer*. The tracer, which is a labeled portion of the same liquid, may be identified by its density and color, by electrical conductivity, etc. Experience shows that during the flow, the tracer mass gradually spreads and occupies an ever increasing portion of the flow domain, beyond the region it is expected to occupy according to the average flow (Fig. 1). This spreading phenomenon is called *hydrodynamic dispersion* (*dispersion*, *miscible displacement*) in a porous medium. It is a nonsteady, irreversible process (in the sense that if the flow is reversed we do not return to the initial tracer distribution) in which the tracer mass mixes with the nonlabeled portion of the liquid. If initially the tracer liquid occupies a separate region, with an abrupt interface separating it from the unmarked liquid, the interface does not remain an abrupt one, the location of which may be determined by the average fluid velocity expressed by Darcy's law. Instead, an ever widening *transition zone* is created, across which the tracer concentration varies from that of the tracer liquid to that of the unmarked liquid.

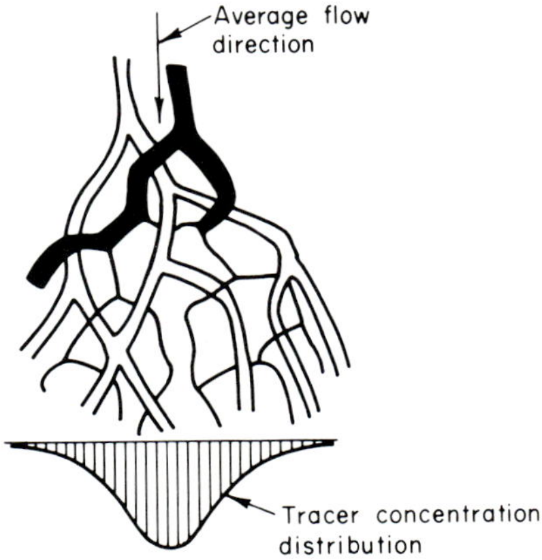

FIG. 1. Mixing of two fluid threads during flow in a porous medium.

One of the earliest observations of these phenomena was reported by Slichter (1905) who used an electrolyte as a tracer in studying ground water movement. Slichter noted that at an observation well downstream of the injection point, the tracer's concentration increases gradually, and that even in a uniform (average) flow field, the tracer advances in the direction of the flow in a pear-like form which becomes longer and wider as it advances.

The dispersion phenomenon may be demonstrated by a simple experiment. Consider steady flow in a cylindrical column of homogeneous sand saturated with water. At a certain instant ($t = 0$), a tracer marked water (e.g., water with NaCl at a low concentration so that the effect of density variations on the flow pattern is negligible), starts to displace the original unlabeled water in the column. Let the tracer concentration $C = C(t)$ at the end of the column be measured and presented in a graphical form (called *breakthrough curve*) as $\varepsilon = \varepsilon(t)$ or $\varepsilon = \varepsilon(U)$ where

$$\varepsilon(t) = [C(t) - C_0]/(C_1 - C_0) \tag{1}$$

is the *relative tracer distribution*, C_0, C_1 are the tracer concentrations of the original and of the displacing water, respectively, and U is the volume of effluent.

In the absence of dispersion, the breakthrough curve might have taken a form similar to that of Fig. 2(a), where U_0 is the pore volume of the column and Q is the constant discharge. Actually, due to hydrodynamic dispersion, it will take the form of the S-shaped curve shown in Fig. 2(b).

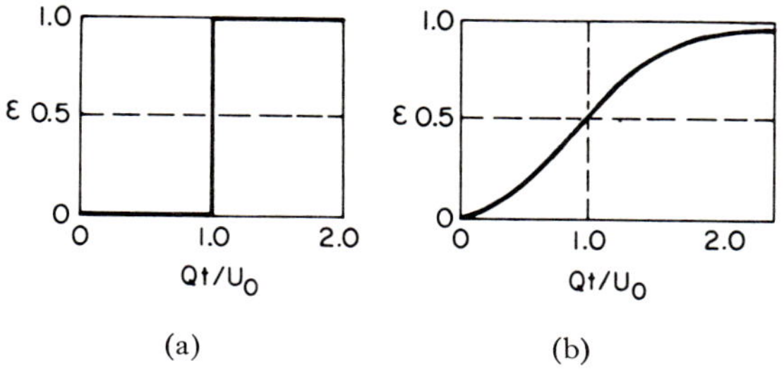

FIG. 2. Breakthrough curves in sand column: (a) without dispersion; (b) actual curve.

Hydrodynamic dispersion is the macroscopic outcome of the actual movements of the individual tracer particles through the pores and the various physical and chemical phenomena which take place within the pores. In general, such movements and phenomena result from (1) external forces acting on the liquid; (2) the microscopic intricate geometry of the pore system; (3) molecular diffusion caused by tracer concentration gradients; (4) variations in liquid properties such as density and viscosity which affect the flow pattern; (5) changes of the tracer's concentration due to chemical and physical processes within the liquid phase; and (6) interactions between the liquid and the solid phases.

There are two basic transport phenomena involved: *convection* and *molecular diffusion.* The complicated system of interconnected passages comprising the microstructure of the medium causes a continuous subdivision of the tracer's mass into finer offshoots. Variations in local velocity, both in magnitude and direction, along the tortuous flow paths and between adjacent flow paths as a result of the velocity distribution within each pore, cause any initial tracer mass within the flow domain to spread and occupy an ever increasing volume of the porous medium (Fig. 3). The two basic elements in this kind of mixing (often called *mechanical dispersion* or *convective diffusion*) are, therefore, the flow and the existence of the pore system through which flow takes place. In addition to inhomogeneity on this microscopic scale (= presence of pores, grains, etc.), we may have also inhomogeneity on a macroscopic scale due to variations in permeability from one portion of the flow domain to the next. This inhomogeneity also contributes to the mechanical dispersion of the tracer.

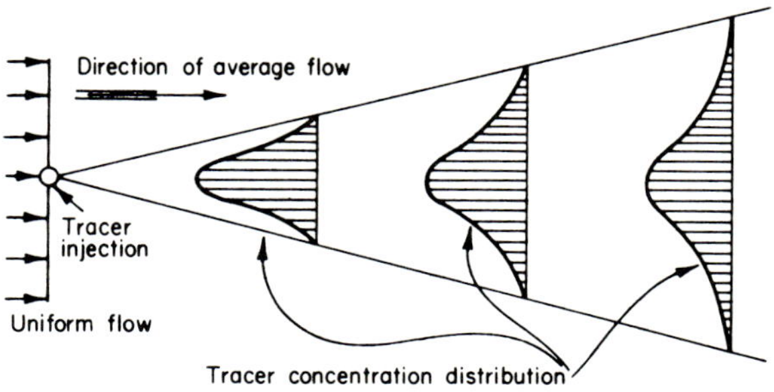

FIG. 3. Variations in tracer concentration (Danel, 1952).

In general we may have a convective mass transport (1) in a laminar flow regime where the liquid moves along definite paths which may be averaged to yield streamlines, and (2) in a turbulent flow regime where the turbulence may cause yet an additional mixing. In what follows, we shall focus our attention mainly to flow of the first type.

An additional mass transport phenomenon which occurs simultaneously with mechanical dispersion is that resulting from variations in tracer concentration within the liquid phase. Actually, the separation between the two processes is somewhat artificial. Hydrodynamic dispersion, as we shall later see, includes *both* processes. However, molecular diffusion does take place also in the absence of both a motion and a pore system. Because molecular diffusion depends on time, its effects on the overall dispersion will be more significant at low flow velocities. The relationships between the two phenomena are considered below.

The interaction between the solid surface of the porous matrix and the liquid may take several forms: adsorption of tracer particles to the solid surface, deposition, solution, ion exchange, etc. All these phenomena cause changes in the concentration of the tracer in the flowing liquid. Radioactive decay and chemical reactions within the liquid also cause tracer concentration changes.

Generally speaking, variations in tracer concentration cause changes in the liquid's density and viscosity. These, in turn, affect the flow regime (i.e., velocity distribution) which depends on these properties. We define an *ideal tracer* as one which is inert with respect to its liquid and solid surroundings and which does not affect the liquid's properties. At relatively low concentrations, the ideal tracer approximation is sufficient for most practical purposes.

1.2. Occurrence of Dispersion Phenomena

Hydrodynamic dispersion phenomena occur in many problems of ground water flow, in chemical engineering processes, in oil reservoir engineering, etc. In ground water flow we encounter it in:

(1) The transition zone between salt water and fresh water in coastal aquifers. The widening of this transition zone as it advances landward as a result of excess pumpage in inland pumping fields causes contamination of wells which are located closer to the coast.

(2) Artificial recharge operations where water of one quality is introduced into aquifers containing water of a different quality. Mixing of water due to hydrodynamic dispersion is among the objectives of artificial replenishment.

(3) Secondary recovery techniques in oil reservoir engineering where the injected fluid dissolves the reservoir's oil.

(4) Radioactive and reclaimed sewage waste disposal into aquifers.

(5) The use of reactors packed with granular material in the chemical industry.

(6) The movement of fertilizers in the soil and the leaching of salts from the soil in agriculture.

Here we shall treat hydrodynamic dispersion mainly from the point of view of ground water hydrology. The discussion will be limited to saturated flow.

1.3. Definition of Concentration and Velocity

To facilitate the discussion, the following definitions are employed:

$$C = \text{tracer concentration} = \frac{\text{mass of tracer}}{\text{volume of solution}}; \qquad (\text{dim: } ML^{-3}) \tag{2}$$

$$\omega = \text{tracer mass concentration} = \frac{\text{mass of tracer}}{\text{mass of solution}}; \qquad (\text{dimensionless}) \tag{3}$$

When an α species ($= \alpha$ component of a solution) is the tracer considered, C in (2) is replaced by $C_\alpha \equiv \rho_\alpha$ and ω in (3) by ω_α.

The movement of fluid through a porous medium may be regarded as a movement of an assemblage of fluid particles. For the purpose of the present discussion, a fluid particle is defined as a volume of fluid smaller than a pore, which consists of a large number of molecules. During the motion, each particle exchanges molecules with neighbouring particles through molecular diffusion. The kinetic properties of the particle are referred to its center of gravity. At a certain instant, a *local* (microscopic) *velocity* at a point is the velocity of the particle for which this point is the center of gravity.

The discussion in the following sections is limited to a single phase–binary liquid system, although, in principle it may be extended to a multiphase liquid system. It will be assumed that the instantaneous true distribution of any component of the liquid may be replaced by a *material continuum*. The entire liquid system will also be replaced by a material continuum. Each one of these continua fills the entire space. Obviously, a continuum approach is justified when the dimensions of the flow domain of interest are much larger than the molecular spacing. This enables liquid properties such as density and viscosity to be continuous functions of location within a given volume of liquid. The spatial distribution and the motion of each component and of the entire liquid. system may be defined by the density and the velocity at each point.

The continuum approach (and we are still dealing with a *liquid system* within solid porous matrix) also facilitates the definition of *velocity* and *flux*.

In a diffusing mixture of n species, the various chemical species are moving at different velocities. Consider a volume dU in the space occupied by the considered liquid system, and let dm_α and dm denote the instantaneous mass of the α species and of the entire liquid system within dU. The volume dU is sufficiently large so as to enable a meaningful statistical averaging of the properties of the various species within it, yet sufficiently small with respect to the flow domain so as to represent a point within it (physically large and mathematically small). The density of each species, or its concentration is defined by

$$\rho_\alpha \equiv C_\alpha = dm_\alpha/dU; \qquad \alpha = 1, 2, \ldots, n \tag{4}$$

For the entire system we have

$$\rho = \sum_{\alpha=1}^{n} \rho_\alpha \equiv \sum_{\alpha=1}^{n} C_\alpha = dm/dU; \qquad \sum_{\alpha=1}^{n} C_\alpha/\rho = \sum_{\alpha=1}^{n} \omega_\alpha = 1 \tag{5}$$

The velocity $\mathbf{V}_\alpha$ (in a fixed coordinate system) of the α species at a point is the statistical average velocity within dU of the individual molecules of the α species (= sum of velocities divided by the number of molecules). With this, we may define for the mixture a local *mass-average velocity* vector $\mathbf{V}^*$ (Bird *et al.*, 1960) (= *velocity of material point* of the system) by

$$\mathbf{V}^* = \sum_{\alpha=1}^{n} \rho_\alpha \mathbf{V}_\alpha \Big/ \sum_{\alpha=1}^{n} \rho_\alpha = \sum_{\alpha=1}^{n} C_\alpha \mathbf{V}_\alpha \Big/ \sum_{\alpha=1}^{n} C_\alpha = \sum_{\alpha=1}^{n} \omega_\alpha \mathbf{V}_\alpha \tag{6}$$

From this definition it follows that $\rho\mathbf{V}^*$ is the (local) rate at which mass passes through a unit area placed perpendicular to $\mathbf{V}^*$. The velocity $\mathbf{V}^*$ is also called the *barycentric velocity*. It is a most useful concept when inhomogeneous liquids are being considered. Another way of defining the velocity of a system is based on a volumetric averaging. Let u_α be the partial specific volume of the α species:

$$u_\alpha = \partial U/\partial m_\alpha$$

$$\sum_{\alpha=1}^{n} u_\alpha \rho_\alpha = \sum_{\alpha=1}^{n} (\partial U/\partial m_\alpha)(dm_\alpha/dU) = 1 \tag{7}$$

The *volume-average velocity* $\mathbf{V}$ is then defined by

$$\mathbf{V} = \sum_{\alpha=1}^{n} \rho_\alpha u_\alpha \mathbf{V}_\alpha \tag{8}$$

In addition to the flow caused by external forces, we have the movement of each species through the mixture because of its concentration gradient. For the sake of simplicity let $n = 2$ (i.e., a binary mixture). We may therefore speak of velocities of the species α with respect to the local motion of the liquid stream: $(\mathbf{V}_\alpha - \mathbf{V}^*)$ = diffusion velocity of the species α with respect to the (mass) average velocity $\mathbf{V}^*$, and $(\mathbf{V}_\alpha - \mathbf{V})$ = diffusion velocity of the species α with respect to the (volume) average velocity $\mathbf{V}$. Similarly, we may define a mass flux $\mathbf{J}_\alpha^* = \rho_\alpha(\mathbf{V}_\alpha - \mathbf{V}^*)$ relative to the mass-average velocity and a mass flux $\mathbf{J}_\alpha = \rho_\alpha(\mathbf{V}_\alpha - \mathbf{V})$ relative to the volume average velocity.

The relationship between $\mathbf{V}^*$ and $\mathbf{V}$ of a nonhomogeneous liquid which is a binary system, is obtained through the use of *Fick's law of diffusion*:

$$\mathbf{J}_\alpha^* = \rho_\alpha(\mathbf{V}_\alpha - \mathbf{V}^*) = -\rho D_d \operatorname{grad} \omega_\alpha \tag{9}$$

where D_d is the coefficient of molecular diffusion of the binary system. Similarly, we have

$$\mathbf{J}_\alpha = \rho_\alpha(\mathbf{V}_\alpha - \mathbf{V}) = -D_d \operatorname{grad} \rho_\alpha \tag{10}$$

From these, by eliminating $\rho_\alpha \mathbf{V}_\alpha$, we obtain

$$\rho_\alpha(\mathbf{V}^* - \mathbf{V}) = -D_d(\text{grad}\ \rho_\alpha - \rho\ \text{grad}\ \omega_\alpha) \tag{11}$$

or

$$\mathbf{V}^* - \mathbf{V} = -(D_d/\rho)\ \text{grad}\ \rho$$

Obviously, for a homogeneous liquid ($\rho =$ const), $\mathbf{V}^* \equiv \mathbf{V}$. In a non-homogeneous liquid they may differ both in magnitude and direction.

1.4. Scope of Summary

Darcy's law and the mass conservation equation describe the macroscopic flow regime within a porous medium; they do not include any information on the dispersion phenomena associated with the flow. Any theory attempting to describe such phenomena in a *macroscopic fashion* must start by considering the behavior of tracer-marked fluid particles in a *microscopic scale.*

The following sections include an attempt to summarize the various theories on hydrodynamic dispersion and their results. It is by no means a complete description or literature survey, and many important works have been omitted. Works have been included only when it was felt that mentioning them would contribute to the presentation.

Section 2 includes a short description of theories and approaches to the dispersion problem. The objective of all theories is to derive a *macroscopic description* of the dispersion. They therefore arrive at *macroscopic parameters* which include all our ignorance as to the exact microscopic nature of the problem. Obviously, different theories may arrive at different parameters. Experiments in the laboratory and in the field are then the only way to test the validity of these theories and to determine the value of the various parameters. Section 3 includes a discussion of the main dispersion parameters. Section 4 gives the partial differential equations which describe dispersion in a macroscopic fashion. Only saturated flow is considered.

2. Hydrodynamic Dispersion Theories

2.1. General

Hydrodynamic dispersion is the macroscopic outcome of the real, microscopic, motion of the tracer particles. One could expect this motion to be described by a suitable solution of the Navier–Stokes equations which describe fluid flow. This could be done if one could formulate correctly the boundary conditions at the highly irregular boundaries. Since it is impossible to define the complicated geometry of the solid surface of the porous matrix in any mathematical form, one cannot treat the problem on hand, and in fact also the general problem of flow through porous media, in any microscopically exact manner.

The above difficulty may be circumvented by replacing, as in many other physical problems, the porous medium by some simplified fictitious model of one kind or another. A model will suit this purpose if it (1) explains the phenomenon on hand, (2) involves parameters which may be measured and related to corresponding properties of the porous medium, and (3) may be treated by available mathematical tools to yield a macroscopic description of the phenomenon with discrepancies which may be neglected for all practical purposes.

Two approaches are commonly employed. The first is to replace the actual medium by a fictitious, greatly simplified, model in which the mixing which takes place can be analyzed by exact mathematical methods. A single capillary tube, a bundle of capillaries, an array of cells, etc., are examples of such models (e.g., Taylor, 1953; Danel, 1952; Bear and Todd, 1960). Models of this kind are also often considered in the flow of a single liquid. The second approach, more commonly used nowadays, is to construct a statistical (conceptual) model of the microscopic motion of the marked fluid particles and to average these motions in order to obtain a macroscopic description of the movement of the marked particles (hydrodynamic dispersion). However, in both approaches one has to investigate the spreading which takes place in an elementary channel. Several examples of both approaches are considered in the following paragraphs.

It must be emphasized that in both approaches, that of the simplified medium model and that of the statistical model, the final test is the experiment. An experiment is also the only way to determine the various coefficients which appear in the equations derived from these models; there is no way to obtain them from the mathematical analysis itself. However, some models may be considered more refined as they relate gross coefficients to more elementary medium properties which still have to be determined experimentally, or they provide a better understanding of the mechanism of the investigated phenomenon.

2.2. Capillary-Tube and Cell Models

One of the simplest porous medium models, is a bundle of capillaries. Taylor (1953) investigated the displacement of a liquid ($C = 0$) from a straight capillary tube of radius R by another liquid ($C = C_0$), miscible with the first. Both liquids have the same density and viscosity. In the absence of molecular diffusion and because of the parabolic velocity distribution across the tube, a group of marked fluid particles initially on a plane which is perpendicular to the tube's axis, will at a later time lie on a paraboloidal surface. This means that the initial tracer mass is continuously spreading along the tube as flow continues.

For any cross section and time, we may define a fractional concentration $C(x, t)$ as the ratio of the area occupied by the displacing fluid to the total area πR^2. The velocity distribution across the tube is $V(r) = 2\bar{V}(1 - r^2/R^2)$, where r is measured from the tube's axis, $\bar{V} = Q/\pi R^2$ is the average velocity, and Q is the discharge through the tube. With the nomenclature of Fig. 4, continuity requires

$$\pi R^2 \bar{V} C(x, t) = C_0 \int_0^{r^*} 2\pi r V\, dr; \qquad r^* = R(1 - x/2\bar{V}t)^{1/2}$$

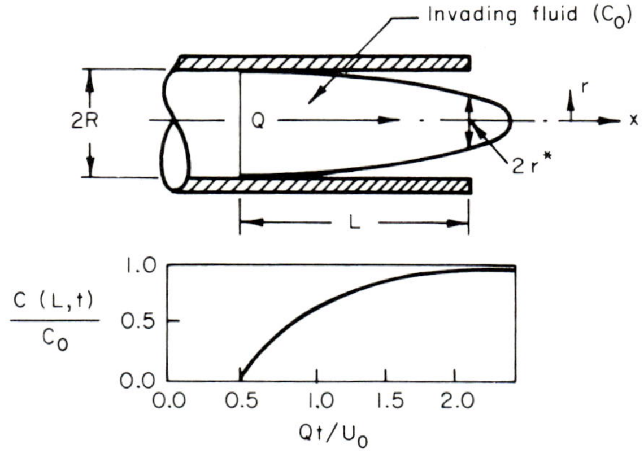

FIG. 4. Breakthrough curve in a capillary tube without molecular diffusion.

Hence, for $Qt/U_0 \geqslant \frac{1}{2}$, and $U_0 = \pi R^2 L$,

$$C(x, t)/C_0 = 1 - x^2/4\bar{V}^2 t^2; \qquad C(L, t)/C_0 = 1 - U_0^2/4Q^2 t^2 \tag{12}$$

If molecular diffusion on the advancing interface is also considered, one obtains an equation for $C = C(x, r, t)$ in the form

$$\frac{\partial C}{\partial t} = D_d \left(\frac{\partial^2 C}{\partial r^2} + \frac{1}{r} \frac{\partial C}{\partial r} + \frac{\partial^2 C}{\partial x^2} \right) - 2\bar{V} \left(1 - \frac{r^2}{R^2} \right) \frac{\partial C}{\partial x} \tag{13}$$

where D_d is the coefficient of molecular diffusion (assumed independent of C). Equation (13) may be simplified by deleting the second derivative

$\partial^2 C/\partial x^2$, because in most cases, radial diffusion predominates over longitudinal diffusion:

$$\partial^2 C/\partial x^2 \ll \partial^2 C/\partial r^2 + (1/r)\,\partial C/\partial r$$

Then, (13) becomes

$$\partial^2 C/\partial \rho^2 + (1/\rho)\,\partial C/\partial \rho = (R^2/D_d)\,\partial C/\partial t + 2(\bar{V}R^2/D_d)(\tfrac{1}{2} - \rho^2)\,\partial C/\partial \xi \quad (14)$$

where $\rho = r/R$ and $\xi = x - \bar{V}t$. Boundary conditions are $x = 0$, $C = C_0$; $\rho = 1$, $\partial C/\partial \rho = 0$. Also, at $\rho = 0$, C is finite. Initial conditions are $t = 0$, $C = 0$ for $x > 0$; $C = C_0$ for $x \leq 0$. Even in this form, the solution is most difficult. Taylor (1953, 1954) obtains approximate solutions corresponding to two extreme cases:

(1) $2L/\bar{V} \ll R^2/14.4\,D_d$; axial convection dominates over radial diffusion.

(2) $2L/\bar{V} \gg R^2/14.4\,D_d$; radial diffusion dominates; i.e., the time required for radial concentration differences to be appreciably reduced by radial diffusion is short relative to the time required for longitudinal convection to cause appreciable radial concentration variations. This assumption is approximately applicable to relatively low velocities.

Case (1) leads to (12). Case (2), described by (14) which may be rewritten as

$$\partial^2 C/\partial \rho^2 + (1/\rho)(\partial C/\partial \rho) = (2\bar{V}R^2/D_d)(\tfrac{1}{3} - \rho^2)(\partial C/\partial \xi) \quad (15)$$

in which $\partial C/\partial \xi$ is taken independent of ρ, is solved by Taylor for $\partial C/\partial \xi = 0$ at $\rho = 1$. He obtains

$$C = C^* + A(\rho^2 - \tfrac{1}{2}\rho^4); \qquad A = (R^2\bar{V}/4D_d)(\partial C/\partial \xi) \quad (16)$$

where C^* is C at $\rho = 0$. The rate of tracer transfer Q_C across the section at ξ is

$$Q_C = 2\pi \int_0^1 CV(\rho)\rho\, d\rho = -\pi R^2 \cdot \frac{R^2\bar{V}^2}{48D_d}\frac{\partial C^*}{\partial \xi} \quad (17)$$

If $\bar{C}$ is the mean concentration over a cross section $\partial \bar{C}/\partial \xi \approx \partial C^*/\partial \xi$ and (17) becomes

$$Q_C = -\pi R^2 \cdot \frac{R^2\bar{V}^2}{48D_d} \cdot \frac{\partial \bar{C}}{\partial \xi} \quad (18)$$

and the mass flux J (per unit area)

$$J = \frac{Q_C}{\pi R^2} = -\frac{R^2\bar{V}^2}{48D_d}\frac{\partial \bar{C}}{\partial \xi} \quad (19)$$

which means that the tracer is dispersed by longitudinal convection and radial molecular diffusion relative to a plane moving with velocity $\bar{V}$ exactly as though it were being diffused by a process which obeys *Fick's law* of molecular diffusion, but with a diffusion coefficient $D' = (R^2/48D_d)\bar{V}^2$. One should notice here the dependence of D' on $\bar{V}^2$. The condition that longitudinal molecular diffusion is negligible with respect to the longitudinal dispersion expressed by D', leading to the above results, is: $(48)^{1/2}(L/R) < L\bar{V}/D_d < 4(L/R)^2$. We shall later discuss the significance of $L\bar{V}/D_d$.

Conservation of mass during this diffusion-like process may be expressed by

$$\partial\bar{C}/\partial t = D' \, \partial^2 C/\partial \xi^2$$

or, in the original coordinate system, by

$$\partial\bar{C}/\partial t = D' \, \partial^2\bar{C}/\partial x^2 - \bar{V} \, \partial C/\partial x \tag{20}$$

which for the initial conditions $C = C_0$ at $x < 0$ and $C = 0$ at $x \geqslant 0$ yields (Von Rosenberg, 1956):

$$\frac{\bar{C}}{C_0} = \frac{1}{2}\left[1 \pm \operatorname{erf}\frac{x - \bar{V}t}{2(D't)^{1/2}}\right], \quad + \text{ for } x - \bar{V}t < 0, \quad - \text{ for } x + \bar{V}t > 0 \tag{21}$$

where $\operatorname{erf} z = (2/\pi) \int_0^z e^{-\alpha^2} \, d\alpha$ (Fig. 5). The point of $\bar{C}/C_0 = 50\%$ moves

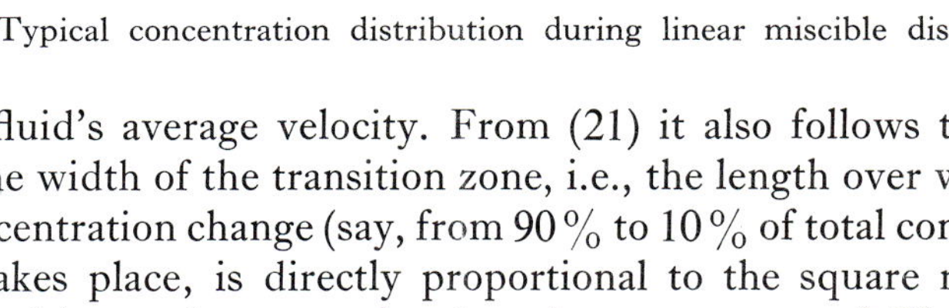

FIG. 5. Typical concentration distribution during linear miscible displacement.

with the fluid's average velocity. From (21) it also follows that at any instant, the width of the transition zone, i.e., the length over which most of the concentration change (say, from 90% to 10% of total concentration change) takes place, is directly proportional to the square root of the velocity and inversely proportional to the square root of D'. At any $\bar{V}$, the length of the transition zone increases as the square root of the average distance traversed. If L denotes the length of the zone along which $\bar{C}$ varies from $0.9C_0$ to $0.1C_0$, we obtain

$$L = 0.52R\bar{V}(t/D_d)^{1/2} \tag{22}$$

which shows that the breakthrough curve flattens when $\bar{V}$ increases and becomes steeper, i.e., tends to a *piston-like* flow, when D_d increases. Van Deemter *et al.* (1955) obtains a more generalized solution of (13). Danel (1952) and others extend the single capillary tube model to bundles of capillary tubes of different diameters. The resulting dispersion is shown to depend both on the parabolic velocity distribution within the tubes and on the distribution of average velocities in the various capillaries. In flow through actual porous media, all the coefficients are, of course, different, but the reason for employing a capillaric model of this kind is obvious. Models of this kind always include an estimate of the relationship between the average diameter $\bar{R}$ of the capillary tubes of the bundle, and the properties k and n of the actual porous medium. For example, by comparing Darcy's law for horizontal flow in the form $q = -(k/\mu)\, \partial p/\partial x$ with Hagen–Poiseuille's law for the average capillary tube $[-(N\pi \bar{R}^4/128\mu)\, \partial p/\partial x$, where N is the number of capillaries per unit area of the bundle of capillaries all in the direction $x]$, one obtains

$$\bar{R} = (32k/n_m)^{1/2}; \qquad n_m = \tfrac{1}{4}N\pi\bar{R}^2 \tag{23}$$

where n_m is the model's porosity. To obtain flow in a three-dimensional space, only one-third of the capillars are placed in each direction. Then

$$\bar{R} = (96k/n_m)^{1/2}$$

Scheidegger (1960) discusses possible definitions for $\bar{R}$. Rifai *et al.* (1956) discuss the application of the capillaric model to dispersion studies.

Aris (1956) extended Taylor's analysis to straight tubes of any cross section. He shows that the above results lead to an effective diffusion coefficient of the form

$$D' = D_d + \delta R^2 \bar{V}^2/D_d\,; \qquad D'/D_d = 1 + (\delta' R\bar{V}/D_d)^2; \qquad \delta' = \delta^{1/2} \tag{24}$$

where δ (restricted to a narrow range; for elliptical cross section $5/12 < 48\delta \leqslant 1$) is a coefficient depending on the shape of the capillary tube's cross section. Aris shows that the dispersion causes an average concentration distribution in the form of a normal (Gauss) distribution about the center of gravity which is displaced at the mean velocity of flow. The coefficient D' of (24) is obtained from the variance of this distribution. D' is usually called the *coefficient of dispersion* (dimensions L^2T^{-1}). Blackwell (1957) performed experiments which verified (24).

As an illustration of a regular type of model, consider the simplified model used by Bear (1960) in studies of one-dimensional dispersion. His model consists of an array of small cells with interconnecting short channels. It is assumed that when a liquid with a certain tracer concentration enters a cell occupied by a liquid of a different concentration, it displaces part of it, while the liquids remaining in the cell immediately mix to form a new homogeneous liquid. We may refer to such a cell as a *perfect mixer*. To obtain perfect mixing, the true movement of the tracer particles (through molecular diffusion or turbulence) must be much faster than the average liquid flow. If U is the volume of a perfect mixer container filled with liquid, $C^{(i)}$, C, and $C^{(o)}$ are the tracer concentrations in the liquid entering the container, in it, and leaving it, and Q is the liquid's discharge, then the variations in tracer concentration in the container are given by

$$\partial C/\partial t = (Q/U)(C^{(i)} - C) \qquad \text{or} \qquad \partial C/\partial t = \tau(C^{(i)} - C) \tag{25}$$

where $\tau = Q/U$ is the average *residence time* of the liquid in the container. For $C^{(i)} = 0$ and $C\,|_{t=0} = C_0$, we obtain

$$\partial C/\partial t = -C/\tau; \qquad C = C_0 e^{-t/\tau} \tag{26}$$

Wentworth (1948) was probably the first to suggest the perfect mixing model in an array of cells as an explanation for creation of a transition zone at a moving interface between fresh water and salt water in a coastal aquifer. One may visualize the pores in a porous medium as such cells.

In the analysis suggested by Bear (1960), the dispersion phenomenon may be regarded as a combination of two processes: (1) complete mixing in the elementary cells, and (2) translation from one cell to the next through the connecting channels. A tracer balance for the jth cell of an array leads to

$$\partial C_j^{(o)}(t)/\partial t + C_j^{(o)}(t)/\tau = C_j^{(i)}(t)/\tau \tag{27}$$

where $C_j^{(i)}$ and $C_j^{(o)}$ are the inflow and outflow concentrations of the jth cell and τ is the residence time in the cell of length $\Delta l'$, $\tau = \Delta l'/\overline{V}$. The solution of (27) is

$$C_j^{(o)}(t) = e^{-t/\tau}\left[\int_0^t \tau^{-1} e^{t/\tau} C_j^{(i)}(t)\,dt + \text{const}\right] \tag{28}$$

If we have an instantaneous inflow which causes an initial concentration of C_0 in the first cell, then $C_j^{(i)}(t) = \overline{C}\,\delta(t)$ where $\delta(t)$ is the *Dirac function* and $\overline{C} = C_0\tau$ may be interpreted as the tracer quantity per unit discharge. We obtain

$$C_j^{(o)}(t) = C_0 \frac{(t/\tau)^{n-1}\, e^{-t/\tau}}{(j-1)!} \tag{29}$$

which is the *Poisson probability distribution* $P(\xi = N) = \lambda^N e^{-\lambda}/N!$ with $N = j - 1$ and $\lambda = t/\tau$. For a constant inflow of concentration C_0 in the first cell, and with $\tau =$ const for all cells, we obtain

$$C_j^{(o)} = C_0\left[1 - e^{-t/\tau}\sum_{i=1}^{j=1}(i!)^{-1}(t/i)^i\right] \tag{30}$$

If we add translation through the cells, with $j\,\Delta t$ the time of translation through j cells, so that of the total time t, $t - j\,\Delta t$ is left for the mixing, and $j\,\Delta t \approx N\,\Delta t$ we obtain

$$\begin{aligned} C_j^{(o)}(t) &= C_0\frac{[(t - N\,\Delta t)/\tau]\exp[-(t - N\,\Delta t)/\tau]}{N!} &&\text{for}\quad t > N\,\Delta t \\ C_j^{(0)}(t) &= 0 &&\text{for}\quad t < N\,\Delta t \end{aligned} \tag{31}$$

By the *central limit theorem* of statistics, as N becomes large, (31) may be approximated by the *normal distribution*

$$C_j^{(o)}(t)/C_0 = (2\pi N)^{-1/2}\exp\{-(T - N)^2/2N\}; \qquad T = (t - N\,\Delta t)/\tau$$

or

$$C(x, t)/C_0 = (4\pi\,D't)^{-1/2}\exp\{-(x - \bar{x})^2/4D't\} \tag{32}$$

where $\bar{x} = \bar{V}t$; $\bar{V} = \Delta l/\tau + \Delta t$; Δl is the distance between centers of cells; $D' = \frac{1}{2}(\Delta l)^2\tau^2/(\tau + \Delta t)^3 = \frac{1}{2}[(\Delta l')^2/\Delta l]\,\bar{V} = a_{\text{I}}\bar{V}$; $a_{\text{I}} = \frac{1}{2}(\Delta l')^2/\Delta l$ is a medium property which may be called the *medium's longitudinal dispersivity* to be considered in detail below. One should note that here D' is proportional to $\bar{V}$ (= average flow velocity), whereas in the capillary tube model, we had $D' \sim \bar{V}^2$. To understand the reason for this difference, one should recall that in a capillary model, dispersion results from transport at different velocities along different streamlines, with exchange of tracer mass between streamlines. The velocity distribution itself is maintained unchanged. In an array-of-cells model, there is no velocity distribution and a perfect mixing is caused only by diffusion which causes a continuous macroscopic mixing of each tracer mass with new masses. The velocity appears just as a parameter with no direct influence on the mixing process itself. In fact, in (32) we have the velocity only in the form $\bar{x} = \bar{V}t$, $\bar{x}$ being the distance traveled by a considered mass, independent of the mass's velocity. We have thus demonstrated how a model of this simple form may lead to results applicable to an actual porous medium. Obviously, before reaching the conclusion that (32) is indeed applicable to actual porous media, one has to perform experiments to verify the validity of (32) and to determine D' experimentally. One has also to check the dependence of D' on $\bar{V}$.

However, an approach of this kind (as also the capillary tube model) fails to yield the transverse dispersion observed in flow through porous media. Let us, therefore, refer to the second approach, the *statistical* one.

All the theories on dispersion considered above were based on regular models which enable an exact analytic treatment. Yet, as a model becomes more simple, less factors affecting dispersion, and especially transverse dispersion, may be taken into account. On the other hand, a complicated model requires a complicated mathematical analysis.

2.3. Statistical Models

The basic idea of the second approach, *the statistical approach*, is that it is impossible to give an exact mathematical description of the motion of a single tracer particle, a description which seemingly should have been necessary for any forecast of spreading or dispersion of a cloud of tracer particles. One should recall at this point that the path of a single tracer particle (which is an ion or a molecule) should not be identified with the path of a liquid particle which is an instantaneous ensemble of a large number of molecules, the identity of which changes as a result of molecular diffusion. One may visualize the path of a tracer particle as the vectorial resultant of two motions: movement along the pathline of a liquid particle, and exchange of such pathlines by motion from one liquid particle to the next as a molecular diffusion process. The nature of both motions, the first determined by the intricate internal geometry of the medium, and the second by the random character of molecular diffusion, prevents any prediction of paths of tracer particles.

The basic postulate of the statistical approach is that, although it is impossible to predict the exact path of each individual tracer particle due to the large number of unknown factors governing its motion, one may apply the rules of probability theory and predict the spatial distribution at any later time of a cloud of many tracer particles which were initially at a close proximity, and which move under the same average conditions. In this approach, one considers the random motion of an anonymous tracer particle which represents an ensemble of many such particles, the paths of which have common statistical characteristics. Obviously, stating that the motion of a tracer particle is a random one is only our way of describing its path which is completely defined by the laws of physics. The probability distribution of the location of a single particle may be interpreted as the spatial relative-concentration distribution of a cloud of tracer particles originating from the neighborhood of a certain point at a certain time and moving under the same average conditions.

This idea conforms to the postulates of probability theory of *random phenomena*. A random phenomenon is one which when observed in repeated experimentes performed under apparently the same conditions, yields each time a different result. These random variations are caused by secondary uncontrolled factors which change from one experiment to the other. Nevertheless, certain trends, or regularities, are observed when a large number of similar experiments is performed. Hence, although the result of each of a series of similar experiments cannot be predicted, the overall behavior, or the mean results, may still be derived.

The conditions for a successful statistical approach are: (1) that we have a (microscopically) *completely disordered* porous medium, (2) that the application of statistical considerations is justified, and (3) that we may actually be able to measure the macroscopic parameters which appear in the solution.

The progress of a liquid particle through a porous medium follows an exact, predetermined path described by the Navier–Stokes equations, and by appropriate boundary conditions. Because the boundary conditions are unknown, it was suggested (Scheidegger, 1958) that instead of considering a specific porous medium, one should investigate a whole *ensemble* of porous media which are assumed macroscopically identical, i.e., have identical macroscopic properties. Each possible path of a liquid particle moving through a porous medium takes place at an ever changing velocity through a system of interconnected pores considered as a particular sample of this ensemble. The entire porous medium domain may be considered as an assembly of a large number of such samples which differ from each other. Obviously, several particles may move through the same sample along different pathlines. Each pathline may be considered as a vector sum of elementary displacements.

Instead of attempting to determine the velocity at a certain time of a definite liquid particle at a certain point in space, one may study the assembly of all possible velocities (or elementary displacements) which many particles, occupying at that time a small volume around the considered point, may have. With this approach, the single-valued velocity of a definite particle at a certain point is replaced by a large number of possible velocities of a large number of anonymous particles. A *probability distribution* may be assigned to this assembly of velocities and referred to this point.

In the case of a homogeneous fluid, the random character of the velocity distribution results only from the nature of the porous matrix. Over a domain including many points, the velocity becomes a *random vector function* of time and space. At any given instant it is a function of space only.

At any instant in a (spatially averaged) uniform flow through a homogeneous porous medium, the velocity (or displacement) is a stationary *random function* of time for a single particle.

Assuming that as a particle moves through a homogeneous medium it encounters all the conditions that are present in the many samples of the flow domain, the *ergodic hypothesis* of interchanging time averages with ensemble averages is applicable. By the *law of large numbers* (Feller, 1958), after a sufficiently long time, the time averaged velocity (or displacement) characteristics of a single particle, may replace averages taken over the assembly of many particles which move under the same flow conditions. Actually this assumption that a tracer particle may reach any point within the flow domain is questionable unless the mechanism of molecular diffusion is also taken into account.

The problem of the spreading of a cloud of labeled particles is thus reduced in the statistical approach to the problem of the random motion of a single anonymous tracer particle through an ensemble of random porous media.

According to Scheidegger (1957a, 1958a), the application of the statistical approach requires: (1) an assumption about the statistical (average) properties of the porous medium (= the ensemble), (2) an assumption on the microdynamics of the flow, i.e., on the relationships between the forces, the liquid properties, and the resulting velocity during each small time step; in general, the flow is assumed laminar, (3) a choice of a type of statistics to be employed, i.e., the probability of occurrence of events during small time intervals within the chosen ensemble. This may take the form of correlation functions between velocities at different points or different times, or joint-probability densities of the local velocity components of the particle as functions of time and space, or a probability of an elementary particle displacement. The chosen correlation function determines the type of dispersion equation derived.

Before describing several typical works based on the statistical approach, certain important general comments seem appropriate here. As the total particle travel time becomes much larger than the time interval during which its successive (local) velocities are still correlated, the total displacement may be considered as a sum of a large number of elementary displacements which are statistically independent. Then, the probability distribution of the particle's total displacement tends to the normal (*Gauss*) *distribution* (Fig. 6). This is based on the *central limit theorem* of probability, which states that no matter what the probability distribution of each elementary displacement (or of the velocity during each elementary displacement) is, as the number of steps increases, the probability distribution of the total displacement tends to normal. In view of the ergodic principle, this distribution also represents the spatial distribution of displacements of a cloud of initially close particles.

In order to express these concepts in a mathematical form, let $\bar{\mathbf{V}}$ denote the average velocity vector (components $\bar{V}_i$) in a field of uniform flow of a homogeneous fluid through a homogeneous porous medium of infinite areal extent. At any instant t, the component $V_i(t)$ of the local

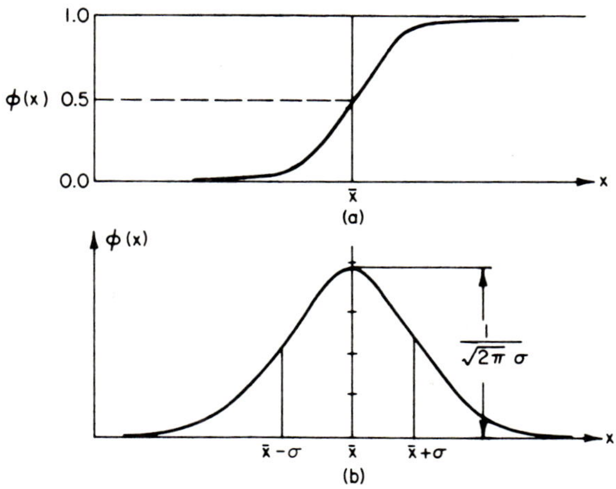

FIG. 6. The normal distribution: (a) distribution function; (b) density function.

velocity vector in the direction i of a marked fluid particle (ideal tracer) may be expressed as

$$V_i(t) = \bar{V}_i + \mathring{V}_i(t) \tag{33}$$

where $\mathring{V}_i(t)$ is the deviation of the particle's instantaneous velocity from the average one. Equation (33) states that the movement of a particle may be considered as a superposition of two motions: an average motion at a velocity $\bar{V}_i$ which obeys some macroscopic, nonrandom law (e.g., Darcy's law), and a random velocity $\mathring{V}_i(t)$.

By definition,

$$x_i(t) = \int_0^t V_i(\theta)\, d\theta$$

$$\bar{x}_i(t) = \int_0^t \bar{V}_i(\theta)\, d\theta = \int_0^t d\theta \int_{-\infty}^{+\infty} V_i(\theta) f(V_i)\, dV_i$$

$$= \int_{-\infty}^{+\infty} x_i(t) f(x_i)\, dx_i$$

where $f(V_i)$ and $f(x_i)$ are the probability distributions of V_i and x_i, respectively; no summation is performed on i. Here and in what follows a bar above a symbol denotes a probability average. By the ergodic property mentioned above, for sufficiently large times, the particle's time-averaged velocity $\overline{V}_i(t)$ approaches $\overline{V}_i$, and therefore $\overline{\mathring{V}_i(t)} = 0$.

To simplify the discussion on the random motion, let the movement be referred to a coordinate system $\mathring{x}_1, \mathring{x}_2, \mathring{x}_3$ which moves at a velocity $\overline{\mathbf{V}}$:

$$\mathring{x}_i(t) = x_i(t) - \bar{x}_i(t); \qquad i = 1, 2, 3 \tag{34}$$

A particle starting from the origin at $t = 0$, will reach a point $\mathring{x}_i(t)$ at some later time t such that

$$d\mathring{x}_i = \mathring{V}_i(t)\, dt; \qquad \mathring{x}_i = \int_0^t \mathring{V}_i(\theta)\, d\theta \tag{35}$$

The spread of the particle's possible displacements around its average displacement $\bar{x}(t)$ at that time is described by the *matrix of convariances* (correlation moments) $\overline{\mathring{x}_i(t)\mathring{x}_j(t)}$. By definition,

$$\begin{aligned}\operatorname{Cov}(x_i, x_j) \equiv \overline{\mathring{x}_i\mathring{x}_j} &= \int_{-\infty}^{+\infty}\int_{-\infty}^{+\infty} (x_i - \bar{x}_i)(x_j - \bar{x}_j) f(x_i, x_j)\, dx_i\, dx_j \\ &= \overline{x_i x_j} - \bar{x}_i\bar{x}_j\end{aligned}$$

where $f(x_i, x_j)$ is the *joint probability density* of x_i and x_j; for $i = j$, the convariances become the variances σ^2 of x_i,

$$\mu_{ii} = \sigma^2(x_i) = \operatorname{Var}(x_i) = \overline{(\mathring{x}_i)^2} \tag{36}$$

where σ is the *standard deviation* of x_i.

Consider now the derivative

$$\begin{aligned}\frac{d}{dt}[\mathring{x}_i(t)\, \mathring{x}_j(t)] &= \frac{d\mathring{x}_i(t)}{dt}\, \mathring{x}_j(t) + \mathring{x}_i(t)\, \frac{d\mathring{x}_j(t)}{dt} \\ &= \mathring{V}_i(t)\, \mathring{x}_j(t) + \mathring{x}_i(t)\, \mathring{V}_j(t) \\ &= \int_0^t \mathring{V}_i(t)\, \mathring{V}_j(\theta)\, d\theta + \int_0^t \mathring{V}_i(\theta)\, \mathring{V}_j(t)\, d\theta\end{aligned} \tag{37}$$

Now, $V_i(t)$ and $V_j(t)$, components in the i and j directions of the velocity vector $\mathbf{V}(t)$, are two *stationary random functions* of time. A random function $V_i(t)$ is called *stationary* if its *correlation coefficient*

$r(t, t')$ is independent of the arguments t, t', but depends only on the interval $(t' - t)$ between them. In general,

$$r_{ij}(t, t') = \frac{\overline{\mathring{V}_i(t)\, \mathring{V}_j(t')}}{\sigma[V_i(t)]\, \sigma[V_j(t')]} \qquad \text{(no summation)} \tag{38}$$

In addition we require that the average, the variance, and the other moments remain constant in time.

$$r_{ij}(t, t') = r_{ij}(t; t + \tau) = r_{ij}(\tau); \qquad \tau = t' - t$$

$$r_{ij}(\tau) = \frac{\overline{\mathring{V}_i(t)\, \mathring{V}_j(t + \tau)}}{\sigma(V_i)\, \sigma(V_j)} \qquad \text{(no summation)}$$

Then, by averaging (37),

$$\frac{d}{dt}\,\overline{[\mathring{x}_i(t)\, \mathring{x}_j(t)]} = 2\sigma(V_i)\, \sigma(V_j) \int_0^t r_{ij}(\tau)\, d\tau$$

$$= 2r_{ij}(0)\, \overline{\mathring{V}_i(t)\, \mathring{V}_j(t)} \int_0^t r_{ij}(\tau)\, d\tau \tag{39}$$

where $r_{ij}(\tau)$ is the *coefficient of correlation* between the velocity V_i at t and V_j at $(t + \tau)$. From (39),

$$\overline{\mathring{x}_i(t)\, \mathring{x}_j(t)} = 2r_{ij}(0)\, \overline{\mathring{V}_i(t)\, \mathring{V}_j(t)} \int_0^t \int_0^t r_{ij}(\tau)\, d\tau\, d\theta$$

$$= 2r_{ij}(0)\, \overline{\mathring{V}_i(t)\, \mathring{V}_j(t)} \int_0^t (t - \tau)\, r_{ij}(\tau)\, d\tau \qquad \text{(no summation)} \tag{40}$$

Because $V_i(t)$ and $V_j(t)$ are stationary random functions, $\overline{\mathring{V}_i(t) \mathring{V}_j(t)}$ is a constant independent of t. If t_0 is the time during which velocities are still effectively correlated (= *autocorrelation time*),

$$t_0 = \int_0^\infty r_{ij}(\tau)\, d\tau \tag{41}$$

We may investigate two cases.

(1) When $t \gg t_0$. Then (40) becomes

$$\overline{\mathring{x}_i \mathring{x}_j} = 2r_{ij}(0)\, \overline{\mathring{V}_i(t)\, \mathring{V}_j(t)}\, t_0 t - 2r_{ij}(0)\, \overline{V_i(t)\, V_j(t)} \int_0^\infty \tau r_{ij}(\tau)\, d\tau$$

$$\approx 2r_{ij}(0)\, \overline{\mathring{V}_i(t)\, \mathring{V}_j(t)}\, t_0 t \tag{42}$$

Introducing the notation D_{ij} which will later be defined as the *coefficient of (mechanical) dispersion*,

$$D_{ij} = \overline{\mathring{x}_i\mathring{x}_j}/2t = r_{ij}(0)\,\overline{\mathring{V}_i\mathring{V}_j}\,t_0 \tag{43}$$

For $i = j$, $r_{ij}(0) = 1$, and

$$D_{ii} = \overline{\mathring{x}_i^{\,2}}/2t = \overline{\mathring{V}_i^{\,2}}\,t_0\,; \qquad \sigma^2(x_i) = 2D_{ii}t \tag{44}$$

(2) When $t \ll t_0$. One obtains (Scheidegger, 1960)

$$\begin{aligned} \overline{\mathring{x}_i\mathring{x}_j} &= r_{ij}(0)\,\overline{\mathring{V}_i\mathring{V}_j}\,t^2; \qquad D_{ij} = \overline{\mathring{x}_i\mathring{x}_j}/2t = \tfrac{1}{2}r_{ij}(0)\,\overline{\mathring{V}_i\mathring{V}_j}\,t \\ D_{ii} &= \overline{\mathring{x}_i^{\,2}}/2t = (\overline{\mathring{V}_i^{\,2}}/2)t; \qquad \sigma^2(x_i) = \overline{\mathring{V}_i^{\,2}}\,t^2 \end{aligned} \tag{45}$$

which expresses the fact that for very short time intervals there is no random process and every particle progresses with its velocity. One should note that (42) and (43) do not prove that the spreading of a cloud of marked particles is given by solutions of the diffusion equation with the effective diffusivity D_{ij}. This is so only if it is shown separately that the probability distribution of x_i is normal. If each elementary displacement $d\mathring{x}_i(t) = \mathring{V}_i(t)\,dt$ is a random variable, the total deviation of the particle's position from its average position tends to a normal distribution only if $t \gg t_0$. As stated above, this is a consequence of the central limit theorem. The general form of the normal density function is (Chandrasekhar, 1943)

$$\phi(x_1, x_2, x_3, t) = (2\pi)^{-3/2}\,|\,\mu\,|^{-1/2} \exp[-\tfrac{1}{2}(\mu'_{ij}/|\,\mu\,|)(x_i - \bar{x}_i)(x_j - \bar{x}_j)] \tag{46}$$

where the summation convention is employed. $|\,\mu\,|$ is the determinant of the correlation matrix $[\mu_{ij}]$

$$[\mu_{ij}] = \begin{bmatrix} \mu_{11} & \mu_{12} & \cdots & \mu_{1n} \\ \mu_{21} & & \cdots & \\ \vdots & & & \\ \mu_{n1} & & \cdots & \mu_{nn} \end{bmatrix} \tag{47}$$

μ_{ij} is the *correlation moment* defined by (36), μ'_{ij} is the cofactor of μ_{ij} in the determinant $|\,\mu_{ij}\,|$. For additional information on the concepts of probability theory, the reader is referred to textbooks on this subject.

In (46), ϕ gives the relative tracer mass concentration in percents of total mass per unit volume of the medium at the considered point; it is to be regarded as an average over a small volume of space around the point. The tracer concentration C, as defined by (2), is related to the relative *concentration* ϕ by

$$C = \phi m/n \tag{48}$$

where n is the medium's porosity and m is the total tracer mass.

The exponent $(\mu'_{ij}/|\mu_{ij}|)\mathring{x}_i\mathring{x}_j$ in (46) is a quadratic form which describes a family of constant ϕ surfaces at time t in the (x_1, x_2, x_3) coordinate system. It is a family of confocal ellipsoids centered at $(\bar{x}_1, \bar{x}_2, \bar{x}_3)$. Let (x'_1, x'_2, x'_3) be a coordinate system which coincides with the major axes of these ellipsoids. In this system, (46) becomes

$$C(x_1, x_2, x_3) = \frac{m/n}{(2\pi)^{3/2}\,\sigma_1\sigma_2\sigma_3} \exp\left(-\sum_{i=1}^{3} \frac{\mathring{x}_i'^2}{2\sigma_i^2}\right) \tag{49}$$

where $\sigma_i = (\mu_{ii})^{1/2}$, defined by (36) in the x'_i system, is the *standard deviation* of x_i. Here, $\mu_{ij} = 0$ for $i \neq j$. This means that the three variables x'_1, x'_2, x'_3 are statistically independent variables. It is easily verified that (49) may be written in the form

$$\phi(x_1, x_2, x_3, t) = \phi_1(x_1, t)\,\phi_2(x_2, t)\,\phi_3(x_3, t)$$

where $\phi_i(x_i, t)$, $i = 1, 2, 3$, is the *normal density function* of the single random variable x'_i at time t. Thus in the one-dimensional case, the density function $\phi(x, t)$ and the distribution function $\Phi(x, t)$ of the normal distribution are

$$\begin{aligned} \phi(x, t) &= (2\pi)^{-1/2}\,\sigma^{-1} \exp\{-(x - \bar{x})^2/2\sigma^2\} \\ \Phi(x, t) &= (2\pi)^{-1/2}\,\sigma^{-1} \int_{-\infty}^{x} \exp\{-(x - \bar{x})^2/2\sigma^2\}\, dx \end{aligned} \tag{50}$$

By (44), $\sigma = (2Dt)^{1/2}$, $D = D_{ii}$. Figure 5 gives $\phi(x)$ and $\Phi(x)$ of the normal distribution.

We have shown here how (46) or (49) could be derived from certain general assumptions. However, these results do not show what are the relationships between σ_i or μ_{ij} and the statistically averaged medium and flow characteristics. This information may be obtained by choosing and analyzing the microdynamics of the flow within each elementary time interval.

As an illustration of the above principles, let us review briefly several works employing the statistical approach. Most of them study dispersion in uniform steady flow through a homogeneous isotropic medium.

The simplest statistical model of our dispersion problem is the one-dimensional *random walk*. In this model, a particle performs displacements along a straight line in the form of a series of steps of equal length, each step being taken either in the forward or in the backward direction with equal probability $\frac{1}{2}$. After N such steps the particle may be at any of the points

$$-N, \quad -N+1, \quad \ldots, \quad -1, \quad 0, \quad +1, \quad \ldots, \quad N-1, \quad N$$

Assuming that each step is equally likely to be taken in either direction, independently of all preceeding displacements, the probability of any sequence of N steps is $(\frac{1}{2})^N$. Hence, the probability $P(M, N)$ that the particle will arrive at a point M after N displacements, is:

$(\frac{1}{2})^N \times$ Number of distinct sequences which will lead to M after N steps

In order to arrive at M after N steps, some $(N + M)/2$ steps must be taken in the positive direction, while the remaining $(N - M)/2$ in the negative one. The total number of distinct sequences is

$$\frac{N!}{[(N+M)/2]!\,[(N-M)/2]!}$$

Hence,

$$P(M, N) = \frac{N}{[(N+M)/2]!\,[(N-M)/2]!}\left(\frac{1}{2}\right)^N$$

which is the *Bernoulli distribution*. The root mean square displacement is $(N)^{1/2}$. As $N \to \infty$ and $M \ll N$,

$$P(M, N) = (2/\pi N)^{1/2} \exp(-M^2/2N) \tag{51}$$

If l is the length of each step and $x = Ml$, the probability element $P(x)\,\Delta x$ that the particle is likely to be in the interval $(x, x + \Delta x)$ after N displacements with $\Delta x > l$, is

$$P(x, N)\,\Delta x = P(M, N) \cdot (\Delta x/2l)$$

or

$$P(x, N) = (1/2\pi N l^2) \exp(-x^2/2Nl^2)$$

If the particle suffers n displacements per unit time (i.e., its velocity is $u = nl$), the probability that it will find itself between x and $x + \Delta x$ at time t is

$$P(x, t)\,\Delta x = [2(\pi\, Dt)^{-1/2} \exp(-x^2/4Dt)]\,\Delta x$$
$$D = (\tfrac{1}{2})\, nl^2 = (l/2)\, u \tag{52}$$

Compare D here with D' in (32). This random walk approach is the basis for molecular diffusion theories, Brownian motion, etc., where l has the meaning of mean free path of a particle.

The result (52) may also be obtained in a different way. Let the duration of each step of length l be Δt and let p be the probability of the particle taking a step in the $+x$ direction, while $q = 1 - p$ is the probability of a step being taken in the $-x$ direction. The probability $P(M, N) \equiv P(x, t)$ of a particle being at a distance $x = Ml$ from the origin at time $t = N\,\Delta t$ is given by the recursive expressions

$$P(M, N + 1) = pP(M - 1, N) + qP(M + 1, N)$$

or

$$P(x, t + \Delta t) = pP(x - \Delta x, t) + qP(x + \Delta x, t)$$

Developing in a Taylor series and neglecting terms beyond the second order yields

$$\Delta t\, \partial P(x, t)/\partial t + O[(\Delta t)^2] = (q - p)\, \Delta x\, \partial P(x, t)/\partial x + [(\Delta x)^2/2]\, \partial^2 P(x, t)/\partial x^2$$

Passing to the limit with $\Delta x \to 0$, $\Delta t \to 0$, but such that the mean and the variance of the displacement will remain finite for all values of t,

$$\bar{x} = t(p - q)\, \Delta x/\Delta t, \qquad \sigma^2 = 4pqt(\Delta x)^2/\Delta t$$

which means that $(\Delta x)^2/\Delta t$ is bounded. With p and q of $O(\Delta x)$, we obtain for P,

$$\partial P/\partial t = D\, \partial^2 P/\partial x^2 - V\, \partial P/\partial x$$

where $D = (\Delta x)^2/2\Delta t$, $V = 2(p - q)D/\Delta x$. This is the differential equation for random walk after a large number of elementary displacements. The elementary solution is

$$P(x, t) = (4\pi\, Dt)^{-1/2} \exp[-(x - Vt)^2/4Dt]$$

which is the same as (52) if $p = q$; V may be interpreted as the *net drift velocity*.

Chandrasekhar (1943) presents a discussion on the general, three-dimensional random walk by using Markov's method. He considers a particle which performs a sequence of elementary steps (each represented by a displacement vector $\mathbf{r}_i : \mathbf{r}_1, \mathbf{r}_2, \ldots$). The displacements are independent of each other. The question is to determine the position of the particle after n displacements. The probability that a step $\mathbf{r}_i$ is in the range $(\mathbf{r}_i + d\mathbf{r}_i\,;\, \mathbf{r}_i)$ is given by

$$\tau_i(\mathbf{r}_i)\, d\mathbf{r}_i = \tau_i(x_i\,, y_i\,, z_i)\, dx_i\, dy_i\, dz_i$$

where $x_i\,, y_i\,, z_i$ are the components of $\mathbf{r}_i$ and τ_i is the joint probability distribution of $x_i\,, y_i\,, z_i$. The location of the particle after n steps is defined by the radius vector $\mathbf{R}(X, Y, Z)$

$$\mathbf{R}(X, Y, Z) = \sum_{i=1}^{n} \mathbf{r}_i\,; \qquad X = \sum_{i=1}^{n} x_i\,; \qquad Y = \sum_{i=1}^{n} y_i\,; \qquad Z = \sum_{i=1}^{n} z_i$$

Chandrasekhar shows that for $n \gg 1$, the probability $W(\mathbf{R})$ that $\mathbf{R}$ will be in the range $(\mathbf{R}, \mathbf{R} + d\mathbf{R})$ is

$$W(\mathbf{R})\, d\mathbf{R} = \frac{dR}{(2\pi)^{3/2}[(n\overline{x^2})(n\overline{y^2})(n\overline{z^2})]^{1/2}} \exp\left\{-\frac{(X - n\bar{x})^2}{2n\overline{x^2}} - \frac{(Y - n\bar{y})^2}{2n\overline{y^2}} - \frac{(Z - n\bar{z})^2}{2n\overline{z^2}}\right\}$$

$$= \frac{1}{(2\pi)^{3/2}\, \sigma_X \sigma_Y \sigma_Z} \exp\left\{-\frac{(X - \bar{X})^2}{2\sigma_X^{\,2}} - \frac{(Y - \bar{Y})^2}{2\sigma_Y^{\,2}} - \frac{(Z - \bar{Z})^2}{2\sigma_Z^{\,2}}\right\}$$

where $\bar{x} = n^{-1} \sum_{i=1}^{n} x_i$, etc. This expression describes a normal distribution in a three-dimensional space. If the particles undergo an average of $\bar{n}$ displacements per unit time $n = \bar{n}t$, we may define the average velocity of the particle by

$$u = \bar{n}\bar{x} = n\bar{x}/t = \bar{X}/t; \qquad v = \bar{n}\bar{y} = \bar{Y}/t; \qquad w = \bar{n}\bar{z} = \bar{Z}/t$$

Then the dispersion coefficients become

$$D_{11} = n\overline{x^2}/2t = \sigma_X^{\,2}/2t; \qquad D_{22} = n\overline{y^2}/2t = \sigma_Y^{\,2}/2t; \qquad D_{33} = \sigma_Z^{\,2}/2t$$

and we obtain for $W(\mathbf{R})$,

$$W(\mathbf{R})\, dR \frac{dX\, dY\, dZ}{(4\pi t)^{3/2}(D_{11}D_{22}D_{33})^{1/2}} \times \exp\left\{-\frac{(X - ut)^2}{4D_{11}t} - \frac{(Y - vt)^2}{4D_{22}t} - \frac{(Z - wt)^2}{4D_{33}t}\right\} \qquad (53)$$

If, instead of considering a single particle, we consider a large number of particles with no interference, all starting from the same origin $\mathbf{R} = 0$ at $t = 0$, $W(\mathbf{R}, t)\, d\mathbf{R}$ may be interpreted at that portion of the total number of particles which will be found at time t in the range $(\mathbf{R}, \mathbf{R} + d\mathbf{R})$.

Equation (53) is an elementary solution of the partial differential equation

$$\frac{\partial W}{\partial t} = D_{11} \frac{\partial^2 W}{\partial X^2} + D_{22} \frac{\partial^2 W}{\partial Y^2} + D_{33} \frac{\partial^2 W}{\partial Z^2} - u \frac{\partial W}{\partial X} - v \frac{\partial W}{\partial Y} - w \frac{\partial W}{\partial Z}$$

In a coordinate system X_1, X_2, X_3 which does not coincide with the principal axes of dispersion, this last equation, using matrix notation, becomes

$$\frac{\partial W}{\partial t} = D_{ik} \frac{\partial^2 W}{\partial X_i \partial X_k} - \bar{V}_i \frac{\partial W}{\partial X_i}$$

where u, v, w are the components of $\bar{\mathbf{V}}$. We shall return to these equations at a later stage.

Scheidegger (1954), who employed the theory of random walk and extended it to three dimensions, based his statistical analysis on: (1) the ensemble is the isotropic homogeneous porous medium; (2) the statistics is such that what occurs in any elemental time interval is entirely independent of what occurs in any other time interval, that the *ergodic hypothesis* is valid, and that the velocity deviates from its mean in a random fashion; (3) the microdynamics of the flow is such that the laminar laws of flow govern the movement in each step, and external forces are uniform and steady.

With these assumptions, and neglecting molecular diffusion, the time interval $(0, t)$ is divided into N equal intervals $\Delta t = t/N$, in each of which the particle undergoes a displacement (vector) $\boldsymbol{\xi}$ with cartesian components (ξ, η, ζ). In Scheidegger's treatment the liquid is homogeneous and a particle is an elementary volume of liquid which remains undivided as it moves through the porous medium. The probability density of a displacement $\boldsymbol{\xi}$ within Δt is denoted by $p(\boldsymbol{\xi})$. By definition,

$$\iiint_{-\infty}^{+\infty} p(\xi, \eta, \zeta)\, d\xi\, d\eta\, d\zeta = 1 \tag{54}$$

A second (primed) coordinate system in which $\boldsymbol{\xi}' = \boldsymbol{\xi} - \bar{\boldsymbol{\xi}}$ (the overbar denoting a probability average over the ensemble), and $\bar{\boldsymbol{\xi}}' = 0$, is introduced.

In the primed system, the probability density is denoted by p' instead of by p. The probability $p'_N(x', y', z')$ that a moving fluid particle, initially at the origin, will have the coordinates (x', y', z') after time t is studied. Scheidegger assumes that since the medium is isotropic and the field of forces is homogeneous, one may split $p'(\xi', \eta', \zeta')$ into identical probabilities

$$p'(\xi', \eta', \zeta') = p'(\xi')\,p'(\eta')\,p'(\zeta')$$

which are, furthermore, identical for every time step. In such a problem of repeated displacements, no matter what the original distribution p' is, the central limit theorem states that if N is sufficiently large, the distribution after N steps is normal (Gaussian). Hence,

$$p'_N(\xi', \eta', \zeta') = (2\pi N\sigma^2)^{-3/2} \exp\{-(x'^2 + y'^2 + z'^2)/2N\sigma^2\}$$

where σ is the standard deviation of p'. With $D = N^2/2N\,\Delta t = \sigma^2/2\Delta t$ and $t = N\,\Delta t$,

$$p'(x', y', z', t) = (4\pi Dt)^{-3/2} \exp\{-(x'^2 + y'^2 + z'^2)/4Dt\}$$

or (55)

$$p(x, y, z, t) = (4\pi Dt)^{-3/2} \exp\{-[(x - \bar{x})^2 + (y - \bar{y})^2 + (z - \bar{z})^2]/4Dt\}$$

where $\bar{x} = u_{(x)}t$, $\bar{y} = u_{(y)}t$, and $\bar{z} = u_{(z)}t$; $u_{(x)}$, $u_{(y)}$, and $u_{(z)}$ are the components of the uniform velocity field.

In view of the ergodic property, (55) describes the spatial tracer concentration at time t of a large number of particles which started from the vicinity of the same origin at $t = 0$, and traveled along independent paths under similar statistical conditions. This ensemble of particles is normally distributed around its center which travels with the average velocity of the flow.

As stated above, Scheidegger (1954) obtains his solution by assuming that there is no correlation between the ith and the $(i - 1)$th step, and that all steps are of equal duration. He also assumes that because the medium is isotropic, the probability is the same in all directions. This leads to a normal distribution and to a D in (55) which is the same in all directions. It will be shown below that this results from an erroneous assumption of isotropy of the spreading. The spreading is not isotropic, that is, it is different in the direction of the mean flow (longitudinal dispersion) and perpendicular to it (transverse dispersion), although the medium is isotropic.

Scheidegger then proceeds to express the relationship between the dispersion average characteristics (mean and variance of the total displacement) by parameters related to the medium, the fluid, and the active forces.

Danckwerts (1953), who studies the efficiency of chemical reactors, also employs the random walk model (in one dimension). He introduces into his model the *residence time* of an elementary step; i.e., the time during which a fluid particle passes through an elementary channel or through a certain length of the medium. In his model, and in other *random residence time models*, the flow takes place through a sequence of cells in each of which a complete mixing takes place. Unlike the deterministic cell theories described above, here there is a certain probability that a fluid particle will leave the cell, the probability being independent of the length of time the particle stayed in the cell. The time of passage of a fluid particle through a sequence of cells simulating the porous medium, which is the sum of the time intervals spent in the individual cells of the sequence, is a random function which tends to normal as the number of cells becomes large. As in (32), the variance of the distribution is proportional to the number of cells. The longitudinal dispersion is then described by the diffusion equation with an effective diffusivity proportional to the product of the average velocity and the length of a cell.

Rifai *et al.* (1956) also considers the time length θ of a step as a random variable. However, in applying the one-dimensional random walk approach to longitudinal dispersion, he assumes that any tracer particle can be either in a moving fluid zone or a stagnant (i.e., immobile) fluid zone, such as the film of water surrounding each sand grain. Consequently, the movement of the tracer particle is reduced to a sequence of "moving" and "stagnant" steps. The length ξ of the single displacement is his second random variable where ξ and θ are independent. By assuming that the length of the step is also a random variable, the random character of the medium is also taken into account. Using a Galton probability distribution $f(\xi, \theta)\, d\xi\, d\theta = e^{-\xi-\theta}\, d\xi\, d\theta$, he obtains for a small number of steps, or small ξ, the probability distribution

$$p(\xi, \theta) = [\exp(-\xi - \theta)]\, J_0[i2(\theta\xi)^{1/2}]$$

where J_0 is a modified Bessel function of first rank and zero order. For large ξ he obtains

$$p(x, t) = (1/2\pi N\sigma^2) \exp[-(x - \bar{x})^2/2N\sigma^2] \tag{56}$$

where σ is the standard deviation of a single displacement, N is the number of elementary displacements, $\bar{x} = ut$ is the average position of the tracer particle, and u is the average velocity.

With $D = N\sigma^2/2t$, (56) becomes

$$p(x, t) = (1/4\pi Dt)^{1/2} \exp[-(x - \bar{x})^2/4Dt] \tag{57}$$

which is again the normal distribution with respect to x.

Another statistical model used by Scheidegger (1958b) is the one involving the effect of *autocorrelation*. Here one assumes that a correlation exists between what happens to a fluid particle in subsequent time intervals. It is similar to the walk of a drunkard who has some memory. For this case, one ends up with a *telegraph equation* instead of the diffusion equation as the equation describing the dispersion. Experiments have yielded the result that the diffusion equation does describe the processes of dispersion qualitatively correctly and that autocorrelation is unimportant.

Scheidegger's (1954) work was probably the first attempt to treat hydrodynamic dispersion as a phenomenon in three-dimensional space. However, he did not take molecular diffusion into account and as a result, his coefficient of disperion was a scalar with no distinction between longitudinal and transversal coefficients of dispersion. Before proceeding to describe other studies in which this distinction was recognized, although not always through the introduction of molecular diffusion (e.g., de Josselin de Jong, 1958), let us examine the mechanism which causes transversal dispersion, i.e., dispersion in a direction which is perpendicular to the average flow. An example of such spreading is given in Figs. 1 and 7(c). A discussion on transversal dispersion is also given by Simpson (1962).

As an example, consider a field of steady uniform average flow in a horizontal plane. In laminar flow, streamlines coincide with pathlines, and once a liquid particle joins a streamline, it will follow the same as the particle preceding it along the same streamline. The exact (microscopic) flow pattern is unknown, yet, as explained above, it is possible to study the statistical characteristics common to a large number of labeled liquid particles starting from the same point: the flow velocity along the streamline which fluctuates around some mean direction such that its amplitude and the nature of the fluctuations remain approximately constant in time. The only mechanism that can cause an ever widening (in a direction perpendicular to the flow) of the zone occupied by the tracer as the flow proceeds is molecular diffusion. This mechanism causes an exchange of tracer particles between adjacent streamlines. This in itself is not sufficient. The second necessary condition for the existence of lateral spreading as observed in experiments is the presence of junctions at which intensive mixing between the fluid streams entering from several channels takes place. Only through such branching can a tracer be transported to points which are at appreciable distances from the average flow axis. The lateral transport through molecular diffusion within a single channel is limited by the dimensions of the channel. The number of particles transported by the diffusion process from one streamline to the next depends on the length of time during which the particles remain within each passage, i.e., on the velocity along the streamlines.

The process of hydrodynamic dispersion is, therefore, also an irreversible one.

de Josselin de Jong (1958), who also used the statistical approach, was probably the first to express in an analytical way the fact that longitudinal dispersion is larger than the lateral one. In Scheidegger's (1954) random walk model, each unit step $r_j(x_j, y_j, z_j)$ has no relationship between its components x_j, y_j, z_j in a cartesian coordinate system, and no preference with respect to the main flow direction. In addition, it is assumed that each step has the same duration. Therefore, after a time interval $T \gg \Delta t$, the number of steps performed by each particle is $N = T/\Delta t$.

In de Josselin de Jong's (1958) work, this number is not a constant. He visualizes a porous medium model composed of a network of interconnected straight channels of equal length oriented at random, but uniformly distributed in all directions, in which an average uniform flow in the z direction takes place (Fig. 7). In this model, the pressure gradient

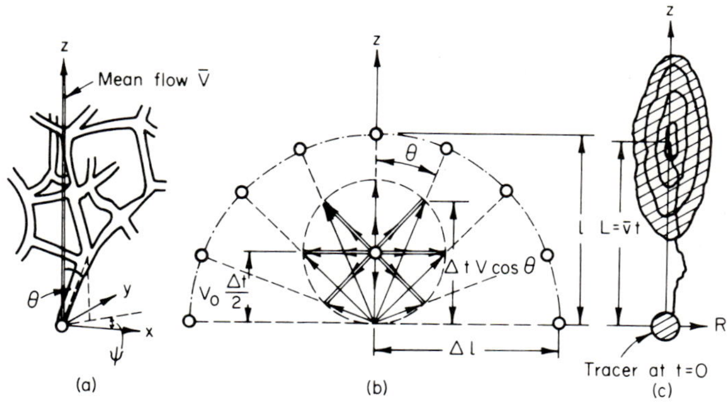

FIG. 7. Path of particle and final distribution of dispersed tracer (de Josselin de Jong, 1958).

in each channel is proportional to cos θ, θ being the angle between the directions of the channel and of the uniform flow. Because of radial diffusion in the narrow channels, one assumes that a tracer particle moves with the mean velocity within each channel. With these assumptions, the residence time t_j in the jth channel is

$$t_j = l/V_0 \cos \theta \tag{58}$$

where V_0 is the velocity in a channel oriented in the z direction and l is the constant length of an elementary channel. The components of the jth displacement are:

$$x_j = l \sin \theta_j \cos \psi_j\,; \qquad y_j = l \sin \theta_j \sin \psi_j\,; \qquad z_j = l \cos \theta_j$$

where θ and ψ are the angles in a spherical coordinate system (Fig. 7).

The probability $p(\theta, \psi)$ that a tracer particle will choose a direction defined by the angular interval $(\theta, \theta + \Delta\theta)$ and $(\psi, \psi + \Delta\psi)$ when arriving at a junction is assumed to be proportional to the ratio between the discharge $Q(\theta, \psi)$ in that direction and the total discharge Q:

$$p(\theta, \psi) = Q(\theta, \psi)/Q = (1/\pi) \cos \theta \sin \theta \qquad (59)$$

The above assumptions are different from those assumed in the random walk theories described above. If the velocity in a channel is $V = V_0 \cos \theta$ [Fig. 6(b)], then in time $\Delta t < l/V$ a particle moves a distance $\Delta t V_0 \cos \theta$ in each channel, and all particles entering a junction simultaneously will be distributed on the circumference of a sphere centered on the z axis at $\Delta t V_0/2$. The difference between the present approach and that of ordinary random walk theories lies in the fact that here all channels have the same length, but the velocity in them varies with their direction. Therefore, particles reach the end of their respective channels at different times and are confronted with new "choices." Each particle, therefore, undergoes a different number of displacements in a given time interval.

Using Markov's method, de Josselin de Jong obtains the probability of a particle to arrive at a given point at time t after a large number of displacements. The result is a normal distribution in three dimensions.

Figure 6(c) shows surfaces of equal concentration resulting from a point injection which are ellipsoids of revolution centered at the point traveling with the mean velocity $\bar{V} = \frac{1}{3}V_0$. The standard deviations σ_L, σ_T of the distribution in the longitudinal and transverse directions, respectively, are

$$\begin{aligned} \sigma_L^2 &= \tfrac{1}{3}(\lambda + 0.173)\,\bar{V}t = \tfrac{1}{3}(\lambda + 0.173)\,L; \qquad & D_L &= \sigma_L^2/2t \\ \sigma_T^2 &= \tfrac{3}{8} l\bar{V}t = \tfrac{3}{8} lL; & D_T &= \sigma_T^2/2t \end{aligned} \qquad (60)$$

where λ is a function related to L by $3L/l = e^{2\lambda}(\lambda - 2.077)$. The ratio of longitudinal to transverse dispersion generally lies in the range from 5 to 7. Both D_L and D_T are proportional to $\bar{V}$. However, D_L increases with the distance traveled. Both are also proportional to a characteristic length l which is of the order of the grain size and depends on the grain-size distribution. Blackwell (1959) mentions ratios of D_L/D_T of up to 24.

The introduction of transverse dispersion is supported by experimental evidence (de Josselin de Jong, 1958; Bear 1961).

Saffman's (1959, 1960) model and statistical method is very similar to that of de Josselin de Jong (1958). However, Saffman introduces molecular diffusion into his model and studies the relationship between mechanical dispersion and molecular diffusion. His model consists of an assembly of randomly oriented and distributed straight circular tubes which are equivalent to a statistically homogeneous, isotropic porous medium. The tubes, of equal length and cross section, are connected to one another at the ends, and several tubes may start or end at these endpoints (= junctions). The dimensions of these tubes (or pores) are comparable with the size of the pores in the real porous medium. In this model, the path of a fluid particle may be regarded as a random one due to variations in direction and velocity.

Saffman (1959) introduces molecular diffusion in determining the residence time of a particle in an elementary tube. He assumes, similar to de Josselin de Jong (1958), that the velocity of a liquid particle in a tube (of radius R) depends on the angle θ between the tube's axis and the direction of the average flow. However, his time of residence also depends on the distance of the particle from the tube's axis:

$$t = l/V = l/2\langle V\rangle(1 - r^2/R^2) = l/6\overline{V} \cos\theta(1 - r^2/R^2)$$

where $\overline{V}$ is average flow velocity and $\langle V\rangle$ is the average velocity in the tube. A tracer particle, on the other hand, also participates in the motion due to molecular diffusion, in addition to the flow of the liquid. This motion will cause the tracer particle to move *across* the tube from one streamline to the next, as well as *along* each streamline relative to the ambient liquid. The difficulty in adding these two motions stems from the fact that the convective motion is described by the Lagrange approach of following a particle, while diffusion is described by Euler's approach of observing particles at a fixed place. Saffman, therefore, suggests an approximate approach, distinguishing between four cases.

(1) $l/\langle V\rangle \ll \tau_1$: Time of passage of an elementary mass through the tube is much shorter than the time τ_1 required for this mass to effectively diffuse across the tube. By analogy with molecular diffusion, Saffman assumes $\tau_1 = (R/2)^2/2D_d$. In this case, Saffman assumes τ = residence time $= l/\langle V\rangle$.

(2) $l/\langle V\rangle \ll \tau_1$, but $r/R \sim 1$, i.e., the tracer particle enters the tube close to the wall. Since here the velocity is very small, there is sufficient time for its diffusion spreading to adjacent streamlines where the velocity is higher. This is equivalent to $\tau = l/\langle V\rangle + \tau_1$.

(3) $\tau_1 < l/\langle V\rangle < \tau_0$: where τ_0 is the time required for effective longitudinal diffusion to take place, $\tau_0 = l^2/2D_d$. For this case, Taylor (1953) shows that the tracer mass spreads with respect to the average motion with a spreading coefficient $R^2\langle V\rangle^2/48D_d$. In a very narrow tube ($R \ll l$) we may then assume $\tau = l/\langle V\rangle$. Otherwise ($R < l$), $\tau = l/\langle V\rangle + \tau_1$.

(4) $\tau_0 \ll l/\langle V\rangle$: Then $\tau = \tau_0$.

In this model, the displacement of a single particle after N (independent) steps is a random variable with components parallel to the x, y, z coordinate system

$$Z_N = \sum_{j=1}^{N} l_j \cos\theta; \qquad X_N = \sum_{j=1}^{N} l_j \sin\theta_j \cos\psi_j; \qquad Y_N = \sum_{j=1}^{N} l_j \sin\psi_j \sin\theta_j$$

where the average flow is in the Z direction and θ and ψ are shown on Fig. 6(a). The time required for the N steps is the random variable

$$T_N = \sum_{j=1}^{N} t_j$$

Saffman's probability function, that a particle will choose a streamline between r and $r + dr$ in a tube of direction between θ and $\theta + d\theta$, ψ and $\psi + d\psi$ is

$$dP = (4/\pi)\sin\theta\cos\theta\cdot(r/R^2)[1 - (r^2/R^2)]\, d\theta\, d\psi\, dr \tag{61}$$

which differs from that of de Josselin de Jong in that it includes the velocity distribution within the tube. Saffman then determines the location of the particle after a large number of steps N and the other statistical parameters for $l_j = l = \text{const}$

$$\bar{x}(N) = \tfrac{2}{3}Nl; \qquad \operatorname{Var} x(N) = [x_j - \bar{x}(N)]^2 = Nl^2\sigma_x^2 = \tfrac{1}{18}Nl^2$$

$$\bar{y}(N) = \bar{z}(N) = 0; \qquad \operatorname{Var} y = \operatorname{Var} z = Nl^2\sigma_y^2 = \tfrac{1}{4}Nl^2$$

$$\bar{t}(N) = \frac{\tfrac{2}{3}Nl}{\bar{V}} = \frac{\bar{x}(N)}{\bar{V}}; \qquad \operatorname{Var} t(N) = \frac{Nl^2}{\bar{V}^2}\sigma_t^2 = \frac{Nl^2}{\bar{V}^2}\left[\frac{2}{9}\ln\frac{3\bar{V}\tau_0}{l} + \frac{1}{18}\left(\ln\frac{6\bar{V}\tau_1}{l}\right)^2 - \frac{1}{6}\ln\frac{6\bar{V}\tau_1}{l} + \frac{1}{4}\right]$$

$$\operatorname{Cov}[x_i(N)\, x_j(N)] = \overline{\mathring{x}_i\mathring{x}_j} = 0 \qquad \text{for } i \neq j;\quad x_1 = x,\quad x_2 = y,\quad x_3 = z$$

$$\operatorname{Cov}[x_i(N)\, t(N)] = -\frac{1}{9}\frac{Nl^2}{\bar{V}} \qquad \text{for } i = 1$$

$$= 0 \qquad \text{for } i \neq 1$$

In performing the integration over time, (61) was employed and the time was divided into three ranges according to the assumptions described above, and terms approaching zero as τ_0 and τ_1 tend to infinity were neglected. The above results assume that X_N, Y_N, Z_N, and T_N are normally distributed.

Assuming that for a long time $\bar{N} = \frac{3}{2}\bar{V}t/l = \frac{3}{2}(L/l)$, Saffman obtains $\bar{x} = L = \bar{V}t$, $\bar{y} = \bar{z} = 0$, and the displacement components perpendicular to the average flow are normally distributed around the mean with a variance which is independent of molecular diffusion

$$\operatorname{Var} y(t) = \operatorname{Var} z(t) = \tfrac{3}{8}\bar{V}lt = (\tfrac{3}{8}l)L$$

In the direction of the flow, $(x - \bar{V}t)/t$ is approximately normally distributed when $P_e \equiv l\bar{V}/D_d \gg 1$. For large times he obtains

$$\overline{\left[\frac{(x-\bar{V}t)^2}{t}\right]} = \tfrac{1}{3}\ln(\tfrac{2}{3}P_e) + \tfrac{1}{12}[\ln(\tfrac{3}{4}(R/l)^2 P_e)]^2 - \tfrac{1}{4}\ln(\tfrac{3}{4}(R/l)^2 P_e) + \tfrac{19}{24} \qquad (62)$$

where

$$\bar{n} \gg \frac{1}{4}\frac{P_e^{\,2}}{\ln(\frac{3}{2}P_e)}; \qquad L/l \gg \frac{1}{6}\frac{P_e^{\,2}}{\ln(\frac{3}{2}P_e)}$$

This is similar to the result obtained by de Josselin de Jong (1958) except for the terms involving $\ln P_e$. In the above expression, $P_e = l\bar{V}/D_d$ is the *Peclet number* which defines the ratio between rate of transport by convection to the rate of transport by molecular diffusion. Here P_e is related to the length l.

Saffman (1959) then defines coefficients of effective longitudinal and transverse diffusivities D_L and D_T which depend on the average Darcy velocity $\bar{V}$, on a pore (or channel) length l, on a pore radius R (related to the intrinsic permeability), on the molecular diffusivity D_d, and on the time t from the initial instant

$$D_L = \frac{1}{2t}\overline{(x-\bar{V}t)^2} = \tfrac{1}{2}\bar{V}lS^2; \qquad S^2 = \ln\frac{3\bar{V}\tau_0}{l} - \frac{1}{12}; \qquad \tau_0 = \frac{l^2}{2D_d} \qquad (63)$$

for

$$\bar{V}\tau_0/l \ll [N\ln(3\bar{V}\tau_0/l)]^{1/2}, \qquad D_T = \frac{1}{2t}\overline{y^2} = \frac{3}{16}\bar{V}l;$$

where N is the number of steps which should be sufficiently large so that the total displacement of the individual tracer particles becomes statistically independent.

The second work (Saffman, 1960), is devoted to the case where molecular diffusion and macroscopic mixing are of the same order of magnitude, i.e., $\bar{V}l/D_d = O(1)$. Again, the model consists of a random network of capillaries. The Lagrangian correlation function defined as $\overline{\mathring{V}_i\mathring{V}_j}$ is introduced, and longitudinal (D_L) and lateral (D_T) dispersion coefficients are derived for all values of $\bar{V}l/D_d$ less than some large value. It is assumed that $l/\bar{V} \gg R^2/8D_d$, i.e., the time scale for molecular diffusion to smooth out concentration variation across the cross section of the capillary tube of radius R is small compared with the mean time spent by a particle in a capillary. Saffman then adapts the analysis of Taylor (1953, 1954) and Aris (1956), who show that in this case the concentration is approximately uniform over the cross section and that the tracer disperses relative to an axis moving at a velocity $\langle V \rangle$, according to the diffusivity given by (24). This means that the dispersion in the network of capillars is the same as if the velocity in each capillar where uniform throughout it, and the material is subject to a diffusion-like process along the capillar with a diffusivity D'. Based on these assumptions, Saffman considers the velocity of a tracer particle at any instant as a sum

$$V = \langle V \rangle + V_{D'} \tag{64}$$

where $\langle V \rangle$ is the mean velocity of motion in the capillar and $V_{D'}$ is a random velocity due to the diffusion process with diffusivity D'. For $V_{D'}$, we have

$$\tilde{V}_{D'} = 0; \qquad \int_0^\infty V_{D'}(0)\, V_{D'}(\theta)\, d\theta = D' \tag{65}$$

where the tilde ($\sim$) denotes an average with respect to the random motion in the tube. In a (statistically) homogeneous medium, the velocity of a tracer particle is a *stationary random function* of time. In a Cartesian coordinate system (x_1, x_2, x_3 or x_i, $i = 1, 2, 3$) with x_1 coinciding with the direction $\bar{\mathbf{V}}$, so that $\bar{V}_1 \neq 0$, $\bar{V}_2 = \bar{V}_3 = 0$; and for large values of t, the tracer spreading is defined by the variances

$$\sigma_{x_i}^2 = \overline{(\mathring{x}_i)^2} \sim 2 \int_0^\infty \overline{\mathring{V}_i(0)\, \mathring{V}_i(\theta)}\, d\theta; \qquad i = 1, 2, 3 \tag{66}$$

where $\mathring{x}_i = x_i - \bar{x}_i$; $\mathring{V}_i = V_i - \bar{V}_i$. The bar above the expression in the integrand denotes an average over an ensemble of realizations, which here means of all possible combinations of the network of tubes.

Saffman then calculates the covariance of velocity components in the direction of the average (Darcy) velocity $\bar{V}$ and perpendicular to it.

Integrating these covariances for a large t, he obtains for longitudinal dispersion

$$D_L = \frac{\overline{\mathring{x}_1^2}}{2t}\frac{1}{3}D_d + \frac{3}{80}\frac{R^2\overline{V}^2}{D_d} + \frac{l^2\overline{V}^2}{4}\int_0^1 (3\alpha^2-1)^2 \frac{M\coth M - 1}{D'M^2}\,d\alpha$$

$$M = \frac{3}{2}\frac{\overline{V}l\alpha}{D'}; \qquad D' = D_d + \frac{3}{16}\alpha^2\left(\frac{R^2\overline{V}^2}{D_d}\right); \qquad \alpha = \cos\theta \tag{67}$$

The first term of D_L is the contribution of molecular diffusion in the longitudinal direction. The second term expresses the contribution of dispersion in a single capillar as derived by Taylor (1953). The third term expresses the resulting convective dispersion.

If the liquid is at rest, $\overline{V} = 0$, we obtain $D_L = D_d/3$. Saffman comments that laboratory experiments in granular materials give $D_L \sim \frac{2}{3}D_d$, which indicates that a granular materials cannot be represented as a random network of capillaries. He therefore suggests replacing $D_d/3$ by some empirical value $(D_d)_m$. For l he uses the value of average grain size, while for R ($\leqslant l/5$), he suggests $R = (24k/n)^{1/2}$ where k is the medium permeability and n is the porosity.

For transverse dispersion he obtains

$$D_T = \frac{\overline{(\mathring{x}_2)^2}}{2t} = \frac{\overline{(\mathring{x}_3)^2}}{2t} \sim (D_d)_m + \frac{1}{80}\frac{R^2\overline{V}^2}{D_d}$$

$$+ \frac{9\overline{V}^2 l^2}{8}\int_0^1 \alpha^2(1-\alpha^2)\frac{M\coth M - 1}{K_s M^2}\,d\alpha \tag{68}$$

For example,

(i) $$P_e \ll 1, \qquad D_L/D_d \sim m + P_e^2/15;$$

$$D_T/D_d = m + P_e^2/40, \qquad \tfrac{1}{3} \leqslant m \leqslant \tfrac{2}{3}$$

(ii) $$1 \ll P_e \ll 8(R/l)^2$$

$$D_L/D_d \sim (P_e/6)\ln(3P_e/2) - (17P_e/72) - (R/l)^2 P_e^2/48 + (m + \tfrac{4}{9}) + O(1/P_e)$$

$$D_T/D_d \sim 3P_e/10 + (R/l)^2 P_e^2/40 + (m - \tfrac{1}{3}) + O(1/P_e)$$

For $P_e < 1$, these results are similar to those obtained by Beran (1955), who obtained for $P_e < 1$, $D = D_d + (\gamma/D_d)\,d^2\overline{V}^2$, where γ is a constant depending on the grain size distribution and d is a characteristic grain dimension.

Hiby (1959) (see Saffman, 1960) obtained for glass beads of diameter l

$$D_L/D_d = 0.67 + 0.65P_e/(1 + 6.7P_e^{-1/2})$$

which agrees with Saffman's results for $R/l = \frac{1}{5}$ in the wide range $10^{-2} \leqslant P_e \leqslant 10^2$. However, for transversal dispersion, Saffman's values are higher than Hiby's. On the other hand, experimental results obtained by Blackwell *et al.* (1959),

$$D_L/D_d = 8.8P_e^{1.17} \qquad \text{for} \quad P_e > 0.5$$

are different.

The works of de Josselin de Jong and Saffman reviewed above, clearly indicate the existence of transversal dispersion. Nikolaevskij (1959) also uses the statistical approach in dealing with an ideal tracer. He regards the velocity of a tracer particle as a random vector function of time. Without referring to the mechanism of molecular diffusion, he applies the central limit theorem of probability, according to which the probability of distribution of the particle's location after a sufficiently long time is approximately normal when the particle's velocities during small time intervals t_0 are independent of each other. In a cartesian coordinate system (x_1, x_2, x_3), which coincides with the principal axes of the normal distribution, the distribution density takes the form

$$P(x_1, x_2, x_3, t) = [1/(2\pi)^{3/2}\,\sigma_1\sigma_2\sigma_3] \exp\left\{-\sum_{i=1}^{3} \frac{\overline{(\mathring{x}_i)^2}}{2\sigma_i^2}\right\} \tag{69}$$

where

$$\mathring{x}_i = x_i(t) - \bar{x}_i(t) = \int_0^t [V_i(\tau) - \bar{V}_i]\, d\tau \tag{70}$$

$$\sigma_i^2 = \overline{(\mathring{x}_i)^2} = 2t \int_0^\infty \overline{\mathring{V}_i(0)\, \mathring{V}_i(\tau)}\, d\tau = 2t\overline{(\mathring{V}_i)^2} \int_0^\infty r(\tau)\, d\tau$$

$$= 2tt_0\overline{(\mathring{V}_i)^2}; \qquad t_0 = \int_0^\infty r(\tau)\, d\tau \tag{71}$$

Here $V_i(\tau)$ is a stationary random function of time having the ergodic property. The average indicated by the bar is a time average of a single realization, or a spatial average over a representative ensemble of porous medium samples. t_0 is the weighted time of correlation of the particle. Nikolaevkij assumes that $t_0 = l/\bar{V}$, which is the residence time of liquid due only to convection in some representative elementary channel, where l is a mixing length characterizing the medium (analogous to the mixing length in turbulent diffusion). He assumes that l is a scalar in an isotropic medium.

P in (69) may be interpreted as the relative tracer mass concentration (percent of the initial mass injected at $t = 0$). It satisfies the differential equation

$$\frac{\partial P}{\partial t} = \sum_{i=1}^{3} \frac{\partial}{\partial x_i}\left(D_i \frac{\partial P}{\partial x_i}\right) \tag{72}$$

where the D_i's are the components of the coefficient of dispersion along the principal spreading axes. With (70) and (71),

$$D_i = \overline{(\mathring{x}_i)^2}/2t = \overline{(\mathring{V}_i)^2}\, l/\overline{V} \tag{73}$$

Various investigators (e.g., Rifai *et al.*, 1956; Bear, 1960) obtained a Poisson distribution for short times, or for a small number of steps (cells, elementary displacements). In a paper on the general form of the partial differential equation describing hydrodynamic dispersion Scheidegger (1965) states that there are two ways by which the differential equation of dispersion may be obtained by purely statistical considerations of a random process. The first is to regard the dispersion process as a *Markov process* (e.g., de Josselin de Jong, 1958). A Markov process is defined as a stochastic process in which the prediction of the probability distribution of a random variable at time t, given its distribution at some time $t_0 < t$, depends only on that given distribution at time t_0. Any additional information regarding this random variable at times t prior to t_0 does not change the resulting probability distribution for $t > t_0$.

The equation describing such a process is the *Kolmogorov equation* which has the form of the diffusion equation.

Let $\xi(t)$ be an ensemble of random variables depending on the parameter t (time), and let $F(t, x; \tau, y)$ describe the probability that at some time τ the random variable $\xi(t)$ will be smaller than y, given that at some earlier time $t < \tau$ we had $\xi(t) = x$

$$F(t, x; \tau, y) \equiv P[\xi(\tau) = y \mid \xi(t) = x]$$

This distribution function also satisfies

$$\lim_{y \to -\infty} F(t, x; \tau, y) = 0; \qquad \lim_{y \to +\infty} F(t, x; \tau, y) = 1$$

We assume that F is continuous with respect to t, τ, x. If a *density* exists, $f(t, x; \tau, y) = \partial F(t, x; \tau, y)/\partial y$, then it satisfies

$$\int_{-\infty}^{y} f(t, x, \tau, \eta)\, d\eta = F(t, x; \tau, y); \qquad \int_{-\infty}^{+\infty} f(t, x, \tau, \eta)\, d\eta = 1$$

We shall also assume that the random process is continuous so that

$$\lim_{\Delta t \to 0} (\Delta t)^{-1} \int_{|y-x| \geqslant \epsilon} F(t - \Delta t; t, \eta)\, d\eta = 0$$

for any positive constant ε.

Under these conditions, and with $\partial F/\partial x$, $\partial^2 F/\partial x^2$ continuous for all t, x, y and $\tau > t$, the *first Kolmogorov equation* is

$$\frac{\partial F}{\partial t} = -a(t, x)\frac{\partial F}{\partial x} - \frac{b(t, x)}{2}\frac{\partial^2 F}{\partial x^2}$$

where we have

$$\lim_{\Delta t \to 0} (\Delta t)^{-1} \int_{|y-x| < \epsilon} (y - x) F(t - \Delta t, x; t, y)\, dy = a(t, x)$$

$$\lim_{\Delta t \to 0} (\Delta t)^{-1} \int_{|y-x| < \epsilon} (y - x)^2 F(t - \Delta t, x; t, y)\, dy = b(t, x)$$

and the convergence is uniform with respect to x.

If in addition to the above conditions, f exists and

$$\partial F/\partial \tau, \quad \partial[a(\tau, y) f(t, x; \tau, y)]/\partial y, \quad \text{and} \quad \partial^2[b(\tau, y) f(t, x; \tau, y)]/\partial y^2$$

exist and are continuous, f will satisfy *the second (or forward) Kolmogorov equation*

$$\frac{\partial f}{\partial \tau} = -\frac{\partial}{\partial y}[a(\tau, y) f] + \frac{1}{2}\frac{\partial^2}{\partial y^2}[b(\tau, y) f]$$

To obtain a physical interpretation of a and b one should notice that

$$\begin{aligned}\int (y - x)\, dy\, F(t - \Delta t, x; t, y) &= \overline{[\xi(t) - \xi(t - \Delta t)]} \\ &= \overline{\xi(t)} - \overline{\xi}(t - \Delta t) = \overline{\Delta \xi}\end{aligned}$$

$$(y - x)^2\, dy\, F(t - \Delta t, x; t, y) = \overline{[\xi(t) - \xi(t - \Delta t)]^2} = \overline{(\Delta \xi)^2}$$

from which we have

$$a(t, x) = d\overline{\xi}/dt; \qquad b(t, x) : \lim_{\Delta t \to 0} \frac{\overline{(\Delta \xi)^2}}{\Delta t}$$

This means that a represents the average rate of change of $\xi(t)$, whereas b represents the average of the square of the change per unit time. If the $\xi(t)$'s are the coordinates of a point moving under the influence of random factors, $a(t, x)$ may be interpreted as its velocity at time t and point x, whereas b denotes the mean of the squares of incremental displacements per unit time (rate of growth of variances). In each case the average is taken over all possible displacements during Δt.

It is thus shown that if the dispersion process is regarded as a Markov process, the average tracer concentration is described by the diffusion equation obtained from the second Kolmogorov equation.

The second way is to use the concept of *constant of motion* (e.g., the energy of a mechanical system) as done in statistical mechanics of irreversible processes. The dispersion is then considered as a stochastic process in which the total tracer mass x remains constant, and in which the interactions between the various parts of the system are weak. It can be shown that when a system consists of a linear combination of a large number of fluctuating subsystems, and when in the larger system a certain values is a constant of motion to which the subsystem adds nonnegative contributions, then in the absence of equilibrium and assuming a linear regression of the fluctuations and microscopic reversibility, the considered value satisfies the diffusion equation with a coefficient which is a symmetric tensor. If we abandon the hypothesis of a linear regression of the fluctuations, we obtain equations of a different type.

2.4. Averaging over a Specified Porous Medium Model

The approach suggested by Bear and Bachmat (1965, 1966) is followed in this section. Although essentially a statistical approach, this work is an attempt to obtain a better insight of the dispersion phenomena and to reveal relationships between the dispersion and the medium characteristics. It will, therefore, be reviewed here at some length.

The basic idea, as in several works described above (whether explicitly stated or only implied) is to replace the porous matrix and the liquid filling the void space with their respective properties by fictituous *interpenetrating continuae*, each of which is occupying the entire flow domain. This is done by averaging point values (or values at the microscopic level) of the various parameters over certain elementary volumes of the porous medium. This is done in the same way that a density of a liquid is defined (Prandtl and Tietjens, 1934). By defining a density (or also viscosity) of a liquid, a passage is made from the microscopic to the macroscopic description of a continuum. Actually, in the same way that the concept of density is a macroscopic description of what happens at the molecular level, the molecule itself with the various parameters describing it is also, in a way, a macroscopic description of a behavior at the submolecular level. In each case, when passing from some microscopic level to the macroscopic one, parameters which express our inability to describe the microscopic level in an exact manner are introduced. Permeability and dispersivity of a porous medium are examples of such parameters. One may refer to them as "ignorance factors."

However, a direct passage from the microscopic level to the macroscopic description of a continuum in the case of a porous medium will not reflect certain media, liquid and flow relationships characterizing a porous medium. We are dealing here with two interpenetrating continuae and some interaction between them (e.g., friction resistance to flow) which require the introduction of a certain porous medium model in which the special behavior of the porous medium will be reflected. *The microscopic level here is the single pore.* The need for averaging over a pore (or channel) cross section seems thus to be essential. These ideas also indicate the advantage of performing the averaging procedure over a specific form of a conceptual model which aims to reflect some of the basic properties of the medium, as far as the phenomena considered here are concerned, rather than to average over a random porous medium directly. In a way, de Josselin de Jong (1958) is doing the same thing, whereas Scheidegger (1960) is not.

The *porous medium model* considered here is composed of a network of interconnected random narrow passages, or *elementary channels*, of varying length, cross section, and orientation. Each channel, bounded by a stable solid surface, defines a fixed streamtube in space. The liquid saturating the channels is incompressible yet nonhomogeneous due to variations in solute concentration. This assumption is valid for practical purposes at low concentrations. The flow in the channels, caused by gravity and pressure differences, is laminar. The resistive force at a point in the channel is due only to viscosity and obeys Poiseulle's law. The liquid is chemically inert and no surface phenomena take place between the solid matrix and the liquid. It is assumed that an *axis* in the form of a smooth curve may be defined along each channel. The orientation of a channel's axis is defined at each point along it by the slope of its elementary length ds at that point.

The cross section normal to the axis at a point is referred to as the *channel's cross section* at that point. Its area is denoted by w. The radius of curvature of the channel's axis is much larger than a characteristic channel length dimension. The angle between the direction of a streamline at a point on a channel cross section and the channel's axis at that cross section, is denoted by θ.

The volumes of *junctions*, where channels meet each other, are much smaller than those of the channels themselves. On entering a junction, the functional relationships relating the liquid's flow parameters, such as velocity and pressure, to the parameters of the particular channel, are disconnected.

Instead of following, as in some of the models described above, a tracer particle moving through the porous medium, the *Eulerian approach* of an observer who is fixed at a point within the medium, is employed. The velocity, pressure, etc., are considered random variables of time and space.

The analysis is carried out in three steps. First, general statements of conservation are stated for a point in a liquid continuum. In the present case, these statements are applicable to points in the liquid filling the void space of the medium. Then, *local transport equations* are derived by considering an infinitesimal displacement of a small liquid volume along a single channel. In developing these local equations (of continuity of mass, conservation of a component of a solution and equation of motion), the "hydraulic" approach is used. Accordingly, the equations, which are in terms of average values over a channel's cross section, include the boundary conditions at the impervious solid walls as random parameters describing channel characteristics.

Finally, making some general assumptions regarding the correlation functions of the various variables involved, without specifying any particular probability distributions, the local equations are averaged at a given instant over an ensemble of channels within a statistically meaningful elementary volume of the porous medium model around the point which serves as its centroid. The resulting macroscopic equations contain average nonrandom variables and parameters assigned to the centroid of this *representative elementary volume*. Since this procedure is applicable to every point in space, the actual medium is thus replaced by a fictitious medium which is continuous with respect to the average variables and parameters.

To determine the representative elementary volume, let the (volumetric) *porosity* n, i.e., ratio between volume of voids and total volume of sample, in the vicinity of the considered point be determined for a sequence of ever decreasing volumes ΔU_1, ΔU_2,... . For large values of ΔU, the resulting porosity values n_1, n_2,... may fluctuate without showing any definite trend, due, for example to macroscopic inhomogeneity. As ΔU becomes smaller, a trend is usually observed, and at a certain ΔU_0, the fluctuations will cease. However, if ΔU is further reduced below ΔU_0, the fluctuations in the resulting n will increase again, this time due to inhomogeneity on a microscopic scale (i.e., presence of grains, pores, etc.). Finally, as $\Delta U \rightarrow 0$, the porosity n will attain the value of 0 or 1, according to whether the considered point is within the solid matrix or inside a void. The volume ΔU_0 is called the *representative elementary volume* (REV).

Because the averaged variables are assigned to the centroid of the representative elementary volume, its largest dimensions must be much smaller than those of the flow domain, yet it must be large enough to contain a representative group of samples of the void space in the neighborhood of the considered point. The representative elementary volume as defined above is the smallest volume around the point such that adding to it or subtracting from it one or several channels has a negligible effect on the value of the volumetric porosity n.

Since the following discussion deals with homogeneous as well as inhomogeneous liquids, we shall distinguish, as in Section 1.3, between:

$\mathbf{V}^*$, barycentric, or mass averaged, velocity;

$\mathbf{V}$, volumetric velocity, $\mathbf{V}^* + (D_d/\rho)$ grad ρ;

$\mathbf{V}_\alpha$, velocity of species α.

The analysis in the following paragraphs is based on a *general conservation statement* (e.g., De Groot and Mazur, 1963). Consider a finite fixed portion of space of volume U bounded by a surface S. *Let* $G_\alpha(\mathbf{x}, t)$ be some *extensive property* (mass, energy) of the species α within U. We may then define an associated *intensive property* $\gamma_\alpha(\mathbf{x}, t)$ by

$$G_\alpha = \int_{(U)} \rho_\alpha(\mathbf{x}, t)\, \gamma_\alpha(\mathbf{x}, t)\, dU \tag{74}$$

γ_α is thus the local density (that is, per unit mass of species α) of the considered property. When G_α is the total mass of species α, $\gamma_\alpha = 1$. If G_α is the total energy, γ_α is the energy per unit mass, etc. We may also introduce the volumetric density $g_\alpha = \rho_\alpha\gamma_\alpha$ which has the meaning of the amount of the property per unit volume of the system.

The rate of change of this property within U at instant t is then expressed by

$$\int_{(U)} \frac{\partial g}{\partial t}\, dU = -\int_{(S)} g_\alpha \mathbf{V}_\alpha \cdot d\mathbf{S} + \int_{(U)} I_\alpha\, dU \tag{75}$$

or by

$$\int_{(U)} \frac{\partial(\rho_\alpha\gamma_\alpha)}{\partial t}\, dU = -\int_{(S)} \rho_\alpha\gamma_\alpha \mathbf{V}_\alpha \cdot d\mathbf{S} + \int_{(U)} I_\alpha\, dU \tag{76}$$

where I_α is a source–sink function expressing the rate of production of $g_\alpha\ (=\rho_\alpha\gamma_\alpha)$ through internal processes (e.g., chemical reactions). Using Gauss' theorem, (75) may be written as

$$\int_{(U)} \frac{\partial g_\alpha}{\partial t}\, dU = -\int_{(U)} \operatorname{div}(g_\alpha \mathbf{V}_\alpha)\, dU + \int_{(U)} I_\alpha\, dU \tag{77}$$

Equations (75) or (77) state that the rate of change of the property of the species α within U is the sum of two terms: the net transport of this property through the surface S and the production rate of this property within U. Since the conservation statement expressed by (77) applies to any volume U, we may rewrite it in the differential form

$$\partial g_\alpha/\partial t + \operatorname{div}(g_\alpha \mathbf{V}_\alpha) = I_\alpha \tag{78}$$

However, the passage from the integral to the differential form is permitted only if the quantities in the integral equation vary smoothly with space and time. One should note here that the product $g_\alpha \mathbf{V}_\alpha$ in (78) is the flux of the property g_α with respect to a fixed coordinate system and hence, the general meaning of $\mathbf{V}_\alpha$ is the velocity at which the property of the α particles advances in the flowing liquid.

Let us now consider some special values of g_α.

(i) *Conservation of mass of a species* α. In this case, $g_\alpha = \rho_\alpha \equiv C_\alpha$, $\gamma_\alpha = 1$. We obtain from (78),

$$\partial C_\alpha/\partial t + \operatorname{div}(C_\alpha \mathbf{V}_\alpha) = I_\alpha \tag{79}$$

(ii) *Conservation of mass of the system.* By adding over all species α we obtain

$$\partial\rho/\partial t + \operatorname{div}(\rho\mathbf{V}^*) = 0; \qquad \sum_{(\alpha)} \int_{(U)} I_\alpha \, dU = 0 \tag{80}$$

which is the *mass conservation equation* for the entire system.

Equation (79) may also be written in the form

$$DC_\alpha/Dt + C_\alpha \operatorname{div} \mathbf{V}_\alpha = 0; \qquad D\rho_\alpha/Dt + \rho_\alpha \operatorname{div} \mathbf{V}_\alpha = 0 \tag{81}$$

where $D(\ )/Dt = \partial(\ )/\partial t + \mathbf{V}_\alpha \cdot \operatorname{grad}(\ )$ is the *hydrodynamic derivative* which expresses the rate of change of the property as it moves. In this case, $\mathbf{V}_\alpha$ is the velocity of propagation of the considered property.

From (81) we obtain for an elementary mass δm_α of species α occupying an element of volume δU,

$$\operatorname{div} \mathbf{V}_\alpha = -\frac{1}{C_\alpha}\frac{D(C_\alpha)}{Dt} = \frac{\delta U}{\delta m_\alpha} \cdot \frac{D(\delta m_\alpha/\delta U)}{Dt} = \frac{1}{\delta U}\frac{D(\delta U)}{Dt} \tag{82}$$

Hence, $\operatorname{div} \mathbf{V}_\alpha$ may be interpreted as the rate of relative growth of the volume occupied by a given mass of species α as it moves. In flow through a porous medium this growth is due to dispersion. We shall indeed see that $\operatorname{div} \mathbf{V}_\alpha$ plays an important role in the dispersion theory.

Equation (81) may now be rewritten as

$$C_\alpha = -(\mathbf{V}_\alpha \cdot \operatorname{grad} C_\alpha + \partial C_\alpha/\partial t)/(1/\delta U)[D(\delta U)/Dt]$$

and for a steady concentration ($\partial C_\alpha/\partial t = 0$) in a cartesian coordinate system x_i, in the form

$$C_\alpha = -(V_\alpha)_j \frac{\partial C_\alpha}{\partial x_j} (\operatorname{div} \mathbf{V}_\alpha)^{-1} \tag{83}$$

We could also write (80) in terms of the volumetric velocity by replacing $\rho\mathbf{V}^*$ by $(\rho\mathbf{V} - D_d \operatorname{grad} \rho)$.

(iii) *Conservation of volume of the system.* Let the property considered be the volume of the liquid system, so that the species α is the liquid as a whole. Then $g_\alpha = 1$ and $\mathbf{V}_\alpha \equiv \mathbf{V}$ (volume average velocity). Also, in an incompressible liquid, volume is not produced. We then have from (77) and (78),

$$\int_{(S)} \mathbf{V} \cdot d\mathbf{S} = \int_{(U)} \operatorname{div} \mathbf{V} = 0; \qquad \operatorname{div} \mathbf{V} = 0; \qquad \partial V_j/\partial x_j = 0 \tag{84}$$

Introducing the mass averaged velocity, we have, by (12),

$$\operatorname{div} \mathbf{V}^* = \operatorname{div} \mathbf{V} - \operatorname{div}[(D_d/\rho) \operatorname{grad} \rho]$$

which means that only for an incompressible liquid, $\rho = \text{const}$, we may write $\operatorname{div} \mathbf{V} = \operatorname{div} \mathbf{V}^* = 0$.

(iv) *Conservation of momentum of a species* α. In this case $g_\alpha = \rho_\alpha \mathbf{V}_\alpha$, and (78) becomes

$$\partial(\rho_\alpha \mathbf{V}_\alpha)/\partial t + \operatorname{div}(\rho_\alpha \mathbf{V}_\alpha \mathbf{V}_{\alpha m}) = I_{\alpha m} \tag{85}$$

where $I_{\alpha m}$ is the rate of production of the momentum density of the α particle, and $\mathbf{V}_{\alpha m}$ is the velocity of propagation of the momentum (in the flowing liquid) of the α particles. One should note here that since the property under consideration is momentum of α particles, the velocity in the divergence term is the velocity of propagation of that property.

(v) *Conservation of momentum of a system.* For the liquid system as a whole, we have $g_\alpha = \rho\mathbf{V}^*$, $\mathbf{V}_\alpha \equiv \mathbf{V}_m$, and (78) becomes

$$\partial(\rho\mathbf{V}^*)/\partial t + \operatorname{div}(\rho\mathbf{V}^*\mathbf{V}_m) = I_m \tag{86}$$

$$\int_{(U)} \frac{\partial(\rho\mathbf{V}^*)}{\partial t}\, dU = -\int_{(S)} \rho\mathbf{V}^*\mathbf{V}_m \cdot d\mathbf{S} + \int_{(U)} I_m\, dU \tag{87}$$

In (86), I_m is the rate of production of the momentum density and $\mathbf{V}_m$ is the velocity of propagation of the momentum in the system.

Equations (86) and (87) are based on the following expressions of momentum fluxes:

$_m\underline{\mathbf{J}} = \rho\mathbf{V}^*\mathbf{V}_m$, momentum flux with respect to fixed coordinates;

$_m\underline{\mathbf{J}}^* = \rho\mathbf{V}^*(\mathbf{V}_m - \mathbf{V}^*)$, momentum flux with respect to the mass average velocity;

$_m\underline{\mathbf{J}}^* = \rho\mathbf{V}^*\mathbf{V}_m - \rho\mathbf{V}^*\mathbf{V}^* = {}_m\underline{\mathbf{J}} - \rho\mathbf{V}^*\mathbf{V}^*$;

$_m\underline{\mathbf{J}}$ and $_m\underline{\mathbf{J}}^*$ are second-rank symmetric tensors. The products $\mathbf{V}^*\mathbf{V}^*$ and $\mathbf{V}^*\mathbf{V}_m$ are dyadic products.

These fluxes may now be expressed in the form

$$_m\underline{\mathbf{J}}^* = p\underline{\boldsymbol{\delta}} + \underline{\boldsymbol{\tau}}, \qquad {}_m\underline{\mathbf{J}} = {}_m\underline{\mathbf{J}}^* + \rho\mathbf{V}^*\mathbf{V}^*$$

where p is pressure, $\underline{\boldsymbol{\delta}}$ is the Kronecker delta, and $\underline{\boldsymbol{\tau}}$ is viscous part (due to shear) of the momentum flux tensor.

We may now rewrite (86) and (87) by applying the second law of mechanics which states that the rate of production of momentum density is equal to the resultant of external forces, in the form

$$\int_{(U)} \frac{\partial(\rho\mathbf{V}^*)}{\partial t}\, dU = -\int_{(S)} \rho\mathbf{V}^*\mathbf{V}_m \cdot d\mathbf{S} + \int_{(U)} \sum_{(\alpha)} \rho_\alpha \mathbf{F}_\alpha\, dU$$

$$\frac{\partial}{\partial t}(\rho\mathbf{V}^*) = -\operatorname{div}(\rho\mathbf{V}^*\mathbf{V}_m) + \sum_{(\alpha)} \rho_\alpha \mathbf{F}_\alpha$$

where $\mathbf{F}_\alpha$ is the external force (per unit mass) acting on the particles of the α component. We may also write

$$\frac{\partial}{\partial t}(\rho\mathbf{V}^*) = -\operatorname{div}(\rho\mathbf{V}^*\mathbf{V}^*) - \operatorname{grad} p - \operatorname{div} \underline{\boldsymbol{\tau}} + \sum_{(\alpha)} \rho_\alpha \mathbf{F}_\alpha$$

Let us now denote div $\underline{\boldsymbol{\tau}}$ (force per unit volume) by $\mathscr{R}$. It represents the density of the force, due to the liquid's viscosity, which resists the motion. Also, in the present case, $\sum_{(\alpha)} \rho_\alpha \mathbf{F}_\alpha = \rho\mathbf{g} = -\mathbf{1z}\rho g$. We may therefore replace (87) by

$$\int_{(U)} \frac{\partial(\rho\mathbf{V}^*)}{\partial t}\, dU = -\int_{(S)} \rho\mathbf{V}^*\mathbf{V}^* \cdot d\mathbf{S} - \mathbf{1z}\int_{(U)} \rho g\, dU - \int_{(S)} p\, d\mathbf{S} + \int_{(U)} \mathscr{R}\, dU \tag{88}$$

where $\mathbf{1z}$ is a unit vector directed vertically upward. Equation (88) is the equation of *momentum conservation* or the *equation of motion*.

All of the above equations are valid for any elementary volume U within the liquid continuum. As explained above, in a flow through a porous medium, the intricate microscopic solid surfaces which bound the liquid continuum are unknown to us. Fortunately, we are not interested in obtaining the local value of the variables involved. Instead, we wish to determine *average values* over representative elementary volumes. We shall use for that purpose the representative elementary volume ΔU_0 (REV) of the model defined above.

Let us first average the various parameters involved over the cross section of the elementary channels of the porous medium model and obtain the *local transport equations*. The second step will be to average over the REV.

Let $f(s)$ be some scalar variable within the pore space with s denoting the distance of the considered cross section along the channel's axis. We shall define the average of $f(s)$ over that cross section by

$$\langle f(s)\rangle = [w(s)]^{-1} \int_{(w)} f(s; \lambda_2, \lambda_3)\, d\lambda_2, d\lambda_3; \qquad d\lambda_2 d\lambda_3 = dw \tag{89}$$

where (λ_2, λ_3) are coordinates of a point on this cross section of area $w(s)$. These coordinates are chosen such that $\mathbf{1\lambda}_1$ is tangent to the channel's axis at its point of intersection with the considered cross section (and in the direction of increasing s), $\mathbf{1\lambda}_2$ is the major normal to the axis, and $\mathbf{1\lambda}_3$ is the binormal vector. If f is differentiable in the void space, we also have

$$\langle \operatorname{grad} f\rangle = \left\langle \frac{\partial f}{\partial s}\right\rangle \mathbf{1\lambda}_1 + \left\langle \frac{\partial f}{\partial \lambda_2}\right\rangle \mathbf{1\lambda}_2 + \left\langle \frac{\partial f}{\partial \lambda_3}\right\rangle \mathbf{1\lambda}_3$$

In what follows, x_i $(i = 1, 2, 3)$ is a cartesian rectangular coordinate system fixed in space. The equation of a channel's axis is $x_i = x_i(s)$, or $\mathbf{R} = \mathbf{R}(s)$, $0 \leqslant s \leqslant L$, where $\mathbf{R}(s)$ is the radius vector of a point on the axis. We shall use the notation ξ_i to indicate the coordinates in the x_i system of points on the channel's cross section, so that

$$\xi_i(\sigma) = x_i(s) + \lambda_j \cos(\mathbf{1\lambda}_j, \mathbf{1x}_i); \qquad ij = 1, 2, 3$$

and the summation convention is used. The parameter σ indicates the length along the particular streamline on which the considered point on the channel's cross section is located, from the cross section with $s = 0$. The two coordinate systems x_i and λ_i are related to each other also by the *Frenet–Serre equations*. With these definitions we have

$$\langle \mathbf{V}\rangle = \langle V_s\rangle\, \mathbf{1\lambda}_1 + \langle V_{\lambda_2}\rangle\, \mathbf{1\lambda}_2 + \langle V_{\lambda_3}\rangle\, \mathbf{1\lambda}_3; \qquad \mathbf{V} = V\mathbf{1\sigma}$$

where $\mathbf{1\sigma}$ is tangent to the streamline and

$$V_s = \mathbf{V} \cdot \mathbf{1\lambda}_1 = V\mathbf{1\sigma} \cdot \mathbf{1\lambda}_1 = V \frac{d\xi_i}{d\sigma} \cdot \frac{dx_i(s)}{ds}$$

If the curvature and tortuousity of the channel's axis are small, we have

$$V_s \approx V/(d\sigma/ds) \tag{90}$$

It can be shown that if $|f - \bar{f}| \ll \tilde{f}$ for a point on the channel's circumference, then

$$\frac{\partial\langle f\rangle}{\partial s} \simeq \left\langle\frac{\partial f}{\partial s}\right\rangle + (\tilde{f} - \langle f\rangle)\frac{\partial w}{w\,\partial s} \tag{91}$$

where $\tilde{f}$ denotes the average of f taken over the circumference of the cross section. We shall use (91), which becomes an exact expression for circular cross sections, in the following discussion.

To obtain average equations for a channel cross section, let us define an elementary volume U_0 of a channel between two cross sections s_1 and $s_2 > s_1$ such that

$$U_0 = \int_{s_i}^{s_2} w(s)\,ds$$

We may now apply (84) to this element of a channel, recognizing that at the solid wall, $\mathbf{V} = 0$. We obtain

$$\int_{(S)} \mathbf{V} \cdot d\mathbf{S} \equiv [\langle V_s\rangle\, w]_{s=s_2} - [\langle V_s\rangle\, w]_{s=s_1} = 0$$

Dividing by U_0 and passing to the limit as $s_2 \to s_1$,

$$\partial(\langle V_s\rangle\, w)/\partial s = 0 \tag{92}$$

In view of (91) and since on the solid surface $\mathbf{V} = 0$, we obtain the volume conservation equation averaged over the cross section in the form

$$\langle \partial V_s/\partial s\rangle = 0 \tag{93}$$

With (90) and (84), (93) may be written as

$$\begin{aligned}
\langle \partial V_s/\partial s\rangle &= \langle \partial V(d\sigma/ds)^{-1}/\partial s\rangle = \langle \partial V/\partial s(d\sigma/ds)^{-1} + V\partial(d\sigma/ds)^{-1}/\partial s\rangle \\
&= \langle(\partial V/\partial\xi_j)(d\xi_j/d\sigma) + V\partial(d\sigma/ds)^{-1}/\partial s\rangle \\
&= \langle V[\partial(d\sigma/ds)^{-1}/\partial s - \partial(d\xi_j/d\sigma)/\partial\xi_j]\rangle = 0
\end{aligned}$$

Hence,

$$\left\langle \frac{\partial}{\partial \xi_j}\left(\frac{d\xi_j}{d\sigma}\right)\right\rangle = \left\langle \frac{\partial}{\partial s}\left(\frac{ds}{d\sigma}\right)\right\rangle = -\left\langle \left(\frac{ds}{d\sigma}\right)^2 \frac{\partial}{\partial s}\left(\frac{d\sigma}{ds}\right)\right\rangle$$

and therefore,

$$\left\langle \frac{\partial}{\partial \xi_j}\left(\frac{d\xi_j}{ds}\right)\right\rangle = \left\langle \frac{\partial}{\partial \xi_j}\left(\frac{d\xi_j}{d\sigma}\frac{d\sigma}{ds}\right)\right\rangle = \left\langle \frac{\partial}{\partial \xi_j}\left(\frac{d\sigma}{ds}\right)\frac{d\xi_j}{d\sigma} + \frac{d\sigma}{ds}\frac{\partial}{\partial \xi_j}\left(\frac{d\xi_j}{d\sigma}\right)\right\rangle$$

$$= \left\langle \frac{\partial}{\partial \sigma}\left(\frac{d\sigma}{ds}\right) - \frac{ds}{d\sigma}\frac{\partial}{\partial s}\left(\frac{d\sigma}{ds}\right)\right\rangle = 0$$

Our next step is to average the mass conservation equation for a species α over a cross section of an elementary porous medium channel. We obtain from (79), with $I_\alpha \equiv 0$ and assuming no interactions at the interface between the solid phase and the liquid,

$$\langle \partial \rho_\alpha / \partial t \rangle = -\langle \partial[\rho_\alpha (V_\alpha)_s] / \partial s \rangle; \qquad (V_\alpha)_s = \mathbf{V}_\alpha \cdot \mathbf{1\lambda}_1 \tag{94}$$

From (10) and (90), we have

$$\rho (V_\alpha)_s = \rho_\alpha[(V_\alpha)_s - V_s] + \rho_\alpha V_s = -D_d \, \partial \rho_\alpha / \partial s + \rho_\alpha V_s$$
$$= -D_d(\partial \rho_\alpha / \partial \sigma)(d\sigma/ds) + \rho_\alpha V(ds/d\sigma) \tag{95}$$

Hence, inserting (95) in (94) and passing to fixed cartesian coordinates x_i, we obtain

$$\left\langle \frac{\partial \rho_\alpha}{\partial t}\right\rangle = \left\langle \frac{\partial}{\partial \xi_i}\left[D_d \left(\frac{\partial \sigma}{ds}\right)^2 T_{ij} \frac{\partial \rho_\alpha}{\partial \xi_j}\right]\right\rangle - \left\langle \frac{\partial}{\partial \xi_i}(\rho_\alpha V_i)\right\rangle$$
$$= \left\langle \frac{\partial}{\partial \xi_i}\left[D_d T_{ij}^* \frac{\partial \rho}{\partial \xi_j}\right]\right\rangle - \left\langle \frac{\partial}{\partial \xi_i}(\rho_\alpha V_i)\right\rangle \tag{96}$$

One should recall that ξ_i are the cartesian coordinates (in the x_i system) of points within the void space. In (96)

$$T_{ij}^* = T_{ij}(d\sigma/ds)^2; \qquad T_{ij} = (d\xi_i/d\sigma)(d\xi_j/d\sigma) \tag{97}$$

The interpretation of this parameter is as follows. No matter what is the direction of the average external driving force (such as pressure gradient, concentration gradient, etc.), the flow at a point inside a channel is restricted by the local configuration of the channel (direction of axis and divergence or convergence of the individual streamlines). Therefore, if any external force **F** is given, we have first to consider its projection along the streamline. This may be expressed as

$$\mathbf{F} \to \mathbf{F} \cdot \mathbf{1}\boldsymbol{\sigma} \quad \text{or} \quad \mathbf{F} \to F \cos(\mathbf{F}, \mathbf{1}\boldsymbol{\sigma}) \quad \text{or} \quad \mathbf{F} \to F_1 \cdot \frac{d\xi_i}{d\sigma}$$

Then the components $F_j^{(\sigma)}$ of this projection along the axes of the x_j system are derived

$$F_i \frac{d\xi_i}{d\sigma} \to F_j^{(\sigma)} = F_i \frac{d\xi_i}{d\sigma} \frac{d\xi_j}{d\sigma} \equiv T_{ij} F_i \qquad \text{(summation convention)}$$

T_{ij} is therefore a linear operator which transforms the components F_i of any vector in the x_i system, acting on a small volume of liquid at a point within a channel, into components $F_j^{(\sigma)}$ (in the x_i system) of the projection of this vector on the path lines. It is a symmetric tensor ($T_{ij} = T_{ji}$) of second rank. The operation $T_{ij}^* = (d\xi_i/ds)(d\xi_j/ds)$ is similar to T_{ij} except that the projection is on the direction of the channel's axis.

In a manner similar to that described above, starting from (75) we obtain the equation of mass conservation of the entire system, averaged over a channel's cross section, in the form

$$\left\langle \frac{\partial \rho}{\partial t} \right\rangle = -\left\langle \frac{\partial}{\partial \xi_i} (\rho V_i) \right\rangle + \left\langle \frac{\partial}{\partial \xi_i} \left[D_d T_{ij}^* \frac{\partial \rho}{\partial \xi_j} \right] \right\rangle \tag{98}$$

One could also obtain (98) by adding (97) for all species α.

Finally, we average over a channel the equation of momentum conservation (88) of the entire system. Written in a differential form for the component of the momentum along the channel's axis we obtain

$$\frac{\partial}{\partial t} (\rho V_s^*) = -\text{div}(\rho V_s^* \mathbf{V}^*) - \rho g \frac{\partial z}{\partial s} - \frac{\partial p}{\partial s} + \mathscr{R}_s \tag{99}$$

At this point we make an *assumption* regarding the resistance force $\mathscr{R}$. It is important to emphasize that this is an assumption which is in a way similar to the constitutive assumptions made by some authors. Here it is made at the "microscopic" level. Obviously, other assumptions are also possible, leading to different macroscopic equations. The only way to verify the validity of any assumption of this kind is the experiment.

Assuming that we deal with laminar flow, we follow the development of Poiseuille's law for flow in a capillary tube and assume

$$\mathscr{R} = -\mu \mathbf{V}^*/B \tag{100}$$

where μ is the dynamic viscosity of the liquid at the considered point and

B is the conductance of the channel (dim L^2) at the point. In Poiseuille's law for a capillary tube, $B = (a^2 - r^2)/4$ where r is the distance of the considered point from the tube's axis and a is the radius of the tube. Therefore, B depends in general on the shape of the channel's cross section and on the location of the considered point on it.

With (100), (99) becomes

$$\frac{\partial}{\partial t}(\rho V_s^*) = -\mathrm{div}(\rho V_s^* \mathbf{V}^*) - \rho g \frac{\partial z}{\partial s} - \frac{\partial p}{\partial s} - \frac{\mu V_s^*}{B} \tag{101}$$

When (101) is combined with the equation of mass conservation (80), and assuming that because of the small channel curvature and cross section variations, $\rho \mathbf{V}^* \cdot \mathrm{grad}\, V_s^* \sim 0$, and introducing $V_s^* = V^*/(d\sigma/ds) = V^*/\sec\theta$, we obtain

$$\mu V_s^* + B\rho \frac{\partial V_s^*}{\partial t} = -B \sec\theta \left(\frac{\partial p}{\partial s} + \rho g \frac{\partial z}{\partial s}\right) \tag{102}$$

Without this assumption, $\partial V_s^*/\partial t$ in (102) should be replaced by dV_s^*/dt.

Multiplying by $d\xi_i/d\sigma$ in order to pass to a cartesian coordinate system x_i $(i = 1, 2, 3)$ and averaging over a cross section, we obtain

$$\left\langle V_i^* + \frac{B}{\nu}\frac{\partial V_i^*}{\partial t}\right\rangle = -\left\langle \frac{B \sec^2\theta T_{ij}}{\mu}\left(\frac{\partial p}{\partial \xi_j} + \rho g \frac{dz}{\partial \xi_j}\right)\right\rangle \tag{103}$$

where

$$\frac{\partial p}{\partial s} = \frac{\partial p}{\partial \xi_j}\frac{d\xi_j}{d\sigma}\frac{d\sigma}{ds}; \qquad \frac{\partial z}{ds} = \frac{\partial z}{\partial \xi_j}\frac{d\xi_j}{d\sigma}\frac{d\sigma}{ds}; \qquad \frac{d\sigma}{ds} = \sec\theta$$

Equation (103) is an expression relating the velocity to the driving external forces; it is therefore the *equation of motion* for the liquid averaged over a channel cross section.

Our final step is to average (93), (97), (98), and (103) over an REV which represents the physical point within the porous medium. With (74) applied to the volume of the REV, we obtain (deleting the subscript α)

$$G = \int_{(n\Delta U_0)} g\, dU_v = \int_{(\Delta U)_{\mathrm{chan}}} g\, dU_{\mathrm{chan}} + \int_{(\Delta U)_{\mathrm{junct}}} g\, dU_{\mathrm{junct}} \tag{104}$$

Neglecting $(\Delta U)_{\mathrm{junct}}$, we obtain

$$G \approx \int_{(\Delta U)_{\mathrm{chan}}} g\, dU_{\mathrm{chan}} = \sum_{(L)} \int_0^L \langle g \rangle\, w(s)\, ds \tag{105}$$

The average over the void space of the REV (considering only the volume of channels U_{chan} and neglecting the volume of junctions U_{junct}), is then expressed for point P by

$$\bar{g}(P) = [n(\Delta U_0)_P]^{-1} \int_{(n\Delta U_0)_P} g\, dU_v \tag{106}$$

In this way, $\bar{g}$ is defined for every point in the porous medium system; this is in accordance with the continuum approach. For the properties g considered here, we assume that $\bar{g}$ is differentiable (up to second order) except at exceptional points, curves, or surfaces. The product $n\bar{g}$ is the macroscopic density, i.e., per unit bulk volume of the medium.

From (106) and (89), it also follows that neglecting the junctions' volume,

$$\overline{\langle g \rangle} \cdot (n\, \Delta U_0) = \int_{(n\Delta U_0)} \langle g \rangle\, dU_v = \int_{(n\Delta U_0)} \langle g \rangle\, w(s)\, ds = \sum_{(L)} \int_0^L \langle g \rangle\, w(s)\, ds$$

$$= \sum_{(L)} \int_0^L \left[w(s)^{-1} \int_{(w)} g\, dw \right] w(s)\, ds = \int_{(n\Delta U_0)} g\, dU_v$$

or

$$\overline{\langle g \rangle} = \bar{g}$$

The value of g at a point within the void space may be represented as the average value of the REV and a local deviation

$$g = \bar{g} + \mathring{g}; \qquad \overline{\mathring{g}} = 0 \tag{107}$$

For products of parameters g and b, we have

$$\overline{gb} = \bar{g}\bar{b} + \overline{\mathring{g}\mathring{b}}; \qquad \overline{g^2} = \bar{g}^2 + \overline{\mathring{g}^2} \tag{108}$$

We have applied here the following definitions from probability theory:

mean,

$$\bar{g} = E(g) = \int_{-\infty}^{+\infty} g f(g)\, dg$$

variance,

$$\operatorname{Var}(g) \equiv \sigma^2(g) = E[(g - \bar{g})^2] = \int_{-\infty}^{+\infty} (g - \bar{g})^2 f(g)\, dg = \overline{g^2} - \bar{g}^2 = \overline{\mathring{g}^2}$$

covariance (or moment of correlation)

$$\text{Cov}(g, b) = \int_{-\infty}^{+\infty}\int_{-\infty}^{+\infty} (g - \bar{g})(b - \bar{b}) f(g, b)\, dg\, db = \overline{\mathring{g}\mathring{b}} = \overline{gb} - \bar{g}\bar{b}$$

The variance $\overline{\mathring{g}^2}$ measures the dispersion of g around $\bar{g}$.

The actual execution of the averaging by performing the integration is impossible, since no information is available as to the exact distribution of the values of g within the void space of the REV. Instead we shall treat the variable g as a random variable and consider only its probability distribution. Moreover, since we are only interested in the (measurable) mean values themselves, there is even no need to specify a particular distribution function. All that is needed is to make some general assumptions regarding the moments of correlation between the various variables represented by g and b. For example, if we assume that g and b are independent of each other, we have $\overline{\mathring{g}\mathring{b}} = 0$.

From the definition of the average it follows that $\overline{\partial g/\partial t} = \partial \bar{g}/\partial t$. By analogy to (91) we may assume

$$\overline{\frac{\partial g}{\partial \xi_j}} = \frac{\partial \bar{g}}{\partial x_j} + (\bar{g} - \bar{\bar{g}})\frac{\partial \ln(n\, \Delta U_0)}{\partial x_j} \tag{109}$$

where $\bar{\bar{g}}$ is the average of g taken over the solid surface of the medium. For $g = \rho, \mu, p$, and ρ_α, we shall assume $\bar{g} = \bar{\bar{g}}$, so that $\overline{\partial g/\partial \xi_j} = \partial \bar{g}/\partial \xi_j$. However, for V we have $V \equiv 0$ on the solid surface and therefore, $\bar{\bar{V}} = 0$. Hence,

$$\left(\overline{\frac{\partial V}{\partial \xi_j}}\right) = \frac{\partial \bar{V}}{\partial x_j} + \bar{V}\frac{\partial \ln(n\, \Delta U_0)}{\partial x_j} = \frac{\partial \bar{V}}{\partial x_j} + \bar{V}\frac{\partial(\ln \Delta U_{0v})}{\partial x_j} \tag{110}$$

where ΔU_{0v} $(= n\, \Delta U_0)$ is the volume of void space within ΔU_0. Since $\partial V_j/\partial \xi_j \equiv 0$, we obtain from (110),

$$\frac{\partial \bar{V}_j}{\partial x_j} = -\bar{V}_j \frac{\partial(\ln \Delta U_{0v})}{\partial x_j}; \qquad \frac{\partial}{\partial x_j}(\bar{V}_j\, \Delta U_{0v}) = 0 \tag{111}$$

In a homogeneous medium, $\Delta U_{0v} =$ const, and therefore, the equation expressing incompressibility of fluid in a porous medium is

$$\partial \bar{V}_j/\partial x_j = 0 \tag{112}$$

From (111), assuming $\Delta U_0 =$ const, we also have

$$\partial(\bar{V}_j n)/\partial x_j = 0 \qquad \text{or} \qquad \partial q_j/\partial x_j = 0; \qquad q_j = n\bar{V}_j \tag{113}$$

If ΔU_{0v} varies, e.g., in an inhomogeneous medium, (111) remains valid.

Next, we apply the averaging procedure described above to (96). We obtain, in view of (109),

$$\frac{\partial \bar{\rho}_\alpha}{\partial t} = \frac{\partial}{\partial x_i}\left[D_d \overline{T^*_{ij}} \frac{\partial \bar{\rho}_\alpha}{\partial x_j}\right] - \frac{\partial}{\partial x_j}(\overline{\rho_\alpha V_i}) - \overline{\rho_\alpha V_i} \frac{\partial \ln(\Delta U_{0v})}{\partial x_i} \tag{114}$$

The product $\rho_\alpha V_i$ represents the instantaneous species α mass flux. It is made up of two parts

$$\rho_\alpha V_i = \rho_\alpha \bar{V}_i + \rho_\alpha \mathring{V}_i \tag{115}$$

where $\rho_\alpha \bar{V}_i$ is the flux carried by the average fluid motion through the REV; $\rho_\alpha \mathring{V}_i$ is the *dispersive flux* resulting from the velocity fluctuations. One should note that the dispersive flux is different from the *diffusive flux* appearing in the square brackets of the first term on the right-hand side of (114). The solute's mass dispersive flux expresses the rate at which mass is transported because of the velocity distribution in a field in which the solute's mass distribution may be taken instantaneously as stationary ($\partial \rho_\alpha / \partial t = 0$). Also for average fluxes we have

$$\overline{\rho_\alpha V_i} = \bar{\rho}_\alpha \bar{V}_i + \overline{\mathring{\rho}_\alpha \mathring{V}_i}$$

Assuming that (1) there is no correlation between the velocity of a liquid particle at a point and the solute's concentration gradient, i.e., $\overline{\mathring{V}_i (\partial \mathring{\rho}_\alpha / \partial \xi_j)} = 0$, and (2) that the mean of the product of three or more deviations is zero, Bear and Bachmat (1965, 1966) obtain from (83)

$$\begin{aligned}\overline{\mathring{\rho}_\alpha \mathring{V}_i} \equiv \overline{\rho_\alpha \mathring{V}_i} &= -\overline{(V_\alpha)_j \frac{\partial \rho_\alpha}{\partial x_j} (\operatorname{div} \mathbf{V})^{-1} \mathring{V}_i} \\ &\approx -\overline{(\operatorname{div} \mathbf{V}_\alpha)^{-1}} \, \overline{\frac{\partial \rho_\alpha}{\partial x_j}} \, \overline{(\mathring{V}_\alpha)_j \mathring{V}_i} - \overline{(\mathbf{V}_\alpha)_j} \, \overline{\frac{\partial \rho_\alpha}{\partial x_j}} \, \overline{(\operatorname{div} \mathbf{V}_\alpha)^{-1} \mathring{V}_i}\end{aligned} \tag{116}$$

Neglecting the effect of the second term on the right-hand side of (116) and introducing (10) in the form

$$(V_\alpha)_j = [(V_\alpha)_j - V_j] + V_j = -\frac{D_d}{\rho_\alpha} \frac{\partial \rho_\alpha}{\partial \xi_j} + V_j \tag{117}$$

we obtain from (116),

$$\overline{\rho_\alpha \mathring{V}_i} \approx \overline{\mathring{V}_i \mathring{V}_j} \, (\operatorname{div} \mathbf{V})^{-1} \frac{\partial \bar{\rho}_\alpha}{\partial x_j} = -D_{ij} \frac{\partial \bar{\rho}_\alpha}{\partial x_j} \tag{118}$$

where D_{ij}, to be considered in detail below, is defined by

$$D_{ij} \approx \overline{\mathring{V}_i \mathring{V}_j} \, \overline{(\text{div } \mathbf{V})^{-1}} \tag{119}$$

By inserting (115) and (119) in (114), we obtain

$$\frac{\partial \bar{\rho}_\alpha}{\partial t} = \frac{\partial}{\partial x_i}\left[(D_{ij} + D_d \overline{T_{ij}^*}) \frac{\partial \bar{\rho}_\alpha}{\partial x_j}\right] - \bar{V}_i \frac{\partial \bar{\rho}_\alpha}{\partial x_i} + D_{ij} \frac{\partial \ln(\Delta U_{0v})}{\partial x_i} \frac{\partial \bar{\rho}_\alpha}{\partial x_j} \tag{120}$$

where $\overline{T_{ij}^*} = T_{ij}(d\sigma/ds)^2$. Equation (120) is the *equation of mass conservation of an* α *species, averaged over an REV of a porous medium.* Equation (120) can also be rewritten in terms of q_i by replacing $\bar{V}_i$ by q_i/n.

In a homogeneous medium, ΔV_{0v} = const and (120) becomes

$$\frac{\partial \bar{\rho}_\alpha}{\partial t} = \frac{\partial}{\partial x_i}\left[(D_{ij} + D_d \overline{T_{ij}^*}) \frac{\partial \bar{\rho}_\alpha}{\partial x_i}\right] - \bar{V}_i \frac{\partial \bar{\rho}_\alpha}{\partial x_i} \tag{121}$$

For a stationary fluid, $\bar{\mathbf{V}} \equiv 0$ and hence, $D_{ij} = 0$; (121) becomes

$$\frac{\partial \bar{\rho}_\alpha}{\partial t} = \frac{\partial}{\partial x_i}\left[D_d \overline{T_{ij}^*} \frac{\partial \bar{\rho}_\alpha}{\partial x_i}\right] \tag{122}$$

which is the *equation of molecular diffusion of a solute in a porous medium.*

Finally, we may rewrite (121) using vector and tensor notation

$$\frac{\partial \bar{\rho}_\alpha}{\partial t} = \text{div}[(\underline{\mathbf{D}} + D_d \overline{\underline{\mathbf{T}}^*}) \text{ grad } \bar{\rho}_\alpha - \bar{\mathbf{V}}\bar{\rho}_\alpha] = \text{div}[(\underline{\mathbf{D}} + \underline{\mathbf{D}}_d^*) \text{ grad } \bar{\rho}_\alpha - \bar{\mathbf{V}}\bar{\rho}_\alpha] \tag{123}$$

In (123), $\underline{\mathbf{D}}$ (second-rank tensor; dim L^2T^{-1}) is the *coefficient of mechanical (or convective) dispersion,* D_d is the coefficient of molecular diffusion (dim L^2T^{-1}), $\underline{\mathbf{T}}^*$ (second-rank tensor, dimensionless) is the *medium's tortuousity*, and $\underline{\mathbf{D}}_d^* = D_d\underline{\mathbf{T}}^*$. They will be discussed in detail in Section 3.

It can be shown that the normal tracer distribution given by (46) or (49) is an elementary solution of the dispersion equation where $\sigma_{x_i}^2 = 2D_{ii}t$.

By adding (120) over all the α's, we obtain

$$\frac{\partial \bar{\rho}}{\partial t} = \frac{\partial}{\partial x_i}\left[(D_{ij} + (D_d^*)_{ij} \frac{\partial \bar{\rho}}{\partial x_j}\right] - \bar{V}_i \frac{\partial \bar{\rho}}{\partial x_i} + D_{ij} \frac{\partial \ln(\Delta U_{0v})}{\partial x_i} \frac{\partial \bar{\rho}}{\partial x_j} \tag{124}$$

This is the *equation of mass conservation of a nonhomogeneous liquid in a porous medium,* averaged over an REV. It can also be obtained by averaging (98).

Turning now to the motion equation, let us continue by averaging (103) over the REV. Assuming that (1) no correlation exists between fluid properties at a point and the matrix geometrical properties at that point [i.e., $\overline{\mathring{B}(\rho/\mathring{\mu})} = 0$, $\overline{(B\mathring{T}^*_{ij})(\rho/\mathring{\mu})} = 0$], (2) no correlation exists between the pressure gradient acting on a liquid at a point in the void space and the direction of the liquid's path at that point [i.e., $\overline{(B\mathring{T}^*_{ij})(\partial p/\partial\mathring{\xi}_i)} = 0$], (3) $(\rho/\mathring{\mu})(\partial\mathring{V}^*_i/\partial t) = 0$, (4) the liquid is a dilute solution such that with $|\mathring{\rho}/\bar{\rho}| \ll 1$ and $|\mathring{\mu}/\bar{\mu}| \ll 1$, we have

$$\overline{(\rho/\mu)} = \bar{\rho}/\bar{\mu} = 1/\bar{\nu} = \mathbf{1}(\bar{\mu}/\bar{\rho}), \qquad \overline{(1/\mu)} = 1/\bar{\mu}$$

we obtain

$$\overline{V}^*_i + \frac{\bar{B}\bar{\rho}}{\bar{\mu}}\frac{\partial \overline{V}^*_i}{\partial t} = -\frac{\overline{BT^*_{ij}}}{\bar{\mu}}\left(\frac{\partial\bar{p}}{\partial x_j} + \bar{\rho}g\frac{\partial z}{\partial x_j}\right) \tag{125}$$

Equation (125) is the *average motion equation for a nonhomogeneous liquid in laminar flow through a porous medium.* When written in terms of the volumetric averaged velocity, we obtain

$$\overline{V}_i - D_d\overline{T^*_{ij}}\left(\overline{\frac{1}{\rho}\frac{\partial\rho}{\partial x_j}}\right) + \frac{\bar{B}\bar{\rho}}{\bar{\mu}}\frac{\partial}{\partial t}\left[\overline{V}_i - D_d\overline{T^*_{ij}}\left(\overline{\frac{1}{\rho}\frac{\partial\rho}{\partial x_i}}\right)\right]$$
$$= -\frac{k_{ij}}{n\bar{\mu}}\left(\frac{\partial\bar{p}}{\partial x_j} + \bar{\rho}g\frac{\partial z}{\partial x_j}\right) \tag{126}$$

where we introduced the *medium's intrinsic permeability*

$$k_{ij} = n\overline{BT^*_{ij}} \tag{127}$$

Assuming that the mass transport (of the total liquid!) by convection is much larger than that due to molecular diffusion, i.e., $|D_d\underline{\mathbf{T}}^* \operatorname{grad}\rho| \ll \rho\mathbf{V}^*$ and $q_i = n\overline{V}^*_i \approx n\overline{V}_i$, we obtain

$$q_i + \frac{\bar{B}}{\bar{\nu}}\frac{\partial q_i}{\partial t} = -\frac{k_{ij}}{\bar{\mu}}\left(\frac{\partial\bar{p}}{\partial x_j} + \bar{\rho}g\frac{\partial z}{\partial x_j}\right) \tag{128}$$

which expresses the *specific discharge* $\mathbf{q}$ of a nonhomogeneous liquid in laminar flow (i.e., neglecting $\mathbf{V}\cdot\operatorname{grad}V_i$) in a non homogeneous non-isotropic porous medium. In most cases we may neglect the entire inertial force (expressed by $\rho\, d\overline{\mathbf{V}}/dt$) with respect to the viscous forces, so that the motion equation becomes

$$q_i = -\frac{k_{ij}}{\bar{\mu}}\left(\frac{\partial\bar{p}}{\partial x_j} + \bar{\rho}g\frac{\partial z}{\partial x_j}\right) \tag{129}$$

which is the extension of Darcy's classical law to inhomogeneous liquids. The relationship between q and $\bar{\mathbf{V}}^*$ requires a proof which is not presented here (see, for example, Bear and Bachmat 1965, 1966).

Had we left $\rho\, dV_s/dt$ in (102), the resulting equation (128) would have included an additional term expressing the convective acceleration.

3. Parameters of Dispersion

3.1. The Coefficients of Mechanical and Hydrodynamic Dispersion

The coefficient of mechanical dispersion $\underline{\mathbf{D}}$ appearing in the dispersion equation (123) has been analyzed by various investigators. It depends on the flow pattern, e.g., through the velocities, on the Peclet number, and on some basic medium characteristics. Many investigators consider the sum $\underline{\mathbf{D}}' = (\underline{\mathbf{D}} + \underline{\mathbf{D}}_d^*)$ as the coefficient of dispersion (or rather, of hydrodynamic dispersion) which depends on the velocity, molecular diffusion, and medium characteristics. Equation (24) may serve as an example. Obviously, works dealing with essentially one-dimensional flow cannot recognize the tensorial nature of $\underline{\mathbf{D}}'$ or of $\underline{\mathbf{D}}$ (when molecular diffusion is neglected).

Taylor (1953) in his one-dimensional analysis obtains D proportional to $\bar{V}^2$. Bear and Todd (1960) in their one-dimensional analysis suggested $D = a_1\bar{V}$, i.e., D proportional to $\bar{V}$, with a_{I} being some characteristic medium length. In the analysis of variances given in Section 2.3, the dispersion, or the spread of particles around their average displacement, is described by the matrix of covariances $\overline{\mathring{x}_i(t)\, \mathring{x}_j(t)}$. There it is shown that these covariances may be related to the coefficient of dispersion D_{ij} for large times in the form $D_{ij} = r_{ij}(0)\, \overline{\mathring{V}_i\mathring{V}_j} t_0$, where t_0 is the time during which velocities are still effectively correlated. If, following Nikolaevskij (1959), we let $r_{ij}(0)t = L/\bar{V}$, where $\bar{V}$ is the absolute value of the average velocity and L is a characteristic medium length (e.g., length of a channel is a porous medium model), we obtain

$$D_{ij} = (\overline{\mathring{V}_i\mathring{V}_j}/\bar{V})L \tag{130}$$

i.e., $\underline{\mathbf{D}}$ is proportional to the first power of the velocity.

Scheidegger (1957) summarizes his analysis on these two possible relationships between D and $\bar{V}$ according to the role played by molecular diffusion.

(a) $D \sim a'\bar{V}^2$, where a', which is a constant of the porous medium alone (dynamical dispersivity), is derived by a dynamic procedure applicable where there is enough time in each flow channel for appreciable mixing by molecular transverse diffusion to take place.

(b) $D \sim a''\bar{V}$, where a'', another constant of the porous medium (geometrical dispersivity), is derived by a geometric procedure applicable where there is no appreciable molecular transverse diffusion from one streamline into another.

Thus, in all models in which the combined effect of a velocity distribution across a channel and transverse molecular diffusion are considered (e.g., Taylor, 1953), the coefficient of dispersion is proportional to $\bar{V}^2$. Where only the mean motion in a channel is considered while mixing occurs at junctions connecting different channels, disregarding molecular diffusion, one obtains $D \sim \bar{V}$. Of course, intermediate cases where D is proportional to some power of the velocity between 1 and 2, lie between these two extremes.

Blackwell (1959), and many others, study the relationship between the coefficient of dispersion, the velocity and the molecular diffusion. Pfannkuch (1962) summarizes 175 experimental values obtained in one-dimensional flow by Rifai *et al.* (1956), Day (1956), Ebach and White (1958), Carberry and Bretton (1958), Blackwell (1959), Blackwell *et al.* (1959), Raimondi *et al.* (1959), and Brigham and Reed (1959). By dimensional analysis one can show that in general the *coefficient of hydrodynamic dispersion* $\underline{\mathbf{D}}' = (\underline{\mathbf{D}} + \underline{\mathbf{D}}_d{}^*)$, appearing in (123), which includes the effects of both mechanical (or convective) dispersion and moledular diffusion, is a function of the *Peclet number* of molecular diffusion $P_e \equiv Vd/D_d = (q/n)\, d/D_d$ where d is the mean grain size or any other characteristic medium length and D_d replaces the coefficient of thermal conductivity in the Peclet number defined in heat conduction studies.

Figure 8 gives a schematic representation of the various experimental results mentioned above. In it, D_L' is the coefficient of hydrodynamic dispersion as obtained from the experimental results. The subscript L denotes here "longitudinal" since we deal with one-dimensional flow. The common experiment is one in which a fluid at constant concentration is introduced at one end of a sand-packed column of length L. The concentration of the effluent is recorded in the form of Fig. 9. From this figure and from the corresponding one-dimensional solution (Section 4.3), we obtain $D_L' = \frac{1}{2}\sigma^2 Lq/n$. In this way, D_L' involves the effects of both dispersion and diffusion in an unseparable form.

Figure 8 may be divided into several zones:

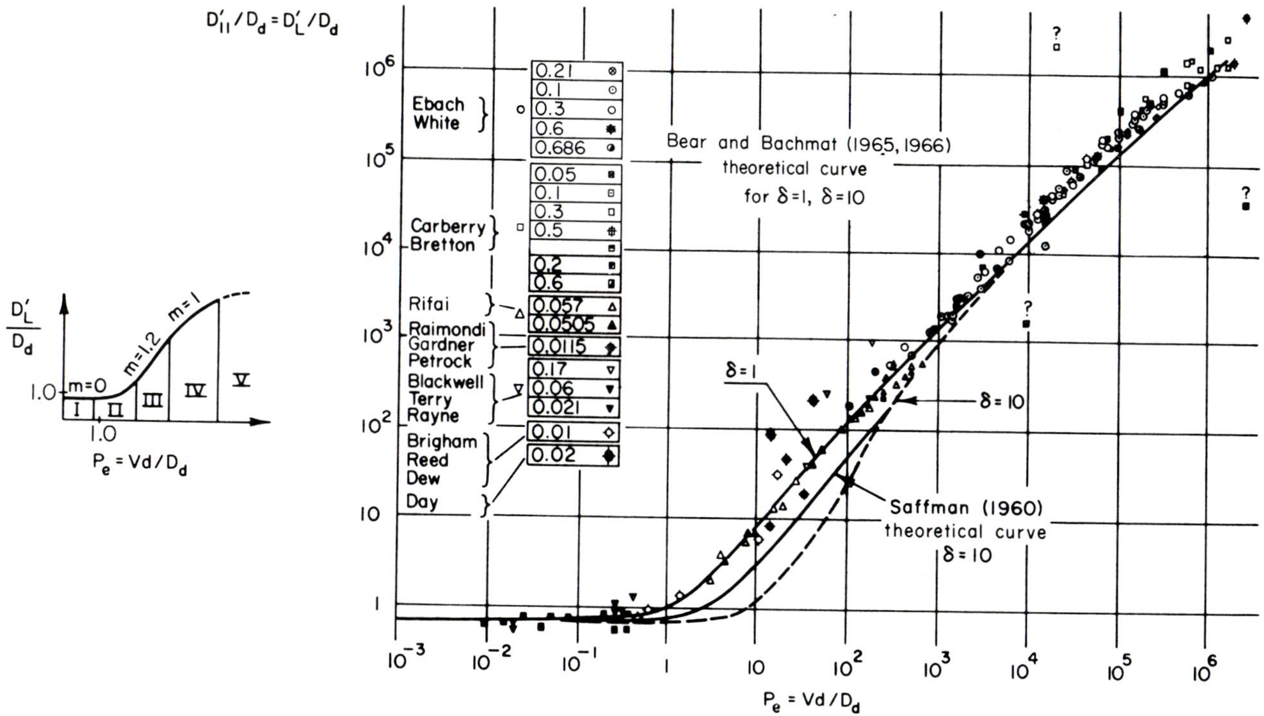

FIG. 8. Relationship between molecular diffusion and convective dispersion (after Pfannkuch, 1962; Saffman, 1960).

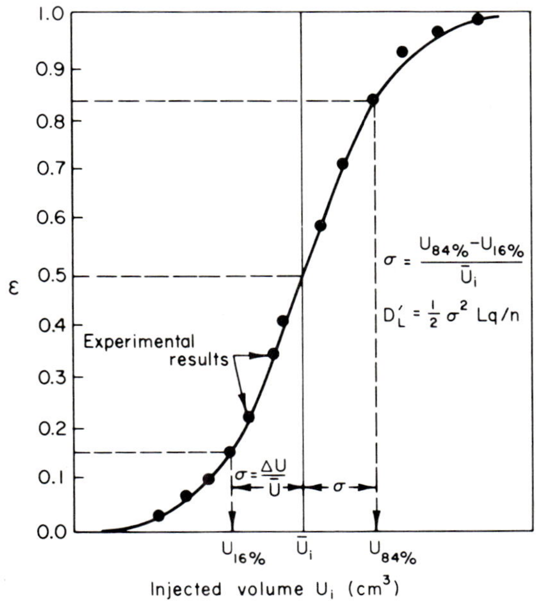

FIG. 9. Computing D_L' from a one-dimensional dispersion experiment.

Zone I: This is a zone where molecular diffusion predominates and $D_L'/D_d = f(P_e) =$ const. Here $a_1q/n \ll T^*D_d$. The effect of the porous medium is only through $T^*(<1)$, which modifies the molecular diffusion in the bulk to yield $D_d^* = T^*D_d$ for the porous medium. To be in this zone, the time of travel through a pore must be equal to or longer than t_0, required for molecular diffusion to affect mass transport in the longitudinal direction

$$t > t_0 ; \qquad t_0 \propto d^2/2D_d$$

An experiment with $q = 0$ may be used to obtain D_d^* and from this the value of T^*. Saffman (1959) obtains in his model for this case $D_L' = T^*D_d = \frac{1}{3}D_d$. However, for an actual porous medium, a value of T^* of approximately $\frac{2}{3}D_d$ (for loose sand) seems a better value.

Zone II: This zone corresponds to a Peclet number of diffusion between 0.4 and 5 (approx.). In this zone the effect of molecular diffusion is of the same order of magnitude as that of mechanical dispersion and the sum of both should be considered.

Zone III: Here the main spreading is caused by mechanical dispersion combined with transversal molecular diffusion. The two mechanisms are no longer additive as they interfere with each other.

In this zone,

$$t_1 < t \ll t_0\,; \qquad t_1 = R^2/8D_d$$

Experimental results yield for this range,

$$D_L'/D_d = \alpha(P_e)^m; \qquad \alpha \approx 0.5; \qquad 1 < m < 1.2 \tag{131}$$

In this zone, transversal molecular diffusion tends to reduce the longitudinal spreading. Obviously, theories on dispersion in which molecular diffusion is neglected, or in which molecular diffusion is considered as an additive term, fail to explain the behavior in this region.

Zone IV: This is a region of dominant mechanical dispersion (as long as we stay in the range of validity of Darcy's law). The effects of molecular diffusion are negligible. In the diagram we obtain a straight line at 45°:

$$D_L'/D_d = \beta P_e\,; \qquad \beta \approx 1.8 \tag{132}$$

Zone V: This is again a zone of pure mechanical dispersion, but the effects of inertia and turbulence may no more be neglected. Their role is equivalent to the role of transversal molecular diffusion in zone III. The slope of the curve in this zone tends to be less than one.

Fewer experiments are available on the coefficient of transversal dispersion. Denoting the coefficient of transversal hydrodynamic dispersion by D_T', one should expect the relationship between D_T'/D_d and P_e to be similar to that of D_L'/D_d (Saffman, 1960).

In the first studies (e.g., Scheidegger 1954), dispersion was taken as isotropic, i.e., D was considered a scalar. However, even in isotropic porous media we observe longitudinal (i.e., in the direction of the flow) and transversal (i.e., in the direction perpendicular to the flow) dispersion which differ from each other. When generalized to anisotropic media, the problem becomes even more complicated. We have seen in this section and in Section 2.3 that the coefficient of mechanical dispersion D_{ij} is a second-rank symmetric tensor which depends on the characteristics of the medium and on the flow. For the sake of simplicity let us consider here only the range where $D \sim V$.

Bear (1960) observed in determining the tracer distribution resulting from a point injection, in a uniform field of flow, that the components V_i of the velocity vector $\mathbf{V}$, or L_i of the displacement $\mathbf{L}$, play no role in the analysis. He suggested that this is probably due to the nature of the porous medium with its interconnected channels distributed at random in all directions, and to the fact that the true velocity in the channels (neglecting molecular diffusion) causes the dispersion, and not the average velocity **V** over the entire cross section and its components (say in a cartesian coordinate system x_i). Since in an isotropic medium the dispersion in the direction of the mean displacement **L** and perpendicular to it [Fig. 10(a)] can be obtained from the longitudinal and the transversal

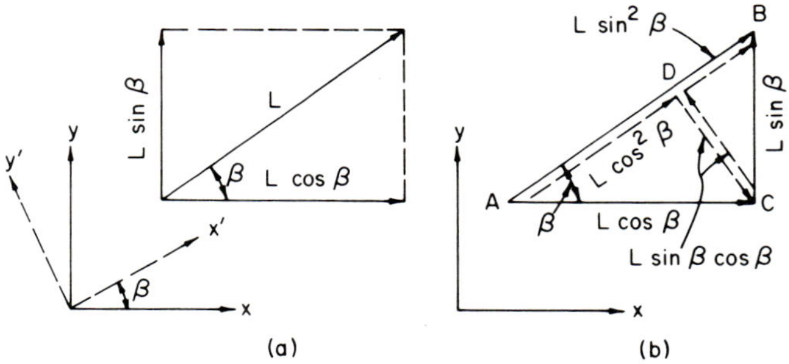

FIG. 10. The resolution of the displacement L.

coefficients of dispersion, $D_L t = a_I L$ and $D_T t = a_{II} L$, where a_I and a_{II} are medium constants, but not from the components L_i of **L** in the x_i system [x, y in the two dimensional case of Fig. 10(a)], Bear suggested to further resolve the components L_i and take their projections L_{ij} on the direction of the flow and perpendicular to it. Thus the displacement is defined as a second-rank tensor which in two dimensions takes the form:

$$[L_{ij}] = \begin{bmatrix} L\cos^2\beta & L\sin\beta\cos\beta \\ L\sin\beta\cos\beta & L\sin^2\beta \end{bmatrix}$$

$$= L\begin{bmatrix} \cos^2\beta & \sin\beta\cos\beta \\ \sin\beta\cos\beta & \sin^2\beta \end{bmatrix} \tag{133}$$

In a three-dimensional space, $i, j = 1, 2, 3$. Then, since in our case the covariance $(\sigma^2)_{kl}$, which defines D_{kl}, e.g., in (44) extended to $l \neq k$, transforms as a second-rank tensor, Bear states as a working hypothesis that

$$(\sigma^2)_{kl} = 2a_{ijkl}L_{ij} = 2D_{kl}t \tag{134}$$

where a_{ijkl} (a fourth-rank tensor) is the *medium's (geometrical) dispersivity*. In general, a_{ijkl} ($i, j, k, l = 1, 2, 3$) contains 81 components, of which all but those with four equal indices or with two pairs of equal indices, are equal to zero.

Equation (133) may be rewritten for a field of uniform flow in the form

$$[L_{ij}] = \frac{L}{V^2}\begin{bmatrix}(V\cos\beta)^2 & V\sin\beta\cdot V\cos\beta \\ V\sin\beta\cdot V\cos\beta & (V\sin\beta)^2\end{bmatrix}$$

$$= \left[\frac{V_iV_j}{V}\right]t; \qquad \frac{L}{V} = t \tag{135}$$

Hence, we obtain from (134) and (135),

$$D_{kl} = a_{ijkl}V_iV_j/V \tag{136}$$

where the summation convention is assumed.

The general form of (136) has also been obtained by Nikolaevskij (1959), starting from an analogy with the statistical theory of turbulence. He states that for an isotropic medium, the point tensor of dispersion must be a tensor of even rank (Batchelor, 1953). He thus obtains (136) as the simplest form which satisfies this requirement.

Poreh (1965) also deals with the tensor form of D_{ij} and its relationship to the components V_i. From macroscopic symmetry considerations following the work of Robertson (1940), he obtains the general forms for an isotropic medium

$$D_{ij} = a\delta_{ij} + bV_iV_j \tag{137}$$

where a and b are considered continuous functions of V^2. Expanding D_{ij} in a power series he obtains

$$D_{ij} = a_0\delta_{ij} + a_1V^2\delta_{ij} + b_0V_iV_j + O(V^4) \tag{138}$$

where a_0, a_1, b_0 are constants depending on the properties of the medium, the fluid, and the tracer; δ_{ij} is the Kronecker delta.

For an anisotropic medium with an axis of symmetry in the direction $\mathbf{1\lambda}$, Poreh (1965) obtains

$$D_{ij} = a'\delta_{ij} + b'V_iV_j + c'\lambda_i\lambda_j + d'(V_i\lambda_j + V_j\lambda_i) \tag{139}$$

where a', b', etc. are functions of V^2 and of $V\cos(\mathbf{V}, \mathbf{1\lambda})$.

3.2. The Medium's Dispersivity

For an isotropic medium, Bear (1960) relates the medium's dispersivity a_{ijkl} to the two constants: a_{I} = *longitudinal dispersivity of the medium*, and a_{II} = *transversal dispersivity of the medium*:

$$a_{1111} = a_{2222} = a_{\mathrm{I}}; \qquad a_{1122} = a_{2211} = a_{\mathrm{II}}; \qquad a_{1112} = a_{1121} = a_{1211}$$

$$= a_{1222} = a_{2111} = a_{2122} = a_{2212} = a_{2221} = 0;$$

$$a_{1212} = a_{1221} = a_{2112} = a_{2121} = \tfrac{1}{2}(a_{\mathrm{I}} - a_{\mathrm{II}}).$$

where subscript 2 denotes any axis perpendicular to the direction of the mean flow. In three dimensions, we have to add similar terms, but with subscript 3 replacing 2.

Scheidegger (1961) studies the symmetry properties of a_{ijkl} and extends the above analysis to the general case of an anisotropic medium. However, the number of constants in specific types of symmetry is not given.

Bachmat and Bear (1964) discuss the general form of D_{ij} in an isotropic medium in any coordinate system. They show that the principal axes of D_{ij} are such that one of them coincides at every point with the direction of the average velocity $\bar{\mathbf{V}}$, and the other two (in a three-dimensional space) are perpendicular to the first. In general, we have for an isotropic medium,

$$D_{ij} = \lambda g_{ij} \bar{V} + 2\mu V_i V_j / \bar{V}$$

$$\lambda = a_{\mathrm{II}}\,; \qquad \mu = (a_{\mathrm{I}} - a_{\mathrm{II}})/2; \qquad \bar{V} = |\,\mathbf{V}\,| \tag{140}$$

In the principal axes of dispersion, D_{ij} takes the form

$$D_{ij} = \begin{bmatrix} h_1{}^2 a_{\mathrm{I}} \bar{V} & 0 & 0 \\ 0 & h_2{}^2 a_{\mathrm{II}} \bar{V} & 0 \\ 0 & 0 & h_3{}^2 a_{\mathrm{II}} \bar{V} \end{bmatrix} \tag{141}$$

where h_i are the scale factors of the curvilinear coordinate system y^i

$$(h_i)^2 = \left(\frac{\partial x}{\partial y^i}\right)^2 + \left(\frac{\partial y}{\partial y^i}\right)^2 + \left(\frac{\partial z}{\partial y^i}\right)^2$$

x, y, z being the coordinates of a cartesian system, and g_{ij} is the fundamental tensor of the Riemanian space

$$g_{ij} = (\partial x/\partial y^i)(\partial x/\partial y^j) + (\partial y/\partial y^i)(\partial y/\partial y^j) + (\partial z/\partial y^i)(\partial z/\partial y^j)$$

Nikolaevskij (1959) discusses the physical meaning of the dispersion constants a_{I} and a_{II} by assuming that the local velocity at each point is linearly dependent on the average flow velocity in the vicinity of that point. Expressed in terms of velocity components, we have

$$V_i = G_{ij} \bar{V}_k \tag{142}$$

where G_{ik} are supposed to be components of a random symmetric second-rank tensor which depends only on the geometry of the porous matrix. By averaging (142) we obtain $\bar{G}_{ik} = \delta_{ik}$.

Then, from (130), and since $\overline{V}_i = \delta_{ik}\overline{V}_k$,

$$D_{ij} = \overline{(G_{ij}\overline{V}_k - \delta_{ik}\overline{V}_k)(G_{jm}\overline{V}_m - \delta_{jm}\overline{V}_m)(L/\overline{V})}$$
$$= [\overline{\mathring{G}_{ik}\mathring{G}_{jm}}\overline{V}_k\overline{V}_m/\overline{V}]L \tag{143}$$

Hence, a_{ijkm} of (136) is related to G_{ik} by

$$a_{ijkm} = \overline{\mathring{G}_{ik}\mathring{G}_{jm}}L; \qquad i, \ , k, m = 1, 2, 3 \tag{144}$$

where L is some characteristic medium length.

Bear and Bachmat (1966) make an attempt (through the analysis of their porous medium model described in Section 2.4) to relate the dispersivity a_{ijkm} to basic medium characteristics by using physical considerations.

In D_{ij}, defined by (119), we have two parts: the correlation $\overline{\mathring{V}_i\mathring{V}_j}$, and $\overline{(\text{div}\,\mathbf{V}_\alpha)^{-1}}$. The latter expresses the relative rate of growth of the volume occupied by an elementary mass of solute as it moves. This growth occurs mainly during the passage through the medium's (model) junctions, where liquid entering from various channels is mixed, say by molecular diffusion. The rate of growth should accordingly be proportional to the number of channels traversed by tracer particles per unit time. If $\overline{L}$ is the average length of a channel and $(\overline{V}_\alpha)_s$ is the average velocity of a particle in a channel (direction s), then $(\overline{V}_\alpha)_s/\overline{L}$ expresses the number of junctions traversed per unit time, while $\overline{L}/(\overline{V}_\alpha)_s$ expresses the average time of passage between two adjacent junctions. A similar expression may be suggested for the relative expansion in the direction normal to the flow. Bear and Bachmat (1966) suggest $(\overline{V}_\alpha)_a/\bar{a}$, where $\bar{a}$ is a characteristic length of the channel's cross section and $(\overline{V}_\alpha)_a$ is the average tracer velocity across the channel due to concentration gradients. Thus they come up with

$$\overline{(\text{div}\,\mathbf{V}_\alpha)^{-1}} \approx \overline{[\text{div}\,\mathbf{V}_\alpha]}^{-1} \div \left[\frac{2\overline{(V_\alpha)_a}}{\bar{a}} + \frac{\overline{(V_\alpha)_s}}{\overline{L}}\right]$$

where $\div$ indicates proportionality.

Taking $\bar{a}^2/2D_d$ to represent the time required for concentrations to equalize over a channel's cross section so that $\overline{(V_\alpha)_a}/\bar{a} \approx 2D_d/\bar{a}$, and $\overline{(V_\alpha)_s} = \overline{(V_\alpha)_s - \overline{V}} + \overline{V} = 2D_d/\overline{L} + \overline{V}$, Bear and Bachmat (1966) end up with

$$\overline{(\text{div}\,\mathbf{V}_\alpha)^{-1}} \div \frac{\overline{L}}{\overline{V}}\frac{P_e}{P_e + 2 + 4\delta^2} = \frac{\overline{L}}{\overline{V}}f(P_e, \delta) \tag{145}$$

where $P_e = \bar{L}\bar{V}/D_d$, $\delta = \bar{L}/\bar{a}$ (i.e., characterizes the shape of the channel), and $f(P_e, \delta)$ is defined by (145). Obviously other estimates of the relative rate of growth and the effect of the interplay between molecular diffusion and convection on it will lead to different forms of the function $f(P_e, \delta)$ [See Fig. 8(a)].

The moment of correlation $\overline{\mathring{V}_i\mathring{V}_j}$ between the velocity components can be obtained by expressing V_i, V_j, $\bar{V}_i$, and $\bar{V}_j$ by (103) and (125)

$$\overline{(V_i - \bar{V}_i)(V_j - \bar{V}_j)} = \overline{\left[-\frac{BT^*_{ij}}{\mu}\left(\frac{\partial p}{\partial \xi_j} + \rho g \frac{\partial z}{\partial \xi_j}\right) + \frac{\overline{BT^*_{ij}}}{\bar{\mu}}\left(\frac{\partial \bar{p}}{\partial x_j} + \bar{\rho} g \frac{\partial z}{\partial x_j}\right)\right]}$$

$$\cdot \overline{\left[-\frac{BT^*_{ij}}{\mu}\left(\frac{\partial p}{\partial \xi_i} + \rho g \frac{\partial z}{\partial \xi_i}\right) + \frac{\overline{BT^*_{ij}}}{\bar{\mu}}\left(\frac{\partial \bar{p}}{\partial x_i} + \bar{\rho} g \frac{\partial z}{\partial x_i}\right)\right]}$$

where we have taken $V^* \approx V$. With the assumptions leading to (125) and

$$\bar{V}_m \overline{(BT^*_{km})^{-1}} = -\frac{1}{\bar{\mu}}\left(\frac{\partial \bar{p}}{\partial x_k} + \bar{\rho} g \frac{\partial z}{\partial x_k}\right)$$

we obtain

$$\overline{\mathring{V}_i\mathring{V}_j} = \overline{(\mathring{BT}^*_{ik})(\mathring{BT}^*_{jl})}\,\overline{(BT^*_{ln})^{-1}}\,\overline{(BT^*_{km})^{-1}}\,\bar{V}_m\bar{V}_n + \frac{\overline{BT^*_{ik}}\;\overline{BT^*_{jl}}}{\bar{\mu}^2}\left(\overline{\frac{\partial \mathring{p}}{\partial \xi_k}\frac{\partial \mathring{p}}{\partial \xi_l}}\right) \tag{146}$$

According to our assumptions of small pressure gradient deviations, the second term on the right-hand side of (146) may be neglected so that (146) becomes

$$\overline{\mathring{V}_i\mathring{V}_j} \approx \overline{\mathring{BT}^*_{ik}\mathring{BT}^*_{j}}\;\overline{(BT^*_{km})^{-1}}\;\overline{(BT^*_{ln})^{-1}}\;\bar{V}_m\bar{V}_n$$

Hence,

$$D_{ij} \div \overline{\mathring{BT}^*_{ik}\mathring{BT}^*_{jl}}\;\overline{(BT^*_{km})^{-1}}\;\overline{(BT^*_{ln})^{-1}}\,\frac{\bar{L}\bar{V}_m\bar{V}_n}{\bar{V}} f(P_e, \delta)$$

$$= a_{ijmn}\frac{\bar{V}_m\bar{V}_n}{\bar{V}} f(P_e, \delta) \tag{147}$$

where

$$a_{ijmn} = [\overline{\mathring{BT}^*_{ik}\mathring{BT}^*_{jl}}/\overline{BT^*_{km}}\;\overline{BT^*_{ln}}]\,\bar{L} \tag{148}$$

is the *medium's dispersivity* (dim L).

From (147) it follows that the coefficient of mechanical dispersion D_{ij} may be expressed as a product of three factors: the medium's dispersivity, the mean flow velocity, and a nondimensional function depending on δ and P_e. The similarity of (145) and (144) or (136) is obvious.

If the above considerations are extended to a general curvilinear coordinate system y^i, we obtain

$$T^{ij} = (dy^i/d\sigma)(dy^j/d\sigma)$$

(again a second-rank tensor), and

$$D^{ij} = a^{ij}_{mn} \frac{\overline{V}^m \overline{V}^n}{\overline{V}} f(P_e, \delta)$$
$$a^{ij}_{mn} = \overline{(B\mathring{T}^{*ik})(B\mathring{T}^{*jl})}\;(\overline{BT^{*}_{km}})^{-1}\,(\overline{BT^{*}_{ln}})^{-1}\,\overline{L} \tag{149}$$

where a superscript and a subscript denote *contravariant* and *covariant* components, respectively. Also, the permeability of (127) becomes $k^{ij} = n\overline{BT^{*ij}}$. Thus, a^{ij}_{mn} is a mixed tensor of fourth rank. The covariance $\overline{(B\mathring{T}^{*ik})(B\mathring{T}^{*jl})}$ appearing in (149) is expressed in terms of the basic medium properties $\overline{B}$, $\overline{T^{*ij}}$, and $\overline{BT^{*ij}}$, in

$$\overline{(B\mathring{T}^{*ik})(B\mathring{T}^{*jl})} = \overline{B}^2[\overline{(\mathring{B}/\overline{B})^2}\;\overline{T^{*ik}}\;\overline{T^{*jl}} + \overline{\mathring{T}^{*ik}\mathring{T}^{*jl}} + \overline{(\mathring{B}/\overline{B})\,\mathring{T}^{*ik}}\;\overline{T^{*jl}} + \overline{T^{*ik}}\;\overline{(\mathring{B}/\overline{B})\,\mathring{T}^{*jl}}] \tag{150}$$

Choosing a coordinate system which coincides with the principal axes of the permeability k^{ij}, we have

$$\overline{BT^{*ij}} = \overline{BT^{*(1)}}\,\mathbf{I}^i\mathbf{I}^j + \overline{BT^{*(2)}}\,\mathbf{J}^i\mathbf{J}^j + \overline{BT^{*(3)}}\,\mathbf{K}^i\mathbf{K}^j \tag{151}$$

where **I**, **J**, **K** are unit vectors in the principal directions of the tensor $\overline{BT^{*ij}}$ and $\overline{BT^{*(1)}}$, $\overline{BT^{*(2)}}$, and $\overline{BT^{*(3)}}$ are the principal values of $\overline{BT^{*ij}}$. The following steps are then undertaken by Bear and Bachmat (1965, 1966): (1) an assumption is made that the principal directions of $\overline{BT^{*ij}}$ coincide with those of $\overline{T^{*ij}}$; (2) they express $\overline{T^{*ij}}$, T^{*ij}, and $\mathring{T}^{*ij}$ through $\alpha^i = (dy^i/d\sigma)(d\sigma/ds)$ and their deviations ($\mathring{\alpha}^i$); and (3) they apply these steps to the various expressions in (150). The resulting expression gives a^{ij}_{mn} in the general case of a medium which is anisotropic with respect to the permeability. From the derived expression it follows that a^{ij}_{mn} possesses in general 21 nonzero terms: 9 of the type $g_m{}^i g_n{}^j$,

6 of the type $g_n{}^i g_m{}^j$, and 6 of the type $g^{ij}g_{mn}$, g_{ij} denotes here the fundamental tensor of the Riemenian space.

For an isotropic medium, all principal values of $\overline{BT^{*ij}}$ are the same and $\overline{\mathring{B}\mathring{T}^{*ij}} = 0$. It is then possible to reduce a^{ij}_{mn} to the form

$$a^{ij}_{mn} = \tfrac{1}{2}(a_{\mathrm{I}} - a_{\mathrm{II}})(g_n{}^i g_m{}^j - g_m{}^i g_n{}^j) + a_{\mathrm{II}} g^{ij} g_{mn} \tag{152}$$

where

$$a_{\mathrm{I}} = [\overline{(\mathring{B}/\bar{B})^2} + \overline{4\mathring{\alpha}^{(1)2}}\;\overline{\alpha^{(1)}}^2 / \overline{\alpha^{(1)2}}^2]\bar{L} \qquad \text{and} \qquad a_{\mathrm{II}} = [\overline{2\mathring{\alpha}^{(1)2}}\;\overline{\alpha^{(1)}}^2 / \overline{\alpha^{(1)2}}^2]\bar{L}$$

The relationship between a_{I} and a_{II} in an isotropic porous medium (with respect to permeability) is given approximately by

$$a_{\mathrm{I}}/a_{\mathrm{II}} = 2 + \overline{(\mathring{B}/\bar{B})^2}\;\overline{\alpha^{(1)2}}^2 / \overline{2\mathring{\alpha}^{(1)2}}\;\overline{\alpha^{(1)}}^2 > 2 \tag{153}$$

Ratios of 6 to 24 are given in the literature by various investigators.

With a^{ij}_{mn} of (152), we obtain

$$\begin{aligned} D^{ij} &= \left[\frac{a_{\mathrm{I}} - a_{\mathrm{II}}}{2}(g_n{}^i g_m{}^j + g_m{}^i g_n{}^j) + a_{\mathrm{II}} g^{ij} g_{mn}\right] \frac{\bar{V}_m \bar{V}_n}{\bar{V}} f(P_e, \delta) \\ &= (a_{\mathrm{I}} - a_{\mathrm{II}}) \frac{\bar{V}^i \bar{V}^j}{\bar{V}} f(P_e, \delta) + a_{\mathrm{II}} g^{ij} \bar{V} f(P_e, \delta) \end{aligned} \tag{154}$$

The molecular diffusivity is introduced by adding the term $D_d \overline{T^*} g^{ij}$ to D^{ij} in (154).

In the special case of an orthogonal system y^i, such that y^1 coincides at every point with the average velocity $\bar{\mathbf{V}}$, while $\bar{V}^{(2)} = \bar{V}^{(3)} = 0$, we obtain

$$D^{ij} = \begin{bmatrix} a_{\mathrm{I}}(h^1)^2\,\bar{V} & 0 & 0 \\ 0 & a_{\mathrm{II}}(h^2)^2\,\bar{V} & 0 \\ 0 & 0 & a_{\mathrm{II}}(h^3)^2\,\bar{V} \end{bmatrix} f(P_e, \delta) \tag{155}$$

If the principal axes of D_{ij} form a cartesian coordinate system, e.g., in the case of a steady uniform flow with $\bar{\mathbf{V}} =$ const, (155) becomes

$$D^{ij} = D_{ij} = \begin{bmatrix} a_{\mathrm{I}}\bar{V} & 0 & 0 \\ 0 & a_{\mathrm{II}}\bar{V} & 0 \\ 0 & 0 & a_{\mathrm{II}}\bar{V} \end{bmatrix} f(P_e, \delta) \tag{156}$$

To conclude this discussion on the parameters of dispersion, the following nomenclature is suggested:

$D'_{ij} = D_{ij} + (D^*_d)_{ij}$	Coefficient of hydrodynamic dispersion
D_{ij}	Coefficient of mechanical (or convective) dispersion

$D_d^* = T_{ij}^* D_d$	Coefficient of molecular diffusion in a porous medium
a_{ijkl}	Dispersivity of the porous medium
a_{I}	Longitudinal dispersivity of an isotropic porous medium
a_{II}	Transversal dispersivity of an isotropic porous medium
T_{ij}^*	Tortuosity, or tensor of directions, of a porous medium

3.3. Dispersivity–Permeability Relationship

Whereas the dispersivity of a medium determines the spreading of a definite fluid portion, the permeability determines the average flux of a fluid through the porous medium. Both are medium geometric characteristics. Although both depend on the fundamental medium property BT_{ij}^*. , they are two distinct characteristics. The permeability k_{ij} depends on the average of BT_{ij}^* ,

$$k_{ij} = n\overline{BT_{ij}^*} \tag{157}$$

If we choose a coordinate system which coincides with the principal directions of k_{ij} , we obtain

$$k_{ij} = k\delta_{ij} , \qquad k_{ij} = k_{ii} \quad \text{for} \quad i = j; \qquad k_{ij} = 0 \quad \text{for} \quad i \neq j.$$

In general, $\overline{BT_{ij}^*}$ (*medium's average directional conductance*) at a point may be expressed as

$$\overline{BT_{ij}^*} = \bar{B}\,\overline{T_{ij}^*} + \overline{\mathring{B}\mathring{T}_{ij}^*} = \bar{B}\,\overline{T_{ij}^*} + \overline{\mathring{B}\mathring{\alpha}_i\alpha_j} + \overline{\mathring{B}\mathring{\alpha}_j\alpha_i} \tag{158}$$

It can be shown that for an isotropic medium, $\overline{\mathring{B}\mathring{T}_{ij}^*} = 0$. Then

$$k_{ij} = n\bar{B}\,\overline{T^*}\delta_{ij}$$

In a general curvilinear coordinate system y^i,

$$k^{ij} = n\overline{BT^{*ij}}; \qquad k^{ij} = k^{ji}$$

In the case of an isotropic porous medium, all principal values of $\overline{BT^{*ij}}$ are the same, and $\overline{\mathring{B}\mathring{T}^{*ij}} = 0$. Hence, $\overline{BT^{*ij}} = g^{ij}\overline{BT^*}$.

The dispersivity a_{ijmn} (or a_{mn}^{ij} in a general curvilinear coordinate system y^i) depends on the variance of BT_{ij}^* (or BT^{*ij}). Through this basic medium characteristic one may therefore find an intrinsic relationship between permeability and dispersivity.

Let us consider the particular case of an *isotropic medium* (with respect to k_{ij}). Then, $k = n\bar{B}\overline{T^*} = (k/3)\, n\bar{B}\overline{T^{*kk}}$, and the dispersivity given by (147) becomes

$$a_{ijmn} = \frac{n^2\bar{L}}{k^2}\overline{\mathring{B}\mathring{T}^*_{im}\mathring{B}\mathring{T}^*_{jn}} \tag{159}$$

from which

$$a_{\mathrm{I}} = a_{1111} = \frac{n^2\bar{L}}{k^2}\overline{(\mathring{B}\mathring{T}^*_{11})^2}; \qquad a_{\mathrm{II}} = a_{2121} = \frac{n^2\bar{L}}{k^2}\overline{(\mathring{B}\mathring{T}^*_{21})^2}$$

If we assume also that $B = \text{const} = \bar{B}$, we obtain

$$a_{\mathrm{I}} = \frac{n^2\bar{L}\bar{B}^2}{k^2}\overline{(\mathring{T}_{11})^2} = \overline{\left(\frac{\mathring{T}_{11}}{\bar{T}}\right)^2}\,\bar{L} \tag{160}$$

As an example, let us consider the special case of a model consisting of straight circular channels of constant radius R and a uniform distribution of directions:

$$B = (r^2 - R^2)/4; \qquad \bar{B} = R^2/8; \qquad \overline{(\mathring{B})^2}/\bar{B}^2 = \tfrac{1}{3};$$

$$\overline{T_{ij}} = \int_{(\Omega)} T_{ij}(\Omega) f(\Omega)\, d\Omega; \qquad \int_{(\Omega)} f(\Omega)\, d\Omega = 1$$

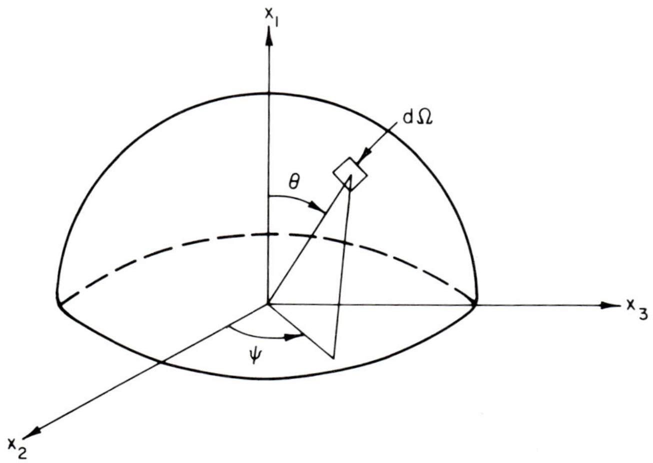

FIG. 11. Definition sketch for the determination of T^*_{ij}.

where $d\Omega$ is a solid angle (Fig. 11):

$$d\Omega = \sin\theta\, d\theta\, d\psi; \qquad f(\Omega) = (2\pi)^{-1} \quad \text{(isotropic medium)}$$

$$T_{11} = \cos^2\theta; \qquad \overline{\mathring{T}_{11}^2} = \overline{T_{11}^2} - (\bar{T}_{11})^2$$

$$\bar{T}_{11} = (2\pi)^{-1}\int_0^{2\pi}\int_0^{\pi/2} \cos^2\theta \sin\theta\, d\theta\, d\psi = \tfrac{1}{3}$$

$$\overline{T_{11}^2} = (2\pi)^{-1}\int_0^{2\pi}\int_0^{\pi/2} \cos^4\theta \sin\theta\, d\theta\, d\psi = \tfrac{1}{5}$$

$$\overline{\mathring{T}_{11}^2} = 4/45, \quad \text{etc.}$$

Assuming no correlation between B and T, we obtain from (149) and (150)

$$a_{1111} \equiv a_{\mathrm{I}} = \frac{\overline{(\mathring{B}T_{11})^2}}{\bar{B}\bar{T}_{11}{}^2}\bar{L} = \left[\overline{\left(\frac{\mathring{B}}{\bar{B}}\right)^2} + \overline{\left(\frac{\mathring{T}_{11}}{\bar{T}_{11}}\right)^2} + \overline{\left(\frac{\mathring{B}}{\bar{B}}\right)^2}\,\overline{\left(\frac{\mathring{T}_{11}}{\bar{T}_{11}}\right)^2}\right]\bar{L}$$

$$= [\tfrac{1}{3} + \tfrac{4}{5} + \tfrac{1}{3}\cdot\tfrac{4}{5}]\bar{L} = \tfrac{7}{5}\bar{L}, \quad \text{etc.}$$

and

$$k = \left(\frac{7}{2880}\right)^{1/2} \frac{nR^2}{(a_{\mathrm{I}}/\bar{L})^{1/2}} = \frac{1}{24}nR^2 = \frac{1}{3}n\bar{B}$$

This expression relates the permeability of the medium to the longitudinal dispersivity, to the porosity, to some characteristic pore cross section, and to some characteristic pore length. Harleman *et al.* (1963) obtained experimentally expressions relating permeability and dispersivity.

3.4. A Note on Other Transport Coefficients

For a homogeneous liquid ($\bar{\rho}$ = const), (129) becomes

$$\mathbf{q} = -\frac{\underline{\mathbf{k}}\bar{\rho}g}{\bar{\mu}}\operatorname{grad}\bar{\varphi}; \qquad \bar{\varphi} = z + \frac{\bar{p}}{\bar{\rho}g} \tag{161}$$

For molecular diffusion of a homogeneous liquid ($\bar{\rho}$ = const), we obtain from (9)

$$\bar{\mathbf{J}}_\alpha^* = -D_d \operatorname{grad} \bar{C}_\alpha$$

For the same liquid, the average flux vector per unit area of voids is, by (122),

$$\bar{\mathbf{J}}_\alpha^* = -\underline{\mathbf{D}}_d^* \operatorname{grad} \bar{C}_\alpha \tag{162}$$

or, per unit area of porous medium,

$$\mathbf{q}_\alpha = n\bar{\mathbf{J}}_\alpha = -nD_d\overline{\underline{\mathbf{T}}^*} \operatorname{grad} \bar{C}_\alpha = -n\underline{\mathbf{D}}_d^* \operatorname{grad} \bar{C}_\alpha \tag{163}$$

where $(D_d^*)_{ij} = \overline{T_{ij}^*}D_d$ (or $(D_d^*)^{ij} = \overline{T^{*ij}}D_d$) is the *coefficient of molecular diffusion in a porous medium.* It is sometimes better to refer to the product $n\underline{\mathbf{D}}_d^*$ as the coefficient of molecular diffusion in a porous medium.

Equations similar to (161) describe also transport phenomena in other branches of physics. For example, Ohm's law in electricity is

$$\mathbf{i} = -\sigma \operatorname{grad} E \tag{164}$$

where σ is the electrical conductivity, E is the electrical potential, and $\mathbf{i}$ is the flux vector. If these transport phenomena take place in a porous medium (e.g., an electrolyte fully saturating an electrically nonconducting porous medium), considerations similar to those leading to (163) will yield

$$\mathbf{q}_i = -n\overline{\underline{\mathbf{T}}^*}\sigma \operatorname{grad} E \tag{165}$$

where $\mathbf{q}_i$ is the average current flux per unit area of medium (voids + solid). Thus, $n\overline{\underline{\mathbf{T}}^*}\lambda$ may be used to define any conductive property of a porous medium, where λ is the conductive property in the bulk (e.g., heat conductivity, or electrical conductivity of the fluid saturating the medium where the porous matrix is assumed nonconductive). Equation (165) also enables an experimental determination of $\overline{T_{ij}^*}$ (Klinkenberg 1951), by comparing the electrical resistivity of the porous material saturated by an electrolyte to the bulk resistivity of the same electrolyte. In reservoir engineering this ratio is often called *formation factor*. Here it is equal to $1/n\overline{T^*}$. Some investigators suggest a formation factor $F = \tau^m/n$ where τ is a *tortuosity* and m is some number.

4. The Differential Equation of Dispersion

4.1. The Fundamental Differential Equation

The tracer concentration distribution $C = C(x, y, z, t)$ within a flow domain is obtained by solving the appropriate partial differential equation of hydrodynamic dispersion subject to a specified initial tracer concentration distribution and to boundary conditions with respect to the tracer's concentration.

In a homogeneous liquid (ρ = const, μ = const) the concentration distribution does not affect the average velocity vector at each point. Hence, the solution of a dispersion problem is made up of two independent subproblems. In the first one the velocity distribution is determined analytically, numerically, or by models. The resulting velocity distribution is inserted in the dispersion equation which is then solved to yield the concentration distribution.

In the case of inhomogeneous liquids, the two problems have to be solved simultaneously, as the instantaneous velocity distribution depends on the instantaneous concentration distribution (affecting ρ and μ). We therefore have to solve simultaneously for the seven dependent variables: $V_i^*(3)$, ρ, μ, p, C. For these variables we have at our disposal the seven equations: the three equations (125), two equations of state for ρ and μ, the mass conservation equation (124), and the dispersion equation (120).

In what follows we shall consider only a homogeneous liquid. For this liquid, the dispersion equation in a cartesian coordinate system x_i is

$$\frac{\partial C}{\partial t} = \frac{\partial}{\partial x_i}\left[(D_{ij} + D_d T_{ij}^*)\frac{\partial C}{\partial x_j} - CV_i\right] \tag{166}$$

where C is the α-species concentration. Here, C, V_i, T_{ij}^* replace $\bar{C}_\alpha$, $\bar{V}_i$, $\overline{T_{ij}^*}$ defined in Section 2.4; they represent average values over a representative elementary volume within the medium. Scheidegger (1961) and de Josselin de Jong and Bossen (1961) also obtained the same equation without the term $D_d T_{ij}^*$, since they included the effect of molecular diffusion already in D_{ij}. Equation (166) may be rewritten in the form

$$\frac{\partial C}{\partial t} = \frac{\partial}{\partial x_i}\left(D'_{ij}\frac{\partial C}{\partial x_j} - CV_i\right); \qquad \underline{\mathbf{D}}' = \underline{\mathbf{D}} + D_d\underline{\mathbf{T}}^* \tag{167}$$

where $\underline{\mathbf{D}}'$ is the coefficient of hydrodynamic dispersion. Two possible extensions of (167) may be mentioned:

(i) The ideal tracer undergoes radioactive decay obeying

$$\partial C/\partial t = -\lambda C \tag{168}$$

where λ is the decay constant equal to the reciprocal of the tracer's mean lifetime.

(ii) Adsorption of the tracer to the solid matrix may occur during the flow. Adsorption here is the tracer mass transfer (in the form of solute ions) from the liquid to the surfaces of the solid phase, controlled by the chemical properties of the solute and the solid matrix, and by the solute concentration in the liquid phase.

The adsorption equation itself may take the form (Lapidus and Amundson, 1952; Bondarev and Nikolaevskij, 1962)

$$\partial A/\partial t = \beta(C - C_{eq}) = \beta(C - A/a_0) \tag{169}$$

where β and a_0 are constants, C_{eq} is the value of C under equilibrium conditions, and A is the tracer concentration on the solid surface (mass of tracer per unit volume of solid phase).

With both adsorption and radioactive decay, $A(\mathbf{r}, t)$ and $C(\mathbf{r}, t)$ are governed by the system of equations (169) and

$$\frac{\partial}{\partial t}\left(C + \frac{1-n}{n} A\right) = \frac{\partial}{\partial x_i}\left(D'_{ij}\frac{\partial C}{\partial x_j}\right) - V\frac{\partial C}{\partial x_i} - \lambda\left(C + \frac{1-n}{n} A\right) \tag{170}$$

In order to solve (169) and (170), we have to know the velocity $\mathbf{V}(\mathbf{r}, t)$ which appears in (170), or in any other form of the dispersion equation as given below, both implicitly and explicitly as a factor of D'_{ij}. For a homogeneous incompressible liquid, $\mathbf{V}(\mathbf{r}, t)$ is determined by Darcy's law, e. g., in the form $V_i = (k_{ij}\rho g/\mu)\, \partial\varphi/\partial x_j$, where $\varphi(\mathbf{r}, t)$ obeys

$$\partial V_i/\partial x_i = \partial[(k_{ij}\rho g/\mu)\, \partial\varphi/\partial x_j]/\partial x_i = 0.$$

The complete solution of an ideal tracer dispersion problem involves, therefore, a simultaneous solution of four equations for the four unknowns A, C, V, φ, provided that all involved medium and fluid parameters are known.

Before discussing boundary and initial conditions for (167), i.e., in the absence of adsorption and radioactive decay, let us review briefly several other forms of this equation in other coordinate systems.

In general curvilinear coordinates y^i, (167) may be written in the form

$$\frac{\partial C}{\partial t} = \nabla_i\left[D'^{ij}\frac{\partial C}{\partial y^j}\right] - \nabla_i(CV^i) \tag{171}$$

where ∇_i is the divergence symbol. In an isotropic medium, let D'^{ij} be represented by (135). Then (171) becomes

$$\frac{\partial C}{\partial t} = \frac{1}{\sqrt{g}}\frac{\partial}{\partial y^i}\left\{\sqrt{g}\, f(P_e, \delta)\left[(a_{\mathrm{I}} - a_{\mathrm{II}})\frac{V^iV^j}{V} + a_{\mathrm{II}}g^{ij}V\right]\frac{\partial C}{\partial y^j} + \sqrt{g}\, D_d T^* g^{ij}\frac{\partial C}{\partial y^j}\right\} - V^i\frac{\partial C}{\partial y^i} \tag{172}$$

where

$$g^{ij} = (\partial y^i/\partial x)(\partial y^j/\partial x) + (\partial y^i/\partial y)(\partial y^j/\partial y) + (\partial y^i/\partial z)(\partial y^j/\partial z)$$

(For additional information the reader is referred to any text on curvilinear coordinates). In an orthogonal system, we have

$$g^{ij} = \begin{cases} (h^{(i)})^2 = 1/h^2_{(i)} & \text{for } i = j \\ 0 & \text{for } i \neq j \end{cases} \tag{173}$$

and (172) becomes

$$\frac{\partial C}{\partial t} = \frac{1}{h_1 h_2 h_3} \frac{\partial}{\partial y^i} \left\{ \frac{h_1 h_2 h_3}{h^2_{(i)}} f(P_e, \delta) \left[(a_{\mathrm{I}} - a_{\mathrm{II}}) \frac{V_i V_k}{V} \frac{1}{h^2_{(k)}} \frac{\partial C}{\partial y^k} + a_{\mathrm{II}} V \frac{\partial C}{\partial y^i} \right] + D_d T^* \frac{\partial C}{\partial y^i} \right\} - \frac{V_i}{h^2_{(i)}} \frac{\partial C}{\partial y^i} \tag{174}$$

where the summation is performed over both i and k, but not over (i), (k). In (174), and everywhere in this section, V is the absolute value of the average velocity.

As explained in Section 3.1, the principal axes of D^{ij} in an isotropic medium are three orthogonal axes of which one is parallel to the average velocity vector. Taking the principal axes of the dispersion as axes of a coordinate system we have $V_1 \neq 0$, $V_2 = V_3 = 0$, and with $V_i = V h_i \cos(\mathbf{V}, \mathbf{1y}^i)$, (174) becomes

$$\begin{aligned} \frac{\partial C}{\partial t} = \frac{1}{h_1 h_2 h_3} \Bigg\{ & \frac{\partial}{\partial y^1} \left[\frac{h_2 h_3}{h_1} (a_{\mathrm{I}} f(P_e, \delta) V + D_d T^*) \frac{\partial C}{\partial y^1} \right] \\ & + \frac{\partial}{\partial y^2} \left[\frac{h_3 h_1}{h_2} (a_{\mathrm{II}} f(P_e, \delta) V + D_d T^*) \frac{\partial C}{\partial y^2} \right] \\ & + \frac{\partial}{\partial y^3} \left[\frac{h_1 h_2}{h_3} (a_{\mathrm{II}} f(P_e, \delta) V + D_d T^*) \frac{\partial C}{\partial y^3} \right] \Bigg\} \\ & + \frac{V}{h_1} \cos(\mathbf{V}, \mathbf{1y}^1) \frac{\partial C}{\partial y^1} \end{aligned} \tag{175}$$

Equation (175) may also be written in terms of lengths along coordinate lines. If ds^i is an element of length along the coordinate line y^i, we have

$$ds^i = h_i \, dy^i \qquad \text{(no summation on } i\text{)}$$

Then, $(1/h_i)\,\partial C/\partial y^i = \partial C/\partial s^i$, and (175) becomes

$$\frac{\partial C}{\partial t} = \frac{1}{h_2 h_3}\frac{\partial}{\partial s^1}\left[h_2 h_3(a_{\mathrm{I}} f(P_e\,,\delta)\,V + D_d T^*)\frac{\partial C}{\partial s^1}\right]$$
$$+ \frac{1}{h_1 h_3}\frac{\partial}{\partial s^2}\left[h_3 h_1(a_{\mathrm{II}} f(P_e\,,\delta)\,V + D_d T^*)\frac{\partial C}{\partial s^2}\right]$$
$$+ \frac{1}{h_1 h_2}\left[h_1 h_2(a_{\mathrm{II}} f(P_e\,,\delta)\,V + D_d T^*)\frac{\partial C}{\partial s^3}\right]$$
$$- V\cos(\mathbf{V}, \mathbf{1s}^1)\frac{\partial C}{\partial s^1} \qquad (176)$$

The family of streamlines may serve as the family of coordinate lines y^i in (175) and (176). If in two-dimensional flow in a homogeneous isotropic medium we use the family of streamlines $\psi =$ const as y^1, and the family of equipotential lines $\Phi =$ const as y^2, where $\Phi = -K\varphi$ is the velocity potential, $(K = k\rho g/\mu)$,

$$\begin{aligned} y^1 &= \Phi = \Phi(x, y); \qquad s^1 = s_\Phi\,; \\ y^2 &= \Psi = \Psi(x, y); \qquad s^2 = s_\Psi\,; \\ dy^1/ds^1 &= d\Phi/ds_\Phi = |\operatorname{grad}\Phi| = V = 1/h_1 \\ dy^2/ds^2 &= d\Psi/ds_\Psi = |\operatorname{grad}\Psi| = 1/h_2 \\ \partial\Phi/\partial y &= -\partial\Psi/\partial x \end{aligned} \qquad (177)$$

Since $\partial\Phi/\partial x = \partial\Psi/\partial y$, we have $|\operatorname{grad}\Phi| = |\operatorname{grad}\Psi|$ and $h_1 = h_2 = 1/V$. With these considerations, (175) becomes

$$\frac{\partial C}{\partial t} = V\left\{\frac{\partial}{\partial s_\Phi}\left[\left(\frac{Va_{\mathrm{I}}}{V+A} + \frac{D_d T^*}{V}\right)\frac{\partial C}{\partial s_\Phi}\right] + \frac{\partial C}{\partial s_\Psi}\left[\left(\frac{Va_{\mathrm{II}}}{V+A} + \frac{D_d T^*}{V}\right)\frac{\partial C}{\partial s_\Psi}\right]\right.$$
$$\left. - \cos(\mathbf{V}, \mathbf{1s}_\Phi)\frac{\partial C}{\partial s_\Phi}\right\} \qquad (178)$$

where $A = (2 + 4\delta^2)D_d/L =$ const.

In the special case of uniform flow in the direction $\mathbf{1x}$ in the (x, y) plane, $ds_\Psi = dy$, $ds_\Phi = dx$; $\cos(\mathbf{V}, \mathbf{1s}_\Phi) = 1$

$$\frac{\partial C}{\partial t} = \left(\frac{V^2 a_{\mathrm{I}}}{V+A} + D_d T^*\right)\frac{\partial^2 C}{\partial x^2} + \left(\frac{V^2 a_{\mathrm{II}}}{V+A} + D_d T^*\right)\frac{\partial^2 C}{\partial y^2} - V\frac{\partial C}{\partial x} \qquad (179)$$

In the symmetrical (or one-dimensional) case of flow in the direction $\mathbf{1x}$, $\partial C/\partial y = 0$, and (179) becomes

$$\frac{\partial C}{\partial t} = \left(\frac{V^2 a_{\mathrm{I}}}{V + A} + D_d T^*\right) \frac{\partial^2 C}{\partial x^2} - V \frac{\partial C}{\partial x} \tag{180}$$

If $D_d T^* \ll (V^2 a_{\mathrm{I}})/(V + A)$, we have

$$\frac{\partial C}{\partial t} = \frac{V^2 a_{\mathrm{I}}}{V + A} \frac{\partial^2 C}{\partial x^2} - V \frac{\partial C}{\partial x} \tag{181}$$

One should recall that in (181), as everywhere else in this section, V represents the absolute value of the average velocity. For flow in the direction $-\mathbf{1x}$, $\cos(\mathbf{V}, \mathbf{1s}_\Phi) = -1$, and the minus sign in (179)–(181) becomes a plus sign. For uniform flow in space, where the direction of x (of the x, y, z cartesian coordinates) coincides with $\mathbf{V}$, $h_1 = h_2 = h_3 = 1$, we have

$$\frac{\partial C}{\partial t} = (f a_{\mathrm{I}} V + D_d T^*) \frac{\partial^2 C}{\partial x^2} + (f a_{\mathrm{II}} V + D_d T^*) \left(\frac{\partial^2 C}{\partial y^2} + \frac{\partial^2 C}{\partial z^2}\right) - V \cos(\mathbf{V}, \mathbf{1x}) \frac{\partial C}{\partial x} \tag{182}$$

where f represents $f(P_e, \delta)$.

A special case of interest is that of radial convergent or divergent plane flow into a sink or from a source at the origin. In the polar coordinate system (r, θ), $ds_\Phi = dr$, $ds_\Psi = r\, d\theta$; $V = G/r$ where G is a constant. We obtain

$$\frac{\partial C}{\partial t} = V \left\{ \frac{\partial}{\partial r} \left[\left(\frac{V a_{\mathrm{I}}}{V + A} + \frac{D_d T^*}{V}\right) \frac{\partial C}{\partial r}\right] + \frac{1}{r} \frac{\partial}{\partial \theta} \left[\left(\frac{V a_{\mathrm{II}}}{V + A} + \frac{D_d T^*}{V}\right) \frac{1}{r} \frac{\partial C}{\partial \theta}\right] \pm \frac{\partial C}{\partial r} \right. \tag{183}$$

where $V = Q/(2\pi B n r)$, Q/B is the strength of the source or the sink, and the plus and minus signs correspond to radially convergent and divergent flow, respectively. If we have axisymmetric conditions, we let $\partial C/\partial \theta = 0$ in (183). Because of variations in V as r varies, the relationship between $a_{\mathrm{I}} V$ and $D_d T^*$ may also depend on r; we may have $a_{\mathrm{I}} V \gg D_d T^*$ near the origin, but $a_{\mathrm{I}} V \ll D_d T^*$ for larger r.

4.2. Boundary and Initial Conditions

Like the solution of any other flow problem from the appropriate differential equations, the determination of $C(\mathbf{r}, t)$, or of $C(\mathbf{r}, t)$ and $A(\mathbf{r}, t)$, requires a knowledge of *boundary and initial conditions.*

The following paragraphs include a brief review of boundary and initial conditions with respect to C.

Specifying initial conditions means specifying C at $t = 0$ for all points inside the considered domain (i.e., excluding the boundary)

$$C(\mathbf{r}, 0) = f_1(\mathbf{r}) \tag{184}$$

In general, the considered porous medium domain may be bounded by either an impervious, noninteracting with the liquid, solid, by a liquid continuum, or by another porous medium domain. The major requirement is the continuity of tracer mass flux across the boundary. If a is the considered domain, and b is the domain outside the boundary, we have

$$J_n^{*(a)} = J_n^{*(b)} \tag{185}$$

where $J_n^{*(a)}$ is the specific (per unit area of boundary surface) tracer mass flux normal to the boundary on its inner side. It is given by

$$J_n^{(a)} = n_a \nu_i \left[-D'^{ij} \frac{\partial C}{\partial x^j} + V^i C\right] \tag{186}$$

where n_a is porosity, ν_i are the covariant components of the outward normal $\mathbf{1\nu}$ to the boundary surface, and D'^{ij} is defined by (167).

Several cases may be considered.

(1) The domain b is also a porous medium domain. Then we have for $J_n^{*(b)}$ an expression similar to (186); (185) takes the form

$$n_a \nu_i \left(-D'^{ij} \frac{\partial C}{\partial x^j} + V^i C\right)_a = n_b \nu_i \left(-D'^{ij} \frac{\partial C}{\partial x^j} + V^i C\right)_b \tag{187}$$

(2) If b is an impervious domain (or practically so) for mass flow, we have

$$\nu_i q^i = 0; \qquad q^{(n)} = 0$$

We then obtain

$$\left[(f^* a_{\text{II}} q + n D_d T^*) \nu^j \frac{\partial C}{\partial x^j}\right]_a = \left[(f^* a_{\text{II}} q + n D_d T^*) \nu^j \frac{\partial C}{\partial x^j}\right]_b \tag{188}$$

where

$$f^* = f(P_e^*, \delta), \qquad P_e^* = q\bar{L}/D_d\,; \qquad n D^{ij} = a_{mn}^{ij} \frac{q^m q^n}{q} f^*(P_e^*, \delta) \tag{189}$$

In this case, only transversal dispersion resulting from the flow which is tangential to the boundary, and molecular diffusion, is active.

(3) If the solid boundary is impervious also to diffusion, (188) reduces to

$$\frac{\partial C}{\partial s^{(n)}} = 0 \quad \text{or} \quad \nu^j \left(\frac{\partial C}{\partial x_j} \right) = 0 \tag{190}$$

where $ds^{(n)}$ is a length element normal to the boundary.

(4) If b is a liquid continuum (sea, lake, river, reservoir), where a given tracer distribution $C(\mathbf{r}_0, t)$ exists,

$$J_n^{(b)} = C(\mathbf{r}_0, t)\, \nu_i q^i - D_d \nu_i \frac{\partial C}{\partial x^i} \tag{191}$$

and $J_n^{(a)}$ is given by the left-hand side of (187). Equations (180) and (191) are then inserted in (185) to yield the boundary condition. The conclusion from this boundary condition is that the concentration distribution on both sides of the boundary are not identical. For the special case of a homogeneous isotropic medium, we obtain

$$[C(\mathbf{r}_0, t) - C]\, \nu_i q^i = D_d \nu^i \frac{\partial C}{\partial x^i} - n D_d T^* \nu^i \frac{\partial C}{\partial x^i}$$
$$- f^* \left[(a_{\text{I}} - a_{\text{II}})\, \nu_i \frac{q^i q^j}{q} + a_{\text{II}} q \nu^j \right] \frac{\partial C}{\partial x^j} \tag{192}$$

When the concentrations on both sides of the boundary become equal and remain so, we have $\partial C/\partial s^{(n)} = 0$ and the boundary condition becomes

$$C(\mathbf{r}_0, t) = C \tag{193}$$

In general, the condition (193) may be obtained only after a sufficiently long time. By comparing (193) with (192), we see that (193) is equivalent to specifying the mass flux of an ideal tracer across the boundary, where the specific discharge vector $\mathbf{q}$ is independent of concentration. Moreover, (193) assumes that the tracer mass transport across the boundary is due to convection alone.

(5) Finally, if b is a void space (i.e., filled with air), C is the same on both sides, i.e., $C(\mathbf{r}_0, t) = C$. There is no molecular diffusion on the outside, and (192) becomes

$$\nu_i (n D^{ij} + n D_d T^{*ij}) \frac{\partial C}{\partial x_j} = 0 \tag{194}$$

For an isotropic homogeneous medium we obtain

$$(f^* a_{\text{I}} q + n D_d T^*)(\partial C/\partial s^{(n)}) = 0, \qquad \text{i.e.,} \quad \partial C/\partial s^{(n)} = 0$$

For an infinite medium without supply at infinity, we require that at $\mathbf{r} \to \infty$, C will be constant and finite.

$$\lim_{r\to\infty} C = C_0\,; \qquad \lim_{x^j\to\pm\infty} \partial C/\partial x^j = 0$$

4.3. Some Solved Problems

Only a few typical examples are given here. The objective is to show typical equations, boundary and initial conditions, and methods of solution. In all cases, the medium is homogeneous and isotropic and we seek $C(\mathbf{r}, t)$. For the sake of simplicity we shall assume $A = 0$ in (179)–(181).

Case (i). Progress of a concentration front in an infinite column of porous medium. Both hydrodynamic dispersion and molecular diffusion are considered; radioactive decay and adsorption are neglected. The specific discharge $q(t)$—the same at any instant along the entire column—is either a constant or may vary with time. Initially, the column is saturated by two (miscible) liquids at different tracer concentrations with an abrupt interface (say at $x = 0$) between them.

The partial differential equation here is (181) with $A \to 0$ or

$$\frac{\partial C}{\partial t} = D' \frac{\partial^2 C}{\partial x^2} \pm \frac{q}{n}\frac{\partial C}{\partial x}\,; \qquad -\infty < x < +\infty \tag{195}$$

with $D' = a_{\text{I}}(|\,q\,|/n) + D_d T^*$, and the minus sign corresponds to flow in the $+x$ direction. For an isotropic medium $T^* = \frac{1}{3}$.

Initial conditions: $t = 0$;

$$\begin{aligned} -\infty < x < 0; &\qquad C = C_0 \\ 0 < x < +\infty; &\qquad C = C_1 \end{aligned} \tag{196}$$

Boundary conditions: $t > 0$;

$$\begin{aligned} x = \pm\infty; &\qquad \partial C/\partial x = 0 \\ x = +\infty; &\qquad C = C_1 \\ x = -\infty; &\qquad C = C_0 \end{aligned} \tag{197}$$

Bear and Todd (1960) solve this problem by applying the Lapalace transform to (195) through (197). The solution is

$$\frac{C(x, t) - C_0}{C_1 - C_0} = \tfrac{1}{2}\,\text{erfc}\left\{- \frac{x - \int_0^t [q(t)/n]\, dt}{2[\int_0^t (a_{\text{I}}\,|\,q\,|/n + D_d T^*)\, dt]^{1/2}}\right\} \tag{198}$$

where $\text{erfc}(z) = 1 - \text{erf}(z)$.

For the special case of negligible diffusion (i.e., $a_I \mid q \mid \gg nD_d T^*$), we set $D_d = 0$ in (198).

For a constant q in the **1x** direction,

$$\frac{C(x, t) - C_0}{C_1 - C_0} = \tfrac{1}{2} \operatorname{erfc} \left\{ - \frac{x - qt/n}{(4D't)^{1/2}} \right\}; \tag{199}$$

Case (ii). An infinite column of porous medium with steady flow q in the **1x** *direction.* At $t = 0$, a very thin slug of tracer-marked fluid is injected at $x = 0$ into the stream.

The tracer concentration as the slug moves downstream is described by (195). For an observer moving with the average flow, this equation becomes

$$\partial C/\partial t' = D'(\partial^2 C/\partial x'^2) \tag{200}$$

where $t' = t$, $x' = x - (q/n)t$. This is the well-known heat conduction equation.

The initial conditions here are in the form of a Dirac delta function $\delta(x)$

$$C(x, 0) = (m/n)\, \delta(x) \tag{201}$$

where m is the total amount of tracer contained in the slug. The Dirac distribution $\delta_m(x)$ is described by

$$\delta_m(x) = 1/m \quad \text{for} \quad 0 < x < m; \qquad \delta_m(x) = 0 \quad \text{elsewhere} \tag{202}$$

where m is a small positive number. Then $\delta(x) = \lim_{m \to 0} \delta_m(x)$. Boundary conditions specified for $C(x', t')$ are

$$\lim C(x', t') = 0, \qquad \mid x' \mid \to 0, \qquad \int_{-\infty}^{+\infty} C(x', t')\, dx' = m/n \tag{203}$$

The solution of (200) under these initial and boundary conditions is (Crank, 1956)

$$C(x, t) = \frac{m/n}{(4\pi D't)^{1/2}} \exp \left[- \frac{x'^2}{4D't'} \right] = \frac{m/n}{(2\pi)^{1/2}\, \sigma} \exp \left[- \frac{(x - \bar{x})^2}{\sqrt{2}\, \sigma} \right] \tag{204}$$

where $\bar{x} = (q/n)t$; $\sigma^2 = 2D't$. The shapes of the curves $C(x', t)$ are given in Fig. 12.

Case (i) may be obtained from this case directly by superposition. Let $C'(x', t')$ denote the slug solution of (198). Remaining in the (x', t') coordinates, the progress of a front is given by integrating the effect of

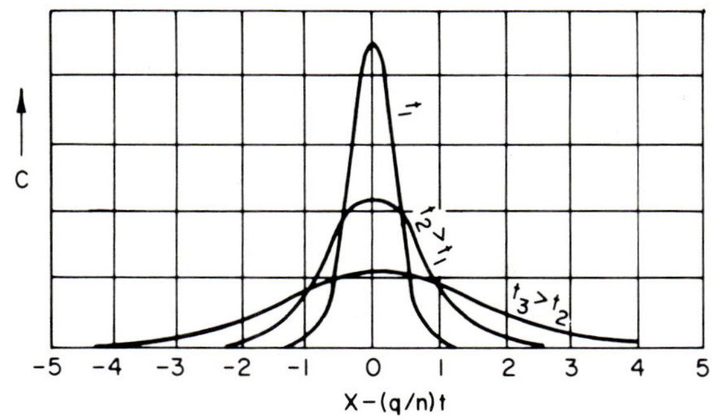

FIG. 12. Progress of a slug.

an infinite number of thin slugs as in case (i), each containing the amount of tracer $m = C_0 n\, d\xi$ (Fig. 13). We obtain (Crank, 1956)

$$\frac{C(x,t) - C_0}{C_1 - C_0} = \int_{-\infty}^{0} C'(x - \xi, t)\, d\xi; \qquad C' = C \text{ of } (204)$$
$$\frac{C(x,t) - C_0}{C_1 - C_0} = \tfrac{1}{2} \operatorname{erfc} \left[\frac{x - qt/n}{(4D't)^{1/2}} \right] \tag{205}$$

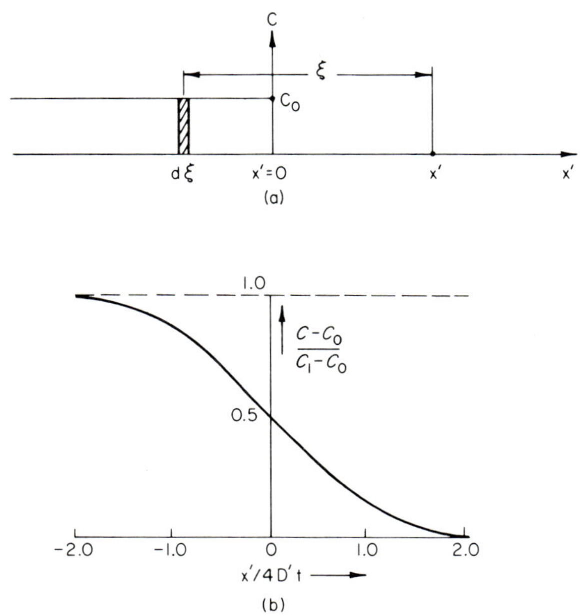

FIG. 13. Progress of a front.

Case (iii). Semi-infinite column of porous medium $x > 0$. The column is adjacent to a reservoir containing a tracer solution of constant concentration C_0. The flow in the column is maintained at a constant rate (specific discharge q in the $\mathbf{1x}$ direction). In addition, the tracer in the column continuously undergoes radioactive decay. We assume that at $x = 0$ the concentration reaches its ultimate value C_0 immediately upon the commencement of the flow. This is equivalent to assuming that $\lim_{t\to 0} (\partial C/\partial x)|_{x=0} = 0$ in (193). The differential equation here is

$$\frac{\partial C}{\partial t} = D' \frac{\partial^2 C}{\partial x^2} - \frac{q}{n}\frac{\partial C}{\partial x} - \lambda C \tag{206}$$

Initial conditions:

$$t = 0, \qquad x \geqslant 0 \qquad C = 0$$

Boundary conditions:

$$\begin{aligned} t > 0, \qquad x &= 0 \qquad C = C_0 \\ x &= \infty \qquad C = 0 \end{aligned} \tag{207}$$

By applying the Laplace transform to (206) and (207) we obtain

$$\begin{gathered} D' \frac{\partial^2 C^*}{\partial x^2} - \frac{q}{n}\frac{\partial C^*}{\partial x} - (\lambda + p)\, C^* = 0, \qquad C^* = \int_0^\infty C \exp(-pt)\, dt \\ x = 0, \qquad t > 0, \qquad C^* = C_0/p \end{gathered} \tag{208}$$

The solution of (208) is

$$C^*(p) = \frac{C_0}{p} \exp\left\{x\left[\frac{q}{2D'n} - \left(\frac{q^2}{4D'n^2} + \frac{\lambda + p}{D'}\right)^{1/2}\right]\right\} \tag{209}$$

which leads to the solution

$$C(x, t) = C_0 \exp\left\{\frac{qx}{2D'n}\right\} \cdot \frac{1}{2\pi i}\int_{\gamma - i\infty}^{\gamma + i\infty} \frac{e^{yt}}{y} \exp\left\{-\left(\beta^2 + \frac{y}{D'}\right)^{1/2}\right\} dy \tag{210}$$

$$\beta^2 = \frac{q^2}{4D'^2 n^2} + \frac{\lambda}{D'}$$

Integrating (210) we obtain (Grobner and Hofreiter, 1949)

$$\begin{aligned} C(x, t) = \tfrac{1}{2} C_0 e^{qx/2D'n} \Big[& e^{-x\beta} \operatorname{erf} \frac{x - [(q/n)^2 + 4\lambda D']^{1/2}\, t}{2(D't)^{1/2}} \\ & + e^{x\beta} \operatorname{erfc} \frac{x + [(q/n)^2 + 4\lambda D']^{1/2}\, t}{2(D't)^{1/2}} \Big] \end{aligned} \tag{211}$$

When $\lambda = 0$ and we neglect molecular diffusion, (211) becomes

$$C(x, t) = \tfrac{1}{2}C_0 \left[\operatorname{erfc} \frac{x - (q/n)\,t}{2(a_I q t/n)^{1/2}} + e^{x/a_I} \operatorname{erfc} \frac{x + (q/n)\,t}{2(a_I q t/n)^{1/2}}\right] \tag{212}$$

derived also by Ogata and Banks (1961). According to them, the second term in (212) may be neglected when x/a_I is sufficiently large, a condition usually satisfied in practice at some distance from the inflow boundary. Equation (212) is an approximation of (211) also when molecular diffusion is negledted and $4\lambda a_I n/q \ll 1$. Thus, this parameter is a criterion for the relative importance of radioactive decay of the considered tracer. It is a combination of the medium's parameters (a_I, n), the flow (q), and the tracer (λ).

If the second term of (211) is neglected and we let $t \to \infty$, we obtain

$$\frac{C(x, t)}{C_0} = \exp \left\{\frac{x}{2a_I} [1 - (1 + 4\lambda n a_I/q)^{1/2}]\right\} \tag{213}$$

Case (iv). An infinite two-dimensional field with uniform flow at $q = \text{const}$ *in the direction* $\mathbf{1x}$. Initially ($t = 0$), an abrupt interface in the form of a straight line $y = x \tan \alpha$ (Fig. 14) separates the tracer fluid region at concentration C_0 from the region free of tracer.

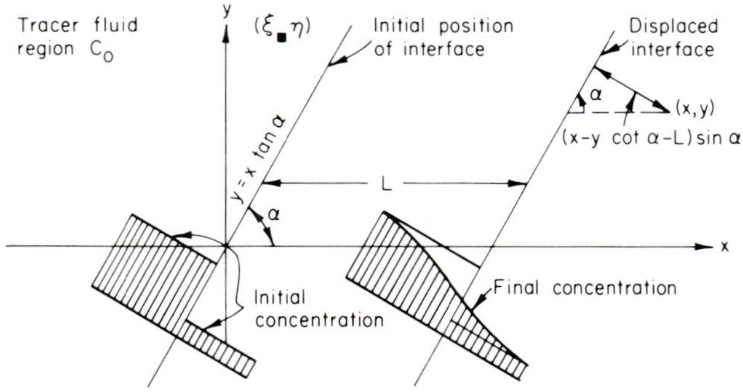

FIG. 14. Transition zone at a moving interface.

The equation defining $C(x, y, t)$ is by (182)

$$\frac{\partial C}{\partial t} = D' \frac{\partial^2 C}{\partial x^2} + D'' \frac{\partial^2 C}{\partial y^2} - \left(\frac{q}{n}\right) \frac{\partial C}{\partial x}; \qquad D' = a_I q/n + D_d T^*; \qquad D'' = a_{II} q/n + D_d T^* \tag{214}$$

where a_I and a_{II} are the longitudinal and the transversal dispersivities of the porous medium.

We may start, as in case (ii), from an elementary solution and integrate over the entire tracer fluid region. If m is the mass of a point source injected at (ξ, η) at $t = 0$, the concentration distribution in the field at any later time is given by

$$m/n^2 = \int_{-\infty}^{+\infty} \int_{-\infty}^{+\infty} C(x', y', t)\, dx'\, dy'$$

$$C(x, y, t) = \frac{m/n^2}{4\pi(D'D'')^{1/2}\, t} \exp\left\{ -\frac{(x - \xi - (q/n)\, t)^2}{4D't} - \frac{(y - \eta)^2}{4D''t} \right\} \tag{215}$$

Lines of equal tracer concentration are in the form of ellipses centered at $[\xi + (q/n)t, \eta]$. To determine the concentration distribution resulting from the moving front, we integrate the effect of an infinite number of small point injections, each with $m = Cn\, d\xi\, d\eta$. The result is

$$\begin{aligned} C(x, y) &= \int_{x \operatorname{tg} \alpha}^{+\infty} d\eta \int_{-\infty}^{+\infty} \frac{m/n^2}{4\pi(D'D'')^{1/2}\, t} \\ &\quad \times \exp\left\{ -\frac{(x - \xi - (q/n)\, t)^2}{4D't} - \frac{(y - \eta)^2}{4D''t} \right\} d\xi \\ &= \frac{C_0}{2} \operatorname{erfc}\left\{ \frac{[x - (q/n)\, t] \sin\alpha - y \cos\alpha}{[4(D' \sin^2\alpha + D'' \cos^2\alpha)\, t]^{1/2}} \right\} \end{aligned} \tag{216}$$

Equation (216) describes a normal distribution perpendicular to the displaced front.

When the motion is parallel to the interface, $\alpha = 0$, and (216) yields

$$C(x, y) = \frac{C_0}{2} \operatorname{erfc}\left\{ \frac{-y}{(4D''t)^{1/2}} \right\} \tag{217}$$

Among other solutions available in the literature we may mention the semi-infinite column with a constant concentration boundary condition (Ogata and Banks, 1961), a semi-infinite column with adsorption (Lapidus and Amundson, 1952; Banks and Ali, 1964), and a finite column with adsorption (Bastian and Lapidus, 1956).

List of Symbols

A	Area; concentration of adsorbed matter; constant
a	Radius of capillary tube; constant
a_{I} ; a_{II}	Longitudinal and transversal dispersivities of a porous medium, respectively (dim L)
a_{ijkl}	Dispersivity tensor of a porous medium (dim L)
B	Coefficient; constant; conductance of an elementary channel in a porous medium model (dim L^2)
b	Variable

C	Concentration (mass of tracer per unit volume of solution)
C_0	Original tracer concentration
C_1	Tracer concentration of displacing water
D	Coefficient of mechanical dispersion (dim L^2T^{-1}, components D_{ij})
D_d	Coefficient of molecular diffusion (dim L^2T^{-1})
D_d^*	Molecular diffusivity of porous medium ($D_d\underline{T}^*$)
D'	Coefficient of dispersion
D_L, D_T	Longitudinal and transversal coefficients of dispersion, respectively
d	Mean grain size
f	Function; scalar variable within pore space
G	Extensive property
g	Gravitational acceleration; intensive property per unit volume
g_{ij}	Fundamental tensor of the Riemannian space
h_i	Coordinate scale
I	Source–sink function
I, J, K	Unit vectors in cartesian coordinates
i	Number; index
$\underline{\mathbf{J}}^*$	Mass flux relative to the mass average velocity)
J_0	Modified bessel function of first rank and zero order
j	Number; integer
K	Hydraulic conductivity (dim LT^{-1})
k	Intrinsic permeability (dim L^2)
L	Length dimension; length; displacement
l	Length
M	Dimension of mass
m	Mass; constant
N	Integer
n	Porosity; number; length along normal
P	Probability
p	Pressure; probability
p_e	Peclet number
Q	Volume discharge
q	Specific flux or discharge (dim LT^{-1}); probability ($= 1 - p$)
R	Radius; radial distance in spherical coordinates; radius vector
REV	Representative elementary volume
$\mathscr{R}$	Density of the force due to liquids viscosity (div τ)
r	Radius; radial distance in polar and cylindrical coordinates; displacement vector
r_{ij}	Correlation coefficient
S	Surface
s	Distance along channel axis
T	Time dimension; aquifer transmissivity
T_{ij}, T_{ij}^*	Components of tortuosity (tensor) of a porous medium
t	Time
t_0	Autocorrelation time
U	Volume
U_0	Pore volume
u	Velocity; partical specific volume
u, v, w	Velocity components in the x, y, z directions, respectively
V	Velocity; volume average velocity
V*	Mass average velocity (barycentric velocity)
W	Probability
w	Cross sectional area of an elementary channel in a porous medium model
x, y, z	Cartesian coordinates
α	Angle; number
β	Angle; constant
γ	Unit weight; density of property per unit mass; constant
δ	Angle; element; Kronecker delta; coefficient of the capillary tube's cross section; Dirac function
ϵ	Relative tracer concentration
ζ, η	Coordinates of a point in a channel in a porous medium model
θ	Angle; time of step; integration variable for time
λ	Function; coefficient; dimensionless time

λ_1 , λ_2 , λ_3	Coordinates in the channel in a porous medium model
μ	Dynamic viscosity
μ_{ii}	Variance; correlation matrix
μ'	Cofactor of determinant
ν	Kinematic viscosity
ξ, η, ζ	Coordinates
ρ	Mass density; specific mass; dimensionless coordinate
σ	Standard deviation; distance along a streamline in a channel in a porous medium model
τ	Viscous stress; average residence time; time; tortuosity
Φ	Normal distribution function
ϕ	Normal density function
φ	Piezometric head
ψ	Angle
Ψ	Streamfunction
Ω	Angle in spherical coordinates
ω	Tracer mass concentration (mass of tracer per unit mass of solution)

Special Symbols

$(\bar{\ })$	Average
$(\underline{\ })$	Second-rank tensor
$(\mathring{\ })$	Deviation from average
$\langle(\ )\rangle$	Average in a tube or channel
$\div$	Proportionality
1i	unit vector in i direction
D/Dt	Hydrodynamic (material, substantive) derivative
	Boldface letter indicates vector

Subscripts

i, j	Component
L	Longitudinal
m	Model
T	Transversal or lateral
$x, y, z, s,; \xi, \eta, \zeta$	Directional components
α	Component of a solution
0	Initial

Acknowledgment

The material presented here was prepared for the Hydrology Institute at Princeton University, July–August 1965 sponsored by the National Science Foundation. Thanks are due to Dr. Y. Bachmat, who reviewed this chapter, for his useful comments and suggestions.

REFERENCES

Aris, R., (1956). On the dispersion of a solute in a fluid flowing through a tube, *Proc. Roy. Soc. (London)* **A235**, 67.

Bachmat, Y., and Bear, J. (1964). The general equations of hydrodynamic dispersion in homogeneous isotropic porous mediums, *J. Geophys. Res.* **69**, 2561–2567.

Banks, R. B., and Ali, I. (1964). Dispersion and absorption in porous media flow, *Proc. Am. Soc. Civil Engrs, J. Hydraulics* **90**, 13–31.

Bastian, W. C., and Lapidus, L. (1956). Longitudinal diffusion in ion exchange and chromatographic columns, *J. Phys. Chem.* **60**, 816.

Batchelor, G. K. (1953). "Theory of Homogeneous Turbulence." Cambridge Univ. Press, London and New York.

Bear, J. (1960). The transition zone between fresh and salt waters in coastal aquifers, Ph.D. thesis, Univ. of California, Berkeley.

Bear, J. (1961). On the tensor form of dispersion in porous media, *J. Geophys Res.* **66**, 1185–1197.

Bear, J. (1961). Some experiments on dispersion, *J. Geophys. Res.* **66**, 2455–2467.

Bear, J., and Bachmat, Y., (1965). A unified approach to transport phenomena in porous media, Underground storage and mixing of water, Progr. Rept., No. 3, Haifa, Technion–Israel Inst. of Technol., Hydraulic Lab., P.N. 1/65.

Bear, J., and Bachmat, Y. (1966). Hydrodynamic dispersion in non-uniform flow through porous media, taking into account density and viscosity differences, Technion-Israel Inst. of Technol., Hydraulic Lab., P.N. 4/66. (in Hebrew with English summary).

Bear, J., and Todd, D. K. (1960). The transition zone between fresh and salt waters in coastal aquifers, Univ. of California, Berkley, Hydraulics Lab., Water Resources Center Contribution No. 29.

Beran, N. J. (1955). Dispersion of soluble matter in slowly moving fluids, Ph.D. thesis, Harvard Univ., Cambridge, Massachusetts.

Bird, R. B., Stewart, W. E., Lightfoot, E. N. (1960). "Transport Phenomena." Wiley, New York.

Blackwell, R. J. (1957). An investigation of miscible displacement processes in capillaries, Paper presented at the local Section Meeting of A.I.Ch.E., Galveston, Texas.

Blackwell, R. J. (1959). Experiments on mixing by fluid flow in porous media, *A.I.Ch.E. S.P.E., 52nd Ann. Meeting, San Francisco,* Reprint No. **29.**

Blackwell, R. J., Rayne, J. R., Terry, W. M. (1959). Factors influencing the efficiency of miscible displacement, *Trans. A.I.M.E.* **216**, 1–8.

Bondarev, E. A., and Nikolaevskij, V. N. (1962). Convective diffusion in porous media with the influence of adsorption phenomena, *Acad. Sci. USSR, J. Appl. Mech. Tech. Phys.* **5**, 128–134.

Brigham, W. E., Reed, P. W., and Dew, J. N. (1959). Experiments on mixing by fluid flow in porous media, *A.I.Ch.E.A.I.M.E., Symp. Oil Recovery Methods, San Francisco, California.*

Carberry, J. J., and Bretton, R. H. (1958). Axial dispersion of mass in flow through fixed beds, *A.I.Ch.E.J.* **4**, 367–375.

Chandrasekhar, S. (1943). Stochastic problems in physics and astronomy, *Rev. Mod. Phys.* **15**, 1–87.

Crank, J. (1956). "The Mathematics of Diffusion." Oxford Univ. Press, London and New York.

Danel, P. (1952). The measurement of ground water flow, *Ankara Symp. Arid Zone Hydrology, Proc. No.* 2, UNESCO, pp. 99–107.

Dankwerts, P. V. (1953). Continuous flow systems (distribution of residence times), *Chem. Eng. Sci.* **2**, 1–13.

Day, P. R. (1956). Dispersion of a moving salt water boundary advancing though a saturated sand, *Trans. Am. Geophys. Union* **37**, 595–601.

De Groot, S. R., and Mazur, P. (1962). "Non-equilibrium Thermodynamics." North-Holland Publ., Amsterdam. (Reprinted 1963.)

De Josselin de Jong, G. (1958). Longitudinal and transverse diffusion in granular deposits, *Trans. Am. Geophys. Union* **39**, 67–74.

De Josselin de Jong, G., and Bossen, M. J. (1961). Discussion of paper by J. Bear, On the tensor form of dispersion, *J. Geophys. Res.* **66**, 3623–3624.

Ebach, E. A., and White, R. R. (1958). Mixing of fluids flowing through beds of packed solids, *A.J.Ch.E.*, **4**, 161–164.

Feller, W. (1958), "An Introduction to Probability Theory and Its Applications," (2nd ed.). Wiley, New York.

Gröbner, W., and Hofreiter, N. (1949). "Integraltafel," Vol. 1 (1949), 166 pp., Vol. 2 (1950), 204 pp. Springer, Vienna.

Harleman, D. R. F., Mehlhorn, P. F., and Rumer, R. R., (1963), Dispersion permeability correlation in porous media, *Am. Soc. Civil Engrs.* HY 2, 67–85.

Hiby, J. W. (1959). Paper presented at conference on displacement in porous media, Imperial College, London, March 1959.

Klinkenberg, L. J. (1951). Analogy between diffusion and electrical conductivity in porous rocks, *Bull. Geol. Soc. Am.* **62**, 559–564.

Lapidus, L., and Amundson, N. R. (1952). Mathematics of adsorption in beds; VI–The effect of longitudinal diffusion in ion exchange and chromatographic columns, *J. Phys. Chem.* **56**, 984–988.

Nikolaevskij, V. N. (1959). Convective diffusion in porous media, *J. Appl. Math. Mech. (P.M.M.)* **23**, 1042–1050.

Ogata, A., and Banks, R. B. (1961). A solution of the differential equation of longitudinal dispersion in porous media, U.S. Geol. Surv., Professional paper 411-A.

Pfannkuch, H. O. (1962). Contribution à l'étude des déplacements des fluides miscibles dans un milieu poreux., Inst. Français du Pétrole.

Poreh, M. (1965). The dispersivity tensor in isotropic and axisymmetric mediums, *J. Geophys. Res.* **70**, 3909–3914.

Prandtl, L., and Tietjens, O. G. (1934). "Fundamentals of Hydro- and Aeromechanics." McGraw-Hill, New York.

Raimondi, P., Gardner, G. H. F., and Petrick, C. B. (1959). Effect of pore structure and molecular diffusion on the mixing of miscible liquids flowing in porous media, *A.I.Ch.E.S.P.E., 52nd Ann. Meeting, San Francisco.* (Preprint 43)

Rifai, M. N. E., Kaufman, W. J., and Todd, D. W. (1956). Dispersion phenomena in laminar flow through porous media, Sanitary Eng. Res. Lab. and Div. of Civil Eng., Univ. of Calif., Berkeley, Progr. Rept. No. 2, Canal Seepage Res., I.E.R. Series No. 93, Issue No. 2.

Robertson, H. P. (1940). The invariant theory of isotropic turbulence, *Proc. Phil. Soc.* **36**, 209–223.

Saffman, P. G. (1959). A theory of dispersion in a porous medium, *J. Fluid Mech.* **6**, 321–349.

Saffman, P. G. (1960). Dispersion due to molecular diffusion and macroscopic mixing in flow through a network of capillaries, *J. Fluid Mech.* **7**, 194–208.

Scheidegger, A. E. (1954). Statistical hydrodynamics in porous media, *J. Appl. Phys.* **25**, 994.

Scheidegger, A. E. (1957, 1960). The physics of flow through porous media (1st ed. 1957; 2nd ed. 1960), Univ. of Toronto Press, Toronto.

Scheidegger, A. E. (1957a). On the theory of flow of miscible phases in porous media, Intern. Assoc. of Scientific Hydrology, IUGG Gen. Assembly of Toronto (1957) Vol. II, pp. 236–242.

Scheidegger, A. E. (1958a). Statistical approach to miscible displacement in porous media, *Bull. Can. Inst. Min. Met.* 26–30.

Scheidegger, A. E. (1958b). The random walk model with autocorrelation on flow through porous media, *Can. J. Phys.* **36**, 649–658.

Scheidegger, A. E. (1961). General theory of dispersion in porous media, *J. Geophys. Res.* **66**, 3273–3278.

Scheidegger, A. E. (1965). Markov processes and dispersion theory, *Bull. IASH*, 38–40.

Simpson, E. S. (1962). Transverse dispersion in liquid flow through porous media, pp. 1–29, USGS, Prof. Paper 411-C, Washington, D. C.

Slichter, C. S. (1905). Field measurements of the rate of movement of underground waters, Water Sup. Paper No. 140, U. S. Geol. Surv.

Taylor, G. I. (1953). Dispersion of soluble matter in solvent flowing slowly through a tube, *Proc. Roy. Soc. (London)* **A219**, 186–203.

Taylor, G. I. (1954). The dispersion of matter in turbulent flow through a pipe, *Proc. Royal Soc. (London)* **A223**, 446.

Van Deemter, J. J., Brader, J. J., and Lauweier, H. A. (1955). Fluid displacement in capillaries, *Appl. Sci. Res.* **A5**, 374–388.

Von Rosenberg, D. U. (1956). Mechanics of steady state single phase fluid displacement from porous media, *A.I.Ch.E.J.* **2**.

Wentworth, C. K. (1948). Growth of the Ghyben-Herzberg transition zone under a rinsing hypothesis, *Trans. Am. Geophys. Union* **29**, 97–98.

5

Velocity and the Longitudinal Dispersion Coefficient in Flow through Porous Media

E. S. Simpson

Hydrology and Water Resources Office
University of Arizona
Tucson, Arizona

1. Introduction

This chapter examines the role and importance of molecular diffusivity in longitudinal dispersion of a fluid flowing through a porous medium. It is universally recognized that molecular diffusivity affects dispersion but many, if not most, authors assume that its effect is small enough to be neglected. Hiby (1962) includes the coefficient of molecular diffusivity in an empirical equation of best fit for his experimental data. Hiby's equation is given later on in this chapter as Eq. (23a) and is compared in Figs. 1 and 2, with this author's equation. Except for Hiby, the author is aware of no other attempt, neither analytical nor empirical, to include molecular diffusivity in equations relating dispersion to flow velocity. In a brief comment on the problem, Biggar and Nielsen (1962) point out that the tendency to neglect molecular diffusivity is often based on the results of crude experiments, and they conclude by saying that "mixing by molecular diffusion cannot be neglected and must be included in any general theory of dispersion in porous media."

In this analysis the following assumptions are made: (a) the fluid is an incompressible liquid at constant temperature, (b) the porous media are homogeneous and isotropic and are composed of randomly oriented grains, and (c) there is no sorption or other interaction between tracer particles and grains of the media. Also, for the most part, the assumption is made that the flow velocity does not exceed the upper limit of validity of Darcy's law as defined by the critical Reynolds number, $R_c \leqslant 1$ (approximately). (For porous media, the Reynolds number may be written $R = ud/\nu$, where d is the average grain diameter instead of the pore or tube diameter.)

A convenient way to approach the mechanism of dispersion is to consider Taylor's (1953) analysis of dispersion in flow through tubes. He verified previous experiments which showed that when the velocity of flow through a straight tube is slow enough, any tracer injected in the fluid (say, some dissolved dye initially introduced into the flow region in the shape of a thin disc), will not disperse solely in response to the velocity distribution over the tube section, which in this case would result in a parabaloid of revolution, but rather will disperse symmetrically about a cross section which moves with the mean velocity of the fluid. The explanation for this phenomenon lies in the microscale or "Brownian" motion of each dissolved particle, as a consequence of the thermokinetic energy of the fluid system. Given sufficient time, the radial component of such motion will have the effect of transferring each tracer particle from its initial fluid flow path or streamline, randomly to every other flow path in the tube section. Hence, even though the initial position of a given particle was, say, in the axial streamline of maximum fluid velocity, the probability for that particle remaining in the same streamline becomes vanishingly small as time increases. And, in addition, as time increases, the probability that a given tracer particle will successively occupy *every* possible position or streamline of the tube cross section approaches certainty [see, for example, Feller (1950, pp. 297–300)]. This mechanism provides a means for giving each tracer particle the possibility of flowing, at one time or another, at all possible velocities from zero to maximum. And, as will be shown later, the average velocity of a large number of elements that randomly occupy all possible streamlines in a tube is the same as the average velocity of the fluid.

Imagine now that the tube is filled with a granular material that satisfies the previous assumption (b), but that the molecular diffusivity of the tracer elements is zero; that is, Brownian motion does not occur. The streamlines are tortuous, but their average direction is parallel to the tube axis. However, individual streamlines no longer represent constant velocities along their length. This arises from the geometry of the flow whereby a streamline which at one point may be near the center of a pore space and at maximum flow velocity, whereas at another point the same streamline may be near the boundary of a pore space and near zero velocity. The medium is homogeneous if the average length of any small bundle of streamlines and the average velocity along such a bundle are the same as that for any other (parallel) small bundle. However, the probability that any two individual streamlines are identical is vanishingly small. Hence, a tracer element flowing along a given streamline would experience all possible flow velocities (except zero velocity in the absence of diffusion) within a flow distance of a few grain diameters. Furthermore, since the grains are randomly oriented, the changes in velocity along each streamline between points more than, say, one grain diameter apart, will also be random. [For a discussion of the random motion of fluid particles see, for example, Beran (1955, pp. 4.1–4.10) or Scheidegger (1959).] As a consequence of the random distribution of streamline lengths and velocities, we should expect dispersion of tracer particles to occur symmetrically about a cross section moving with average fluid velocity. However, the molecular diffusivity of tracer particles is never zero; therefore, in a porous medium both mechanisms for dispersion are present: (a) that resulting from molecular diffusivity as outlined by Taylor, where each pore space is here considered as a tiny tube, and (b) that resulting from the geometry of the flow as outlined in the previous paragraph. An important difference between them exists: the former is velocity dependent, whereas the latter, within the range of Darcy's law, is not velocity dependent. The present problem is to mathematically describe the simultaneous effect on dispersion of both mechanisms as a function of velocity, and in a manner that will represent the essential physical reality.

2. Analysis

2.1. Average Velocity of Tracer Particles in Flow through a Tube

By virtue of the radial component of Brownian motion, and neglecting short-term "memory," a tracer particle may randomly occupy all positions in the flow cross section. Short-term "memory" is the condition that from a given position and within a time period that is short compared to the total time of flow, a particle is restricted to positions only in its immediate vicinity. Suppose the cross section of a fluid flowing through a circular tube is divided into a nest of small concentric areas. The average residence time of a particle in any small concentric area compared to the total time of flow equals the size of that area compared to the total cross section area. Or, in differential form,

$$dt/t_0 = dA/A_0 \tag{1}$$

where t_0 is the total time of flow and A_0 is the cross section area. For a finite distance of flow at sufficiently low flow velocity, a tracer particle might randomly occupy all possible small concentric areas; more realistically, this statement is valid for a particle repeatedly introduced into the same flow field a large number of times, but at different points in the initial cross section. The average axial distance of flow of such a particle would be

$$\bar{x} = \int_0^{t_0} u\,dt = (t_0/A_0)\int_0^{A_0} u\,dA = (2\pi t_0/A_0)\int_0^{r_0} ur\,dr \tag{2}$$

where $\bar{x}$ is the average axial motion of tracer particle, u is the fluid velocity, and r_0 is the tube radius. For laminar flow through a circular tube, the velocity is a function of the radius

$$u/u_{\max} = 1 - (r/r_0)^2 \tag{3a}$$

Differentiation of (3a) gives

$$r\,dr = -(r_0^2/2u_{\max})\,du \tag{3b}$$

Substitution of Eq. (3b) into Eq. (2) and integration lead to

$$\bar{x} = (-t_0/u_{\max})\int_{u_{\max}}^{0} u\,du = \tfrac{1}{2}(t_0 u_{\max}) = \bar{u}t_0 \tag{4}$$

where $\bar{u}$ is the average fluid velocity and, for a circular tube, $\bar{u}$ is the $\frac{1}{2}u_{\max}$.

Hence, $\bar{x}/t_0 = \bar{u}$, or the average axial velocity of tracer elements equals the average fluid velocity.

Since the random motion of a single particle repeated a large number of times is equivalent to the simultaneous but independent motion of a large number of particles, Eq. (4) also describes the average velocity of a thin disc of tracer particles introduced into a section of the flow in a straight circular tube.

It will be noted that in the previous analysis, the axial component of Brownian motion is neglected. The average fluid velocity, while small, is nevertheless large compared to the axial component of thermokinetic motion. In any case, the effect of the latter on *average* axial velocity sums to zero.

2.2. Dispersion of Tracer Particles in Flow through a Tube

In Taylor's (1953) analysis, the dispersion of tracer particles about a cross section moving with average fluid velocity through a straight tube was described by the following differential equation (symbols altered):

$$D_x(\partial^2 C_m/\partial x_1^2) = \partial C_m/\partial t \tag{5}$$

where $D_x = (r_0^2\bar{u}^2/48D_M)$ is the coefficient of dispersion in axial direction; $x_1 = x - \bar{u}t$; C_m is the mean concentration of the tracer over a given tube cross section; D_M is the coefficient of molecular diffusivity of the tracer; and r_0 is the tube radius. Equation (5) is of the same form as Fick's equation describing molecular diffusion. Solutions to this equation for semi-infinite regions in a straight tube may be found in the literature [see, for example, Crank (1956, pp. 9–15)] and are given by Taylor. The solutions usually are given in terms of the error function, but it is equally correct to write the solutions in terms of the normal distribution function. The normal distribution function, or, more precisely, its derivative, the normal density function, is the limiting function of the binomial distribution, and the binomial distribution is a description of the distribution of the total number of successes in Bernoulli trials [see, for example, Feller (1950, Chapters 6–9)]. That is, the normal density function is a good approximation of the distribution of the total successes in Bernoulli trials, becoming better as the number of trials increases, and in the limit as the number approaches infinity the two distributions are identical. It seems valid, therefore, to describe the observed dispersion in terms of a large number of Bernoulli trials.

Suppose the tube cross section is divided into two concentric regions of equal area A_1 and A_2. The radius of the circle separating the two areas is $1/\sqrt{2}$ times the tube radius. Let A_2 be the inner area and A_1 the outer area. By the previous argument, after sufficient time, the probability for an average tracer particle to be found in region A_1 is the same as for region A_2, no matter what the initial position of the particle might have been. Furthermore, by the nature of the fluid velocity distribution in region A_1, the axial velocity of the particle ranges between zero and $\bar{u}$; in region A_2 the axial velocity ranges between $\bar{u}$ and u_{max}.

Imagine that the time of flow is divided into a large number of equal intervals, each one short compared to the total time, but longer than the short-term "memory." At the end of each interval, each tracer particle is imagined to undergo a Bernoulli trial, where success S means that the particle in question is in flow region A_2 where $u > \bar{u}$, and failure F means it is in region A_1 where $u < \bar{u}$. For purposes of analysis, the velocity everywhere in region A_1 is considered zero, and in region A_2 it is everywhere equal to maximum velocity. Since the two regions are equal in area, the probability p for success is the same as the probability q for failure. That is,

$$p = q = \tfrac{1}{2} \tag{6}$$

After n trials, the average number of successes $\bar{S}$ will be

$$\bar{S} = np = \tfrac{1}{2}n \tag{7a}$$

A distance β_d is designated (hereinafter called the *diffusion beta-length*) to represent the distance of axial flow associated with each success. Then

$$\bar{x} = \beta_d \bar{S} = \tfrac{1}{2}\beta_d n \tag{7b}$$

where $\bar{x}$ is the average axial distance of flow after n trials.

If the molecular diffusivity is assumed to be constant, then for a given tube radius, the number n of trials will vary directly with average flow distance and inversely with average velocity:

$$n = M(\bar{x}/\bar{u}) \tag{8}$$

where M is a constant of proportionality.

Substitution of Eq. (8) into Eq. (7b) and rearrangement of terms gives

$$\beta_d = 2\bar{u}/M \tag{9}$$

The standard deviation σ_s of the successes in n Bernoulli trials is

$$\sigma_s = (npq)^{1/2} = \tfrac{1}{2}\sqrt{n} \tag{10a}$$

Hence, the standard deviation of the dispersion in the axial direction is

$$\sigma_x = \beta_d \sigma_s = \tfrac{1}{2}\beta_d\sqrt{n} \tag{10b}$$

Substitution of Eq. (9) into Eq. (10b) gives

$$\sigma_x = (\bar{u}/M)\sqrt{n}$$

or

$$\sigma_x^2 = (\bar{u}^2/M^2)\, n \tag{10c}$$

Substitution of Eq. (8) into Eq. (10c) gives

$$\sigma_x^2 = \bar{u}\bar{x}/M \tag{10d}$$

As previously stated, solutions of Eq. (5) for dispersion in a straight tube may be obtained either in terms of the error function, in which case the coefficient of dispersion D_x appears as a variable, or in terms of the normal distribution function, in which case the standard deviation of the dispersion σ_x appears as a variable. The two are related by the following equation

$$\sigma_x = (2D_x t)^{1/2} = [2D_x(\bar{x}/\bar{u})]^{1/2} \tag{11a}$$

or

$$D_x = \sigma_x{}^2\bar{u}/2\bar{x} = (2M)^{-1}\,\bar{u}^2 \tag{11b}$$

Combining this result with definition of D_x in Eq. (5) gives

$$(2M)^{-1} = r_0{}^2/48D_{\mathrm{M}} \qquad \text{or} \qquad M = 24D_{\mathrm{M}}/r_0{}^2 \tag{12}$$

2.3. Dispersion of Tracer Particles in Flow through a Porous Medium, Assuming Zero Molecular Diffusivity

The standard deviation of dispersion in fluid flow through porous media is in part velocity dependent and in part nondependent. To examine the nondependent part, molecular diffusivity is, for the moment, assumed zero. The porous medium is understood to be homogeneous. Thus, by definition, for any small cross section of flow δA_m such that δA_m is small compared with the total cross section, but large compared with an average pore section, we have

$$n\bar{u} = (\delta A_1)^{-1}\int_0^{\delta A_1} u\,dA = (\delta A_2)^{-1}\int_0^{\delta A_2} u\,dA = \cdots = (\delta A_m)^{-1}\int_0^{\delta A_m} u\,dA$$

where $A_0 = \delta A_1 + \delta A_2 + \cdots + \delta A_m$ is the total cross section of flow; n is the porosity; and $\bar{u}$ is the average pore velocity.

In the absence of molecular diffusivity, the path of each tracer particle will remain fixed in its initial streamline and its average velocity will be the same as that of the fluid particles in the same streamline. The particles will experience a succession of changes in velocity both in magnitude and direction because each streamline is tortuous in shape and passes through regions of maximum velocity at or near a pore axis to regions of minimum velocity at or near a pore boundary. Because of the random packing of the medium, the changes in velocity also are random, neglecting the effect of short-distance "memory." The argument here is the same as in Section 2.1, except that here the random motion is space dependent, whereas in the previous case, it is time dependent.

Let a streamline be divided into a large number of sections of equal length where the section length is small compared to the total length of the streamline, but the same or slightly larger than the short-distance "memory." For each section let there be associated a magnitude of velocity (rounded off, say, to two or three significant figures), equal to the axial component of the average velocity along that section. Such component magnitudes would range from zero to the flow maximum and would constitute a long sequence of random numbers. The expected average of the sequence would equal one-half the maximum, or $\bar{u} = \frac{1}{2}u_{\max}$. Furthermore, the probability p for any magnitude to be greater than average is the same as the probability q for it to be less than average, or $p = q = \frac{1}{2}$. Motion along the streamline is imagined to be governed by a sequence of Bernoulli trials whereby success S occurs when $u > \bar{u}$ and failure F occurs when $u < \bar{u}$. Success means that the fluid or tracer particle flows an axial distance β_{h} (hereinafter called the *hydrodynamic beta-length*), and failure means no motion. The average axial distance $\bar{x}$ of flow after n trials is

$$\bar{x} = \beta_{\mathrm{h}} np = \tfrac{1}{2}\beta_{\mathrm{h}} n \tag{13}$$

Implicit in the above discussion is the consequence that the length of β_{h} is more than twice the length of the sections into which the streamline is subdivided.

In considering flow patterns from one streamline to another, there will be differences in the number of sections where $u > \bar{u}$. These differences will result in a longitudinal dispersion of fluid or tracer particles. The standard deviation of this dispersion is

$$\sigma_x = \beta_{\mathrm{h}}(npq)^{1/2} = \tfrac{1}{2}\beta_{\mathrm{h}}\sqrt{n} \tag{14}$$

The number n of trials in a given average flow distance $\bar{x}$ is constant no matter what the velocity is within the range of validity of Darcy's law, as a consequence of the assumed constancy of the flow geometry. Hence, by Eq. (13), β_{h} also must be constant. Combination of Eqs. (13) and (14) gives

$$\sigma_x^2 = \tfrac{1}{2}\beta_{\mathrm{h}}\bar{x} \tag{15a}$$

and

$$D_x = \sigma_x^2\bar{u}/2\bar{x} = (\tfrac{1}{4}\beta_{\mathrm{h}})\,\bar{u} \tag{15b}$$

Hence, in flow through a tube with nonzero molecular diffusivity [Eq. (11b)], the coefficient of dispersion is proportional to the square of the average velocity, whereas in flow through a porous medium with zero molecular diffusivity [Eq. (15b)], dispersion is proportional to the first power of the average velocity.

2.4. Dispersion of Tracer Particles in Flow through a Porous Medium, Assuming Nonzero Molecular Diffusivity

Within the framework of the Bernoulli trials analogy, the effect of nonzero molecular diffusivity on longitudinal dispersion in flow through porous media is to increase the number of trials of tracer particles per unit flow distance. Because of Brownian motion, the tracer particles "jump" from streamline to streamline as flow proceeds, often changing from a velocity $u > \bar{u}$ to a velocity $u < \bar{u}$ and vice-versa. The number of such jumps per unit flow distance in a given medium will very inversely with velocity. The total number of trials will then be the sum of those resulting from the geometry of flow n_{h} and those resulting from molecular diffusivity n_{d}, each assumed independent of the other. (Subscripts added to distinguish between the two.) We have then

$$\bar{x} = \beta_{\mathrm{r}}(n_{\mathrm{h}} + n_{\mathrm{d}})\, p = \tfrac{1}{2}\beta_{\mathrm{r}}(n_{\mathrm{h}} + n_{\mathrm{d}}) \tag{16a}$$

or

$$\beta_{\mathrm{r}}^2 = 4\bar{x}^2/(n_{\mathrm{h}} + n_{\mathrm{d}})^2 \tag{16b}$$

where β_{r} is the now reduced beta-length associated with a successful trial. The standard deviation of the dispersion is

$$\sigma_x = \beta_{\mathrm{r}}[(n_{\mathrm{h}} + n_{\mathrm{d}})\, pq]^{1/2} = \tfrac{1}{2}\beta_{\mathrm{r}}(n_{\mathrm{h}} + n_{\mathrm{d}})^{1/2} \tag{17a}$$

$$\sigma_x^2 = \tfrac{1}{4}\beta_{\mathrm{r}}^2(n_{\mathrm{h}} + n_{\mathrm{d}}) = \bar{x}^2/(n_{\mathrm{h}} + n_{\mathrm{d}}) \tag{17b}$$

If each pore space is regarded as a small section of a tube, the use of Eq. (8) gives a value of n_{d} :

$$n_{\mathrm{d}} = M(\bar{x}/\bar{u})$$

Also, Eq. (13) renders a value of n_{h} :

$$n_{\mathrm{h}} = 2\bar{x}/\beta_{\mathrm{h}}$$

Substitution of these values of n_{d} and n_{h} in Eq. (17b) gives

$$\sigma_x^2 = \beta_{\mathrm{h}}\bar{x}\bar{u}/(2\bar{u} + M\beta_{\mathrm{h}}) \tag{17c}$$

and

$$D_x = \sigma_x^2\bar{u}/2\bar{x} = \tfrac{1}{2}[\beta_{\mathrm{h}}\bar{u}/(2\bar{u} + M\beta_{\mathrm{h}})]\,\bar{u} \tag{18}$$

The hydrodynamic beta-length β_{h} is an intrinsic property of the porous medium since it depends on flow geometry only, and within the range of Darcy's law, flow geometry depends on medium geometry only.

Since this beta-length is associated with the characteristics of a streamline, it seems reasonable to suppose that it will be some linear function of a characteristic length, say, of average grain diameter. Hence,

$$\beta_{\mathrm{h}} = \alpha_1 d \tag{19}$$

where d is the average grain diameter and α_1 is the dimensionless coefficient depending on porosity, grain shape, packing, etc. According to Eq. (12), for flow through a straight tube,

$$M = 24(D_{\mathrm{M}}/r_0^2) = 96(D_{\mathrm{M}}/d_0^2)$$

where d_0 is the diameter of a tube.

With the additional assumption that the pore spaces may be approximated by a network of interconnecting short tubes, and that average pore diameter is a function of average grain diameter,

$$d_0 = \sqrt{\alpha_2}\, d \tag{20}$$

where α_2 is the dimensionless coefficient.

Substitution of Eqs. (12), (19), and (20) into (18) renders

$$D_x = \frac{\sigma_x^2 \bar{u}}{2\bar{x}} = \frac{1}{2}\left(\frac{\alpha_1 d\bar{u}}{2\bar{u} + 96(D_{\mathrm{M}}\alpha_1/\alpha_2 d)}\right)\bar{u} \tag{21a}$$

Division of the top and bottom of the right side of Eq. (21a) by the kinematic viscosity ν of the fluid finally gives

$$D_x = \frac{\sigma_x^2 \bar{u}}{2\bar{x}} = \frac{d}{4}\left(\frac{\alpha_1 R}{R + 48(\alpha_1 S/\alpha_2)}\right)\bar{u} \tag{21b}$$

where R is the Reynolds number $d\bar{u}/\nu$ and S is the Schmidt number D_{M}/ν.

3. Comparison with Experiment

In 1962, Biggar and Nielsen pointed out that published experiments were too few and inadequate to test the validity of the various theories concerning dispersion. This statement is still essentially correct. The bulk of published experiments lack a systematic examination of the effect on dispersion of molecular diffusivity. In much of the published work the results are given in graphic form only, or with insufficient data to make possible a determination of the Schmidt and Reynolds numbers, or of the standard deviation of the dispersion for individual runs. Fortunately, the report of Harleman *et al.* (1963) provides data in tabular form making possible the calculation of the points shown on Figs. 1 and 2. The Schmidt number is calculated from handbook values of molecular diffusivity of 0.1 M NaCl in water, divided by the kinematic viscosity of water at 20°C, and equals 1.05×10^{-3}.

Rearranging Eq. (21b), we get

$$\frac{2D_x}{\bar{u}d} = \frac{\sigma_x^2}{\bar{x}d} = \frac{\frac{1}{2}(\alpha_1)\,R}{R + 48\,S(\alpha_1/\alpha_2)} \tag{22a}$$

and for $S = 1.05 \times 10^{-3}$,

$$\frac{2D_x}{\bar{u}d} = \frac{\sigma_x^2}{\bar{x}d} = \frac{\frac{1}{2}(\alpha_1)\,R}{R + 0.050(\alpha_1/\alpha_2)} \tag{22b}$$

For each of their experiments, Harleman *et al.* tabulate the value of each variable in the term on the left side of Eq. (22b) plus the value of R based on the average grain diameter of the medium. Hence, the middle term of Eq. (22b) is immediately known. The remaining unknowns are the two dimensionless constants α_1 and α_2.

As previously mentioned, the hydrodynamic beta-length β_h is more than twice the length of the sections into which each streamline is imagined to be divided. The length of each section, in turn, is about the same as the short-distance "memory" of the tracer particles. The short-distance "memory" may be expected to be on the order of one or two grain diameters. Hence, by Eq. (19), α_1 should have a value of about 3 or 4.

The expected value of α_2 is more difficult to estimate. Replacing a porous medium by a network of interconnecting short tubes does not provide a direct estimate of tube diameter. Furthermore, the estimate of M, given in Eq. (12) is based on an analysis of flow through straight tubes. Assuming that the form of Eq. (12) remains the same for the more complex boundery conditions of a network or of the porous medium itself, the constant 24 will not be the same. By the development leading to Eq. (21b), any change in this constant will be contained in the value of α_2. Hence, it does not seem useful to attempt an *a priori* estimate of α_2.

Figure 1 is a plot of the data of Harleman *et al.* for flow through packed spheres compared with Eq. (22b) in which $\alpha_1 = 3.0$ and $\alpha_2 = 1.5$. The fit, although far from perfect, is not unreasonable. Figure 2 is a plot of the data for flow through sand compared with Eq. (22b) in which $\alpha_1 = 3.6$ and $\alpha_2 = 3.6$. Here the fit is much better.

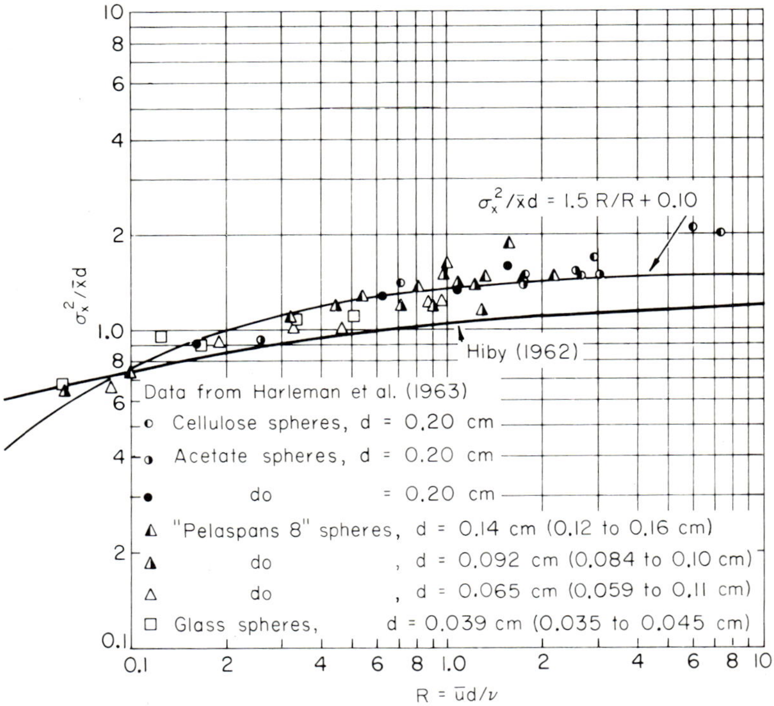

FIG. 1. Longitudinal dispersion versus Reynolds number for flow through spheres.

The empirical equation of Hiby [1962, Eq. (13)], with a change of symbols, is

$$\frac{D_x}{\bar{u}d} = \frac{0.65}{1 + 7.0(D_{\mathrm{M}}/\bar{u}d)^{0.5}} + \frac{0.67\,D_{\mathrm{M}}}{\bar{u}d} \tag{23a}$$

The term on the right side of Eq. (23a) accounts for the effect of the *axial* component of molecular diffusivity when flow velocity is of the same order or less than this component. Hiby's experiments show this to occur at a Reynolds number of about 0.005 or less for aqueous solutions of NaCl. Since the data considered herein are for experiments where $0.04 < R < 10$, the last term of Eq. (23a) may be neglected. The term on the left side, when multiplied by 2, is identical with that of Eq. (23b). Hence, we get

$$\frac{2D_x}{\bar{u}d} = \frac{1.30}{1 + 7.0(D_{\mathrm{M}}/\bar{u}d)^{0.5}} = \frac{\frac{1}{2}(\alpha_1)R}{R + 48\,S(\alpha_1/\alpha_2)} = \frac{\sigma_x{}^2}{\bar{x}d} \tag{23b}$$

Hiby's equation was fitted to data derived from experiments in flow through spheres only. Hence, it is not surprising that his equation provides a better fit to the data in Fig. 1 than in Fig. 2.

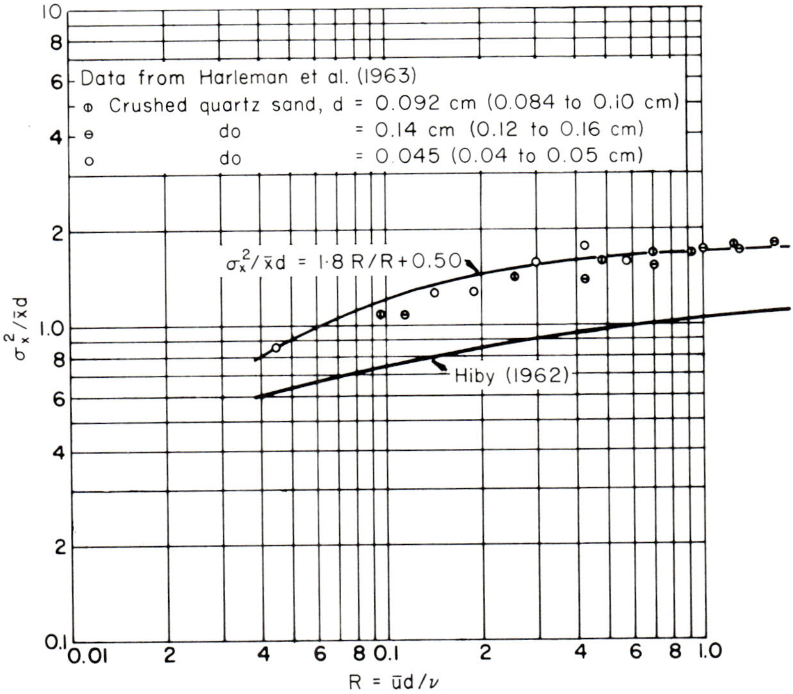

FIG. 2. Longitudinal dispersion versus Reynolds number for flow through quartz sand.

If we divide Eq. (22a) by R, we get

$$\frac{2D_x}{\bar{u}d} = \frac{\frac{1}{2}(\alpha_1)}{1 + 48(\alpha_1/\alpha_2)(D_M/\bar{u}d)} \tag{24}$$

Thus, the principal difference between Hiby and the present writer is that in Hiby's results, the term containing molecular diffusivity is raised to the 0.5 power.

Undoubtedly, data of other authors might also be used. But the magnitude of the variation of dispersion with variation in flow velocity is small. In fact, it is not much larger than the experimental error in the work of a single investigator. There is no standard "dispersometer," nor is there agreement on methodology of measurement; hence, it is not surprising that results vary somewhat from investigator to investigator. To attempt to compare the results of various investigators would add little value to the present analysis. The critical experiments remain to be performed: a systematic investigation of the effect on dispersion of changes in the molecular diffusivity of the tracer particles.

REFERENCES

Beran, N. J. (1955). Dispersion of soluble matter in slowly moving fluids. Thesis, Harvard Univ., Cambridge, Massachusetts.

Biggar, J. W., and Nielsen, D. R. (1962). Some comments on molecular diffusion and hydrodynamic dispersion in porous media, *J. Geophys. Res.* **67**, 3636–3637.

Crank, J., 1956, "The Mathematics of Diffusion." Oxford Univ. Press, London.

Feller, W. (1950). "An Introduction to Probability Theory and Its Application." Wiley, New York.

Harleman, D. R. F., Mehlhorn, P. F., and Rumer, R. R. (1963). Dispersion-permeability correlation in porous media, *J. Hydraulics Div., Am. Soc. Civil Engrs.* **89**, 67–85.

Hiby, J. W. (1962). Longitudinal and transverse mixing during single-phase flow through granular beds, *Symp. Interaction between Fluids and Particles, London, June, 1962*, pp. 312–325. Inst. Chem. Engr.

Scheidegger, A. E. (1959). Statistical approach to miscible displacement in porous media, *Can. Mining Met. Bull.* **52**, 26–30.

Taylor, G. I. (1953). Dispersion of soluble matter in solvent flowing slowly through a tube, *Proc. Roy. Soc. (London)* **A219**, 186–203.

6 The Flow of Water in Unsaturated Soils

Dale Swartzendruber

Department of Agronomy
Purdue University
Lafayette, Indiana

1. Introduction

When the interconnected free pore space of a porous medium is no longer completely filled with a given homogeneous liquid, the porous medium is considered to be unsaturated with respect to that liquid. The character of the unsaturation is, however, dependent upon the nature of the second fluid which occupies the partial pore space vacated by the homogeneous liquid. If the second fluid is another liquid, the resulting overall liquid saturation is heterogeneous if the two liquids are imperfectly mixed or are immiscible. If the second fluid is a gas, or a mixture of gases such as air, then the resulting composite system of solid, liquid, and gaseous phases becomes the one which is frequently meant when one speaks of unsaturated flow. Particularly has this been so if the ultimate concern is the hydrologic one related to soil water, especially when the soil region involved is far above the water table. The process of water flow in such an unsaturated system could also be construed as a two-fluid problem in immiscible displacement, where water and air are the fluids.

Even if the second fluid is air, a further distinction can be made between an entrapped gas phase and an interconnected gas phase. In the entrapped phase, isolated individual bubbles or groups of bubbles exist throughout the system composed otherwise of water and porous medium. As the unsaturation increases, that is, water saturation decreases, the various bubbles coalesce with each other until a reasonably continuous interconnected region of uniform gas pressure exists throughout the wetted system, more or less intimately distributed with the water–solid phases. It is this latter type of unsaturation which forms the primary concern of the present chapter, although admittedly a sharp delineation between the two types may not always be possible. Generally speaking, entrapped air or gas tends to be removed by or through the permeating water (Christiansen, 1944; Gupta and Swartzendruber, 1964; Bloomsburg and Corey, 1964) unless some provision is made to maintain the dissolved air or gas at a suitable level of supersaturation (Wyckoff and Botset, 1936).

Whereas in a water-saturated porous medium the process of water transport is one of liquid flow, the existence of the gaseous phase in unsaturated flow allows also for the possibility of water transport in vapor form. It is generally felt that the component of vapor-phase transport is relatively small compared with the liquid-phase component, unless the soil water content is very low or appreciable vapor-pressure gradients are induced by temperature variations or salts. The primary concern of the present chapter will be with liquid transport under isothermal conditions in salt-free soils.

The subject of unsaturated flow is supported by a rapidly growing literature, all of which cannot be presented within the confines of a single chapter. Certain aspects of the intake and release of water from soil will be treated in the most detail, because these processes are important hydrologic concerns. Readers wishing further coverage, or detail on topics treated briefly or not at all, are directed to a recent review (Swartzendruber, 1966b) and the references cited therein.

2. Liquid-Phase Transport for Flow Response Proportional to Gradient

2.1. The Equation for Flow Velocity

It has become customary to describe unsaturated flow by beginning with the Darcy proportionality for saturated flow and modifying it to take into account the effects of unsaturation. The Darcy equation for saturated flow is

$$\mathbf{q} = -K\,\nabla h \tag{1}$$

where $\mathbf{q}$ is the vector flow velocity, K is the hydraulic conductivity, ∇ is the standard del operator of vector notation, and h is the driving potential expressed as a hydraulic head. In turn, $h = z + p/\rho g$, where z is the elevation (gravitational) head, $p/\rho g$ is the pressure head, p is the gauge pressure, ρ is the density of liquid water, and g is the acceleration of gravity; z is also the vertical position coordinate, positive upward.

2.1.1. *Unsaturated Conductivity*

The first unsaturation modification of Eq. (1) involves the recognition that the air-filled pores will reduce the effective cross section for liquid flow and will increase the tortuosity of the remaining liquid flow paths. Hence, K for unsaturated flow is less than for saturated flow and becomes a function of the water content, expressible as $K = K(\theta)$, where θ is the volumetric water content of the porous medium. Arguments of this type were first set forth by Buckingham (1907), on the basis of analogy with heat and electrical flow, completely without mention of Darcy's equation. The connection with Darcy's equation was noted later by Israelson (1927), explicitly discussed by L. A. Richards (1928, 1931), and utilized by Childs and Collis-George (1948) in relation to diffusion and diffusivity concepts. Buckingham called $K(\theta)$ the capillary conductivity, but, since the term capillary is somewhat overworked and not as descriptive as it might be, $K(\theta)$ will here be called the *unsaturated conductivity*, or simply *conductivity* if the unsaturated connotation is clear from the context.

2.1.2. *Unsaturation Effects on Hydraulic Head*

The second unsaturation modification of Eq. (1) involves the potential function h. The gravitational component z should contribute in the same way as it does for saturated flow, but the component corresponding to the pressure head $p/\rho g$ will be different. To measure this component, it is no longer adequate to conceive of an ordinary piezometer being inserted into the soil to determine the height to which water will rise in it.

Instead, a tensiometer must be envisioned as inserted at the point in question, as shown in Fig. 1. In general, a tensiometer consists of a pressure-measuring device (manometer standpipe or otherwise) filled

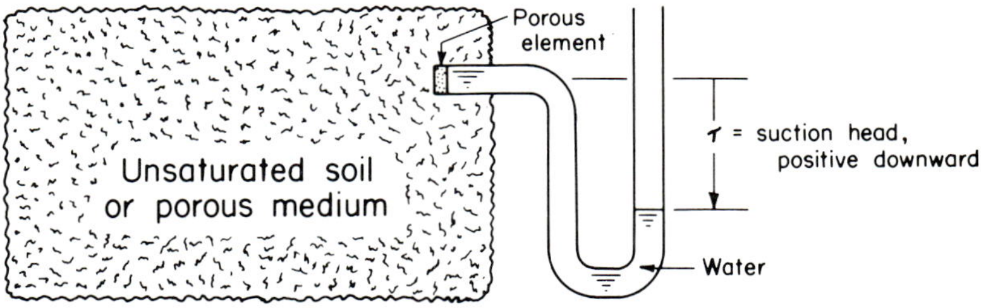

FIG. 1. Schematic representation of the measurement of matric suction head τ with a tensiometer. The water in the tensiometer is at equilibrium with the water in the unsaturated soil, hydraulic communication being obtained through the porous element.

with water which communicates through a porous element with the soil water. The porous element, in contact with the moist soil, must have pores sufficiently small to prevent air from entering the tensiometer.

That the water level in the tensiometer of Fig. 1 stands below the position of the porous element may be construed as evidence for considering the energy of the soil water to be less than the energy of free water positioned at the same level as the porous element. The magnitude of this change in energy might be expressed as the product of an infinitesimal mass of water multiplied by $g\tau$. Or, a potential could be defined as the energy per unit mass of water, that is, simply $g\tau$, which is positive since τ is positive downward. In essence, this is the capillary potential of Buckingham (1907), defined as the energy required to move unit mass of water from the soil to the free-water reference, and identified physically with porous-element apparatus by W. Gardner *et al.* (1922). L. A. Richards (1928), however, has pointed out that the usual convention for potentials would be the reverse; namely, the energy required to move the unit quantity of water from the free-water reference to the soil, a convention followed in recent definitions (Aslyng, 1963; Lutz, 1965). Under this convention, the potential would be $-g\tau$.

L. A. Richards (1961) has also pointed out that if Buckingham's original definition of capillary potential were expressed as energy per unit volume (rather than per unit mass), the result would be $\rho g\tau$. This has units of force per unit area, or pressure, and Richards proposes that it be called the *matric suction* (see also Marshall, 1959). Previously, $\rho g\tau$ has been called the soil moisture tension. The present writer, however, agrees with L. A. Richards (1961) that the term *tension* has already been sufficiently preempted in the soil-water related concepts of surface tension and vapor tension, so that a new term such as matric suction is necessary. If this argument were followed consistently, it would also require a new name for the instrument of Fig. 1, since the term *tensiometer* has also been previously used for instruments which measure surface tension. However, because of euphemistic shortcomings, it may be doubtful whether some instrument-naming prefix derived from the term "suction" would gain general acceptance. What is needed is a generally acceptable substitute for the term suction, and one which can also be adapted for naming the measuring instrument. But, lacking such a term at present, we shall use matric suction (or suction) as needed, without renaming the tensiometer. Strictly speaking, if $\rho g \tau$ is called the *suction*, then τ itself should be called the *suction head*. However, a consistent effort will not necessarily be made to preserve this minor distinction, except in instances where confusion might otherwise result.

2.1.3. *Soil-Water Diffusivity*

The suction head τ is also a function of water content θ. Using both $K = K(\theta)$ and $\tau = \tau(\theta)$ in Eq. (1), with $h = z - \tau(\theta)$, yields

$$\mathbf{q} = -K(\theta)[\nabla z - \nabla \tau(\theta)] \tag{2}$$

Since $\partial\tau/\partial x = (d\tau/d\theta)(\partial\theta/\partial x)$, and $\partial\tau/\partial y$ and $\partial\tau/\partial z$ can be evaluated similarly, Eq. (2) can be expressed as

$$\mathbf{q} = -K(\theta)\ \nabla z - D(\theta)\ \nabla\theta \tag{3}$$

where, since both K and τ are functions of θ,

$$D(\theta) = K(\theta)(-d\tau/d\theta) \tag{4}$$

The right side of Eq. (4) was introduced by Buckingham (1907) but not named by him. It was identified as a diffusion coefficient by Childs and Collis-George (1948), and is now generally called the *soil-water diffusivity*, with units of area per unit time. The physical transport process, however, is not one of diffusion in the classical sense, since the hydrodynamic basis of Eqs. (1) and (2) has not been changed in the process of arriving at Eq. (3). Rather, the introduction of $K(\theta)$ and $\tau(\theta)$ merely enables the grouping of $K(\theta)$ and $-d\tau/d\theta$ into a product that is also a function of θ, which then is conveniently identified as a diffusivity function since $\nabla\theta$ represents a concentration gradient.

L. A. Richards (1961) has suggested that Eq. (2) be named after Buckingham. The present writer is in essential agreement, but would alter the suggestion to the extent of adding Darcy's name. Hence, both Eqs. (2) and (3) could be designated as forms of the Buckingham–Darcy equation. This would not only credit Buckingham for his very material contributions to both equations, but would also reflect the practice of the last four decades in citing the connective analogy with the Darcy equation [see, for example, Israelsen (1927), L. A. Richards (1928, 1931, 1936, 1952), W. Gardner (1936), Christensen (1944), Childs and Collis-George (1948), Childs and Collis-George (1950), Klute (1952, 1965), E. E. Miller and R. D. Miller (1956), W. R. Gardner (1956), and Philip (1957b)]. That Buckingham (1907) did not even mention Darcy (1856) seems somewhat surprising, since analogy with porous media, even if saturated, would appear to be at least as relevant as analogies with heat and electrical flow.

2.1.4. *Nature of Conductivity and Suction Functions*

Whether the Buckingham–Darcy relationship is expressed by either Eq. (2) or (3), it is clear that K and τ are functions of very basic import for describing the flow process. The general nature of these functions is illustrated in Fig. 2 for both a coarse- and a fine-textured soil. The range of the water content θ is from zero to the total porosity, which is n_1 for the coarse soil and n_2 for the fine, with $n_2 > n_1$. K increases with θ [Fig. 2(A)] and reaches its maximum (saturated) value at θ equal to the porosity, so that K_{s1} and K_{s2} are the hydraulic conductivities of the saturated coarse and fine soils, respectively, with $K_{s1} > K_{s2}$. For both soils, the change of K with θ is most pronounced over the higher water-content range.

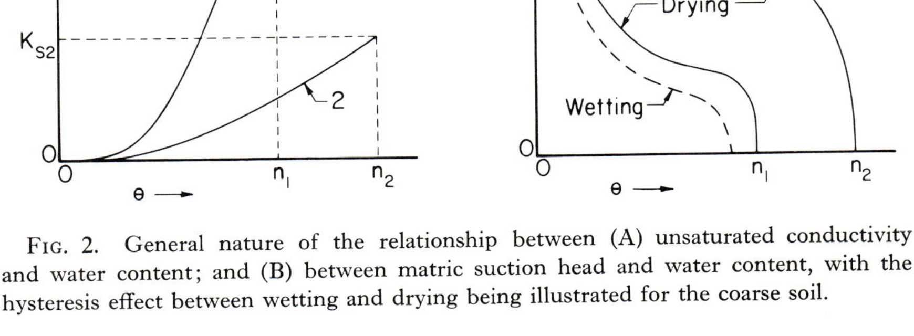

FIG. 2. General nature of the relationship between (A) unsaturated conductivity and water content; and (B) between matric suction head and water content, with the hysteresis effect between wetting and drying being illustrated for the coarse soil.

In contrast with K, τ in general increases as θ decreases [Fig. 2(B)], since progressively more energy is required to extract water as the water content decreases. Thus, $d\tau/d\theta$ is inherently negative, so that the negative sign in Eq. (4) serves to make $D(\theta)$ positive. That θ at a given τ is generally less for the coarse soil than for the fine soil is the consequence of the larger pores and smaller internal surface area in the coarse soil. This difference in θ for soils at the same τ can also be used to demonstrate Eq. (2) as a more general criterion of flow than Eq. (3). Consider that single samples of each soil, both at the same value of any $\tau > 0$, are brought into contact in such a way that the gravitational gradient is zero. With $\nabla z = 0$, and, since $\nabla\tau$ also equals zero by virtue of the same τ, Eq. (2) correctly predicts that the flow velocity for water transport between the samples is zero. Equation (3), on the other hand, gives the flow velocity as $-D(\theta)\,\nabla\theta$. But this product is not zero at the junction between the soil samples, since $D(\theta)$ is not zero at the junction and since Fig. 2(B) shows the two soils to have different θ at the same τ; therefore, $\nabla\theta \neq 0$ at the junction. We are thus left with the conclusion that Eq. (3) can hold only in a spacially uniform soil of given internal geometry, whereas Eq. (2) would hold even in stratified or nonuniform soils. This observation for stratified soils has also been noted by Hallaire (1958).

Another important aspect of τ behavior is brought out by the broken-line curve in Fig. 2(B). Whereas the solid-line curves are for θ being decreased from n_1 or n_2 (drying), the broken line is $\tau(\theta)$ for the coarse soil when θ is increased from zero (wetting). The separation between the two curves (wetting and drying, coarse soil) at $\tau = 0$ is caused by entrapped air. The general hysteresis between the wetting and drying curves for the same soil causes considerable difficulty in mathematical analysis, since it means that τ is not really a unique function of θ if the water movement involves wetting and drying simultaneously. The most common way of avoiding this difficulty in the past has been to consider wetting only or drying only but not both together, so that $\tau(\theta)$ might reasonably be taken as unique. Strictly speaking, a broken-line curve for $K(\theta)$ somewhat different from the solid-line curve in Fig. 2(A) should also be shown, to correspond with the wetting curve of Fig. 2(B). Some reports (Nielsen and Biggar, 1961; Elrick and Bowman, 1964) indicate hysteresis in $K(\theta)$ to be relatively small, whereas others (Youngs, 1964; Staple, 1965) indicate it to be more sizable. If K were to be expressed as a function of τ, since $K(\theta) = K[\theta(\tau)]$, then any hysteresis in $\tau(\theta)$ would be imposed on K in conjunction with that already existing in $K(\theta)$.

Up to this point, the development of suction (that is, $\tau > 0$) has been particularly associated with desaturation ($\theta < n$). Even in saturated soils, however, suction can develop for vertically downward flow in a stratum of high hydraulic conductivity overlain by a stratum of lower conductivity. For steady-state flow, this was demonstrated experimentally and described theoretically by Day and Luthin (1953), while experimental extension to transient flow was provided by Behnke and Bianchi (1965). Zaslavsky (1964a), with an extensive theoretical analysis of nonuniform systems in some respects similar to Takagi's (1960), has described the development of suction under both saturated and unsaturated conditions, as related to the transition between saturated and unsaturated flow.

2.2. The Partial Differential Equation of Flow

The law of conservation of mass is embodied in the equation of continuity already derived in Chapter 1 [Eq. (30)]. Only a slight modification is needed for unsaturated flow. For the infinitesimal cartesian-coordinate parallelopiped of soil bulk volume $dx\,dy\,dz$, one expression of the total net accumulation of water mass per unit time is still $-dx\,dy\,dz\,\nabla\cdot(\rho\mathbf{q})$. The alternative expression when the soil is unsaturated is $\partial(\rho\theta\,dx\,dy\,dz)/\partial t = dx\,dy\,dz\,\partial(\rho\theta)/\partial t$. Equating these two expressions and considering ρ constant yields

$$-\nabla\cdot\mathbf{q} = \partial\theta/\partial t \tag{5}$$

as the equation of continuity for unsaturated flow. Combining Eqs. (2) and (5), observing that $\nabla\cdot(K\nabla z) = \partial K/\partial z$ for cartesian coordinates, and using $\theta = \theta(\tau)$ [simply the inverse expression of $\tau = \tau(\theta)$] to write $\partial\theta/\partial t = (d\theta/d\tau)\,\partial\tau/\partial t$, yields

$$-\nabla\cdot\mathbf{q} = \frac{\partial K}{\partial z} - \frac{\partial}{\partial x}\left(K\frac{\partial\tau}{\partial x}\right) - \frac{\partial}{\partial y}\left(K\frac{\partial\tau}{\partial y}\right) - \frac{\partial}{\partial z}\left(K\frac{\partial\tau}{\partial z}\right) = \frac{d\theta}{d\tau}\frac{\partial\tau}{dt} \tag{6}$$

It is convenient here to consider K as a function of τ, and this is permissible since $K(\theta) = K[\theta(\tau)]$. Also, $-d\theta/d\tau$, which may be defined as the specific water capacity, may also be considered a function of τ. Equation (6) is then essentially the governing partial differential equation first derived by L. A. Richards (1931).

If, instead of Eq. (2), we elect to combine Eq. (3) with Eq. (5), the result is

$$-\nabla\cdot\mathbf{q} = \frac{\partial K}{\partial z} + \frac{\partial}{\partial x}\left(D\frac{\partial\theta}{\partial x}\right) + \frac{\partial}{\partial y}\left(D\frac{\partial\theta}{\partial y}\right) + \frac{\partial}{\partial z}\left(D\frac{\partial\theta}{\partial z}\right) = \frac{\partial\theta}{\partial t} \tag{7}$$

where $D = D(\theta)$ of Eq. (4). Or, Eq. (6) can be changed to Eq. (7) simply by assuming the functional forms to be $K = K(\theta)$ and $\tau = \tau(\theta)$, and then making use of Eq. (4). Equation (7) is essentially that given by Klute (1952) and recast in the present form by Philip (1957c).

Continuity principles for unsaturated flow were first used by W. Gardner (1919) in one dimension, and in three dimensions by W. Gardner and Widtsoe (1921). In these treatments, however, the authors used only special-case functional forms for both $K(\theta)$ and $\tau(\theta)$, and did not leave the forms general until much later (W. Gardner, 1936). Hence, the first continuity analysis to leave both K and τ in general form was that of L. A. Richards (1931). He also recognized explicitly that the use of τ and $K(\tau)$ [Eq. (6)] or of θ and $K(\theta)$ [Eq. (7) by implication] is strictly a matter of mathematical expediency, provided $\tau(\theta)$ is unique. Hence, Eqs. (6) and (7) could appropriately be referred to as forms of the Richards equation.

2.3. Solution of the Partial Differential Equation for One-Dimensional Flow

Problems illustrating one-dimensional unsaturated flow will now be solved on the basis of the Buckingham–Darcy equation as the governing relationship. In every case it will be assumed that $\tau(\theta)$ and $K(\theta)$, and hence $D(\theta)$, are unique.

2.3.1. *Absorption of Water by Soil*

2.3.1.1. HORIZONTAL DIRECTION. Consider a uniform horizontal soil column of constant cross section and infinite length to have constant initial water content θ_0 (Fig. 3). At time $t = 0$, a higher constant water

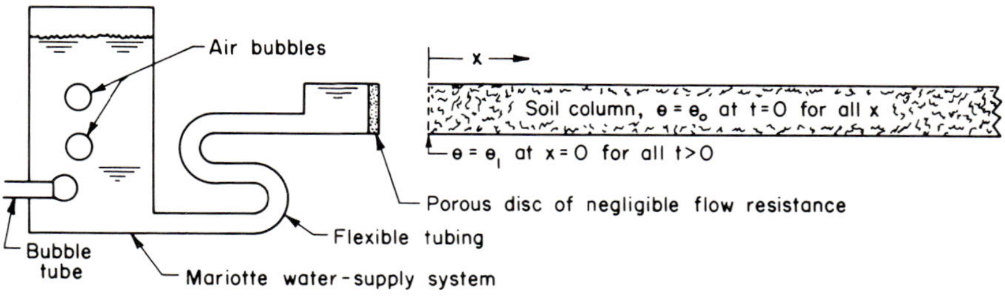

FIG. 3. Schematic diagram of apparatus for measuring the horizontal absorption of water by soil.

content θ_1 is applied and maintained at one end of the column chosen as $x = 0$. In Fig. 3, this is accomplished by bringing the porous disc of the water supply system into contact with the soil at $t = 0$. A θ_1 near saturation is provided by having the Mariotte bubble tube at a constant position only slightly below the bottom edge of the soil column, whereas a smaller θ_1 is provided if the bubble tube is positioned still lower. For such a horizontal flow problem, Eq. (7) for the x direction becomes

$$\frac{\partial \theta}{\partial t} = \frac{\partial}{\partial x}\left(D \frac{\partial \theta}{\partial x}\right) \tag{8}$$

subject to the conditions

$$\theta = \theta_0, \qquad x > 0, \qquad t = 0 \tag{9}$$

$$\theta = \theta_1, \qquad x = 0, \qquad t > 0 \tag{10}$$

where $\theta = \theta(x, t)$ and $D = D(\theta)$. A variables-separable solution can be obtained if x is made to be the dependent variable, that is, $x = x(\theta, t)$. Carrying out the transformation from $\theta(x, t)$ to $x(\theta, t)$, and making use of the standard relationships (see Zemansky, 1957),

$$\frac{\partial \theta}{\partial t} = -\frac{\partial x/\partial t}{\partial x/\partial \theta} \tag{11}$$

$$\frac{\partial \theta}{\partial x} = \frac{1}{\partial x/\partial \theta} \tag{12}$$

Eq. (8) then takes the form

$$-\frac{\partial x}{\partial t} = \frac{\partial}{\partial \theta}\left(\frac{D}{\partial x/\partial \theta}\right) \tag{13}$$

The problem has thus been changed to Eq. (13) subject to conditions (9) and (10), with x considered dependent on θ and t.

Assume that the solution to Eq. (13) can be written in product form as

$$x = \lambda(\theta)\, S(t) \tag{14}$$

where λ is a function only of θ, and S is a function only of t. Partial differentiation of Eq. (14) provides $\partial x/\partial t = \lambda\, dS/dt$ and $\partial x/\partial \theta = S\, d\lambda/d\theta$; substituting these into Eq. (13) and rearranging yields

$$S \frac{dS}{dt} = -\frac{1}{\lambda} \frac{d}{d\theta}\left(\frac{D}{d\lambda/d\theta}\right) \tag{15}$$

But, since the left side of Eq. (15) depends only on t and the right side only on θ, each side must be equal to a constant, say a. Doing this and rearranging gives the two ordinary differential equations

$$S\, dS = a\, dt \tag{16}$$

$$\frac{d}{d\theta}\left(D \frac{d\theta}{d\lambda}\right) = -a\lambda \tag{17}$$

Integrating Eq. (16) and solving for S gives

$$S = [2a(t + C_1)]^{1/2} \tag{18}$$

where C_1 arises from the constant of integration. Combining Eq. (18) with (14) provides

$$x = (t + C_1)^{1/2}\eta(\theta) \tag{19}$$

where $\eta = \lambda(2a)^{1/2} = \eta(\theta)$, since λ depends on θ only. Hence, $d\eta = (2a)^{1/2}\, d\lambda$, and Eq. (17) takes the form

$$\frac{d}{d\theta}\left(D\frac{d\theta}{d\eta}\right) = -\frac{\eta}{2} \tag{20}$$

Equation (19) must also accommodate the boundary conditions if it is to be a solution to the problem. Condition (10) in Eq. (19) gives

$$\eta(\theta_1) = 0 \tag{21}$$

Condition (9) in Eq. (19) yields

$$\eta(\theta_0) = x/C_1^{1/2} \tag{22}$$

If C_1 were finite and nonzero, θ_0 would vary with x, in violation of condition (9). Infinity and zero are the only choices of C_1 which would make $\eta(\theta_0)$, and hence θ_0, constant and independent of finite nonzero x. But $C_1 \to \infty$ would make $\eta(\theta_0) = 0$, so that from Eq. (21) we would have the trivial static case of $\theta_0 = \theta_1$. Hence, for a nontrivial solution we are left with $C_1 = 0$, or

$$\eta(\theta_0) \to \infty \tag{23}$$

The problem of Eq. (13) subject to conditions (9) and (10) is thus reduced to Eq. (20) subject to conditions (21) and (23), but this is precisely the problem of Crank and Henry (1949) and Klute (1952), and for which Philip (1955b) has given a rapid numerical solution. Furthermore, the use of $C_1 = 0$ in Eq. (19) leads to

$$\eta(\theta) = xt^{-1/2} \tag{24}$$

which is known as the Boltzmann (1894) transformation, and is seen to be an ultimate consequence of Eq. (13) and condition (9). The foregoing variables-separable analysis is taken from Swartzendruber (1966a), who has also considered the separation of variables by an additive form.

Philip's (1955b) solution of Eq. (20) subject to conditions (21) and (23) utilizes an integration of Eq. (20) to obtain

$$\int_{\theta_0}^{\theta} \eta \, d\theta = -2D \, (d\theta/d\eta) \tag{25}$$

The integration between limits yields the term $D(\theta_0)[d\theta/d\eta]_{\theta=\theta_0}$ to be added on the right side of Eq. (25), but this term vanishes because condition (23) implies

$$d\theta/d\eta = 0 \qquad \text{at} \quad \theta = \theta_0 \tag{26}$$

Thus, Eq. (25) remains subject only to condition (21). For given θ_0, θ_1, and a specified diffusivity function $D(\theta)$, Philip's (1955b) finite-difference numerical solution of Eq. (25) subject to condition (21) enables a fairly rapid evaluation of $\eta(\theta)$, which completes the solution to the problem of horizontal flow. Note that $D(\theta)$ needs only to be specified in numerical form—an analytical expression is not necessarily required. Nevertheless, an analytical expression is convenient for illustrative purposes and will be so used here. Assuming $D(\theta) = D_1(\theta/\theta_1)^m$, where $D_1 = D(\theta_1)$ and m is positive, Philip's method was used to obtain $\eta(\theta)$ for $m = 1$ and $m = 5$, and $\theta_1/\theta_0 = 7$. The results are shown in Fig. 4(A), in the form $\eta/D_1^{1/2}$ versus θ/θ_0. The values of m were chosen arbitrarily to demonstrate what happens when D varies rapidly ($m = 5$) with water content, and when it varies much more slowly ($m = 1$). W. R. Gardner and Mayhugh (1958) have used the analytical form $D(\theta) = D_1 \exp \beta(\theta - \theta_1)$,

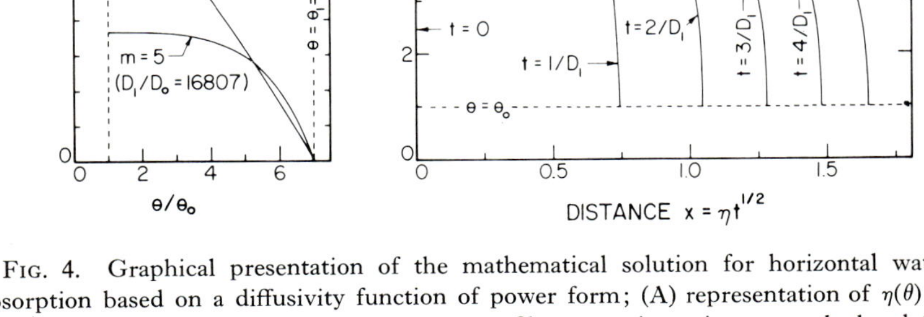

FIG. 4. Graphical presentation of the mathematical solution for horizontal water absorption based on a diffusivity function of power form; (A) representation of $\eta(\theta)$ as $\eta/D_1^{1/2}$ versus θ/θ_0, and (B) water-content profiles at various times as calculated for $m = 5$ of part (A).

where β is a constant, with results qualitatively similar to those in Fig. 4(A).

With $\eta(\theta)$ known, the water content in the soil can be determined at any desired time or position. Or, Eq. (24) can be solved to give

$$x = [\eta(\theta)/D_1^{1/2}](D_1 t)^{1/2} \tag{27}$$

which, since $\eta(\theta)$ is unique by virtue of $D(\theta)$ being unique, asserts the position x of any given water content θ to be directly proportional to the square root of the time t. Or, at a given t, the function $\eta t^{1/2} = x(\theta)$ is $\eta(\theta)/D_1^{1/2}$ multiplied by $(D_1 t)^{1/2}$. The function $x(\theta)$, or $\theta(x)$, at a given t may be called a water-content profile. Such profiles are shown in Fig. 4(B) as calculated from $\eta/D_1^{1/2}$ for $m = 5$ in Fig. 4(A). At $t = 0$, the profile simply coincides with the θ/θ_0 axis. As t increases linearly, the profile advances into the soil, but the progress of the advance is retarded by virtue of the $(D_1 t)^{1/2}$ factor. Another interesting feature is the steep slope (large $-\partial\theta/\partial x$) in the forward portion of each profile.

The cumulative volume U of water absorbed by the horizontal soil column at any given time t is given by the area under the water-content profile from the integration

$$U = A \int_{\theta_0}^{\theta_1} x \, d\theta \tag{28}$$

where A is the cross-sectional area of the soil column. By Eq. (24) we can express x as $\eta t^{1/2}$ and substitute it into Eq. (28) to get

$$U = A t^{1/2} \int_{\theta_0}^{\theta_1} \eta \, d\theta \tag{29}$$

where $t^{1/2}$ is brought outside the integral sign since the integration takes place at a given (constant) t. The definite integral in Eq. (29) is a constant and can be evaluated as $\theta_0 D_1^{1/2}$ times the total area under such a curve as in Fig. 4(A). Hence, the cumulative volume of water absorbed is proportional to the square root of the time.

It is pertinent to consider how well these various facets of the mathematical solution are in accord with experimental findings. For small θ_0 on the order of air dryness, Kirkham and Feng (1949) found generally good linearity of U versus $t^{1/2}$ and of wet-front advance versus $t^{1/2}$, although these linear plots did not always pass through the origin as required for proportionality. Since the optically visual difference between wet and dry soil marks the position of the wet front, it might be argued that no one value of $\theta_{\text{wet front}}$ can be identified, so that

Eq. (27) might not be expected to hold for the wet front. If, however, the water-content profile is very steep in its forward portion, $\theta_{\text{wet front}}$ could actually vary progressively through a range of values near θ_0, and the resulting slight effect on wet-front advance would be extremely difficult to detect experimentally. In Fig. 4(B), this range in θ would be from θ_0 to about $2.5\,\theta_0$. The forward-portion steepness in actual water-content profiles was very clearly evident in Kirkham and Feng's experimental results, and demonstrated conclusively that constant-diffusivity theory was inadequate to describe the profiles. The signal accomplishment of the nonconstant diffusivity approach (Klute, 1952), embodied here in Eqs. (8)–(29), is the prediction of profiles of the correct qualitative shape, while retaining the square-root-of-time dependencies [Eqs. (27) and (29)].

Equation (24) implies that all $\eta(\theta)$ curves determined from water-content profiles $\theta(x)$ at various given times should coalesce into a single curve then identifiable as the unique $\eta(\theta)$. Such data obtained for Adelanto loam by Jackson (1963a) are shown in Fig. 5 for profiles at the two

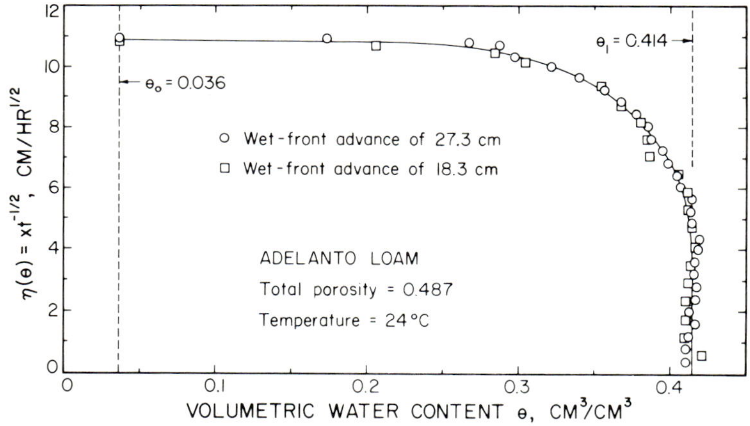

FIG. 5. Coalescing experimental curves of $\eta(\theta)$ determined at two different positions of the visible wet front in horizontal water absorption by soil [data supplied by Jackson (1963a) and used by permission].

times corresponding to wet-front advances of 18.3 and 27.3 cm, and do coalesce nicely into a single curve. However, Nielsen *et al.* (1962) and Davidson *et al.* (1963b) found such behavior only when θ_1 was near saturation, while Ferguson and W. H. Gardner (1963) and Rawlins and W. H. Gardner (1963) were unable to verify Eq. (27) at constant θ even when θ_1 was near saturation. Other experiments which demonstrate one or more aspects of profile shape or $t^{1/2}$ behavior are those of Bruce and Klute (1956), W. R. Gardner and Mayhugh (1958), and Stockinger *et al.* (1965).

2.3.1.2. VERTICALLY DOWNWARD INFILTRATION. We shall next consider the problem in which one-dimensional water infiltration occurs vertically downward, but with the same boundary conditions as for horizontal flow. Physically, this may be visualized similarly to Fig. 3, except that the right end of the soil column is turned downward to obtain a vertical position, and the porous disc is appropriately reoriented to acommodate the top end of the vertical column. Or, if $\theta_1 = n$ (saturation) is desired, apparatus to apply and maintain free water on the soil surface from $t = 0$ on may be used, but the depth of ponded water must not be more than a very thin film. Only derivatives with respect to z and t need be retained from Eq. (7). Dealing with a space coordinate Z as positive downward, we set $Z = -z$, and the governing partial differential equation thus becomes

$$\frac{\partial\theta}{\partial t} = \frac{\partial}{\partial Z}\left(D\frac{\partial\theta}{\partial Z}\right) - \frac{\partial K}{\partial Z} \tag{30}$$

Again we make Z the dependent variable by transforming from $\theta = \theta(Z, t)$ to $Z = Z(\theta, t)$, using Eqs. (11) and (12) written with Z in place of x to obtain

$$-\frac{\partial Z}{\partial t} = \frac{\partial}{\partial\theta}\left(\frac{D}{\partial Z/\partial\theta}\right) - \frac{dK}{d\theta} \tag{31}$$

which is then taken subject to

$$\theta = \theta_0, \qquad Z > 0, \qquad t = 0 \tag{32}$$

$$\theta = \theta_1, \qquad Z = 0, \qquad t > 0 \tag{33}$$

It is granted that condition (32) may be more difficult to realize experimentally for the vertical case than for the horizontal [condition (9)], unless θ_0 is sufficiently small so that the gravitational field will not cause it to vary appreciably with Z.

Let us first examine the nature of the solution for very small t. In general, Eq. (31) may also be expressed as

$$-\frac{\partial Z}{\partial t} = \frac{\partial}{\partial\theta}\left(\frac{D}{\partial Z/\partial\theta} - K\right) = \frac{\partial}{\partial\theta}\left(D\frac{\partial\theta}{\partial Z} - K\right) \tag{34}$$

For very small t, the magnitude of $\partial\theta/\partial Z$ is extremely large, having been theoretically infinite at $t = 0$ as in Fig. 4(B); in magnitude, $D(\partial\theta/\partial Z)$ thus greatly exceeds K, so that K can safely be neglected. But with K

neglected in Eq. (34), the solution takes the form of that for horizontal flow; that is, on the basis of Eq. (24) we can write

$$Z = \eta t^{1/2} \tag{35}$$

where η is the solution of Eq. (20) subject to conditions (21) and (23), or, alternatively, the solution of Eq. (25) subject to condition (21).

Considering Eq. (31) subject to conditions (32) and (33) at other than small times, we find that a product form such as Eq. (14) written in Z will not separate the variables. An alternative formulation which can be tried, and which has abundant precedent in the pragmatic approach to the solution of differential equations, is to assume a solution in series form

$$Z(\theta, t) = f_0 + f_1 t^m + f_2 t^{2m} + f_3 t^{3m} + \cdots = \sum_{i=0}^{\infty} f_i t^{im} \tag{36}$$

where m is a positive constant, and

$$f_i = f_i(\theta) \tag{37}$$

Note that the general term in the series is assumed to be the product of a function of θ alone and a function of t alone, the t function being the power form t^{im}. Equation (36) must also reduce to Eq. (35) at very small times. For very small t in Eq. (36) we may write

$$Z = f_0 + f_1 t^m \tag{38}$$

assuming for all $i > 1$ that the $f_i(\theta)$ remain properly bounded at small t, so that the $f_i t^{im}$ for $i > 1$ are negligible compared with the first two terms of the series of Eq. (36). Comparing Eq. (38) with Eq. (35), we choose $f_0 = 0$ and $m = \frac{1}{2}$, so that a more specific form of the proposed solution is written as

$$Z(\theta, t) = f_1 t^{1/2} + f_2 t + f_3 t^{3/2} + f_4 t^2 + \cdots = \sum_{i=1}^{\infty} f_i t^{i/2} \tag{39}$$

Expanding the derivative with respect to θ in Eq. (31), and rearranging, yields

$$D \frac{\partial^2 Z}{\partial \theta^2} + K' \left(\frac{\partial Z}{\partial \theta}\right)^2 - \frac{\partial Z}{\partial t}\left(\frac{\partial Z}{\partial \theta}\right)^2 - D' \frac{\partial Z}{\partial \theta} = 0 \tag{40}$$

where the primes designate ordinary first derivatives with respect to θ.

The partial derivatives in Eq. (40) are obtained by partial differentiation of Eq. (39), which yields

$$\partial Z/\partial\theta = f_1't^{1/2} + f_2't + f_3't^{3/2} + f_4't^2 + \cdots = \sum_{i=1}^{\infty} f_i't^{i/2} \tag{41}$$

$$\partial^2 Z/\partial\theta^2 = f_1''t^{1/2} + f_2''t + f_3''t^{3/2} + f_4''t^2 + \cdots = \sum_{i=1}^{\infty} f_i''t^{i/2} \tag{42}$$

$$\begin{aligned}\partial Z/\partial t &= (f_1/2)\, t^{-1/2} + f_2 + (3f_3/2)\, t^{1/2} + 2f_4 t + \cdots \\ &= \sum_{i=1}^{\infty} (i/2)\, f_i t^{(i-2)/2}\end{aligned} \tag{43}$$

$$(\partial Z/\partial\theta)^2 = (f_1')^2 t + 2f_1'f_2't^{3/2} + [2f_1'f_3' + (f_2')^2]\, t^2 + 2[f_1'f_4' + f_2'f_3']\, t^{5/2} + \cdots \tag{44}$$

where again the single and double primes designate respectively the ordinary first and second derivatives with respect to θ. A general series term is not deduced for Eq. (44), which was obtained by multiplying Eq. (41) by itself.

If Eqs. (41)–(44) are substituted back into the governing partial differential equation [Eq. (40)], and like powers of t are collected, manipulation and rearrangement eventually enable the result to be expressed as

$$Y_1 t^{1/2} + Y_2 t + Y_3 t^{3/2} + Y_4 t^2 + \cdots = \sum_{i=1}^{\infty} Y_i t^{i/2} = 0 \tag{45}$$

where the first four Y_i are given by

$$Y_1 = Df_1'' - (f_1/2)(f_1')^2 - D'f_1' \tag{46}$$

$$Y_2 = Df_2'' + K'(f_1')^2 - (f_1')^2 f_2 - f_1 f_1' f_2' - D'f_2' \tag{47}$$

$$\begin{aligned}Y_3 = {} & Df_3'' + 2K'f_1'f_2' - (3f_3/2)(f_1')^2 - 2f_2 f_1' f_2' \\ & - f_1 f_1' f_3' - (f_1/2)(f_2')^2 - D'f_3'\end{aligned} \tag{48}$$

$$\begin{aligned}Y_4 = {} & Df_4'' + K'[2f_1'f_3' + (f_2')^2] - 2(f_1')^2 f_4 - 3f_1'f_2'f_3 \\ & - 2f_1'f_3'f_2 - (f_2')^2 f_2 - f_1'f_4'f_1 - f_2'f_3'f_1 - D'f_4'\end{aligned} \tag{49}$$

Since Eq. (45) is an identity that must hold for all positive square roots and powers of $t > 0$, each coefficient of a power of t in this identity must therefore be zero, that is

$$Y_i = 0 \qquad \text{for} \quad i = 1, 2, 3, \ldots \tag{50}$$

This formulation expresses an ordinary differential equation for each value of i.

To determine the corresponding boundary conditions holding for the differential equations that arise out of condition (50), we begin by substituting condition (33) into Eq. (39), with the result

$$f_1(\theta_1)\, t^{1/2} + f_2(\theta_1)\, t + f_3(\theta_1)\, t^{3/2} + f_4(\theta_1)\, t^2 + \cdots = \sum_{i=1}^{\infty} f_i(\theta_1)\, t^{i/2} = 0 \quad (51)$$

But the only way that Eq. (51) can hold for positive square roots and powers of $t > 0$ is for the coefficient of each power of t to vanish, that is,

$$f_i(\theta_1) = 0 \qquad \text{for} \quad i = 1, 2, 3, \ldots \quad (52)$$

To handle condition (32) we first rewrite Eq. (39) as

$$Z/t^{1/2} = f_1 + f_2 t^{1/2} + f_3 t + f_4 t^{3/2} + f_5 t^2 + \cdots = f_1 + \sum_{i=2}^{\infty} f_i t^{(i-1)/2} \quad (53)$$

Since it was assumed in obtaining Eq. (38) that the f_i for $i \geqslant 2$ remained suitably bounded for very small t, it seems reasonable to assume tentatively that the same would be true of Eq. (53) at $t = 0$, so that the summation sign on the right side of Eq. (53) vanishes at $t = 0$. Thus, when condition (32) is used in Eq. (53) the result is

$$f_1(\theta_0) \to \infty \quad (54)$$

When $Y_1 = 0$, Eq. (46) becomes an ordinary differential equation in f_1 and θ, and can easily be manipulated to give, in ordinary derivative notation,

$$\frac{d}{d\theta}\left(D \frac{d\theta}{df_1}\right) = -\frac{f_1}{2} \quad (55)$$

But we now observe that if $f_1 = \eta$, then Eqs. (55), (54), and (52) for $i = 1$ become identical respectively with Eqs. (20), (23), and (21) for horizontal flow; note that $f_1 = \eta$ also completes the identification of Eq. (38) with Eq. (35) on the basis of $f_0 = 0$ and $m = \frac{1}{2}$ as previously chosen. Hence, the first term in the series of Eq. (39) is the solution for horizontal flow [Eq. (35)], in which $f_1 = \eta$ is given by the solution of Eq. (20) subject to conditions (21) and (23) [or condition (26)], or, alternatively, by Eq. (25) subject to condition (21).

When $i = 2$ in condition (50), the use of $Y_2 = 0$, $f_1 = \eta$, and

Eq. (55) in Eq. (47) ultimately yields, after some manipulation and rearrangement,

$$f_2 = \frac{d}{d\theta}\left[D\left(\frac{d\theta}{d\eta}\right)^2 \frac{df_2}{d\theta}\right] + \frac{dK}{d\theta} \tag{56}$$

Integrating Eq. (56) with respect to θ between limits of θ_0 and θ, and making use of condition (26), which ultimately reflects condition (32) by way of $f_1 = \eta$ and conditions (54) and (23), we find

$$\int_{\theta_0}^{\theta} f_2 \, d\theta = D\left(\frac{d\theta}{d\eta}\right)^2 \frac{df_2}{d\theta} + (K - K_0) \tag{57}$$

where $K_0 = K(\theta_0)$. Equation (57) is subject to

$$f_2(\theta_1) = 0 \tag{58}$$

obtained by setting $i = 2$ in condition (52), and which ultimately reflects condition (33).

When $i = 3$ in condition (50), the use of $Y_3 = 0, f_1 = \eta$, and Eqs. (55) and (56) in Eq. (48) provides, after considerable manipulation and rearrangement,

$$\frac{3f_3}{2} = \frac{d}{d\theta}\left[D\left(\frac{d\theta}{d\eta}\right)^2 \frac{df_3}{d\theta} - D\left(\frac{d\theta}{d\eta}\right)\left(\frac{df_2}{d\eta}\right)^2\right] \tag{59}$$

which can be integrated between limits of θ_0 and θ in conjunction with condition (26) to obtain

$$\frac{3}{2}\int_{\theta_0}^{\theta} f_3 \, d\theta = D\left(\frac{d\theta}{d\eta}\right)^2 \frac{df_3}{d\theta} - D\left(\frac{d\theta}{d\eta}\right)\left(\frac{df_2}{d\eta}\right)^2 \tag{60}$$

subject to

$$f_3(\theta_1) = 0 \tag{61}$$

obtained by setting $i = 3$ in condition (52). Similarly, when $i = 4$ in condition (50), the use of $Y_4 = 0, f_1 = \eta$, and Eqs. (55), (56), and (59) in Eq. (49) yields, after extensive manipulation and rearrangement, a differential equation which again can be integrated between limits of θ_0 and θ in conjunction with condition (26) to obtain

$$2\int_{\theta_0}^{\theta} f_4 \, d\theta = D\left(\frac{d\theta}{d\eta}\right)^2 \frac{df_4}{d\theta} - D\left(\frac{d\theta}{d\eta}\right)\left(\frac{df_2}{d\eta}\right)^2 \left[2\frac{df_3}{df_2} - \frac{df_2}{d\eta}\right] \tag{62}$$

subject to

$$f_4(\theta_1) = 0 \tag{63}$$

obtained by setting $i = 4$ in condition (52). The procedure of this and the preceding paragraph can be continued as often as necessary to obtain differential equations corresponding to $i \geqslant 5$.

The solutions of Eqs. (57), (60), and (62) subject respectively to conditions (58), (61), and (63) yield the functions f_2, f_3, and f_4, and are obtained numerically by the finite-difference methods of Philip (1957a), which are applicable for $i \geqslant 2$. Both $D(\theta)$ and $K(\theta)$ must be known or assumed, and $\eta(\theta)$ must first be evaluated by the solution of Eq. (25) subject to condition (21) as previously discussed. The results of such numerical solutions are shown in Fig. 6(A) for f_1 ($= \eta$), f_2, f_3,

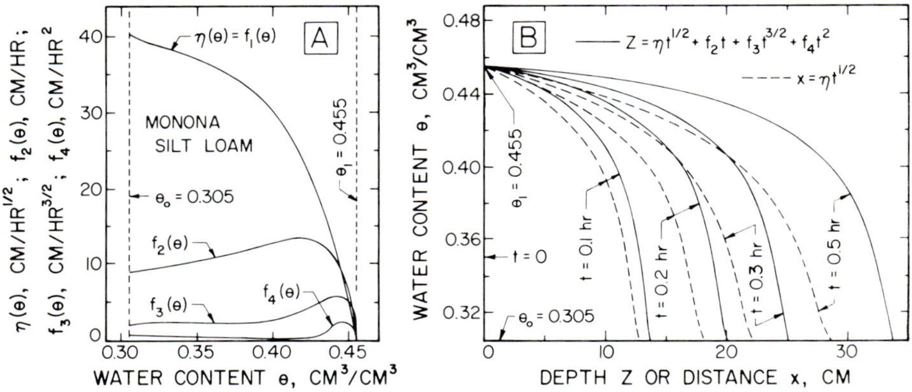

FIG. 6. (A) Infiltration functions [see Eq. (39)] $f_1(= \eta)$, f_2, f_3, and f_4 [data of Nielsen *et al.* as adapted from *Soil Sci. Soc. Am. Proc.* **25**, 167 (1961), by permission of publisher and authors]. (B) Water-content profiles at various times for both vertical infiltration and horizontal absorption, as computed from Eqs. (39) and (24), respectively, using the f_1, f_2, f_3, and f_4 curves of part (A).

and f_4 as calculated by Nielsen *et al.* (1961) for Monona silt loam. The relative importance of each $f_i(\theta)$ decreases rapidly as i increases, and would appear to be essentially negligible for this soil for $i > 4$. In Fig. 6(B), the water-content profiles for vertically downward infiltration are shown (solid-line curves) at various times as calculated from Fig. 6(A) on the basis of the first four terms of the series in Eq. (39). For comparison, the broken-line curves computed from the first term of Eq. (39) are also shown, since these would be the corresponding water-content profiles for horizontal flow. The profiles advance more rapidly for vertical flow, as would be expected from the influence of gravity. Also, for vertical flow the forward portion of the profile is more steep and the steepness persists to higher θ than for horizontal flow, but, in the region near to θ_1, the vertical-flow profile is actually less steep than that for horizontal flow.

At a given time t, the cumulative volume U of water infiltrated into the vertical soil column is

$$U = AK_0t + A\int_{\theta_0}^{\theta_1} Z\, d\theta \tag{64}$$

The component AK_0t arises from the transmission of water through the soil column by the gravitationally induced unit hydraulic gradient acting at the initial soil-water content θ_0. Substituting Eq. (39) into Eq. (64), using $f_1 = \eta$, and remembering that t is held constant during the integration with respect to θ, we find

$$U/A = t^{1/2}\int_{\theta_0}^{\theta_1} \eta\, d\theta + t\left[K_0 + \int_{\theta_0}^{\theta_1} f_2\, d\theta\right] + t^{3/2}\int_{\theta_0}^{\theta_1} f_3\, d\theta + t^2\int_{\theta_0}^{\theta_1} f_4\, d\theta + \cdots \tag{65}$$

The definite integrals in Eq. (65) are constants which can be evaluated as the areas under the curves of the various $f_i(\theta)$ between limits θ_0 and θ_1 [see Fig. 6(A)]. Note that the first term on the right side of Eq. (65) is that for horizontal absorption [Eq. (29)]. The infiltration velocity $v_0 = d(U/A)/dt$ may be calculated from Eq. (65) as

$$v_0 = F_1/2t^{1/2} + (K_0 + F_2) + 3F_3t^{1/2}/2 + 2F_4t + \cdots \tag{66}$$

where the F_i are the constants given by

$$F_i = \int_{\theta_0}^{\theta_1} f_i\, d\theta \tag{67}$$

For time t not too large, it may be possible to neglect all but the terms in $t^{1/2}$ and t in Eq. (65) (Philip, 1957e), yielding

$$U/A = F_1t^{1/2} + (K_0 + F_2)\, t \tag{68}$$

and, for the infiltration velocity

$$v_0 = F_1/2t^{1/2} + (K_0 + F_2) \tag{69}$$

As time t becomes very large, however, the use of enough terms in such series as Eqs. (39), (65), and (66) becomes inefficient and practically impossible, since the water-content profile becomes constant in shape (Philip, 1957d; Youngs, 1957). Without giving details, it will simply be noted that the solution becomes linear in t as expressed by Philip (1964)

$$Z(\theta, t) = \Omega(\theta) + [(K_1 - K_0)/(\theta_1 - \theta_0)]\, t \tag{70}$$

where $\Omega(\theta)$ denotes the constant-shape profile, and is the solution of an ordinary differential equation which again may be found by numerical methods (Philip, 1957d). The cumulative volume of infiltrated water per unit cross-sectional area at given time t now becomes, using Eq. (70) in Eq. (64),

$$U/A = \int_{\theta_0}^{\theta_1} \Omega(\theta)\, d\theta + K_1 t \tag{71}$$

and $v_0 = d(U/A)/dt$ becomes

$$v_0 = K_1 \tag{72}$$

Note that although Eq. (69) will also predict a constant v_0 as t becomes large, the constant value is erroneously asserted to be $(K_0 + F_2)$ rather than the correct value K_1 given by Eq. (72). This should serve as a reminder that Eqs. (68) and (69) are approximations not holding for large t.

Experimental verifications of the vertical infiltration solution were first given by Youngs (1957), who reported good laboratory confirmation of the theoretical predictions of profile shape and movement for slate dust and glass beads. Nielsen *et al.* (1961) found fair experimental validation in field studies of soil-water-content profiles. Davidson *et al.* (1963b), in laboratory experiments with disturbed field soils, reported good agreement between theoretical and experimental profiles for θ_1 near saturation, as illustrated in Fig. 7 for Columbia silt loam, but could not even make comparisons for smaller values of θ_1 because of nonuniqueness encountered in $\eta(\theta)$. D. E. Miller and W. H. Gardner (1962) did not find

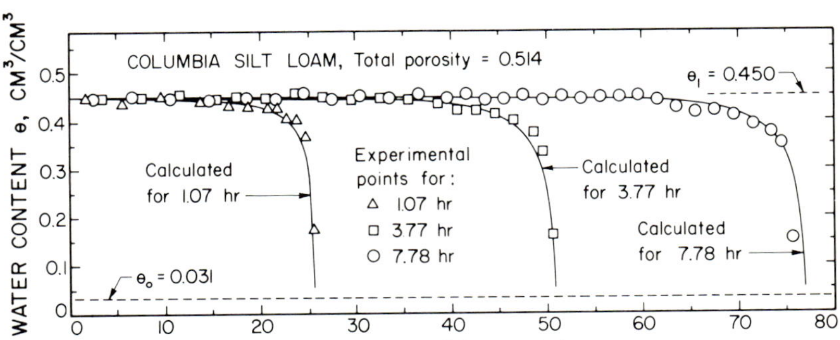

FIG. 7. Experimental and theoretical water-content profiles for Columbia silt loam [data of Davidson *et al.* (1963b) as adapted from *Hilgardia* **34**, 614 (1963), by permission of the publisher and authors].

linearity of v_0 with $t^{-1/2}$ [Eq. (69)] for t larger than about 3 min. However, Stockinger *et al.* (1965), for times as high as 300 min., reported F_1 as given by an experimental test of Eq. (68) to be the same for both horizontal and vertical flow, but no attempt was made to determine whether the coefficient of t was $(K_0 + F_2)$ or K_1.

The use of electronic computers is suggested rather naturally by the numerical nature of the various solutions for water intake by soil. Hanks and Bowers (1962) developed for vertical infiltration a computer solution which can accommodate stratified soils, finite depth, and nonuniform initial water content. Whisler and Klute (1965) obtained a computer solution for infiltration into a uniform finite soil column, but in which the initial water-content distribution is that of the drying $\tau(\theta)$ attained by the soil in equilibrium above a water table. Computer solutions and related experimental studies have also been provided for infiltration from constant rainfall (Rubin and Steinhardt, 1963, 1964; Rubin *et al.*, 1964; Rubin, 1966), both for the unponded surface-water case of the saturated hydraulic conductivity $K_s \geqslant v_0$, and for the case $K_s < v_0$, in which a nonzero depth of water eventually will pond on the soil surface.

For a constant nonzero depth of ponded water applied to the soil surface from time $t = 0$ on, Philip (1958b) has a given solution obtained by methods similar to those used in solving Eq. (30) or (31) subject to conditions (32) and (33). This ponded-water solution, however, was not found altogether adequate experimentally by Gupta and Staple (1964), who then utilized a computer approach to achieve agreement between experimental and theoretical water-content profiles.

2.3.2. *Soil-Water Movement to an Evaporating Soil Surface*

In ultimate detail, the evaporation of water from soil involves numerous interactions between the evaporative energy, the soil water, and the energy and water transport processes. This complexity prevents an unequivocal interpretation of the evaporative flow mechanism, as noted by Wiegand and S. A. Taylor (1961). The process would not be expected to be truly isothermal. Nevertheless, the assumption of isothermality permits enormous simplification of the transport problem, and does provide useful mathematical solutions and descriptions. Using the isothermal assumption, Philip (1957b) was able to give some quantitative characterization of both the constant-rate and falling-rate stages of evaporative soil-water loss. Several aspects of these two stages will now be considered, assuming isothermality and liquid-phase transport.

2.3.2.1. STEADY-STATE LOSS FROM A CONSTANT-POSITION WATER TABLE. The simplest constant-rate stage of soil-water evaporation is shown

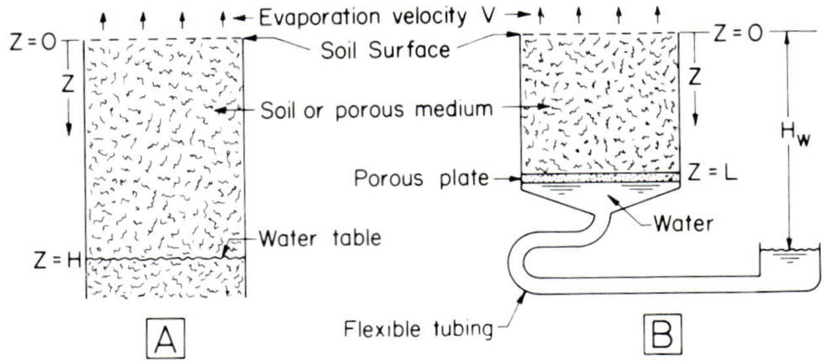

FIG. 8. Schematic diagrams of steady-state evaporation through vertical soil columns, (A) from an actual constant-position water table at depth H below the evaporating soil surface, and (B) from a constant-position simulated water table at depth H_w below the evaporating soil surface. Devices for maintaining a constant H or H_w are not shown, but could utilize a Mariotte system as in Fig. 3.

in Fig. 8(A) by considering a uniform vertical column of soil, of constant cross-sectional area A, to be subject to steady-state evaporation at its top end, while a water table is constantly maintained at a depth H. Let this steady-state evaporation velocity (constant volumetric flow rate divided by A) be V. Using the scaler form of Eq. (2) for the z direction, and again taking $-z = Z$ as positive downward, we have

$$v = V = -K(1 - \partial\tau/\partial z) = -K(1 + d\tau/dZ) \tag{73}$$

which can be rearranged and expressed in integral form as

$$Z = -\int \frac{d\tau}{1 + V/K} + C_2 \tag{74}$$

where C_2 is the constant of integration. With V being constant, it is expedient to take $K = K(\tau)$. An empirical expression for this has been suggested by W. R. Gardner (1958) in a form equivalent to

$$K(\tau) = K_S/(c\tau^m + 1) \tag{75}$$

holding only for $\tau \geqslant 0$. K_s is the saturated hydraulic conductivity [$K(\tau)$ at $\tau = 0$], and c and m are empirical constants. For $m = 1, \frac{3}{2}, 2, 3,$ or 4, the use of Eq. (75) enables Eq. (74) to be integrated (W. R. Gardner, 1958). For illustrative purposes, we present here the result for $m = 2$, using Eq. (75) and formula No. 49 of Peirce (1929) to perform the integration. In the integrated result we use the boundary condition at the water table, $\tau = 0$ at $Z = H$, to evaluate C_2, with the final result being

$$Z = H - (K_S/[cV(K_S + V)]^{1/2}) \arctan \tau[cV/(K_S + V)]^{1/2} \tag{76}$$

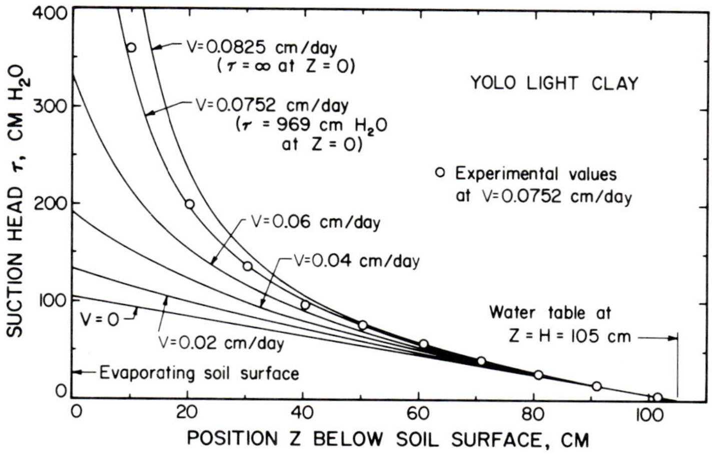

FIG. 9. Suction-head distributions at various steady-state evaporation velocities, for Yolo light clay (R. E. Moore, 1939). Solid-line curves were calculated from Eq. (76) using $c = 0.0025$ (cm $H_2O)^{-2}$ and $K_s = 1$ cm/day as given by W. R. Gardner and Fireman (1958).

For a given V, Eq. (76) implicitly specifies the distribution of τ with Z. Such curves are shown in Fig. 9 for various V for the Yolo light clay of R. E. Moore (1939), using Eq. (76) with $c = 2.5 \times 10^{-3}$ (cm $H_2O)^{-2}$ and $K_s = 1$ cm/day as given by W. R. Gardner and Fireman (1958). When $V = 0$, we have the linear static distribution $\tau = H - Z$ as yielded from Eq. (74), taking $V = 0$ and again evaluating C_2 from $\tau = 0$ at $Z = H$. As V becomes nonzero and increases, the suctions in the upper part of the column increase rapidly, while those near the water table are but slightly affected. The experimental values (R. E. Moore, 1939) are in good conformity with the corresponding theoretical curve.

Equation (76) implies, and Fig. 9 demonstrates, that the evaporation velocity rises to a maximum as $\tau \to \infty$ at the evaporating soil surface. Letting $V = V_m$ when $\tau \to \infty$ at $Z = 0$ in Eq. (76), we obtain, after rearrangement,

$$V_m(K_s + V_m) = K_s^2\pi^2/4cH^2 \tag{77}$$

which, if desired, could be solved explicitly for V_m by the use of the standard quadratic formula. Equation (77) thus provides for a given soil the maximum evaporation velocity as a function of depth to water table, and is shown in Fig. 10 by the solid-line curve computed for Chino clay using $K_s = 1.95$ cm/day and $c = 1.77 \times 10^{-3}$ (cm $H_2O)^{-2}$ as given by W. R. Gardner and Fireman (1958). With increasing water-table depth, V_m falls rapidly to only several percent of K_s when H reaches 300 cm,

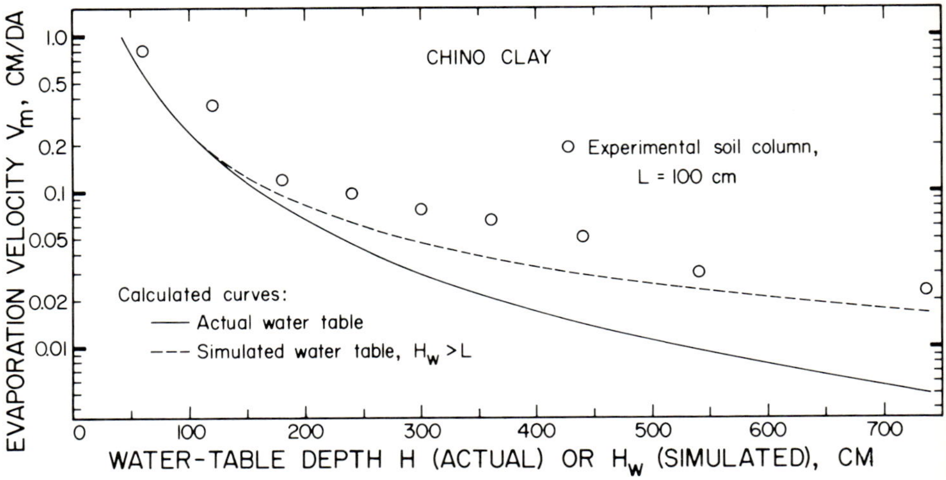

FIG. 10. Maximum steady-state evaporation velocity versus actual [Eq. (77)] or simulated [Eqs. (79) and (80)] water-table depth, along with experimental values for a simulated water table in Chino clay [adapted from W. R. Gardner and M. Fireman, *Soil Sci.* **85**, 246 (1958), by permission of the publisher and authors].

and thus demonstrates the importance of deep water tables in the reduction of evaporative soil-water loss.

Since direct experimental assessment of the relationship between V_m and H becomes inconvenient for large H, W. R. Gardner and Fireman (1958) employed the experimental arrangement of Fig. 8(B). A column of fixed length L is subjected to suction at its bottom end by means of the porous plate and a hanging water column of length $(H_w - L)$, where H_w is the depth of the simulated water table as maintained constant for a given steady V. Because of flow resistance in the porous plate (plus that in the saturated soil if $H_w < L$), H_w and H of Fig. 8 cannot agree precisely when $V > 0$, but will be quite close to each other if V and the porous-plate flow resistance are small. For the simulated water table, Eq. (74) on the basis of Eq. (75) integrates as before, but yields the final form

$$Z = L + \frac{K_s}{[(K_s + V)\,cV]^{1/2}} \left[\arctan(H_w - L)\left(\frac{cV}{K_s + V}\right)^{1/2} - \arctan \tau \left(\frac{cV}{K_s + V}\right)^{1/2}\right] \tag{78}$$

The arctan term involving $(H_w - L)$ in Eq. (78) comes out of the evaluation of C_2 from Eq. (74), as a result of applying $\tau = H_w - L$ at $Z = L$ as the lower boundary condition in Fig. 8(B); this assumes a negligible head loss for flow through the porous plate. As before, the maximum evaporation velocity is obtained from Eq. (78) by setting $V = V_m$ for $\tau \to \infty$ at $Z = 0$, with the rearranged result that

$$H_W - L = [(K_s + V_m)/cV_m]^{1/2}/(\tan L[(K_s + V_m)\, cV_m]^{1/2}/K_s) \qquad (79)$$

Note that Eqs. (78) and (79) are good only for $H_w \geqslant L$; to apply them for $H_w < L$ would violate the condition $\tau \geqslant 0$ attached to Eq. (75). Instead, to evaluate C_2 for $H_w < L$, the lower boundary condition in Fig. 8(B) is taken as $\tau = 0$ at $Z = H_w$. The resulting expression for Z is then the same as Eq. (76) except that H_w replaces H. Taking $V = V_m$ as $\tau \to \infty$ at $Z = 0$, we get

$$V_m(K_s + V_m) = K_s^2\pi^2/4cH_w^2 \qquad (80)$$

which is the same as Eq. (77) except that H_w replaces H. At $H_w = L$, Eq. (79) will also reduce to Eq. (80). Thus, the solid-line curve of Fig. 10 also holds for the simulated water table when $H_w \leqslant L = 100$ cm; for $H_w > L$, the broken-line curve of the figure was plotted from Eq. (79) with the same values of K_s and c as used previously for Chino clay in Eq. (77). The circles in Fig. 10 are experimental values obtained by W. R. Gardner and Fireman (1958) on a column of Chino clay 100 cm long. The agreement between theory and experiment is considered satisfactory, since the values of K_s and c were obtained from measurements on a separate sample of Chino clay. An even better fit could be attained if the K_s used in Eqs. (77) and (79) were increased about 45%.

If V_m can be neglected in comparison with K_s, Eq. (77) becomes

$$V_m = K_s\pi^2/4cH^2 \qquad (81)$$

so that V_m becomes inversely proportional to the square of H. Similarly, for $m = \frac{3}{2}$, 3, and 4 in Eq. (75), V_m turns out to be inversely proportional to $H^{3/2}$, H^3, and H^4, respectively (W. R. Gardner, 1958), if $V_m << K_s$. Hence, this provides some theoretical basis for a log-log plot of evaporation velocity versus water-table depth to be a straight line of negative slope, as reported experimentally for a number of cases by Wiegand and S. A. Taylor (1961).

Whether V reaches its maximum value of V_m for a given soil (K_s and c) and water-table depth would appear to depend upon the evaporative capability of the atmosphere. A measure of this capability used by Schleusener and Corey (1959) was V_f, the velocity of evaporation from a free water surface. It would seem reasonable that V should equal V_f as long as $V_f < V_m$, for under such circumstances the suction at the soil surface could remain finite. However, as V_f becomes large and equals or exceeds V_m, it would seem that τ should approach infinity at the soil surface, as used to obtain Eqs. (77) and (79). Hence, a plot of V versus V_f should rise to and remain at the constant limiting value V_m for given K_s, c, and H. Experiments by Schleusener and Corey (1959), however, showed V to rise to a maximum and then to decrease as V_f was continually increased. They suggested hysteresis as a possible cause of such behavior, but this is somewhat difficult to conceive if V_f and the suction at the soil surface increase monotonically. The anomaly might be at least partially related to the oversimplifications arising out of the assumption of isothermality, or to the development of slight salt crusts at the evaporating soil surface (W. R. Gardner and Hillel, 1962) so as to reduce the vapor pressure of the soil water (Cary, 1965b).

Steady-state evaporation analysis for a stratified soil column is also possible. The effect of stratification inhomogeneities can in some instances be handled by a weighted-average expression for $K(\tau)$ (Willis, 1960).

Liakopoulos (1966) has recently presented a computer analysis for a transient case somewhat like that of Fig. 8(A), but wherein the soil at time $t = 0$ is taken as saturated everywhere above the water table. The upper boundary condition is taken as $\tau/(H - Z) = \text{const}$ at $Z = 0$ for $t > 0$. But this would require τ to be constant at $Z = 0$, whereas the final results (his Figs. 10 and 12) show τ to vary with time at $Z = 0$. It therefore appears that the upper boundary condition would be more correctly expressed as $\partial\tau/\partial Z = \text{const}$ at $Z = 0$ for $t > 0$.

2.3.2.2. CONSTANT-RATE LOSS WITH TRANSIENT REDUCTION OF SOIL-WATER CONTENT. When water is no longer supplied steadily to the bottom of an evaporating soil column, Eq. (6) or (7) will apply with $\partial\tau/\partial t$ or $\partial\theta/\partial t$ being nonzero. Even when τ or θ changes with time, however, an initial period exists during which the evaporation velocity at the upper end of the column remains constant (see Philip, 1957b; Wiegand and S. A. Taylor, 1961, and references therein), provided the evaporative capability of the atmosphere remains constant. Covey (1963) has attacked this problem on the basis of one-dimensional flow ignoring gravity, thus reducing Eq. (7) to $\partial\theta/\partial t = \partial(D\partial\theta/\partial z)/\partial z$, which, if $-z = Z$ is positive downward, becomes

$$\frac{\partial\theta}{\partial t} = \frac{\partial}{\partial Z}\left(D\frac{\partial\theta}{\partial Z}\right) \tag{82}$$

and is taken subject to

$$\theta = \theta_b, \qquad Z > 0, \qquad t = 0 \tag{83}$$

$$\partial\theta/\partial Z = 0, \qquad Z = L, \qquad t > 0 \tag{84}$$

$$D\,\partial\theta/\partial Z = V, \qquad Z = 0, \qquad t > 0 \tag{85}$$

where θ_b is the constant initial soil-water content, and V once again is the constant evaporation velocity at the top of the soil column. The physical arrangement may be visualized as similar to Fig. 8(B), except that θ is initially constant, and the porous plate at $Z = L$ is replaced with an impermeable barrier so that the zero flow across it gives rise to condition (84). The constant-rate aspect is embodied in condition (85). The diffusivity function is taken as the exponential form of W. R. Gardner and Mayhugh (1958), namely,

$$D = D_b \exp \beta(\theta - \theta_b) \tag{86}$$

where D_b is the diffusivity at θ_b, and β is a constant. Equations (82) and (86) subject to conditions (83), (84), and (85) are then solved by numerical methods.

For finite L, Covey (1963) tabulates $\beta(\theta_b - \theta)$ versus Z/L and $\beta Vt/L$ for various values of the characteristic parameter $\beta VL/D_b$, which is small for slow rates of drying, short column lengths, and large initial water contents (which produce large D_b). To illustrate his solution we use Pachappa sandy loam, taking $\beta = 18.5$, $\theta_b = 0.387$, $\theta_a = 0.014$, and $D_a = 1.5$ cm²/day, where θ_a and D_a are the water content and diffusivity, respectively, at air-dryness. These numerical values are taken from W. R. Gardner and Hillel (1962), θ_a being the product of the bulk density of 1.4 gm/cm³ and a gravimetric air-dry water content of 0.01 (W. R. Gardner, 1959a). Setting $D = D_a$ for $\theta = \theta_a$ in Eq. (86) gives $D_b = D_a \exp \beta(\theta_b - \theta_a) = 1489$ cm²/day. For $L = 16.1$ cm and V of 0.05, 0.5, and 5 cm/day, the results are shown in Fig. 11 by the solid-

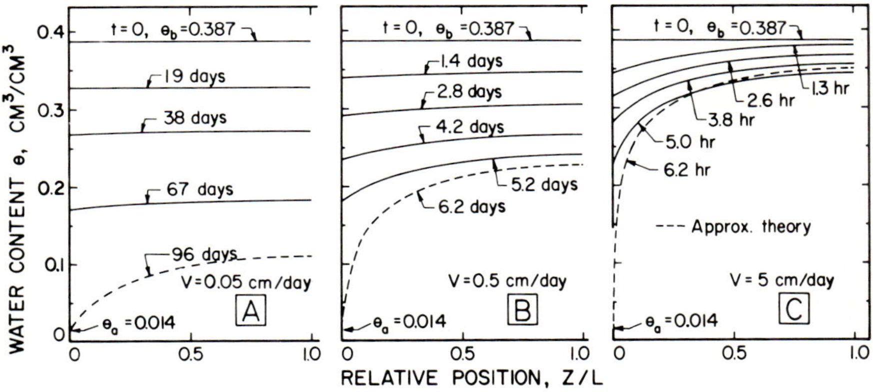

FIG. 11. Transient water-content profiles for constant-rate soil-water evaporation; solid-line curves from Covey's (1963) solution, and broken-line curves from the approximate theory [Eq. (92)] of W. R. Gardner and Hillel (1962). Soil and other parameters used in calculations are for Pachappa sandy loam (W. R. Gardner, 1959a; W. R. Gardner and Hillel, 1962).

line curves, which are water-content profiles (θ versus Z/L) at the times indicated by the numbers on the curves. For the smallest V [0.05 cm/day, Fig. 11(A)], the profiles are essentially straight horizontal lines which shift proportionally with time for the first 38 days. Even at 67 days, the highest time available from Covey's tables, the left end of the profile is but slightly curved toward θ_a. As V increases, however, the region of proportionally shifting, horizontally linear profiles is sharply reduced, in terms of both time and the amount of water removed. At the highest times available from Covey's tables [5.2 days and 5.0 hr in Figs. 11(B) and (C), respectively], the curvature of the profiles toward θ_a at $Z = 0$ is progressively increased.

An important aspect of the constant-rate transient problem is the determination of the time t_c during which the constant evaporation velocity will be sustained. To do this requires the use of some criterion to mark the point at which the soil column is no longer able to maintain the evaporation velocity at V, such as, for instance, the water content of the soil surface being reduced to the air-dry value θ_a (W. R. Gardner, 1959a). One approach utilizing this criterion would be to extend Covey's (1963) results to larger values of $\beta Vt/L$, thus enabling determination of the time for which $\theta = \theta_a$ at $Z = 0$. An approximate approach first outlined by W. R. Gardner and Hillel (1962) may also be used, and is developed here as follows:

The magnitude of the overall volumetric evaporation loss rate per unit volume of soil is $AV/AL = V/L$. Whereas condition (85) implies the volumetric soil-water loss rate to be constant only at $Z = 0$, the approximate approach introduces the further assumption that it shall be constant at all positions Z in the soil column; that is,

$$\partial(A\theta\, dZ/A\, dZ)/\partial t = \partial\theta/\partial t = -V/L$$

By Eq. (82) we then write

$$\frac{\partial}{\partial Z}\left(D\frac{\partial\theta}{\partial Z}\right) = -\frac{V}{L} \tag{87}$$

which integrates directly to

$$D\frac{\partial\theta}{\partial Z} = -\frac{VZ}{L} + f(t) \tag{88}$$

where $f(t)$ is an arbitrary function of t. Using condition (84) in Eq. (88) yields $f(t) = V$, which makes Eq. (88) become

$$D\frac{\partial\theta}{\partial Z} = -\frac{VZ}{L} + V \tag{89}$$

Direct integration of Eq. (89) may proceed if D can be expressed as an integrable function of θ. Using Eq. (86) for this purpose, by rearranging it first as $D = D_a \exp \beta(\theta - \theta_a)$ and then substituting it into Eq. (88) and integrating, we obtain

$$(1/\beta) \exp \beta(\theta - \theta_a) = (V/D_a)(Z - Z^2/2L) + g(t) \tag{90}$$

where $g(t)$ is another arbitrary function of t. Letting θ_c be the water content as a function of Z at the end of the constant-rate period, we use $\theta = \theta_c$ at $t = t_c$ in Eq. (90) to obtain

$$(1/\beta) \exp \beta(\theta_c - \theta_a) = (V/D_a)(Z - Z^2/2L) + g(t_c) \tag{91}$$

But, at t_c we also have $\theta_c = \theta_a$ at $Z = 0$, which when used in Eq. (91) evaluates $g(t_c) = 1/\beta$, hence enabling Eq. (91) to be written in the rearranged form

$$\theta_c = \theta_a + (1/\beta) \ln[1 + (\beta VL/2D_a)(1 - [1 - Z/L]^2)] \tag{92}$$

Equation (92) thus expresses θ_c as a function of Z at t_c; that is, the water-content profile at the end of the constant-rate period. That all operations on Eq. (87) and those following have been rigorous must not obscure the fact that Eq. (92) is approximate by virtue of Eq. (87) itself being approximate. This may be further emphasized by noting that in obtaining Eq. (87) from Eq. (82) we used $\partial\theta/\partial t = -V/L$, which, if itself integrated, would have yielded $\theta = -Vt/L + f(Z)$, where $f(Z)$ is an arbitrary function of Z. But, by condition (83), $f(Z) = \theta_b$, so that $\theta = \theta_b - Vt/L$, which is independent of Z and is rather drastically different from either Eq. (90) or (92).

The mean water content $\bar{\theta}_c$ at time t_c may be determined by using Eq. (92) in the expression

$$\bar{\theta}_c = (1/L) \int_0^L \theta_c \, dZ = \theta_a + (1/\beta L) \int_0^L \ln[1 + b_1(1 - [1 - Z/L]^2)] \, dZ \tag{93}$$

where $b_1 = \beta VL/2D_a$. We twice transform the integrand in Eq. (93), first by $\xi = L - Z$, and second by setting a new variable equal to the resulting $(1 + b_1 - b_1\xi^2/L^2)$. Then integrating by parts and using formulas No. 94 and No. 98 of Peirce (1929), we obtain finally

$$\bar{\theta}_c = \theta_a - 2/\beta + (b_2/\beta) \ln[(b_2 + 1)/(b_2 - 1)] \tag{94}$$

where $b_2 = [(b_1 + 1)/b_1]^{1/2}$. Then, since the total amount of water lost from the column per unit cross-sectional area is $\theta_b L - \bar{\theta}_c L = Vt_c$, we solve this for t_c to get

$$t_c = (\theta_b - \bar{\theta}_c) L/V \tag{95}$$

For the Pachappa sandy loam of Fig. 11, the broken-line curves are the approximate water-content profiles at the end of the constant-rate period, as calculated from Eq. (92). The time values on these approximate profiles are calculated from Eq. (95) with $\bar{\theta}_c$ given by Eq. (94), using the same θ_a, β, V, L, D_a, and θ_b as for the solid-line profiles. For the two smaller evaporation velocities [0.05 and 0.5 cm/day, Fig. 11(A) and (B)], the approximate profiles are seen to form reasonable-appearing advanced stages of their solid-line predecessors, with respect to shape of curve, time, and the graph area swept out from the preceding profile between the lines $Z/L = 0$ and $Z/L = 1$. The same cannot be said for the V of 5 cm/day in Fig. 11(C), but it should be remembered that this high velocity might induce substantial departure from the assumption of a constant volumetric loss rate ($\partial\theta/\partial t = -V/L$). On the whole, it may be somewhat surprising that the approximate approach works as well as it seems to. W. R. Gardner and Hillel (1962), while not assessing the experimental validity of approximate profiles such as the broken-line curves of Fig. 11, did make the further approximation that the mean water content at t_c was given by $[\theta_a + (1/\beta)\ln(1 + \beta VL/2D_a)]$, which is the right side of Eq. (92) when $Z = L$. The estimates of t_c on this basis were reasonably well verified by experiments with several soils and various V.

Fig. 11 may also be used to demonstrate the behavior for $V = 0.05$ cm/day, but in which the column length L is increased to 161 and 1610 cm in parts (B) and (C), respectively. However, for these same two parts, the time values on the water-content profiles must be multiplied by 10^2 and 10^4, respectively.

2.3.2.3. FALLING-RATE LOSS WITH TRANSIENT REDUCTION OF SOIL-WATER CONTENT. As the evaporative capability of the atmosphere increases relative to the ability of the soil system to supply water, Covey's (1963) parameter $\beta VL/D_b$ increases. Then, Fig. 11 illustrates that the constant-rate time period t_c decreases, the water-content drop from θ_b to θ_a at the soil surface is more rapid, and the water content at the bottom of the column ($Z = L$) is progressively less reduced from θ_b. Let us now consider the limiting case in which θ at the soil surface drops instantaneously from θ_b to θ_a at time $t = 0$, thus implying a very large or infinite evaporative capability of the atmosphere. For simplicity we take the column length to be infinite, and consider Eq. (82) to be subject to

$$\theta = \theta_b, \qquad Z > 0, \qquad t = 0 \tag{96}$$

$$\theta = \theta_a, \qquad Z = 0, \qquad t > 0 \tag{97}$$

The solution of this problem may again be approached by the separation-of-variables technique used for the horizontal absorption case [Eq. (8) subject to conditions (9) and (10)], with the result

$$\frac{d}{d\theta}\left(D\frac{d\theta}{d\eta}\right) = -\frac{\eta}{2} \tag{98}$$

subject to

$$\eta(\theta_b) \rightarrow \infty \tag{99}$$

$$\eta(\theta_a) = 0 \tag{100}$$

where $\eta(\theta) = \eta = Zt^{-1/2}$ is the Boltzmann variable as defined in Eq. (24). Equation (98) subject to conditions (99) and (100) has been solved numerically by W. R. Gardner (1959a) and Scott and Hanks (1962), using the exponential diffusivity function of Eq. (86).

From the results of Scott and Hanks (1962), $\eta(\theta)$ is illustrated graphically by the solid-line curves of Fig. 12, where $\eta/D_a^{1/2}$ is plotted

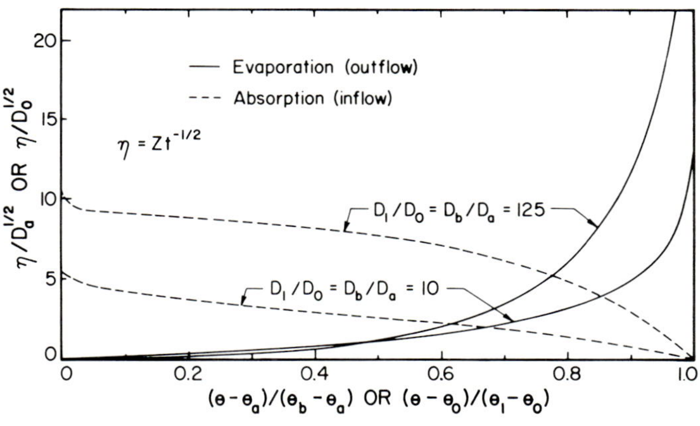

FIG. 12. Theoretical solutions for falling-rate evaporative soil-water loss (solid-line curves) and horizontal water absorption (broken-line curves), developed and adapted from the results of Scott and Hanks (1962) for the diffusivity function of Eq. (86).

against a relative water content $(\theta - \theta_a)/(\theta_b - \theta_a)$. Although Eq. (98) subject to conditions (99) and (100) is the reverse of Eq. (20) subject to conditions (21) and (23), Fig. 12 demonstrates that the two solutions are by no means mirror images of each other. This is seen by comparing the solid-line curves with their respective broken-line counterparts for horizontal absorption, again as determined from the results of Scott and Hanks (1962).

Figure 12 also provides a starting point for an interesting comparison between water absorption and evaporation, which was first noted by

W. R. Gardner (1959a). If $\theta_0 = \theta_a$ and $\theta_1 = \theta_b$, so that in both flow cases the water-content changes take place between the same endpoints, then $D_0 = D_a$ and $D_1 = D_b$, and the curves of the figure become quantitatively comparable. It appears that the area under a given absorption curve, which area is a measure of the cumulative volume U of water absorbed [Eq. (29)], exceeds the area under the corresponding evaporation curve, which area is a measure of the cumulative volume of water U_e lost by evaporation. In similar vein to Eq. (28), U_e is given by

$$U_e = A \int_{\theta_a}^{\theta_b} Z\, d\theta = At^{1/2} \int_{\theta_0}^{\theta_1} \eta_e\, d\theta = At^{1/2} F_{1e} \tag{101}$$

where η_e simply denotes the Boltzmann variable for the evaporation process, and F_{1e} designates the integral of η_e. From Eqs. (28) and (67), $U = At^{1/2}F_1$, so that at any given time t the ratio of U to U_e becomes

$$U/U_e = F_1/F_{1e} \tag{102}$$

Scott and Hanks (1962) have presented graphical values of a variable ω^*, and it turns out that $F_1/F_{1e} = |\, \omega^*_{\text{inflow}}/\omega^*_{\text{outflow}} \,|$, thus enabling the calculation of U/U_e from Eq. (102). A plot of U/U_e versus D_b/D_a (or D_1/D_0) is given in Fig. 13, and shows indeed that U exceeds U_e and

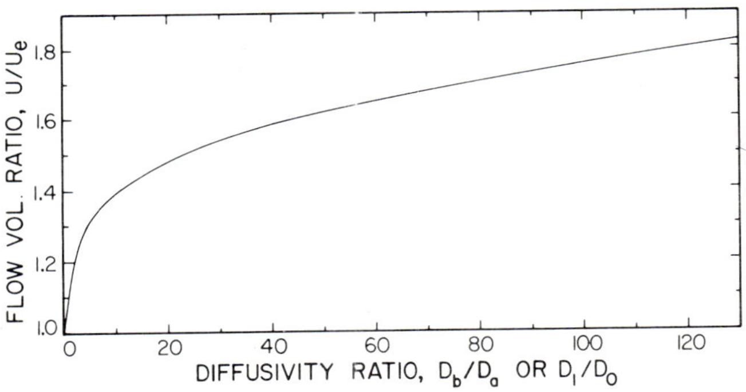

FIG. 13. Excess of water inflow over water outflow for similar initial and boundary soil-water contents ($\theta_0 = \theta_a$ and $\theta_1 = \theta_b$), as expressed by flow-volume ratio versus diffisivity ratio, developed and adapted from the results of Scott and Hanks (1962) for the diffusivity function of Eq. (86).

that U/U_e increases with D_b/D_a. Thus, for any given time and the conditions here considered ($\theta_0 = \theta_a$ and $\theta_1 = \theta_b$), more water is absorbed by the soil than is lost by evaporation. This behavior is of relevance for the conservation of soil water. The magnitude of the effect shown in Fig. 13 is due solely to the decrease of diffusivity with decreasing water content. If D were constant, so that $D_b = D_a = D_1 = D_0 = D$, the absorption solution would be the classical form

$$\theta = \theta_b - (\theta_b - \theta_a)\,\mathrm{erf}(\eta/2D^{1/2})$$

and the evaporation solution would be the mirror image

$$\theta = \theta_a + (\theta_b - \theta_a)\,\mathrm{erf}(\eta_e/2D^{1/2})$$

where "erf" is the error function; U_e can then be shown to equal U.

U/U_e can normally be expected to exceed unity by even more than that indicated in Fig. 13, for two reasons. First, since absorption is a wetting process and evaporation is one of drying, the hysteresis effect, as considered for $K(\theta)$ and $\tau(\theta)$ in Fig. 2, will cause the wetting diffusivity to exceed the drying diffusivity (W. R. Gardner, 1959b; Staple, 1965), whereas for Fig. 13, an identical diffusivity function [Eq. (86) with given D_b and β] was used for both processes. Second, gravity was ignored in Fig. 13, an omission which does not appear serious for evaporation (and hence U_e) when theory is compared with experiment (W. R. Gardner, 1959a; W. R. Gardner and Hillel, 1962; Rose, 1967), but is of consequence for downward infiltration, and means that U should be increased in accord with Eq. (65).

Questions also might be raised about the effect of assuming infinite column length. Preliminary computations by the author, based on numerical finite-column solutions of Klute *et al.* (1965), indicate the curve of Fig. 13 to be virtually unaffected by this factor. This conclusion must remain tentative, however, since the results presented by Klute *et al.* (1965) are not in a form conveniently suited to the precise evaluation of U/U_e.

From Eq. (101), the evaporation velocity $v_e = d(U_e/A)/dt$ becomes

$$v_e = 2F_{1e}t^{-1/2} \tag{103}$$

which, since F_{1e} is a constant, means that the falling-rate evaporation velocity decreases as the inverse square root of the time. Equation (101) was nicely verified by W. R. Gardner (1959a) on a 100-cm column of Chino clay during the time it would reasonably be expected to behave as infinitely long. Ultimately, however, the finite aspect of the column must be included in the theoretical solution, particularly when the water content at the far end of the column decreases from θ_b, and then the evaporation velocity falls off even more rapidly than the inverse square root of the time. Again, however, using a finite-column theoretical solution, W. R. Gardner (1959a) was able to describe the experimental cumulative evaporation from short columns of soil of differing length and initial water content. By combining certain features of the finite-column solution with the approximate constant-rate analysis already discussed [Eqs. (87)–(92)], W. R. Gardner and Hillel (1962) were able to describe the complete time course of the cumulative evaporation loss over both the constant-rate and falling-rate stages of the transient drying process. During the falling-rate stage, the evaporation velocity was found to approach closely a function of the water content of the soil and to be essentially independent of the potential evaporative capability of the atmosphere. A relatively simple equation was suggested for the evaporation rate. Hanks and H. R. Gardner (1965), however, encountered some substantial discrepancies upon comparing this simple equation with a computer analysis of falling-rate stage evaporation.

2.3.3. *Vertical Drainage to a Water Table*

One-dimensional gravity drainage to a constant-position water table has been considered by a number of investigators. The early studies of Lambe (1951) and Luthin and Miller (1953) were followed by a finite-difference numerical solution of the flow equation (Day and Luthin, 1956), but this solution is laborious unless a computer is employed (Liakopoulos, 1964; Remson *et al.*, 1965). Interest in the problem has often arisen out of concern with the nature of the drainable porosity or specific water-yield of a porous medium (Childs, 1960; G.S. Taylor, 1960; Youngs, 1960; Childs and Poulovassilis, 1962; Ligon *et al.*, 1962; Prill *et al.*, 1965). The approach we will employ here is that of using certain simplifying assumptions to enable solution of the boundary-value problem, as done by W. R. Gardner (1962) and Fujioka and Kitamura (1964), although the particular details will differ somewhat from those used in these two previous studies.

Since gravity is directly involved, we begin with Eq. (30) but rewrite it on the basis of Eq. (4) to obtain

$$\frac{\partial\theta}{\partial t} = \frac{\partial}{\partial Z}\left[-K\left(\frac{\partial Z}{\partial\theta} + \frac{d\tau}{d\theta}\right)\frac{\partial\theta}{\partial Z}\right] \tag{104}$$

Introduce the assumption that

$$-K\left(\frac{\partial Z}{\partial\theta} + \frac{d\tau}{d\theta}\right) = N \tag{105}$$

where N is a constant similar to but not identical with a diffusivity. Note that taking N constant does not necessarily require K to be constant in Eq. (105); also, since $h = -(Z + \tau)$, the coefficient of K may be interpreted as $\partial h/\partial \theta$, where h is the hydraulic head. Use of Eq. (105) in Eq. (104) then yields

$$\frac{\partial \theta}{\partial t} = N \frac{\partial^2 \theta}{\partial Z^2} \tag{106}$$

To describe the physical problem and the boundary conditions, we will refer to Fig. 8(A), assuming evaporation to be absent. Consider that the water table is first at the top of the column ($Z = 0$), and, at time $t = 0$, is lowered instantaneously to its constant position at $Z = H$. Mathematically,

$$\theta = n, \qquad 0 \leqslant Z \leqslant H, \qquad t = 0 \tag{107}$$

$$\theta = n, \qquad Z = H, \qquad t > 0 \tag{108}$$

where n, the constant porosity, is also the saturated water content. Since there is no flow across the upper soil boundary, we there have $\partial h/\partial Z = 0$, but this yields $\partial \theta/\partial Z = -d\theta/d\tau$, since $h = -(Z + \tau)$. If we introduce the further linearizing approximation

$$\theta = n - b_3 \tau \tag{109}$$

where b_3 is a positive constant, the upper boundary condition finally becomes

$$\partial \theta/\partial Z = b_3 , \qquad Z = 0, \qquad t > 0 \tag{110}$$

since $d\theta/d\tau = -b_3$ on the basis of Eq. (109). The solution of Eq. (106) subject to conditions (107), (108), and (110) can then be synthesized from variables-separable particular solutions of Eq. (106) appropriately expanded into a Fourier cosine series to meet condition (107). The final form is

$$\theta = n + b_3(Z - H) + \frac{8b_3H}{\pi^2} \sum_{i=1}^{\infty} \frac{\cos[(2i - 1)\,\pi Z/2H]}{(2i - 1)^2} \times \exp\left\{-\left[\frac{(2i - 1)\,\pi}{2H}\right]^2 Nt\right\} \tag{111}$$

The total volume of water U drained from the soil at time t is given by

$$U = A \int_0^H (n - \theta)\, dZ \tag{112}$$

Using Eq. (111) in Eq. (112) and performing the integration at given (constant) t yields

$$U = \frac{Ab_3H^2}{2} - \frac{16Ab_3H^2}{\pi^3} \sum_{i=1}^{\infty} \frac{(-1)^{i+1}}{(2i - 1)^3} \exp\left\{-\left[\frac{(2i - 1)\,\pi}{2H}\right]^2 Nt\right\} \tag{113}$$

The total outflow volume $U = U_\infty$, which obtains when $t \to \infty$ in Eq. (113), is $U_\infty = Ab_3H^2/2$, so that Eq. (113) can finally be written

$$\frac{U}{U_\infty} = 1 - \frac{32}{\pi^3} \sum_{i=1}^{\infty} \frac{(-1)^{i+1}}{(2i-1)^3} \exp \left\{ - \left[\frac{(2i-1)\,\pi}{2H} \right]^2 Nt \right\} \tag{114}$$

If only the first term of the series in Eq. (114) is retained, the approximate result is

$$U/U_\infty = 1 - (32/\pi^3)\, e^{-(\pi/2H)^2 Nt} \tag{115}$$

which, since $32/\pi^3$ is near unity and N is constant, is an expression similar to that which Youngs (1960) derived on the basis of a capillary-tube model, and which he found to hold reasonably well experimentally. W. R. Gardner (1962) also found his approximate equation, which differs from Eq. (115) only in that $32/\pi^3$ is replaced by $8/\pi^2$, to have fair experimental validity. The discrepancies which did occur implied failure in the assumption equivalent to $N =$ const.

If a portion of the soil immediately above the water table does not drain, that is, a so-called "saturated capillary fringe" exists, the lower boundary condition (108) will be altered. Analysis should still be possible along the lines given by W. R. Gardner (1962). This additional refinement, however, may not always lead to better agreement with experimental data, unless the length of the saturated capillary fringe comprises an appreciable portion of H.

2.3.4. *Measurement of Unsaturated Flow Characteristics*

In principle, any solution to a water-flow boundary-value problem could serve as the basis for a method of measuring K, τ, or D if these quantities appear in the solution. Ideally, a successful method should achieve some optimum blend of both physical and mathematical feasibility and convenience. As yet, however, the matter of experimental measurement is hardly to be considered as satisfactorily resolved, even though a degree of progress has been realized.

2.3.4.1. STEADY-STATE UNSATURATED CONDUCTIVITY. For the system of Fig. 14 (Olson, 1966), wherein the steady flow is created by the difference between the inflow and outflow hanging-water-column levels, the constant flow velocity V is the steady volumetric flow rate divided by the cross-sectional area A. Within the region $x_1 \leqslant x \leqslant x_3$, the suction gradient $\partial\tau/\partial x$ is determined as the slope of the curve established by plotting τ_1, τ_2, and τ_3 against their respective x_1, x_2, and x_3. From the scalar form of Eq. (2) for horizontal flow, the unsaturated conductivity

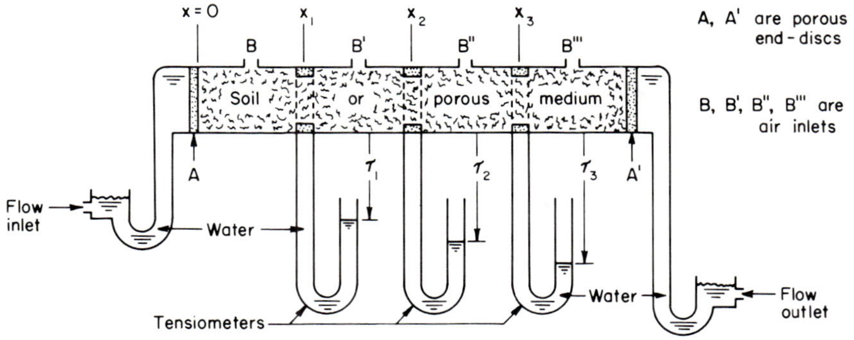

FIG. 14. Schematic representation of steady-state unsaturated water flow through a soil column. The tensiometers at the three internal positions have porous elements shown in section view as portions of circumferential rings.

is calculated as $K = V/(\partial\tau/\partial x)$ at a position in the column such as x_2, for which the corresponding τ is known (in this case, τ_2). Normally, $\tau(x)$ between x_1 and x_3 is curvilinear. However, if the distance such as $(x_2 - x_1)$ is small, the suction gradient may be taken as $(\tau_2 - \tau_1)/(x_2 - x_1)$ at $(x_1 + x_2)/2$, so that the computed K corresponds to a mean suction $(\tau_1 + \tau_2)/2$. The vertical thickness of the horizontal soil column should be small relative to τ, so that the change of τ with vertical position in the column will be negligible.

The principle of the steady-state method was first introduced by L. A. Richards (1931), using a vertical-flow column of relatively short length and large cross section, and with two interior tensiometers located just inside the porous end-discs. The gradient is then $(h_1 - h_2)/(x_2 - x_1)$, where $(h_1 - h_2)$ is the difference in hydraulic head between the tensiometers at x_1 and x_2. Later, L. A. Richards and D. C. Moore (1952) modified the system by raising the pressure of the soil air to induce changes in suction. In Fig. 14 this would be accomplished by connecting the air inlets to a source of air at gauge pressure P. The suction at any tensiometer position would then be the height of the hanging water column in that tensiometer plus $P/\rho g$, where ρg is the unit weight of water.

After K for a given steady state has been obtained, a new steady state and mean suction are achieved by shifting the inlet and outlet levels or by changing P. $K(\tau)$ can thus be determined for either drying or wetting, depending on whether τ is being progressively increased or decreased. Fig. 15 shows examples of such measurements of the drying $K(\tau)$, as obtained by L. A. Richards and D. C. Moore (1952). At saturation $(\tau = 0)$, the order in hydraulic conductivity is: sand $>$ loamy fine sand $>$ silty clay loam, which is to be expected on the basis of larger pore

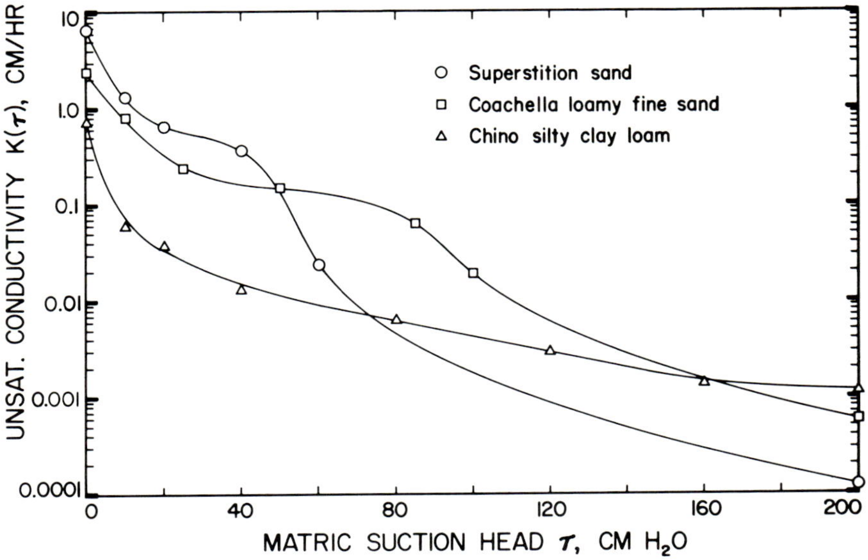

FIG. 15. Experimental curves of unsaturated conductivity versus matric suction head for three of the soils of L. A. Richards and D. C. Moore (1952).

sizes in the coarser textured soils. As τ increases (θ decreases), the curves cross each other in a systematic fashion until at $\tau = 200$ cm H_2O the order in unsaturated conductivity is the reversed one of: silty clay loam > loamy fine sand > sand. This can be accounted for on the basis of the greater number of undrained small pores existing in the finer textured soils at high suction.

Modifications of the steady-state method but which still employ the two porous end-discs have been given by Corey (1957), Nielsen and Biggar (1961), and Elrick and Bowman (1964). Childs and Collis-George (1950), however, eliminated the porous end-discs by subjecting long soil columns to downward flow at various velocities less than the saturated hydraulic conductivity. They used a capacitance method to measure the water content θ, so that K was determined as a function of θ; tensiometers, however, can be installed along the column to provide $K(\tau)$ (Luthin and Day, 1955). R. E. Moore (1939) used long vertical soil columns, without porous end-discs, by immersing the lower end of the column in a constant-level water supply and then subjecting the upper end of the column to steady-state evaporation. Nielsen *et al.* (1960) modified R. E. Moore's system by applying suction through a porous end-disc to the bottom end of a short vertical column, but continuing to use evaporation at the top of the column without a porous end-disc.

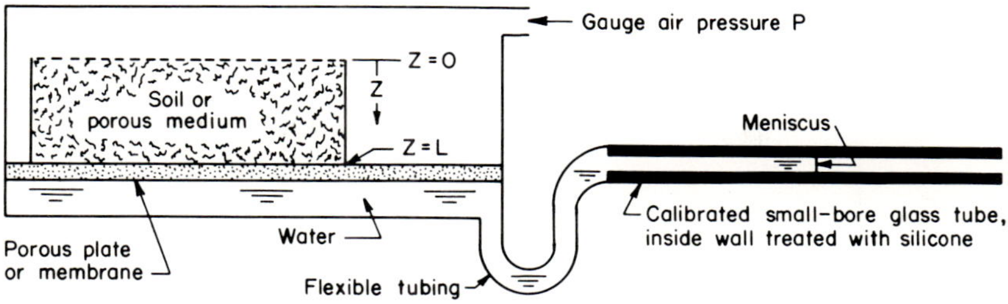

FIG. 16. Schematic diagram of apparatus for measuring changes of soil-water content in response to applied matric suctions obtained from gauge air pressure $P > 0$ or from lowering of the horizontal small-bore glass tube.

2.3.4.2. EQUILIBRIUM SUCTION AND WATER CONTENT. In the apparatus shown in Fig. 16, the water in the soil sample of total height L communicates through the porous plate or membrane with the water in the horizontal calibrated glass tube, which, with appropriate provision for draining or refilling, serves to measure the volume of water given up or taken in by the soil. The suction applied in terms of water column is $P/\rho g$, plus the distance of the horizontal glass tube below the porous plate or membrane if this tube is located lower than indicated in the figure. L should be small relative to the applied τ, so that at equilibrium the change of τ with Z is negligible.

To determine $\tau(\theta)$ for drying, the soil is initially saturated with water, and the first increment of τ is applied either by imposing a constant $P > 0$ or by lowering the horizontal glass tube to a new constant position. Outflow equilibrium is marked by a cessation of movement of the meniscus in the glass tube, at which time the total outflow volume of water is recorded. The next increment of τ is then applied, and the procedure is repeated as often as needed. After equilibrium at the final applied τ, the water content of the soil is determined gravimetrically. By proper accumulation of the various increments of outflow, the water content of the soil at each suction is determined by back calculation. The $\tau(\theta)$ for wetting is determined similarly except that the maximum suction is applied initially, followed by τ values which become progressively smaller. Detailed specifications for both apparatus and method have been given by L. A. Richards (1947, 1949), Tanner and Elrick (1958), and Reginato and van Bavel (1962).

Examples of actual data from several sources (L. A. Richards, 1953; Bruce and Klute, 1963; Elrick and Bowman, 1964) are shown in Fig. 17, and include a hysteresis loop as outlined by the broken-line rewetting and solid-line drying curves for Bloomfield sand. If the drying curves are compared as the soil texture shifts from sand to silty clay, the Guelph loam fails to fall into perfect sequence with the other soils, but several

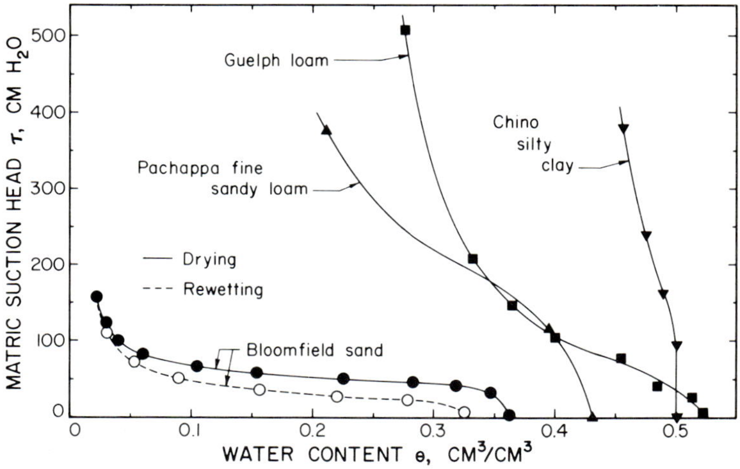

FIG. 17. Experimental curves of matric suction head versus water content for various soils. [Pachappa and Chino curves adapted from L. A. Richards, *Intern. Symp. Des. Res. Proc.*, p. 537 (1953); Bloomfield curves adapted from Bruce and Klute, *Soil Sci. Soc. Am. Proc.* **27**, 19 (1963); Guelph curves adapted from Elrick and Bowman, *Soil Sci. Soc. Am. Proc.* **28**, 450 (1964). All data used by permission of the publishers and authors.]

reasonably distinct tendencies can be noted nevertheless. Not only does θ at $\tau = 0$ increase, in accord with the greater porosity often attributed to finer texture, but the reduction of θ with increasing τ is also more gradual, along with a less pronounced inflection behavior. The difficulty of removing water from Chino silty clay is further illustrated by pointing out that even at $\tau = 15{,}500$ cm H_2O (15-atm suction) the value of θ has been reduced only to 0.34 (L. A. Richards, 1953).

2.3.4.3. TRANSIENT DIFFUSIVITY AND CONDUCTIVITY BY THE POROUS-PLATE OR MEMBRANE METHOD. Consider the soil in the porous-plate apparatus of Fig. 16 to have come to outflow equilibrium with a constant gauge air pressure $P = P_1$. Hence, for $0 \leqslant Z \leqslant L$, τ is constant at $\tau_1 = P_1/\rho g$, and so $\theta = \theta_1$ is constant. This is only true with acceptable accuracy if τ is always $>> L$, so that gravity effects can be ignored. At time $t = 0$, the air pressure is raised instantaneously from P_1 to a new constant $P_2 > P_1$, but τ throughout the soil sample will not reach the new value $\tau_2 = P_2/\rho g$ until sufficient water outflow has occurred to reduce θ to its new constant value $\theta_2 < \theta_1$. To attack this transient problem, ignoring gravity, Eq. (82) is considered subject to

$$\theta = \theta_1, \qquad 0 \leqslant Z \leqslant L, \qquad t = 0 \tag{116}$$

$$\partial\theta/\partial Z = 0, \qquad Z = 0, \qquad t > 0 \tag{117}$$

$$K\,\partial\tau/\partial Z = (\tau_2 - \tau)/I_p, \qquad Z = L, \qquad t > 0 \tag{118}$$

Condition (117) represents the absence of flow across the upper soil surface. Condition (118) arises from considerations at the lower soil surface in contact with the porous plate (or membrane), where the flow velocity from the soil is $v_s = K\,\partial\tau/\partial Z$, and that through the porous plate is $v_p = (\tau_2 - \tau)/(L_p/K_p)$. L_p and K_p are the porous-plate thickness and hydraulic conductivity, respectively. Setting $v_s = v_p$ and designating $L_p/K_p = I_p$ as the hydraulic impedance of the porous plate then results in condition (118).

The solution of Eq. (82) subject to conditions (116), (117), and (118), proceeds by two further simplifying conditions originally proposed by W. R. Gardner (1956). First, restrict the pressure step

$$P_2 - P_1 = \rho g(\tau_2 - \tau_1)$$

to be small enough that the water content is but slightly altered, so that D can be taken constant during the time of outflow. This then makes Eq. (82) become

$$\frac{\partial\theta}{\partial t} = D\,\frac{\partial^2\theta}{\partial Z^2} \tag{119}$$

Second, with $\tau_2 - \tau_1$ being small, take θ to be linearly related to τ during the period of outflow, that is,

$$\theta = a_1 - a_2\tau \tag{120}$$

where a_1 and a_2 are positive constants. Use of Eq. (120) permits condition (118) to be written

$$K\,\partial\theta/\partial Z = -(\theta - \theta_2)/I_p\,, \qquad Z = L, \qquad t > 0 \tag{121}$$

Note that the use of Eq. (120) and of $D =$ const forces K also to be constant, by virtue of Eq. (4).

The solution of Eq. (119) subject to conditions (116), (117), and (121) proceeds by separation of variables in Eq. (119) to obtain in product form a particular solution which meets conditions (116) and (117). This is then expanded into a Fourier-series-type solution capable of meeting condition (121) (E. E. Miller and Elrick, 1958; Rijtema, 1959), the final form being

$$\theta = \theta_2 + 2(\theta_1 - \theta_2)\sum_{i=1}^{\infty}\frac{\cos(\alpha_i Z/L)\exp(-\alpha_i^2\,Dt/L^2)}{(R + \csc^2\alpha_i)\,\alpha_i\sin\alpha_i} \tag{122}$$

where α_i is the ith solution of

$$R\alpha_i = \cot\alpha_i \tag{123}$$

and $R = I_p/I_s$, where $I_s = L/K$ is the hydraulic impedance of the unsaturated soil sample and is constant for any given pressure step $P_2 - P_1$ since K is then constant. Hence, R also is constant.

The mean soil-water content $\bar{\theta}$ at any time t is given by

$$\bar{\theta} = (1/L)\int_0^L \theta\, dZ$$
$$= \theta_2 + 2(\theta_1 - \theta_2)\sum_{i=1}^{\infty} \frac{\exp(-\alpha_i^2\, Dt/L^2)}{\alpha_i^2(R + \csc^2 \alpha_i)} \tag{124}$$

If U and U_∞ are the total volumes of outflow at time t and infinite time, respectively, we can write $U = AL(\theta_1 - \bar{\theta})$ and $U_\infty = AL(\theta_1 - \theta_2)$. Using these expressions in conjunction with Eq. (124) yields finally,

$$\frac{U}{U_\infty} = 1 - \sum_{i=1}^{\infty} \frac{2\exp(-\alpha_i^2\, Dt/L^2)}{\alpha_i^2(R + \csc^2 \alpha_i)} \tag{125}$$

Experimental values of U/U_∞ are plotted against time on log-log overlay graphs, and are then matched with an appropriate member of the family of theoretical curves obtained from Eq. (125) for various R. Figure 18 illustrates the matching procedure of Kunze and Kirkham (1962), using their data for Colo clay loam. The resulting evaluation of D thus takes into account the impedance of the porous plate plus any impedance arising at the soil-plate contact, even though such impedance need not be directly evaluated or measured. The resulting D is multiplied

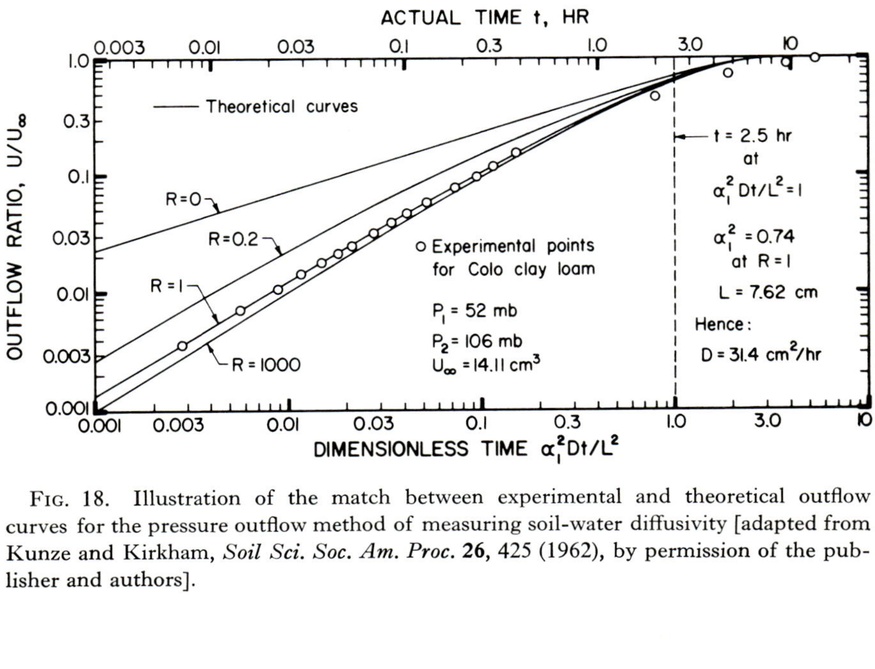

FIG. 18. Illustration of the match between experimental and theoretical outflow curves for the pressure outflow method of measuring soil-water diffusivity [adapted from Kunze and Kirkham, *Soil Sci. Soc. Am. Proc.* **26**, 425 (1962), by permission of the publisher and authors].

by $-d\theta/d\tau$ to get K; by Eq. (120), $-d\theta/d\tau = a_2 = (\theta_1 - \theta_2)/(\tau_2 - \tau_1)$, where $\theta_1 - \theta_2 = U_\infty/AL$, and $\tau_2 - \tau_1 = (P_2 - P_1)/\rho g$. The D and K may then be associated with either $(\tau_2 + \tau_1)/2$ or $(\theta_1 + \theta_2)/2$, depending on whether suction or water content is needed as the argument variable. The desired overall range in water content or suction is covered simply by repeating the measurement process at successively higher P_1; frequently, the P_2 of a given pressure step may be used as the P_1 of the subsequent step. Over the range encompassing a number of pressure steps, the experimental water contents are obtained by back calculation from the final gravimetric water-content determination as described previously (Section 2.3.4.2.). Finally, as well as being applicable for outflow (drying) as described herein, the method is also adaptable for inflow (wetting).

Experimental values of D and K obtained for drying (Kunze and Kirkham, 1962) are plotted against water content [that is, $(\theta_1 + \theta_2)/2$ for each pressure step] in Fig. 19. Results for the two soils are comparable in that over the total water-content range, $D(\theta)$ changes 10^3–10^4-fold while $K(\theta)$ changes 10^7-fold. Curve shapes too are similar, and, in view of the scatter of points for diffusivity, a straight line on the semilog graph will provide a fairly reasonable expression of the data, in accord with Eq. (86). It appears that the same type of expression would work reasonably well for $K(\theta)$ of the Webster clay loam, but not for $K(\theta)$ of the Ida silt loam, except over restricted ranges of the water content.

The method does have several attractive features. In principle, D, K, and τ are all obtainable as functions of θ by one set of experimental

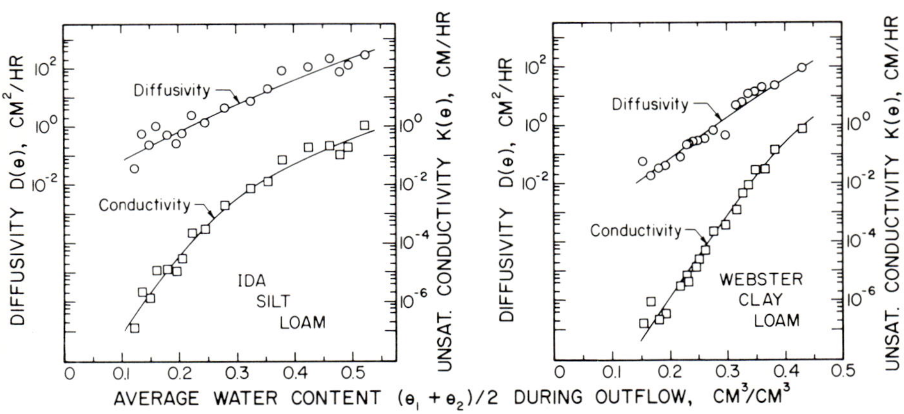

FIG. 19. Experimental measurements of diffusivity and unsaturated conductivity versus water content, as determined for two soils by the pressure outflow method [adapted from Kunze and Kirkham, *Soil Sci. Soc. Am. Proc.* **26**, 425 (1962), by permission of the publisher and authors].

operations. Use of high pressure and appropriate porous plates or membranes allows measurements at greater suctions and lower water contents than does the steady-state setup (Fig. 14). Extension also has been made to a radial flow geometry (L. A. Richards and P. L. Richards, 1962). Nevertheless, the method has not stood up well in the light of continuing and more stringent experimental tests. Jackson *et al.* (1963) and Bazargani (1964) reported experimental errors of from 7- to 10-fold in D, which is supported by Fig. 19. Also, it is now well documented (Bruce and Klute, 1963; Elrick, 1963; Jackson *et al.*, 1963; Bazargani, 1964; Elrick and Bowman, 1964; Miklich and L. A. Richards, 1965; Kunze and Peters, 1966; Davidson *et al.*, 1967) that the experimentally determined outflow curves (such as U/U_∞ versus t) often cannot be matched to any of the theoretical curves [Eq. (125)] over the complete outflow period. The nature of the deviation is such that the experimental curves do not approach equilibrium as rapidly as theory [Eq. (125)] implies. This may be illustrated even in Fig. 18, where, with the experimental values of U/U_∞ for early times being matched to the appropriate theoretical outflow curve, the later experimental values lag the theoretical curve. The discrepancy is not removed even for pressure steps progressively reduced to small values (Bazargani, 1964; Miklich and L. A. Richards, 1965), and raises serious questions on the setting of $D =$ const to obtain Eq. (119). Ohmstede (1964) has obtained a computer solution in which D of Eq. (82) is allowed to vary during outflow in accord with Eq. (86), but this can lead to difficulty in devising an objective fitting procedure. Doering (1965) proposed a one-step pressure procedure which allows D to vary during outflow, but the porous plate or membrane impedance is not taken into account. Kunze and Peters (1966) modified Eq. (125) by introducing an empirical relationship between D and U/U_∞, and thus were able to fit the resulting theoretical outflow curves to the experimental. But then the diffusivity was implied to undergo greater change within a pressure step than between successive steps, as was also observed later by Davidson *et al.* (1967).

Further comment on a possible cause of intrapressure-step diffusivity reduction will be made in a later section of this chapter. But until the matter is more completely understood and resolved, composite evidence from the literature (W. R. Gardner, 1956; Bruce and Klute, 1963; Elrick, 1963; Elrick and Bowman, 1964; and Doering, 1965) suggests that a researcher is probably well advised to use some form of the pressure outflow method for D or K only at large suctions, perhaps for $\tau > 1000$ cm H_2O. At low suction (high water content), steady-state methods for K are likely preferable.

2.3.4.4. DIFFUSIVITY BY TRANSIENT INFLOW. The horizontal absorption of water by dry soil (Fig. 3) affords a direct method of determining $D(\theta)$ for the wetting process, and was first utilized for soil by Bruce and Klute (1956). Equation (25) solved for $D(\theta)$ yields

$$D(\theta) = -\frac{d\eta}{2\,d\theta}\int_{\theta_0}^{\theta} \eta \, d\theta \tag{126}$$

The necessary experimental data are either a determination of a water-content profile $\theta(x)$ at a given time t, or water content θ as a function of t at a given position x; this second procedure is naturally adapted to nondestructive measurement of water content by gamma or neutron radiation (Ferguson and W. H. Gardner, 1962, 1963; Stewart, 1962; Davidson *et al.*, 1963a; Rawlins and W. H. Gardner, 1963). Equation (24) is used to calculate η, which is then plotted against θ. At any θ, the slope of the $\eta(\theta)$ curve is measured graphically to evaluate $d\eta/d\theta$, the area under the curve between θ_0 and θ is evaluated, and D is then calculated from Eq. (126). Curves of D versus θ obtained in this manner for Adelanto loam by Jackson (1963a) are shown in Fig. 20, the values for a porosity of 0.487 having been obtained from the $\eta(\theta)$ plot of Fig. 5. At the higher water contents the diffusivity curves of Fig. 20 essentially merge for the different porosities, but separate as the water content decreases, the effect of a lower porosity being to accentuate the reduction of diffusivity with water content. Once again, the data would conform reasonably well to a linear semilog relationship in accord with Eq. (86),

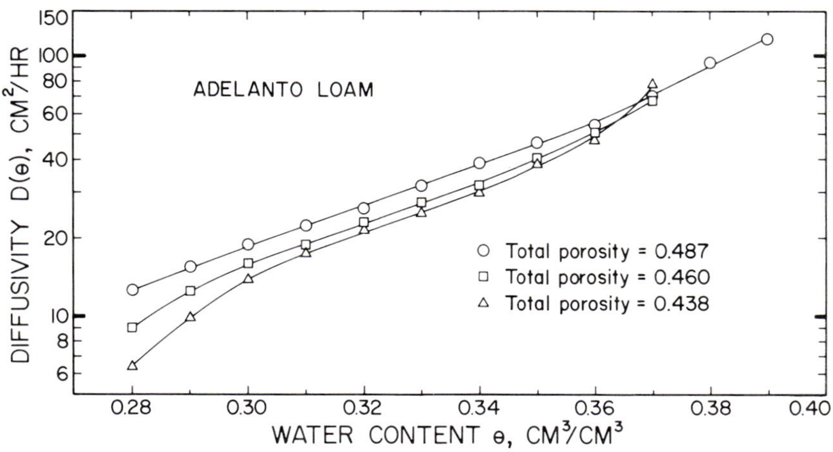

FIG. 20. Diffusivity versus water content for soil at three porosities (Jackson, 1963a), as determined by the transient horizontal inflow technique. [Adapted from *Soil Sci. Soc. Am. Proc.* **27**, 124 (1963), by permission of the publisher and author].

but the total range in experimental water content admittedly is small. Later data of Jackson (1963b) for Adelanto loam at a porosity of 0.486, and included subsequently in Fig. 27 of the present chapter, indicate for $\theta > 0.39$ that D increases much more rapidly with increasing θ than would be indicated by a straight line extrapolated from the data points of Fig. 20 for a porosity of 0.487.

The small range of water content in Fig. 20 is caused by the shape of the $\eta(\theta)$ curve, which, as may be noted from Fig. 5, yields a steep slope near θ_1, and a flat slope over a considerable range of θ as θ_0 is approached. Both extremely steep or flat slopes must be avoided if the graphical determination of curve slope (Swartzendruber and Olson, 1963) is to be reasonably accurate. Nielsen *et al.* (1962) have pointed out that this difficulty can be partially alleviated by taking experimental data at several successively smaller values of θ_1, since a less steep slope near θ_1 will result when θ_1 is reduced.

2.3.4.5. OTHER PROCEDURES AND METHODS. A transient method for $K(\tau)$ has been given by S. J. Richards and Weeks (1953). Several procedures for evaluating diffusivity and conductivity by flow into or out of a cylindrical tensiometer cup in contact with soil have been given by W. R. Gardner (1961). A one-dimensional constant-flux method, reportedly capable of giving considerable detail in $D(\theta)$, has been described by W. R. Gardner and Miklich (1962). Davidson *et al.* (1963b) have used vertical infiltration into long soil columns to determine $K(\theta)$ on the basis of Eq. (72), while Youngs (1964) has presented an infiltration determination of $K(\theta)$ which does not require the use of long columns. Jackson *et al.* (1965), in view of some of the difficulties associated with direct measurement of K, have reconsidered the question of calculating $K(\theta)$ from $\tau(\theta)$. Slight modifications in the application of previously proposed calculational schemes appeared to have promise.

Details of procedure and apparatus construction for many of the methods of measurement described herein may be found in a recent monograph (Black, 1965).

2.4. Solution of the Partial Differential Equation for Flow in More than One Dimension

2.4.1. *Flow above a Water Table*

If flow occurs in the x and z directions, and is steady so that $\partial\tau/\partial t = 0$, Eq. (6) becomes

$$\frac{\partial}{\partial x}\left(K\frac{\partial\tau}{\partial x}\right) + \frac{\partial}{\partial z}\left(K\frac{\partial\tau}{\partial z}\right) - \frac{\partial K}{\partial z} = 0 \tag{127}$$

Luthin and Day (1955) solved the equivalent of Eq. (127) by numerical relaxation methods, for boundary conditions appropriate to a rectangular sand model in which upward, lateral, and downward flow could occur in the unsaturated region above a gently sloping water table. $K(\tau)$ had been determined experimentally for the sand, and was introduced into the analysis by considering τ in the argument of $K(\tau)$ to be given by the height above the water table, although this is rigorously true only for equilibrium conditions. Below the water table, K was taken as the saturated hydraulic conductivity. Theoretical total flow rates were nicely validated experimentally, and theoretical hydraulic heads at various positions in the sand model agreed well with experimental heads determined from $h = z - \tau$, where τ was measured by tensiometers located at the various positions in the model.

For water tables above tile lines, involving drainage or subirrigation, Bouwer and Little (1959) used an electrical resistance network to analyze steady-state flow in the composite system containing both saturated and unsaturated soil. For the unsaturated part, the conductivity was again specified by $K(\tau)$. The analysis was extended to open-ditch drains (Bouwer, 1959a), and was also used to show (Bouwer, 1959b) that the flow region immediately above the water table could increase the drainage rate for tile lines in shallow soils by as much as 24–57%, depending upon water-table height.

For two-dimensional nonsteady-state drainage to tile lines or open ditches, W. R. Gardner (1962) introduced an approximation equivalent to the Dupuit–Forchheimer assumption. The position and shape of the water table were not given by the analysis, but the time course of the cumulative soil-water outflow was nicely validated with experimental data from the literature.

2.4.2. *Water Movement to Plant Roots*

Although the movement of soil water to plant roots is conceivably some combination of one-, two-, and three-dimensional flow, it seems reasonable to suppose that the major features of the problem might be described if the plant-root system were idealized to an equivalent two-dimensional line sink, through which water is removed from the soil by so-called passive plant action (Philip, 1957b; W. R. Gardner, 1960). In polar coordinates ignoring gravity, Eq. (7) is written

$$\frac{\partial \theta}{\partial t} = \frac{1}{r}\frac{\partial}{\partial r}\left(rD\frac{\partial \theta}{\partial r}\right) \tag{128}$$

where r is the radial distance coordinate. Because of the great difficulty attendant to letting diffusivity vary with water content in Eq. (128),

D is assumed constant (Philip, 1957b). In addition, W. R. Gardner (1960) has taken $K =$ const, so that by Eq. (4), $d\theta/d\tau$ is also constant. This permits Eq. (128) to be written

$$\frac{\partial \tau}{\partial t} = \frac{D}{r} \frac{\partial}{\partial r} \left(r \frac{\partial \tau}{\partial r} \right) \tag{129}$$

which is then considered subject to

$$\tau = \tau_0 , \qquad r > r_1 , \qquad t = 0 \tag{130}$$

$$K \, \partial\tau/\partial r = V, \qquad r = r_1 , \qquad t > 0 \tag{131}$$

where τ_0 is the constant initial suction head (at constant initial water content), and V is a constant water inflow velocity at the cylindrical root surface of radius r_1. On the basis of solutions given by Carslaw and Jaeger (1959, p. 262) for large t, an approximate solution for Eq. (129) subject to conditions (130) and (131) is written as

$$\tau - \tau_0 = (\pi r_1 V/K)[\ln(4Dt/r^2) - 0.57722] \tag{132}$$

where 0.57722 is Euler's constant.

In spite of the many approximations used in obtaining Eq. (132), W. R. Gardner (1960) was able to use it to good advantage in accounting for plant-response phenomena as related to matric suction. Then, applying Eq. (132) to the root surface by setting $r = r_1$, he introduced additional considerations and approximations to arrive eventually at the van den Honert (1948) resistance form (W. R. Gardner and Ehlig, 1962a,b)

$$W = (\tau_L - \tau)/I \tag{133}$$

where W is the rate of water uptake per unit volume of soil, τ_L is the suction head of the leaf (sometimes called diffusion pressure deficit), and I is a total resistance factor for water movement in both plant and soil. The soil component of I was taken as inversely proportional to $K(\theta)$. Again, this simplified form was found useful and effective in describing plant–soil–water relationships (W. R. Gardner and Ehlig, 1962a, b, 1963; W. R. Gardner, 1964).

2.4.3. *Other Examples*

The use of two-dimensional solutions in relation to water flow into and out of a cylindrical tensiometer (W. R. Gardner, 1961) has already been mentioned (Section 2.3.4.5.). For two- and three-dimensional seepage from ponds to underlying water tables, Reisenauer (1963) has given computer solutions to the equivalent of Eq. (6) for steady-state flow ($\partial\tau/\partial t = 0$).

For transient two-dimensional water absorption into soil initially at a constant water content θ_0, Singh (1965) has provided a computer analysis which begins with the solution of the radial flow problem without gravity; that is, Eq. (128) subject to

$$\theta = \theta_0, \qquad r > r_0, \qquad t = 0 \tag{134}$$

$$\theta = n, \qquad r = r_0, \qquad t > 0 \tag{135}$$

where r_0 is the constant radius of the cylindrical water source, n the porosity is the constant saturated water content, and the diffusivity function is an exponential form similar to Eq. (86). The effect of gravity was then approximated by an implicit finite-difference method, taking K as proportional to a power function of the water content. The analogous approach for three-dimensional flow was also introduced. Philip (1968), however, has pointed out an error in the Singh analysis.

3. Nonproportional Liquid-Phase Flow Response to Gradient

One of the basic premises upon which the foregoing unsaturated flow theory and analyses depend is that of proportionality between flow velocity and hydraulic gradient [Eq. (1)] or of linearity between velocity and water-content gradient at a given water content [Eq. (3)], this latter case reducing to proportionality if flow is in the horizontal direction only. So far in the present chapter, questions on the validity of these basic equations have only been implied indirectly, but will now be considered in some detail. For background, we will begin with saturated flow, even though this case *per se* is not the primary concern of the present chapter.

The Darcian proportionality of Eq. (1) may be written in scalar form as

$$V = Ki \tag{136}$$

where V is the steady-state flow velocity, K is the saturated hydraulic conductivity considered constant, and $i = -\Delta h/L$ is the hydraulic gradient. In turn, $\Delta h = h_2 - h_1$, where h_1 and h_2 are the inlet and outlet hydraulic heads, respectively, separated by the length L. If desired, the volumetric flow rate Q and the cross-sectional area A may be used to write $V = Q/A$.

In addition to Darcy's (1856) original data, which have been presented graphically by Hubbert (1956), one of the strongest traditionally cited evidences for the validity of Eq. (136) is the work of Fancher and Lewis (1933) with sands and sandstones; they made special note, however, of the general absence of clay in these materials. Stearns (1927) found fair adherence to Eq. (136) for sands, as did Bodman and Edlefsen (1934) for sands and soils, and Leonards (1962) credits Terzaghi (1925a,b) with some validation of Darcian proportionality for clays. Recently, Olsen (1965, 1966) has reported excellent conformity for compacted kaolinite.

3.1. Deviations from Darcian Proportionality in Saturated Soils

Darcy (1856) recognized explicitly that Eq. (136) would fail at high gradients if kinetic effects and turbulence were present. Under such circumstances, V would increase less than proportionally with i. We will here be concerned, however, with the opposite type of deviation, in which V increases more than proportionally with i, or, alternatively, $V/i = K$ of Eq. (136) increases with i rather than remaining constant. One of the earliest reports of experimental data exhibiting this type of deviation is that of King (1898), but seems to have been largely ignored during the first half-century thereafter. More recently, such behavior has been found in a number of materials (von Engelhardt and Tunn, 1955; Lutz and Kemper, 1959; Hansbo, 1960; Paez, 1962; R. J. Miller and Low, 1963; Hadas, 1964; Kutilek, 1965), although Lutz and Kemper and Paez did find proportionality for certain hydrogen- and calcium-saturated clays. Terzaghi (1925a,b) also had mentioned more-than-proportional behavior in clays. In the data of Stearns (1927), often cited in support of proportionality, deviations were found by Swartzendruber (1962b) for certain sands with clay contents in excess of 5%.

3.1.1. *Modified Saturated-Flow Equations*

Attempts to develop a modified flow equation capable of accommodating the various more-than-proportional deviations reported up to that time (Swartzendruber, 1962a,b) culminated in the empirical form (Swartzendruber, 1962b)

$$V = B[i - J(1 - e^{-Ci})] \tag{137}$$

where B, J, and C are constants. As i increases from zero at $V = 0$, Eq. (137) approaches the linear asymptote $V = B(i - J)$, from which J is seen to be the gradient-axis intercept of the linear asymptote. If J, C, or both are zero, Eq. (137) reduces to the Darcian form, Eq. (136), with B replacing K; the quantity J^2C serves as a combination parameter for expressing non-Darcian behavior, being zero for Darcian flow. Also, since the intrinsic permeability k is generally defined by $k = \mu K/\rho g$ (L. A. Richards, 1952), where μ is the viscosity, a corresponding quantity termed the *permeance* b (Swartzendruber, 1962b) was defined by

$$b = \mu B/\rho g \tag{138}$$

As with intrinsic permeability, the permeance would be construed as independent of the properties of the liquid.

Equation (137) was found to fit the various nonproportional flow data quite adequately (Swartzendruber, 1962b), as illustrated by a typical example in Fig. 21, in which the solid-line curves were computed from

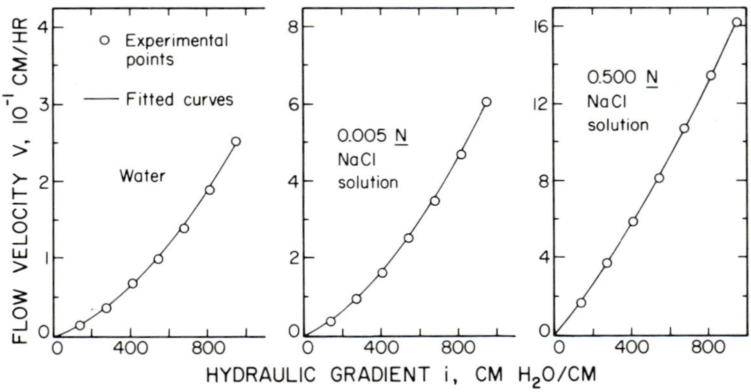

FIG. 21. Flow velocity versus hydraulic gradient for saturated flow through the sodium Utah bentonite of Lutz and Kemper (1959); fitted curves as calculated from Eq. (137).

the equation and the fitted constants. As the salt concentration of the flowing water increases, the behavior in Fig. 21 becomes more Darcian, as reflected in the J^2C values of 4560, 2130, and 484, corresponding to the three salt concentrations of 0, 0.005, and 0.500 N, respectively. Such maximal non-Darcian behavior for pure water, followed by a decrease with increasing concentration of salt or other electrolyte, was found as a general trend for the various media which exhibited non-Darcian behavior with pure water. The behavior of b as calculated from Eq. (138) remained qualitatively similar to that generally expected of k for various electrolytes. That is, b would generally increase with ionic concentration, conceivably reflecting the better flocculation status of the porous medium as usually associated with higher electrolyte concentration.

Kutilek (1965) has presented a modified flow equation

$$V = (B'/C')[\ln(e^{C'J'} + e^{C'i} - 1) - C'J'] \tag{139}$$

where the constants B', C', and J' are analogous to their unprimed counterparts in Eq. (137). Swartzendruber (1967) has found very similar characterizations of non-Darcian behavior to be given by Eqs. (137) and (139), so that a choice between them may usually be made on the basis of convenience.

Although the magnitude of non-Darcian behavior may be expressed in terms of the empirical constants of Eqs. (137) or (139), such as J^2C or some appropriate combination of J' and C', another approach is to cast the phenomenon into a Darcian framework by computing $V/i = K = K(i)$, where the saturated hydraulic conductivity $K(i)$ becomes a function of gradient i. One then examines the relative change in $K(i)$ over any desired gradient range, or formulates a K_{max}/K_{min} on the basis of Eq. (137) or (139). An analysis of this type has been carried out (Swartzendruber, 1968) for the various data available. Actually measured variations in $K(i)$ of 2–4-fold were found to be common, with possibilities of 5–15-fold variations being inferred from K_{max}/K_{min} calculations based on Eqs. (137) and (139). Variations of such magnitude, in a quantity supposed to remain constant, are sufficient to justify concern with the cause of non-Darcian deviations, along with some effort to assess the effects of the deviations on flow problems.

3.1.2. *Possible Causes of Non-Darcian Behavior*

3.1.2.1. MODIFIED WATER PROPERTIES. Low (1961) has set forth in detail his hypothesis of how the properties of soil water could be different from those of normal bulk water, as the result of interaction between clay particles and water. Studies interpreted in support of such an hypothesis have been those on water density (Anderson and Low, 1958), ion movement (Low, 1962), suspension behavior (Kolaian and Low, 1962; Leonard and Low, 1964), thin water films (Oster and Low, 1963), heat capacity (Oster and Low, 1964), and water flow in clays (Low, 1960) and argillaceous rocks (Young *et al.*, 1964). Such modified water was postulated to possess a non-Newtonian viscosity decreasing with increasing rate of shear, and, in its extreme development, to exhibit a rigid-type resistance to shear and flow. Hence, instead of just being initially curvilinear with gradient as in Fig. 21, the flow velocity in this extreme case would be zero over some range $0 \leqslant i \leqslant i_0$, and would exceed zero only when $i > i_0$, where i_0 is the threshold gradient. Such threshold gradients have been reported experimentally by R. J. Miller and Low (1963) and Li (1963), but Olsen (1965) has severely criticized Li's study on the basis of major errors in procedure.

In one form or another, non-Newtonian explanations have been recognized or offered by von Engelhardt and Tunn (1955), Lutz and Kemper (1959), and Hansbo (1960). Non-Newtonian explanations in the light of Low's modified-water hypothesis are those of Low (1961), Swartzendruber (1962a,b), and Kutilek (1965). The non-Newtonian explanation is a logical one if the velocity-gradient relationship is viewed as an expression of the rheological consistency diagram (Swartzendruber, 1962a), and at present seems to be about the only way of accounting for a threshold gradient (R. J. Miller and Low, 1963).

Martin (1962) avoided the rigid-phase aspects of Low's modified water hypothesis by proposing a two-dimensional fluid concept, in which the soil water would resist normal stresses but would shear at least as readily as ordinary bulk water. Intrinsically, this concept has difficulty in rationalizing a theshold gradient, but may have utility in describing other facets of non-Darcian flow. Nonlinearity of the velocity-gradient relationship is such a facet, but does have several alternative explanations which will now be considered.

3.1.2.2. CHANGES IN POROUS-MEDIUM FABRIC. If the component particles of the porous medium are asymmetrically shaped, it is conceivable that the greater velocities of flow associated with increased gradients might progressively reorient the particles to produce a larger *K*. A hypothetical illustration of how this might occur in a highly idealized assemblage of clay platelets is given in Fig. 22, which shows qualitatively

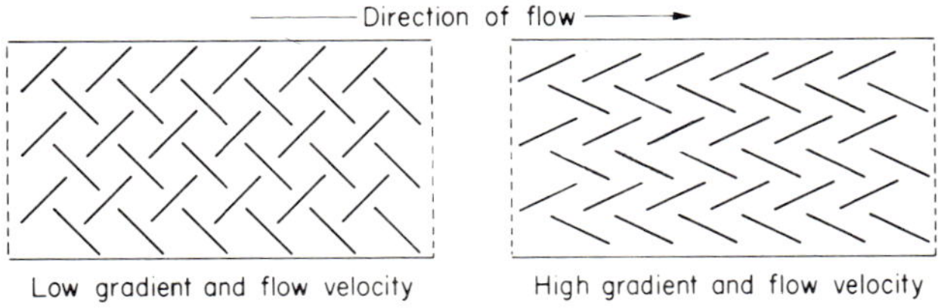

FIG. 22. Schematic diagram of how a highly idealized assemblage of clay platelets might reorient at higher water velocity, thus resulting in a greater hydraulic conductivity at high gradient than at low gradient.

how the greater velocity might simultaneously enlarge the cross section for flow and reduce the tortuosity of the flow path. Such particle reorientation, however, would need to be essentially reversible if the flow velocity and gradient were then reduced, since the careful experiments of von Engelhardt and Tunn (1955) included only results wherein flow measurements at high gradients could be followed by a low-gradient check that agreed with a previous value at low gradient.

If the porous-medium particles are of a relatively symmetrical shape, it may be conceivable that the increased internal pressure normally associated with higher gradient might induce certain localized elastic compressions of clusters of particles. This might decrease the tortuosity and increase the cross section for flow, perhaps by decreasing the volume of dead-end pores if such exist in the medium. This might be comparable to Terzaghi's (1925a,b) analogy with a flow system composed of easily expansible rubber tubes, and might also bear some compatibility with Martin's (1962) concept of plugging and unplugging of voids (pores). Perhaps the deforming-media approaches of Zaslavsky (1964b) and Raats (1965) will ultimately give some guidance in this respect.

3.1.2.3. ELECTRICAL STREAMING POTENTIAL. If an electrical streaming potential (Abramson, 1934; Overbeek, 1952) is developed in an otherwise Darcian system, it seems logical to represent the flow effect as a correction on Eq. (136), that is,

$$V = Ki - c_e i_e \tag{140}$$

where c_e is the constant transport coefficient for electroosmotic water flow (Winterkorn, 1955), and $i_e = E/L$ is the electrical streaming-potential gradient, where E is the streaming potential. Rearrangement of Eq. (140) yields

$$V = (K - c_e i_e / i)\, i \tag{141}$$

If i_e/i is a constant as assumed classically for membranes and capillary tubes (Abramson, 1934; Overbeek, 1952), then Eq. (141) is still Darcian in that V is proportional to i (Kemper, 1960). However, data of Bull and Gortner (1932) for powdered quartz and Michaels and Lin (1955) for kaolinite, as presented by Swartzendruber (1966b), indicate a decrease in $i_e/i = E/(h_1 - h_2)$ as i increases, so that the parenthesis quantity of Eq. (141) would increase with i to cause more-than-proportional behavior. Fortunately, Bull and Gortner also measured rates of liquid flow, reporting them as mass flow rate ρVA. Multiplying Eq. (141) by ρA, and using $i_e = E/L$ and $i = (h_1 - h_2)/L$, enables the rewriting of Eq. (141) as

$$\rho VA/(h_1 - h_2) = \rho KA/L - (c_e \rho A/L)[E/(h_1 - h_2)] \tag{142}$$

which asserts $\rho VA/(h_1 - h_2)$ to vary linearly with $E/(h_1 - h_2)$ but with a negative slope. Such a plot (Swartzendruber, 1967) prepared from Bull and Gortner's data is shown in Fig. 23, and confirms Eq. (142) reasonably

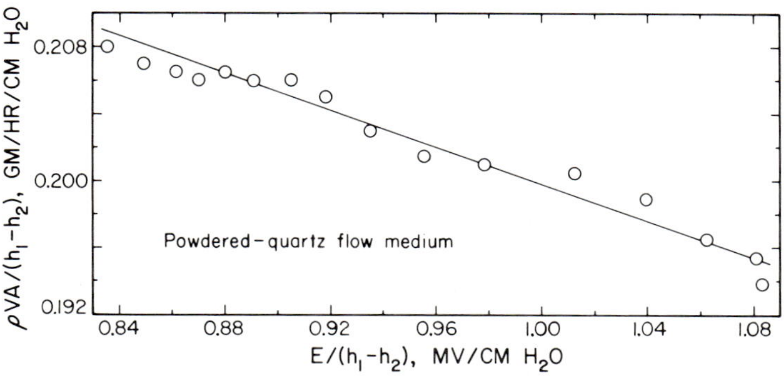

FIG. 23. Test of Eq. (142), using the electrical streaming potentials and flow rates of Bull and Gortner (1932) as measured for 10^{-4} N sodium chloride solution flowing through powdered quartz.

well. It therefore seems that streaming-potential effects might merit further study for assessing their possible role in causing non-Darcian behavior.

3.1.2.4. SALT AND OSMOTIC EFFECTS. The analysis just described for streaming potential is closely related to the approach of irreversible thermodynamics (S. A. Taylor and Cary, 1964), and could well be extended to include additional potentials such as osmotic. The elucidation of the role of salt and osmotic effects in non-Darcian behavior, however, is not likely to be easy, in view of such complicated phenomena as ion-sieving and membrane leakage (Kemper and Evans, 1963; Tschapek, 1963; Kemper and Maasland, 1964; Young and Low, 1965; Kemper and Rollins, 1966) along with miscible displacement and diffusion mixing of salts (Nielsen and Biggar, 1963; Biggar and Nielsen, 1962). The general problem of salt and osmotic flow effects, however, appears worthy of considerable research attention.

3.1.2.5. EXPERIMENTAL ERROR. Olsen (1965) reported that the measurement of flow rates with an air bubble in a contaminated capillary tube can cause a false gradient dependence in K, if it is assumed that there is no loss of hydraulic head across the moving bubble. Invoking such error, he could account for the non-Darcian deviations of Hansbo (1960), for a small part of R. J. Miller and Low's (1963) threshold gradients, and for practically none of von Engelhardt and Tunn's (1955) and Lutz and Kemper's (1959) deviations. Furthermore, Olsen did not indicate how his bubble-error explanation could account for a threshold gradient that would decrease with increasing temperature and decreasing clay concentration, as observed by R. J. Miller and Low, and which can be explained on the basis of modified water properties.

Since the hydraulic conductivity can be strongly affected by such factors as entrapped air, swelling and shrinking of colloids and clay, and microbial growth (see Gupta and Swartzendruber, 1962, 1964, and references therein), there exists the possibility that one or some of these factors might somehow interact with the flow measuring procedure to produce a false gradient dependence in K as an artifact. To dismiss non-Darcian behavior on the basis of a vague appeal to some unspecified trend of possible experimental error, however, hardly does justice to the several careful investigators who have been cognizant of the attendant difficulties and have made special effort to eliminate or minimize them.

3.1.3. *Flow in More than One Dimension*

The use of such as Eqs. (137) and (139) in the equation of continuity leads to partial differential equations very difficult to solve even for steady-state flow. Hence, to gain some insight into the manner in which more-than-proportional effects are manifest when flow is multidimensional, we shall consider only three-dimensional radial flow, using an approach previously applied to two-dimensional flow (Swartzendruber, 1968). We approximate the curvilinear regions of Eqs. (137) or (139) by expanding the exponential or logarithmic expressions in series, neglecting terms of higher degree than i^2 to obtain

$$V = a_3 i + a_4 i^2 \tag{143}$$

where a_3 and a_4 are constants (Swartzendruber, 1962b).

Assume that steady-state flow takes place radially through a spherical shell containing uniform non-Darcian material describable in one-dimensional flow by Eq. (143). The spherical inlet surface, of constant radius r_1 and hydraulic head h_1, is concentric with the spherical outlet surface of constant radius r_2 and hydraulic head h_2, that is

$$h = h_1, \qquad r = r_1 \tag{144}$$

$$h = h_2, \qquad r = r_2 \tag{145}$$

where $r_2 > r_1$ and $h_1 > h_2$. Since the flow is radial, only total derivatives with respect to the radial coordinate r will appear in the three-dimensional equation of continuity which then is $(1/r^2)\, d(r^2 V)/dr = 0$. This integrates directly to $r^2 V = C_3$, where C_3 is the constant of integration, and is evaluated by taking $V = V_1$ when $r = r_1$, thus yielding

$$r^2 V = r\ {}^2 V_1 \tag{146}$$

Setting $i = -dh/dr$ in Eq. (143), Eqs. (146) and (143) are combined and solved for dh/dr by the standard quadratic formula to obtain

$$dh/dr = (a_3/2a_4)[1 - (r^2 + 4a_4V_1r_1^2/a_3^2)^{1/2}/r] \tag{147}$$

Separating variables and using formula No. 130 of Peirce (1929) to assist in the integration provides

$$h = (2/a_3G)[rf(r, GV_1) + r_1(GV_1)^{1/2} \ln g(r, GV_1)] + C_4 \tag{148}$$

where $G = 4a_4/a_3^2$, the constant of integration is C_4, and

$$f(r, GV_1) = 1 - (1 + GV_1r_1^2/r^2)^{1/2} \tag{149}$$

$$g(r, GV_1) = (GV_1r_1^2/r^2)^{1/2} + (1 + GV_1r_1^2/r^2)^{1/2} \tag{150}$$

Using conditions (144) and (145) in Eq. (148) enables elimination of C_4 by subtraction, so that

$$h_1 - h_2 = (2/a_3G)\left[r_1f(r_1, GV_1) - r_2f(r_2, GV_1) + r_1(GV_1)^{1/2} \ln \frac{g(r_1, GV_1)}{g(r_2, GV_1)}\right] \tag{151}$$

Use of condition (145) in Eq. (148) also enables elimination of C_4 by forming the difference $(h - h_2)$, which is then employed with Eq. (151) to form the ratio

$$\frac{h - h_2}{h_1 - h_2} = \frac{rf(r, GV_1) - r_2f(r_2, GV_1) + r_1(GV_1)^{1/2} \ln[g(r, GV_1)/g(r_2, GV_1)]}{r_1f(r_1, GV_1) - r_2f(r_2, GV_1) + r_1(GV_1)^{1/2} \ln[g(r_1, GV_1)/g(r_2, GV_1)]} \tag{152}$$

Equation (151) gives the relationship between the total hydraulic head difference $(h_1 - h_2)$ and the flow velocity V_1 at the spherical inlet surface, while Eq. (152) provides the pattern of head dissipation as r varies between r_1 and r_2. As GV_1 becomes small, Eq. (152) reduces to

$$\frac{h - h_2}{h_1 - h_2} = \left(\frac{1}{r} - \frac{1}{r_2}\right)\Big/\left(\frac{1}{r_1} - \frac{1}{r_2}\right) \tag{153}$$

which is the well-known reciprocal form for Darcian flow (Muskat, 1946, p. 260). A small GV_1 results not only from small flow velocity but also from Darcian behavior, since G is proportional to a_4. Clearly, Eq. (143) is reduced to Darcian form if $a_4 = 0$, or even if i (and hence V) is small.

Several aspects of the solution will be illustrated by choosing $r_1 = 1$ cm and $r_2 = 5$ cm, using $a_3 = 1.16$ cm/hr and $a_4 = 0.482$ cm/hr as determined by Swartzendruber (1968) from the data of Hadas (1964) for Gilat loess. The broken-line curve in Fig. 24(A), calculated from Eq. (143), shows this equation to be a good fit to the experimental data

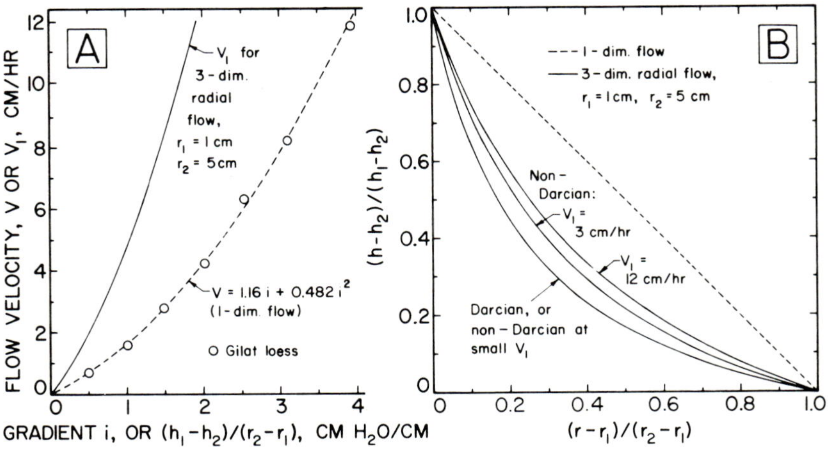

FIG. 24. One- and three-dimensional non-Darcian flow behavior for essentially saturated [$\tau = 10$ cm H_2O] Gilat loess (Hadas, 1964). (A) Velocity-gradient relationships showing V_1 as calculated from Eq. (151) [solid-line curve], and Eq. (143) [broken-line curve] as fitted to the experimental points [data taken from A. Hadas, *Israel J. Agr. Res.* **14**, 164 (1964), by permission of the publisher and author]. (B) Head dissipation patterns, solid-line curves calculated from Eqs. (152) [non-Darcian] and (153) [Darcian].

for one-dimensional flow. A plot of V_1 versus $(h_1 - h_2)/(r_2 - r_1)$, calculated from Eq. (151), is shown by the solid-line curve of Fig. 24(A). This three-dimensional radial-flow curve is of more-than-proportional type, and rises more rapidly than its one-dimensional counterpart. Figure 24(B) shows head dissipation patterns, the broken-line linear pattern being that which obtains in a uniform one-dimensional flow medium regardless of the nature of the velocity-gradient relationship (Wyckoff *et al.*, 1933). The greatest departure from linearity is that exhibited by the solid-line, reciprocal-form Darcian pattern as calculated from Eq. (153). The remaining solid-line curves, calculated from Eq. (152) for V_1 of 3 and 12 cm/hr, show that non-Darcian behavior in three-dimensional radial flow induces velocity dependence in a head-dissipation pattern that shifts progressively toward the one-dimensional pattern as the velocity increases. In comparison with the velocity-independent Darcian pattern, the non-Darcian pattern is more steep near the flow outlet and less steep near the flow inlet. This is a reasonable consequence of $K(i)$ being greater at large gradients than at small gradients, and agrees qualitatively with the results found previously for two-dimensional radial flow (Swartzendruber, 1968). Note that $V_1 = 12$ cm/hr is the maximum velocity which can be chosen without unduly exceeding the range of the data in Fig. 24(A).

3.2. Nonproportionality in Unsaturated Soils

For one-dimensional unsaturated flow in the x-direction, a scalar expression of Eq. (2) can ultimately be written as

$$v = K(\tau) \frac{\partial \tau}{\partial x} \tag{154}$$

where v is the flow velocity, and the form $K(\tau)$ is chosen since $K(\theta) = K[\theta(\tau)] = K(\tau)$ on the basis of the assumed unique relationship between θ and τ. Or, alternatively, the corresponding horizontal scaler expression of Eq. (3) is

$$v = -D(\theta) \frac{\partial \theta}{\partial x} = D(\theta)\, w \tag{155}$$

where $w = -\partial\theta/\partial x$ is the water-content gradient. At a given τ, Eq. (154) asserts proportionality between flow velocity v and suction gradient $\partial\tau/\partial x$, while Eq. (155) asserts proportionality between v and w at given θ. Although foundational for the nonlinear diffusivity theory of unsaturated soil-water movement described in Section 2, the only early experimental validations of proportionality were for coarse sand (Childs and Collis-George, 1950) and a loam soil (S. J. Richards and Weeks, 1953).

3.2.1. *Tests and Modifications of Unsaturated-Flow Equations*

More stringent and refined experimental tests of the mathematical solutions for water absorption by soil (Section 2.3.1) have revealed persistent discrepancies. Nielsen *et al.* (1962) and Davidson *et al.* (1963b) found that whereas a unique $\eta(\theta)$ could be found for a given soil if the applied water content θ_1 were near n (saturation), a θ_1 reduced appreciably below saturation would cause nonuniqueness. That is, with η computed from Eq. (24), the $\eta(\theta)$ curves at various times would not coalesce into a single curve as in Fig. 5. For Columbia silt loam (Davidson *et al.*, 1963b) at $\theta_1 = 0.325$ (n of 0.514), this nonuniqueness is illustrated in Fig. 25(A) by a progressive retardation-type shift in the $\eta(\theta)$ curves at increasing times. Even for a θ_1 near saturation, the same behavior was present in Salkum silty clay loam (Rawlins, 1961; Stewart, 1962; Ferguson and W. H. Gardner, 1963), as illustrated in Fig. 25(B) for the data of Rawlins and W. H. Gardner (1963). A detailed velocity-gradient analysis of

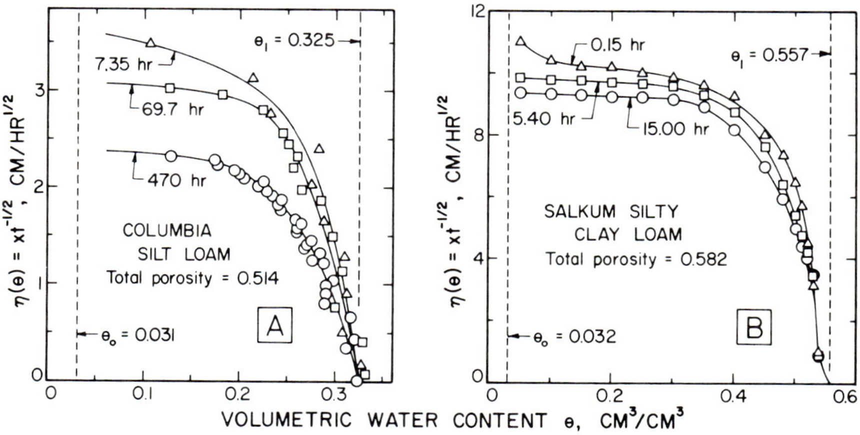

FIG. 25. Nonunique $\eta(\theta)$ at various times during horizontal water absorption. (A) Columbia silt loam at θ_1 substantially less than the total porosity [Davidson *et al.*, 1963b, adapted from *Hilgardia* **34**, 608 (1963), by permission of the publisher and authors]. (B) Salkum silty clay loam at θ_1 near the total porosity (Rawlins and W. H. Gardner, 1963. Data supplied by the authors and used by permission).

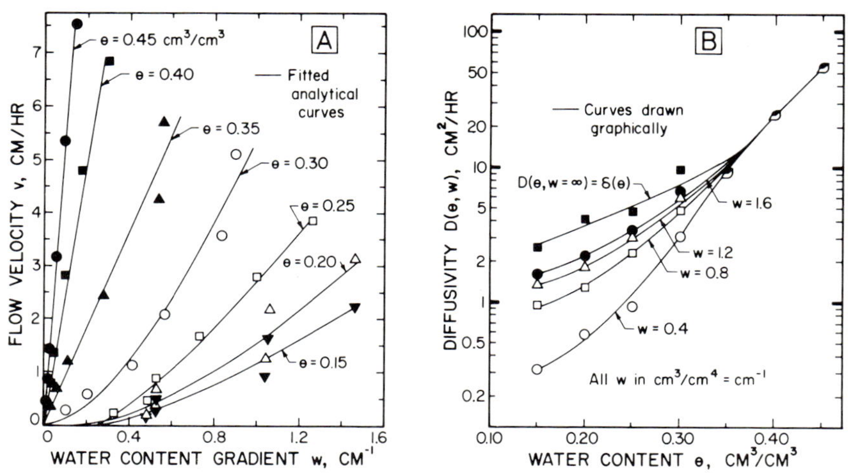

FIG. 26. (A) Flow velocity versus water-content gradient at selected water contents; fitted curves determined from Eq. (156), while experimental points for Salkum silty clay loam (Rawlins and W. H. Gardner, 1963) were evaluated from data supplied by the authors and used by permission. (B) Diffusivity versus water content at specified water-content gradients; points calculated from Eq. (157) using δ, α, and γ as fitted (Swartzendruber, 1963) in part (A).

these same data was carried out by Swartzendruber (1963), and is shown graphically in Fig. 26(A). For $\theta \geqslant 0.35$, proportionality of v with w at given θ holds fairly well, in conformity with Eq. (155). For $\theta \leqslant 0.30$, the velocity increases more than proportionally with gradient, in fashion similar to Fig. 21. Again, the data in Fig. 26(A) are reasonably well represented by the fitted curves calculated from the modified flow equation

$$v = \delta[w - \alpha(1 - e^{-\gamma w})] \tag{156}$$

Equation (156) is derived from Eq. (137) by considering B, J, and C as functions of θ, along with $i = -\partial h/\partial x$, $h = z - \tau(\theta)$, and $w = -\partial\theta/\partial x$. Hence, $\delta = \delta(\theta) = B(\theta)(-d\tau/d\theta)$, $\alpha = \alpha(\theta) = J(\theta)(-d\theta/d\tau)$, and $\gamma = \gamma(\theta) = C(\theta)(-d\tau/d\theta)$, so that δ, α, and γ are constants at given θ.

The data of Fig. 26(A) may be cast in a diffusivity framework in the sense of Eq. (155) if D is allowed to depend on gradient w as well as water content θ; that is, $D = v/w = D(\theta, w)$. Applied to Eq. (156) this yields

$$D(\theta, w) = \delta[1 - \alpha(1 - e^{-\gamma w})/w] \tag{157}$$

Using the fitted values of δ, α, and γ for Salkum silty clay loam (Swartzendruber, 1963), $D(\theta, w)$ as calculated from Eq. (157) is plotted against θ for different w [Fig. 26(B)]. The coalescing curves for $\theta > 0.35$ demonstrate that Eq. (155) holds in that region. However, the curves for different gradients diverge markedly as the water content decreases, with the maximum discrepancy approaching 8-fold at $\theta = 0.15$.

For vertically downward one-dimensional flow, the counterpart of Eq. (155) is the scaler version of Eq. (3) expressible as

$$v = -D(\theta)\frac{\partial\theta}{\partial Z} + K(\theta) = D(\theta)\,w + K(\theta) \tag{158}$$

where v and Z $(= -z)$ are positive downward and $w = -\partial\theta/\partial Z$. At given θ, Eq. (158) asserts v to increase linearly with w, while both the slope $D(\theta)$ and the v intercept $K(\theta)$ are positive. In a test of Eq. (158) for Rustan sandy loam and Providence silt loam, Thames (1966) did find the v intercept to be positive and to decrease with θ, but the plots of v against w were curvilinear upward in much the same fashion as the curves in Figs. 21 and 26(A). At given θ, he found for both soils that v minus the v intercept varied as w raised to slightly less than the third power, rather than as the first power implied by Eq. (158).

In assessing the steady-state validity of Eq. (154) for several soils, Hadas (1964) in every case found more-than-proportional increases of velocity v with gradient $\partial\tau/\partial x$ at given τ. Presented in terms of a gradient-dependent $K(\tau, \partial\tau/\partial x) = v/(\partial\tau/\partial x)$, 2–4-fold variations were common, for τ not in excess of 70 cm H_2O. In a similar test with a silt loam soil, Abd-el-Aziz and S. A. Taylor (1965) found a short gradient range of more-than-proportional behavior located between two gradient ranges of proportionality, with a maximum variation in $K(\tau, \partial\tau/\partial x)$ of about 8-fold. Olson (1966), however, in steady-state experiments particularly designed to avoid errors arising from hysteresis and nonuniform packing, has been able to obtain generally good verification of Eq. (154) for sand–silt and sand–kaolinite mixtures, for τ as high as 275 cm H_2O. The nonproportionalities of largest magnitude were actually of less-than-proportional nature.

3.2.2. *Possible Causes of Nonproportional Behavior*

3.2.2.1. NON-DARCIAN BEHAVIOR. Since by Eq. (4) one of the components of diffusivity D is the unsaturated conductivity $K(\theta)$, such nonuniqueness as in Fig. 25 and nonproportionality as in Fig. 26 could arise through $K(\theta)$ from non-Darcian factors as previously discussed for saturated flow (Section 3.1.2). The close resemblance of the nonproportionalities of Fig. 26(A) to those of Fig. 21 have already been noted. The non-Newtonian viscosity aspect of Low's (1961) modified-water hypothesis has been suggested by Swartzendruber (1963, 1964) as a tentative explanation for the results of Nielsen *et al.* (1962), Rawlins and W. H. Gardner (1963), and Davidson *et al.* (1963a), while Hadas (1964) has also invoked it to account for his results. Unsaturated conditions would indeed be expected to accentuate the flow effect of any particle-modified water, since the water remaining in the soil would be existing in closer proximity to particle surfaces. This could account for a soil which behaves proportionally when relatively wet but which exhibits nonproportionality as the water content is reduced (Nielsen *et al.*, 1962; Davidson, *et al.*, 1963b), as is also demonstrated in Fig. 26(A). Desaturation, however, does not necessarily pose difficulties for the other previously listed non-Darcian explanations. Salt effects, for instance, might also be expected to become more pronounced with reduced water content, in view of the salt-sieving and electrolyte redistribution phenomena as discussed by Bolt and Lagerwerff (1965).

Transient entry of water into soil may, however, be accompanied by additional effects on $K(\theta)$ due to air or to soil swelling. Included in the air effect would be the removal of initially entrapped air as well as any back-pressure of the displaced air. But both of these factors would appear to operate in the direction of less-than-proportional behavior, since flow velocities would be too high (Wilson and Luthin, 1963) at the greater times associated with low gradients. Soil swelling, however, could induce a false more-than-proportional effect by tending to reduce $K(\theta)$ at greater times. Rawlins and W. H. Gardner (1963) avoided the air effect by working at near-vacuum, and minimized swelling by using a soil previously found to be macroscopically nonswelling. Furthermore, if some internally compensating microswelling did occur, it would seem that the effect should have been detectable at all water contents and possibly even maximally manifest at high water contents. Yet, it is at high water contents that Eq. (155) holds (Fig. 26). For soils with pronounced swelling characteristics, however, recent work of Christenson and Ferguson (1966) does imply that gradient dependence in $K(\theta)$ might be attributable to swelling.

3.2.2.2. ANOMALIES IN THE SUCTION–WATER-CONTENT RELATIONSHIP. In place of or in conjunction with non-Darcian effects in $K(\theta)$ as just discussed, Swartzendruber (1963, 1966b) has noted that nonproportionality in unsaturated flow could be introduced by way of the $-d\tau/d\theta$ factor [Eq. (4)], or, alternatively, by nonuniqueness in $\tau(\theta)$ even for a monotonic wetting process. Such kind of nonuniqueness was first suggested explicitly by W. H. Gardner and W. Gardner (1951) as conceivably arising from a dependence of τ on water-content gradient w in addition to the dependence on θ. Bazargani (1964), in pressure outflow studies of sand–kaolinite mixtures and calcined clay, found a fluctuating behavior in $-d\tau/d\theta$ when determined from small increments of τ. He suggested that the fluctuations might somehow have been caused by threshold gradients or discreteness in the pore-size distribution. Mokady and Low (1964) pointed out that the modified-water hypothesis implies a different τ for flowing water than for static water, but were not able to measure such a difference for a sand–kaolinite mixture, perhaps because their experimental arrangement did not provide sufficient desaturation for a definitive test. Recently, Davidson *et al.* (1966) have reported small but distinct tendencies for $\tau(\theta)$ to depend on the size of the τ increment. They proposed a number of suggested causes for the effect, including immiscible displacement of air, salt sieving, and contact angle.

In a diagrammatic sketch of water in unsaturated soil, L. A. Richards (1965) explicitly represented the soil water as different from bulk liquid water, and advised considerable caution in identifying the properties of bulk liquid water with those of soil water. Such a view has much in common with Low's (1961) modified-water hypothesis. Richards kept open the possibility that the soil-particle surface might exert such a profound influence on the soil water that the hydrostatic pressure within it might actually be positive and large near the particle surface. Equilibrium of soil water would therefore need to be specified by a criterion of potential broader than pressure alone (Low and Deming, 1953; Bolt and R. D. Miller, 1958). This is in contrast with the more frequent explanations based on the classic concept of pressure deficiencies induced by curved air–water interfaces. While the curved-interface mechanism can be operative in the relatively large soil pores at low suctions, the results of Waldron *et al.* (1961) and Wilkinson and Klute (1962) suggest that retention mechanisms other than air–water interfaces may become sensible and appreciable even at relatively low suctions.

The difficulty of separating the individual effects of $K(\theta)$ and $\tau(\theta)$ on nonproportionality is not avoided by returning to Eq. (154), since, in effect, $\tau = \tau(\theta)$ has been employed in shifting from $K(\theta)$ to $K(\tau)$. Rather, the validity of $v = K(\theta)\ \partial\tau/\partial x$ should be assessed by independent measurements of v, τ, and θ to provide corresponding values of v and $\partial\tau/\partial x$ at given θ. Similar measurements for vertical-flow systems would test the validity of $v = -K(\theta)\ \partial h/\partial z$, which was the form assessed by the early steady-state experiments of Childs and Collis-George (1950), although their gravitationally induced hydraulic gradients could not exceed unity. But this limitation was not overcome even in the recent refined experimentation of Watson (1966). With the exception of having been obtained under transient-flow conditions, Watson's work was primarily a sophisticated corroboration of Childs and Collis-George's earlier work, since the sand was still relatively coarse (particle diameter 0.15–0.30 mm) and the suctions relatively small ($\leqslant$50 cm H_2O).

3.2.2.3. VAPOR AND TEMPERATURE EFFECTS. If appreciable water can be transported in vapor form through an unsaturated soil, it is conceivable that some feature of this process might somehow couple with an aspect of the overall water-transport process to induce nonproportionality. The results of Rose (1963a) suggest that vapor transport may not truly be isothermal even if temperature gradients cannot be detected experimentally. If sizable temperature gradients are present, the likelihood of appreciable vapor transfer is greatly enhanced, and Anderson and Linville (1960, 1962) have shown that an increase in temperature is often produced in the soil ahead of an advancing wet front. Whether the resulting counter gradient of temperature would have had sufficient magnitude and the necessary transient features to cause nonproportionality between velocity and gradient in the various inflow experiments (Nielsen *et al.*, 1962; Davidson *et al.*, 1963b; Ferguson and W. H. Gardner, 1963; Rawlins and W. H. Gardner, 1963) is difficult to assess, but has been suggested as a possibility by Stewart (1962).

3.3. Implications of Nonproportional Behavior

The rather widespread existence of nonproportional behavior in both saturated and unsaturated flow implies that the Darcy [Eq. (1)] and the Buckingham–Darcy [Eq. (2) or (3)] relationships should be viewed as first approximations somewhat in the sense of working hypotheses. This would be especially true for the Buckingham–Darcy equation where both $K(\theta)$ and $\tau(\theta)$ [or $-d\tau/d\theta$] are involved. A more strict experimental validity should be claimed only when actually demonstrated for the porous material in question. This does not mean that scientists should be discouraged from attempting to use the proportional assumption, although they should be cognizant of its possible limitations. Obviously, there would be little point in worrying about nonproportionalities for such as a practical field problem in which the overall experimental error was great enough to mask the nonproportional effects. Also, certain flow aspects may not always be critically sensitive even to appreciable nonproportionality. For example, from the data of Rawlins and Gardner (1963), Swartzendruber (1968) has found the overall velocity of horizontal water absorption to be affected only 40% as a result of the 8-fold gradient-induced diffusivity variation in Fig. 26(B). Apparently, the unique (proportional) wet-end diffusivity for $\theta > 0.35$ largely dominates the nonunique dry-end discrepancies, but even the 40% error is more than would have been anticipated from the similar type of proportionally-based analysis of Hanks and Bowers (1963).

Nonproportionality should be kept in mind as a possible explanation for experimental flow aspects which do not agree with proportionally based theoretical solutions and concepts. The present dilemma with the pressure outflow method of measuring diffusivity (Section 2.3.4.3) may be a case in point, although the cause of error is not yet finally known, nor is it even yet clear whether the diffusivity discrepancy lies with $K(\theta)$ or $-d\tau/d\theta$. Another example is the possibility of threshold-gradient behavior in porous plates (Hillel and Mottes, 1966), a phenomenon which could have important bearing on the accuracy of measured $\tau(\theta)$ relationships.

One of the more difficult theoretical aspects of nonproportionality is that of extension to multidimensional flow and more general boundary and initial conditions. Present-day mathematical emphases on computers and nonlinear differential equations, however, render such difficulties less formidable than would have been true several decades ago. Nevertheless, in cases where these difficulties still persist, they constitute ample justification for invoking the proportionality assumption as a means of forcing progress and enabling insight. For, as has been aptly noted by Muskat (1946, pp. 68–69), it may be self-defeating to insist on highly precise one-dimensional velocity-gradient equations, if their mathematical form seriously impedes their further use in extension and application to different or more complicated flow problems of interest.

From a scientific and academic viewpoint, nonproportionality is not altogether a disadvantage. Potentiality for new knowledge is provided by investigation of the phenomenon, unless, of course, it is ultimately shown to be merely an artifact of experimental procedure. If eventually shown to arise from such as modified-water properties, porous-medium fabric changes, electrical streaming potential, or salt-osmotic effects, nonproportionality would likely become in its own right an interesting interaction property of the flow liquid and the porous medium.

4. Temperature and Vapor Aspects of Soil-Water Transport

4.1. Different Isothermal Temperatures

If the saturated hydraulic conductivity of Eq. (136) is partitioned in the customary manner (L. A. Richards, 1952), then

$$K = k\rho g/\mu \tag{159}$$

where k is the intrinsic permeability and μ is the viscosity. On this basis, it is rather generally held that temperature effects on saturated water flow in sands are essentially attributable to and describable by the temperature dependence of the viscosity (see Gustafsson, 1940; Pillsbury, 1950, and references therein). The matter may also be approached in terms of the activation-energy concept (Glasstone *et al.*, 1941), with the viscosity being expressed as

$$\mu = M \exp(E_a/R_g T) \tag{160}$$

where M is an empirical constant, E_a is the activation energy, R_g is the molar gas constant, and T is the absolute temperature. For saturated flow, Low (1960) in effect combined Eqs. (159) and (160) to obtain

$$K = k\rho g M \exp(-E_a/R_g T) \tag{161}$$

By obtaining K from flow experiments conducted at various temperatures, Eq. (161) can be used to evaluate E_a for soil-water flow, which can then be compared with E_a for pure water as determined from Eq. (160) applied to viscosity-temperature data. Using this approach, Low (1960) found E_a for water flow through sodium bentonite to range from 3.8 to 4.4 kcal/mole, compared with $E_a = 3.9$ kcal/mole for pure water. He concluded from this that a more ordered structure had been induced in the liquid water as a result of clay-water interaction.

For unsaturated flow, activation-energy methods were first used by Biggar (1956) and Biggar and S. A. Taylor (1960). They combined a diffusion equivalent of Eq. (160) with Eq. (29) for horizontal absorption. This assumes that the definite integral from Eq. (29) [that is, F_1 from Eq. (67) when $i = 1$ and $f_1 = \eta$] can fill the role of K in Eq. (161), and led to E_a values ranging from 1.3 to 2.8 kcal/mole. Anderson *et al.* (1963) criticized the use of F_1 as an inappropriate characterization of the rate coefficient, and suggested the use of F_1^2 as preferable. This criticism may be justified, since F_1^2 can be shown proportional to a mean diffusivity $\bar{D}$ as utilized by W. R. Gardner (1959a,b). Proceeding on this basis to a modified analysis of Biggar's (1956) data, Anderson *et al.* (1963) found E_a to be as high as 6 kcal/mole, a value commensurate with their own results as obtained from temperature-sensed measurements of wet-front position. Jackson (1965) employed activation-energy methods to infer an E_a of 4.2 kcal/mole at water contents above that corresponding to an estimated monolayer of water. From an earlier study (Jackson, 1963b), he had concluded for much higher water contents that the temperature dependence of the diffusivity was essentially that exhibited by the ratio of the surface tension to the viscosity.

While the activation-energy approach can probably furnish further insight into the water transport process, its use is not without certain limitations and complications. Low (1960) has pointed out that quantitative comparison of viscosity with activation energy is impossible, since M of Eq. (160) involves an unknown entropy of activation. Young *et al.* (1964) have pointed out that appreciable curvature may sometimes occur in a plot of ln K versus $1/T$ as used to evaluate E_a from Eq. (161), and implies the complication of a temperature-dependent activation energy. Another requirement for unambiguous interpretation would seem to be that the transport coefficient as in Eq. (161) should be gradient independent (proportional transport), or else the gradient must be standardized to avoid extraneous effects. Finally, question also might be raised in regard to the two-component diffusivity of Eq. (4) being a suitable rate coefficient, or whether only $K(\theta)$ qualifies, since the net temperature effect in $D(\theta)$ depends somehow on the resultant of the individual effects in $K(\theta)$ and $-d\tau/d\theta$ (W. R. Gardner, 1959b). The temperature dependence of $\tau(\theta)$ alone has been studied (R. Gardner, 1955; S. A. Taylor, 1958; Wilkinson and Klute, 1962), and, although the conclusions have not always been definitive, it appears that the observed effects are not wholly attributable to the temperature dependence of the surface tension.

4.2. Vapor-Phase Transport of Soil Water

The derivation of Eq. (3) was presented (Section 2.1.3) in terms of only hydrodynamic liquid flow, but Philip (1955a, 1957c, 1958a) has shown that the mathematical form is not changed if isothermal vapor- and adsorbed-phase transport are included. $D(\theta)$, however, would then be an overall diffusivity given by the sum

$$D(\theta) = D_{\text{liq}}(\theta) + D_{\text{vap}}(\theta) + D_{\text{ads}}(\theta) \tag{162}$$

where $D_{\text{liq}}(\theta)$, $D_{\text{vap}}(\theta)$, and $D_{\text{ads}}(\theta)$ are the true liquid-, vapor-, and adsorbed-phase diffusivities, respectively. Both $D_{\text{vap}}(\theta)$ and $D_{\text{ads}}(\theta)$ are here conceived as arising from classical diffusion mechanisms, while $D_{\text{liq}}(\theta)$ retains the hydrodynamic liquid-flow mechanism discussed previously in connection with Eqs. (3) and (4).

As indicated in the introduction (Section 1), $D_{\text{liq}}(\theta)$ is generally presumed to dominate the sum in Eq. (162), unless the soil-water content is extremely low. To examine such behavior experimentally, Jackson (1964a) investigated the one-dimensional water-absorption process (Section 2.3.1) at water contents less than air-dry. The initial water content θ_0 was established by vacuum desiccation methods, and the applied θ_1 was obtained by constant water-vapor density as regulated by saturated salt solutions (Wexler and Brombacher, 1951). At various times after the application of θ_1, water-content profiles were determined by gravimetric sectioning, and were found to specify a unique $\eta(\theta)$ as determined from Eq. (24), although the shapes of the curves were somewhat different from those of Fig. 5. By the method of Section 2.3.4.4., diffusivities were calculated from Eq. (126) and plotted against water content. The typical results shown in Fig. 27 are for Adelanto loam and exhibit a distinct peak maximum at a θ of about 0.035. Such a maximum had been predicted by Philip (1955a), along with a nonzero minimum at a slightly higher water content as reported previously by Staple and Lehane (1954) and recently by Rose (1967). Philip associated the minimum diffusivity with a water content not far removed from that at which complete liquid continuity breaks down and $K(\theta)$ is essentially zero. He associated the peak maximum with vapor transport, on the basis of previous observations (Gurr *et al.*, 1952, and references therein) that reported an optimum low water content for vapor movement. The existence of the maximum was further confirmed by Jackson (1964b,c) with both steady-state and

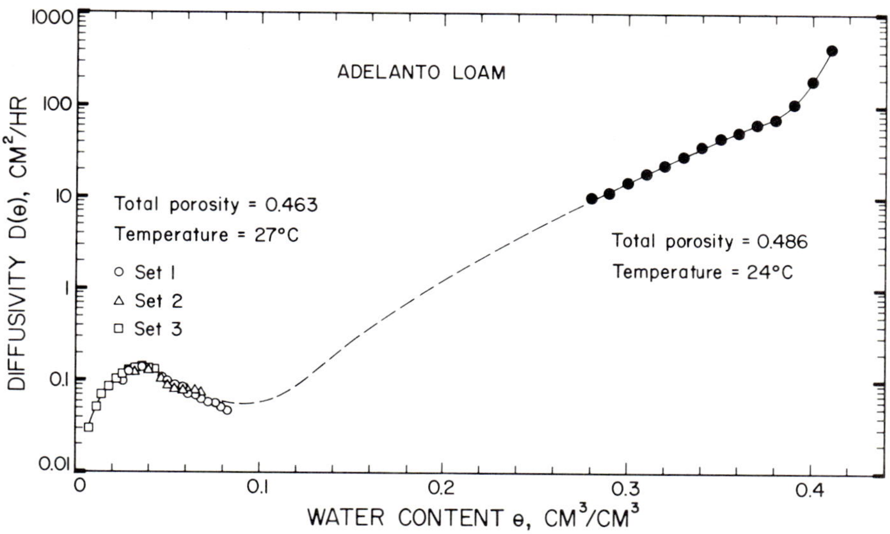

FIG. 27. Soil-water diffusivity for Adelanto loam at low water contents [open symbols] (Jackson, 1964a) and at higher water contents [solid symbols] (Jackson, 1963b). [Adapted from *Soil Sci. Soc. Am. Proc.* **28**, 174 (1964); **27**, 365 (1963); by permission of the publisher and the author.]

drying experiments. Finally, Fig. 27 also shows data of Jackson (1963b) for Adelanto loam at much higher water contents, to illustrate the diffusivity function in the context of a broad range of water content.

Rose (1963a), using measurements of water-vapor transport to evaluate $K(\theta)$ at low water contents, did find that K increased very rapidly with small θ but then leveled off to a value as little as 10^{-6}–10^{-7} of the saturated hydraulic conductivity, even for θ substantially below n (saturation). This was attributed to flow-induced counter gradients of temperature that were too small to be detected experimentally as sensible departures from isothermality. Utilizing gaseous hydrogen diffusion, Rose (1963b) also attempted to separate the components of vapor and liquid transport, and concluded that the vapor-phase contribution decreased with increased θ and became negligible in the water-content range normally associated with plant growth. Jackson (1965) attempted a separation of vapor- and liquid-transport components by flow experiments at different total pressures. Where measurable, this pressure-determined liquid-flow component increased with θ and comprised up to 40% of the total transport, over the dry-end water-content range associated with the peak maximum as in Fig. 27.

For conditions that are no longer isothermal, the diffusivity approach can be generalized to accommodate flow responses to temperature gradients (Philip and de Vries, 1957; de Vries, 1958). In principle, the broadest approach is that of irreversible thermodynamics (Winterkorn, 1958; Cary and S. A. Taylor, 1962a,b; S. A. Taylor and Cary, 1964), and which has been used to compare temperature and suction gradients for equal water transport (Cary, 1965a), as well as to unify a recent review of practical aspects of thermal gradients (Cary, 1966). Readers wishing to explore in more detail the literature on temperature gradients are directed to the foregoing citations and the references contained therein.

5. Concluding Comments

The many unsaturated-flow problems which have been attacked and solved by combining the equation of continuity with the Buckingham–Darcy equation should leave no doubt concerning the utility and fruitfulness of such an approach. Also, the development of this approach owes a great debt to its predecessor methods developed from the use of potential-theory analysis of saturated Darcian flow, foremost among them being the early comprehensive treatment of Muskat (1946). Furthermore, there is little reason to suppose that the usefulness of proportionally based approaches has yet been completely exhausted.

Nevertheless, there likewise is little good reason to suppose that soil and water scientists should avoid or flee the complexities of nonproportional behavior, if and when it is encountered. Rather, it may be viewed as a complex but interesting challenge. In approaching it, there is room for bringing to bear the background knowledge from proportional conductivity and diffusivity analyses, plus that from classical and irreversible thermodynamics, miscible and immiscible displacement, continuum mechanics, or any other relevant discipline. Mathematically, support is provided from the currently live and vital areas of nonlinear differential equations and computer technology. Experimentally, the modern techniques of rapid nondestructive water-content measurement (W. H. Gardner, 1965; Mack and Brach, 1966), rapid-response recording tensiometers (Klute and Peters, 1962; Watson, 1965; Peck and Rabbidge, 1966), and thermocouple psychrometers (L. A. Richards, 1965) should be of increasing utility. To assess and separate the various possible causative factors of nonproportionality will likely require a subtle blend of both experiment and theory. Even if a final definite assessment of nonproportionality and its causes is different from what might presently be envisioned, the formulation and testing of hypotheses in this regard should of themselves yield results of information and value to soil and water scientists and their disciplines.

REFERENCES

Abd-el-Aziz, M. H., and Taylor, S. A. (1965). *Soil Sci. Soc. Am. Proc.* **29**, 141–143.

Abramson, H. A. (1934). "Electrokinetic Phenomena and Their Application to Biology and Medicine." Chem. Catalog Co., New York.

Anderson, D. M., and Linville, A. (1960). *Science* **131**, 1370–1371.

Anderson, D. M., and Linville, A. (1962). *Soil Sci. Soc. Am. Proc.* **26**, 14–18.

Anderson, D. M., and Low, P. F. (1958). *Soil Sci. Soc. Am. Proc.* **22**, 99–103.

Anderson, D. M., Linville, A., and Sposito, G. (1963). *Soil Sci. Soc. Am. Proc.* **27**, 610–613.

Aslyng, H. C. (1963). *Intern. Soc. Soil Sci. Bull.* **23**, 7–10.

Bazargani, J. I. (1964). Ph.D. Thesis, Purdue Univ., Lafayette, Indiana.

Behnke, J. J., and Bianchi, W. C. (1965). *Water Resources Res.* **1**, 557–562.

Biggar, J. W. (1956). Ph.D. Thesis, Utah State Univ., Logan, Utah.

Biggar, J. W., and Nielsen, D. R. (1962). *J. Geophys. Res.* **67**, 3636–3637.

Biggar, J. W., and Taylor, S. A. (1960). *Soil Sci. Soc. Am. Proc.* **24**, 81–85.

Black, C. A., ed. (1965). "Methods of Soil Analysis," Part 1. Am. Soc. Agron., Madison, Wisconsin.

Bloomsburg, G. L., and Corey, A. T. (1964). *Colorado State Univ. Hydrology Paper* No. 7.

Bodman, G. B., and Edlefsen, N. E. (1934). *Soil Sci.* **38**, 425–444.

Bolt, G. H., and Lagerwerff, J. V. (1965). *Soil Sci.* **99**, 147–153.

Bolt, G. H., and Miller, R. D. (1958). *Trans. Am. Geophys. Union* **39**, 917–928.

Boltzmann, L. (1894). *Ann. Physik Chemie (Leipzig)* **53**, 959–964.

Bouwer, H. (1959a). *Agr. Eng.* **40**, 395–400.

Bouwer, H. (1959b). *Soil Sci. Soc. Am. Proc.* **23**, 260–263.

Bouwer, H., and Little, W. C. (1959). *Soil Sci. Soc. Am. Proc.* **23**, 91–96.

Bruce, R. R., and Klute, A. (1956). *Soil Sci. Soc. Am. Proc.* **20**, 458–462.

Bruce, R. R., and Klute, A. (1963). *Soil Sci. Soc. Am. Proc.* **27**, 18–21.

Buckingham, E. (1907). *U. S. Dept. Agr., Bur. Soils Bull.* **38**.

Bull, H. B., and Gortner, R. A. (1932). *J. Phys. Chem.* **36**, 111–119.

Carslaw, H. S., and Jaeger, J. C. (1959). "Conduction of Heat in Solids," 2nd ed. Oxford Univ. Press, London.

Cary, J. W. (1965a). *Soil Sci.* **100**, 168–175.

Cary, J. W. (1965b). *Agron. J.* **57**, 607–610.

Cary, J. W. (1966). *Soil Sci. Soc. Am. Proc.* **30**, 428–433.

Cary, J. W., and Taylor, S. A. (1962a). *Soil Sci. Soc. Am. Proc.* **26**, 413–416.

Cary, J. W., and Taylor, S. A. (1962b). *Soil Sci. Soc. Am. Proc.* **26**, 417–420.

Childs, E. C. (1960). *J. Geophys. Res.* **65**, 780–782.

Childs, E. C., and Collis-George, N. (1948). *Discussions Faraday Soc.*, No. 3, pp. 78–85.

Childs, E. C., and Collis-George, N. (1950). *Proc. Roy. Soc. (London)* **A201**, 392–405.

Childs, E. C., and Poulovassilis, A. (1962). *J. Soil Sci.* **13**, 271–285.

Christensen, H. R. (1944). *Soil Sci.* **57**, 381–391.

Christenson, D. R., and Ferguson, H. (1966). *Soil Sci. Soc. Am. Proc.* **30**, 549–553.

Christiansen, J. E. (1944). *Soil Sci.* **58**, 355–365.

Corey, A. T. (1957). *Soil Sci. Soc. Am. Proc.* **21**, 7–10.

Covey, W. (1963). *Soil Sci. Soc. Am. Proc.* **27**, 130–134.

Crank, J., and Henry, M. E. (1949). *Trans. Faraday Soc.* **45**, 636–650.
Darcy, H. (1856). "Les fontaines publique de la ville de Dijon," pp. 570, 590–594. Victor Dalmont, Paris.
Davidson, J. M., Biggar, J. W., and Nielsen, D. R. (1963a). *J. Geophys. Res.* **68**, 4777–4783.
Davidson, J. M., Nielsen, D. R., and Biggar, J. W. (1963b). *Hilgardia* **34**, 601–617.
Davidson, J. M., Nielsen, D. R., and Biggar, J. W. (1966). *Soil Sci. Soc. Am. Proc.* **30**, 298–304.
Davidson, J. M., Biggar, J. W., Nielsen, D. R., Warrick, A. W., and Cassel, D. K. (1967). *Proc. Symp. Water Unsat. Zone, Wageningen, Netherlands,* 1966 (in press).
Day, P. R., and Luthin, J. N. (1953). *Soil Sci. Soc. Am. Proc.* **17**, 87–91.
Day, P. R., and Luthin, J. N. (1956). *Soil Sci. Soc. Am. Proc.* **20**, 443–447.
de Vries, D. A. (1958). *Trans. Am. Geophys. Union* **39**, 909–916.
Doering, E. J. (1965). *Soil Sci.* **99**, 322–326.
Elrick, D. E. (1963). *Australian J. Soil Res.* **1**, 1–8.
Elrick, D. E., and Bowman, D. H. (1964). *Soil Sci. Soc. Am. Proc.* **28**, 450–453.
Fancher, G. H., and Lewis, J. A. (1933). *Ind. Eng. Chem.* **25**, 1139–1147.
Ferguson, H., and Gardner, W. H. (1962). *Soil Sci. Soc. Am. Proc.* **26**, 11–14.
Ferguson, H., and Gardner, W. H. (1963). *Soil Sci. Soc. Am. Proc.* **27**, 243–246.
Fujioka, Y., and Kitamura, T. (1964). *J. Geophys. Res.* **69**, 5249–5255.
Gardner, R. (1955). *Soil Sci.* **79**, 257–265.
Gardner, W. (1919). *Soil Sci.* **7**, 313–317.
Gardner, W. (1936). *J. Agr. Res.* **53**, 57–60.
Gardner, W., and Widtsoe, J. A. (1921). *Soil Sci.* **11**, 215–232.
Gardner, W., Israelson, O. W., Edlefsen, N. E., and Clyde, H. (1922). *Phys. Rev. (Series 2)* **20**, 196.
Gardner, W. H. (1965). *In* "Methods of Soil Analysis" (C. A. Black, ed.), Part 1, pp. 82–127. Am. Soc. Agron., Madison, Wisconsin.
Gardner, W. H., and Gardner, W. (1951). *Soil Sci. Soc. Am. Proc.* **15**, 42–50.
Gardner, W. R. (1956). *Soil Sci. Soc. Am. Proc.* **20**, 317–320.
Gardner, W. R. (1958). *Soil Sci.* **85**, 228–232.
Gardner, W. R. (1959a). *Soil Sci. Soc. Am. Proc.* **23**, 183–187.
Gardner, W. R. (1959b). *Soil Sci. Soc. Am. Proc.* **23**, 406–407.
Gardner, W. R. (1960). *Soil Sci.* **89**, 63–73.
Gardner, W. R. (1961). *Trans. 7th Intern. Congr. Soil Sci., Madison, Wisconsin, 1960,* **1**, pp. 300–305. Elsevier, Amsterdam.
Gardner, W. R. (1962). *Soil Sci. Soc. Am. Proc.* **26**, 129–132.
Gardner, W. R. (1964). *Agron. J.* **56**, 41–45.
Gardner, W. R., and Ehlig, C. F. (1962a). *Science* **138**, 522–523.
Gardner, W. R., and Ehlig, C. F. (1962b). *Agron. J.* **54**, 453–456.
Gardner, W. R., and Ehlig, C. F. (1963). *J. Geophys. Res.* **68**, 5719–5724.
Gardner, W. R., and Fireman, M. (1958). *Soil Sci.* **85**, 244–249.
Gardner, W. R., and Hillel, D. I. (1962). *J. Geophys. Res.* **67**, 4319–4325.
Gardner, W. R., and Mayhugh, M. S. (1958). *Soil Sci. Soc. Am. Proc.* **22**, 197–201.
Gardner, W. R., and Miklich, F. J. (1962). *Soil Sci.* **93**, 271–274.
Glasstone, S., Laidler, K. J., and Eyring, H. (1941). "The Theory of Rate Processes," pp. 477–551. McGraw-Hill, New York.
Gupta, R. P., and Staple, W. J. (1964). *Soil Sci. Soc. Am. Proc.* **28**, 729–732.
Gupta, R. P., and Swartzendruber, D. (1962). *Soil Sci. Soc. Am. Proc.* **26**, 6–10.
Gupta, R. P., and Swartzendruber, D. (1964). *Soil Sci. Soc. Am. Proc.* **28**, 9–12.

Gurr, C. G., Marshall, T. J., and Hutton, J. T. (1952). *Soil Sci.* **74**, 335–345.
Gustafsson, Y. (1940). *Lantbrukshogskolans Annaler (Ann. Agr. Coll. Sweden)* **8**, 425–456.
Hadas, A. (1964). *Israel J. Agr. Res.* **14**, 159–168.
Hallaire, M. (1958). *In* "Water and Its Conduction in Soils" (H. F. Winterkorn, ed.), pp. 88–105. Highway Res. Board, Spec. Rept. 40.
Hanks, R. J., and Bowers, S. A. (1962). *Soil Sci. Soc. Am. Proc.* **26**, 530–534.
Hanks, R. J., and Bowers, S. A. (1963). *Soil Sci. Soc. Am. Proc.* **27**, 263–265.
Hanks, R. J., and Gardner, H. R. (1965). *Soil Sci. Soc. Am. Proc.* **29**, 495–498.
Hansbo, S. (1960). *Swed. Geotech. Inst. Proc. No.* 18. Publ. in Stockholm by the Institute.
Hillel, D. I., and Mottes, J. (1966). *Soil Sci.* **102**, 135–139.
Hubbert, M. K. (1956). *Trans Am. Inst. Mining Met. Eng.* **207**, 222–239.
Israelson, O. W. (1927). *Hilgardia* **2**, 479–528.
Jackson, R. D. (1963a). *Soil Sci. Soc. Am. Proc.* **27**, 123–126.
Jackson, R. D. (1963b). *Soil Sci. Soc. Am. Proc.* **27**, 363–366.
Jackson, R. D. (1964a). *Soil Sci. Soc. Am. Proc.* **28**, 172–176.
Jackson, R. D. (1964b). *Soil Sci. Soc. Am. Proc.* **28**, 464–466.
Jackson, R. D. (1964c). *Soil Sci. Soc. Am. Proc.* **28**, 467–470.
Jackson, R. D. (1965). *Soil Sci. Soc. Am. Proc.* **29**, 144–148.
Jackson, R. D., van Bavel, C. H. M., and Reginato, R. J. (1963). *Soil Sci.* **96**, 249–256.
Jackson, R. D., Reginato, R. J., and van Bavel, C. H. M. (1965). *Water Resources Res.* **1**, 375–380.
Kemper, W. D. (1960). *Soil Sci. Soc. Am. Proc.* **24**, 10–16.
Kemper, W. D., and Evans, N. A. (1963). *Soil Sci. Soc. Am. Proc.* **27**, 485–490.
Kemper, W. D., and Maasland, D. E. L. (1964). *Soil Sci. Soc. Am. Proc.* **28**, 318–322.
Kemper, W. D., and Rollins, J. B. (1966). *Soil Sci. Soc. Am. Proc.* **30**, 529–534.
King, F. H. (1898). U. S. Geol. Surv., 19th Ann. Rept., Part 2, pp. 59–294.
Kirkham, D., and Feng, C. L. (1949). *Soil Sci.* **67**, 29–40.
Klute, A. (1952). *Soil Sci.* **73**, 105–116.
Klute, A. (1965). *In* "Methods of Soil Analysis" (C. A. Black, ed.), Part 1, pp. 253–261. Am. Soc. Agron., Madison, Wisconsin.
Klute, A., and Peters, D. B. (1962). *Soil Sci. Soc. Am. Proc.* **26**, 87–88.
Klute, A., Whisler, F. D., and Scott, E. J. (1965). *Soil Sci. Soc. Am. Proc.* **29**, 353–358.
Kolaian, J. H., and Low, P. F. (1962). *Proc. 9th Natl. Conf. Clays Clay Minerals, Lafayette, Indiana, 1960*, pp. 71–84. Pergamon Press, Oxford.
Kunze, R. J., and Kirkham, D. (1962). *Soil Sci. Soc. Am. Proc.* **26**, 421–426.
Kunze, R. J., and Peters, D. B. (1966). *Soil Sci. Soc. Am. Proc.* **30**, 286–288.
Kutilek, M. (1965). *Sci. Sol* **1**, 3–14.
Lambe, T. W. (1951). *Am. Soc. Civ. Engrs. Trans.* **116**, 401–423.
Leonard, R. A., and Low, P. F. (1964). *Proc. 12th Natl. Conf. Clays Clay Minerals, Atlanta, Georgia, 1963*, pp. 311–325. Macmillan, New York.
Leonards, G. A., ed. (1962). "Foundation Engineering," pp. 66–240. McGraw-Hill, New York.
Li, S. P. (1963). *Soil Sci.* **95**, 410–413.
Liakopoulos, A. C. (1964). *J. Hydraulic Res., Intern. Assoc. Hydraulic Res.* **2**, 50–74.
Liakopoulos, A. C. (1966). *Water Resources Res.* **2**, 227–240.
Ligon, J. T., Johnson, H. P., and Kirkham, D. (1962). *J. Geophys. Res.* **67**, 5199–5204.
Low, P. F. (1960). *Proc. 8th Natl. Conf. Clays Clay Minerals, Norman, Oklahoma, 1959*, pp. 170–182. Pergamon Press, Oxford.
Low, P. F. (1961). *Advan. Agron.* **13**, 269–327.

Low, P. F. (1962). *Proc. 9th Natl. Conf. Clays Clay Minerals, Lafayette, Indiana, 1960,* pp. 219–228. Pergamon Press, Oxford.

Low, P. F., and Deming, J. M. (1953). *Soil Sci.* **75**, 187–202.

Luthin, J. N., and Day, P. R. (1955). *Soil Sci. Soc. Am. Proc.* **19**, 406–410.

Luthin, J. N., and Miller, R. D. (1953). *Soil Sci. Soc. Am. Proc.* **17**, 329–333.

Lutz, J. F. (1965). *Soil Sci. Soc. Am. Proc.* **29**, 330–351.

Lutz, J. F., and Kemper, W. D. (1959). *Soil Sci.* **88**, 83–90.

Mack, A. R., and Brach, E. J. (1966). *Soil Sci. Soc. Am. Proc.* **30**, 544–548.

Marshall, T. J. (1959). *Tech. Communication No.* 50, *Commonwealth Bur. Soils.* Harpenden, England.

Martin, R. T. (1962). *Proc. 9th Natl. Conf. Clays Clay Minerals, Lafayette, Indiana, 1960,* pp. 28–70. Pergamon Press, Oxford.

Michaels, A. S., and Lin, C. S. (1955). *Ind. Eng. Chem.* **47**, 1249–1253.

Miklich, F. J., and Richards, L. A. (1965). *Soil Sci. Soc. Am. Proc.* **29**, 485–487.

Miller, D. E., and Gardner, W. H. (1962). *Soil Sci. Soc. Am. Proc.* **26**, 115–119.

Miller, E. E., and Elrick, D. E. (1958). *Soil Sci. Soc. Am. Proc.* **22**, 483–486.

Miller, E. E., and Miller, R. D. (1956). *J. Appl. Phys.* **27**, 324–332.

Miller, R. J., and Low, P. F. (1963). *Soil Sci. Soc. Am. Proc.* **27**, 605–609.

Mokady, R. S., and Low, P. F. (1964). *Soil Sci. Soc. Am. Proc.* **28**, 583–584.

Moore, R. E. (1939). *Hilgardia* **12**, 383–426.

Muskat, M. (1946). "The Flow of Homogeneous Fluids Through Porous Media," Edwards, Ann Arbor, Michigan.

Nielsen, D. R., and Biggar, J. W. (1961). *Soil Sci.* **92**, 192–193.

Nielsen, D. R., and Biggar, J. W. (1963). *Soil Sci. Soc. Am. Proc.* **27**, 10–13.

Nielsen, D. R., Kirkham, D., and Perrier, E. R. (1960). *Soil Sci. Soc. Am. Proc.* **24**, 157–160.

Nielsen, D. R., Kirkham, D., and van Wijk, W. R. (1961). *Soil Sci. Soc. Amer. Proc.* **25**, 165–168.

Nielsen, D. R., Biggar, J. W., and Davidson, J. M. (1962). *Soil Sci. Soc. Am. Proc.* **26**, 107–111.

Ohmstede, W. D. (1964). *J. Geophys. Res.* **69**, 633–642.

Olsen, H. W. (1965). *Soil Sci. Soc. Am. Proc.* **29**, 135–140.

Olsen, H. W. (1966). *Water Resources Res.* **2**, 287–295.

Olson, T. C. (1966). Ph.D. Thesis, Purdue Univ., Lafayette, Indiana.

Oster, J. D., and Low, P. F. (1963). *Soil Sci. Soc. Am. Proc.* **27**, 369–373.

Oster, J. D., and Low, P. F. (1964). *Soil Sci. Soc. Am. Proc.* **28**, 605–609.

Overbeek, J. T. G. (1952). *In* "Colloid Science" (H. R. Kruyt, ed.), Vol. I, pp. 194–207. Elsevier, Amsterdam.

Paez, J. A. (1962). M. S. Thesis, Univ. of Calif., Davis, California.

Peck, A. J., and Rabbidge, R. M. (1966). *Science* **151**, 1385–1386.

Peirce, B. O. (1929). "A Short Table of Integrals." Ginn and Company, New York.

Philip, J. R. (1955a). *Proc. Natl. Acad. Sci. India* **A24**, 93–104.

Philip, J. R. (1955b). *Trans. Faraday Soc.* **51**, 885–892.

Philip, J. R. (1957a). *Australian J. Phys.* **10**, 29–42.

Philip, J. R. (1957b). *Trans. 3rd Congr. Irrigation Drainage, San Francisco, California, 1957,* Vol. 3, Question 8, pp. 8. 125–8.154. Intern. Commission Irrigation and Drainage, New Delhi, India.

Philip, J. R. (1957c). *Soil Sci.* **83**, 345–357.

Philip, J. R. (1957d). *Soil Sci.* **83**, 435–448.

Philip, J. R. (1957e). *Soil Sci.* **84**, 257–264.

Philip, J. R. (1958a). *In* "Water and Its Conduction in Soils" (H. F. Winterkorn, ed.), pp. 147–163. Highway Res. Board, Spec. Rept. 40.
Philip, J. R. (1958b). *Soil Sci.* **85**, 278–286.
Philip, J. R. (1964). *Natl. Symp. Water Resources Use Management, Canberra, Australia, 1963*, pp. 257–275. Melbourne Univ. Press, Melbourne.
Philip, J. R. (1968). *J. Geophys. Res.* **73**, 3968–3970.
Philip, J. R., and de Vries, D. A. (1957). *Trans. Am. Geophys. Union* **38**, 222–232.
Pillsbury, A. F. (1950). *Soil Sci.* **70**, 299–300.
Prill, R. C., Johnson, A. I., and Morris, D. A. (1965). U. S. Geol. Surv., Water-Supply Paper 1662B.
Raats, P. A. C. (1965). Ph.D. Thesis, Univ. of Illinois, Urbana, Illinois.
Rawlins, S. L. (1961). Ph.D. Thesis, Washington State Univ., Pullman, Washington.
Rawlins, S. L., and Gardner, W. H. (1963). *Soil Sci. Soc. Am. Proc.* **27**, 507–511.
Reginato, R. J., and van Bavel, C. H. M. (1962). *Soil Sci. Soc. Am. Proc.* **26**, 1–3.
Reisenauer, A. E. (1963). *J. Geophys. Res.* **68**, 5725–5733.
Remson, I., Drake, R. L., McNeary, S. S., and Wallo, E. M. (1965). *J. Hydraulics Div., Am. Soc. Civ. Engrs.* **91**, 55–74.
Richards, L. A. (1928). *J. Agr. Res.* **37**, 719–742.
Richards, L. A. (1931). *Physics* **1**, 318–333.
Richards, L. A. (1936). *J. Amer. Soc. Agron.* **28**, 297–300.
Richards, L. A. (1947). *Agr. Eng.* **28**, 451–454.
Richards, L. A. (1949). *Soil Sci.* **68**, 95–112.
Richards, L. A. (1952). *Soil Sci. Soc. Am. Proc.* **16**, 85–88.
Richards, L. A. (1953). *Proc. Intern. Symp. Desert Res., Jerusalem, Israel, 1952*, pp. 523–546. Res. Council of Israel, Jerusalem.
Richards, L. A. (1961). *Trans. 7th Intern. Congr. Soil Sci., Madison, Wisconsin, 1960* Vol. 1, pp. 67–79. Elsevier, Amsterdam.
Richards, L. A. (1965). *In* "Methods of Soil Analysis" (C. A. Black, ed.), Part 1, pp. 128–152. Am. Soc. Agron., Madison, Wisconsin.
Richards, L. A., and Moore, D. C. (1952). *Trans. Am. Geophys. Union* **33**, 531–540.
Richards, L. A., and Richards, P. L. (1962). *Soil Sci. Soc. Am. Proc.* **26**, 515–518.
Richards, S. J., and Weeks, L. V. (1953). *Soil Sci. Soc. Am. Proc.* **17**, 206–209.
Rijtema, P. E. (1959). *Neth. J. Agr. Sci.* **7**, 209–215.
Rose, D. A. (1963a). *Brit. J. Appl. Phys.* **14**, 256–262.
Rose, D. A. (1963b). *Brit. J. Appl. Phys.* **14**, 491–496.
Rose, D. A. (1967). *Proc. Symp. Water Unsat. Zone, Wageningen, Netherlands, 1966* (in press).
Rubin, J. (1966). *Water Resourses Res.* **2**, 739–749.
Rubin, J., and Steinhardt, R. (1963). *Soil Sci. Soc. Am. Proc.* **27**, 246–251.
Rubin, J., and Steinhardt, R. (1964). *Soil Sci. Soc. Am. Proc.* **28**, 614–619.
Rubin, J., Steinhardt, R., and Reiniger, P. (1964). *Soil Sci. Soc. Am. Proc.* **28**, 1–5.
Schleusener, R. A., and Corey, A. T. (1959). *J. Geophys. Res.* **64**, 469–475.
Scott, E. J., and Hanks, R. J. (1962). *Soil Sci.* **94**, 314–322.
Singh, R. (1965). Stanford Univ., Dept. Civil Eng., Tech. Rept. 54.
Staple, W. J. (1965). *Can. J. Soil Sci.* **45**, 78–86.
Staple, W. J., and Lehane, J. J. (1954). *Can. J. Agr. Sci.* **34**, 329–342.
Stearns, N. D. (1927). U. S. Geol. Surv., Water-Supply Paper 596, pp. 121–176.
Stewart, G. L. (1962). Ph.D. Thesis, Washington State Univ., Pullman, Washington.
Stockinger, K. R., Perrier, E. R., and Fleming, W. D. (1965). *Soil Sci.* **100**, 124–129.
Swartzendruber, D. (1962a). *Soil Sci.* **93**, 22–29.

Swartzendruber, D. (1962b). *J. Geophys. Res.* **67**, 5205–5213.
Swartzendruber, D. (1963). *Soil Sci. Soc. Am. Proc.* **27**, 491–495.
Swartzendruber, D. (1964). *J. Geophys. Res.* **69**, 1676–1677.
Swartzendruber, D. (1966a). *Soil Sci. Soc. Am. Proc.* **30**, 7–11.
Swartzendruber, D. (1966b). *Advan. Agron.* **18**, 327–370.
Swartzendruber, D. (1967). *Proc. Intern. Symp. Soil Water, Prague, Czechoslovakia, 1967*, pp. 207–221. Czechoslovak Sci.-Tech. Soc. for Water Management, Prague.
Swartzendruber, D. (1968). *Soil Sci. Soc. Am. Proc.* **32**, 11–18.
Swartzendruber, D., and Olson, T. C. (1963). *Soil Sci. Soc. Am. Proc.* **27**, 108–110.
Takagi, S. (1960). *Soil Sci.* **90**, 98–103.
Tanner, C. B., and Elrick, D. E. (1958). *Soil Sci. Soc. Am. Proc.* **22**, 575–576.
Taylor, G. S. (1960). *Soil Sci.* **90**, 338–343.
Taylor, S. A. (1958). *Soil Sci.* **86**, 83–90.
Taylor, S. A., and Cary, J. W. (1964). *Soil Sci. Soc. Am. Proc.* **28**, 167–172.
Terzaghi, K. (1925a). "Erdbaumechanik," p. 127. Franz Deuticke, Vienna.
Terzaghi, K. (1925b). *Eng. News-Record* **95**, 832–836.
Thames, J. L. (1966). Ph.D. Thesis, Univ. of Arizona, Tucson, Arizona.
Tschapek, M. (1963). *Z. Pflanzenernaehr. Dueng. Bodenk.* **102**, 193–203.
van den Honert, T. H. (1948). *Discussions Faraday Soc.* **3**, 146–153.
von Engelhardt, W., and Tunn, W. L. M. (1955). *Illinois State Geol. Surv., Circ. 194.*
Waldron, L. J., McMurdie, J. L., and Vomocil, J. A. (1961). *Soil Sci. Soc. Am. Proc.* **25**, 265–267.
Watson, K. K. (1965). *Water Resources Res.* **1**, 577–586.
Watson, K. K. (1966). *Water Resources Res.* **2**, 709–715.
Wexler, A., and Brombacher, W. G. (1951). *Natl. Bur. Standards, Circ. 512.*
Whisler, F. D., and Klute, A. (1965). *Soil Sci. Soc. Am. Proc.* **29**, 489–494.
Wiegand, C. L., and Taylor, S. A. (1961). *Utah Agr. Exp. Sta. Spec. Rept. 15.*
Wilkinson, G. E., and Klute, A. (1962). *Soil Sci. Soc. Am. Proc.* **26**, 326–329.
Willis, W. O. (1960). *Soil Sci. Soc. Am. Proc.* **24**, 239–242.
Wilson, L. G., and Luthin, J. N. (1963). *Soil Sci.* **96**, 136–143.
Winterkorn, H. F. (1955). *Am. Soc. Testing Mater. Spec. Tech. Publ.* **163**, 27–35.
Winterkorn, H. F., ed. (1958). "Water and its Conduction in Soils," pp. 324–338. Highway Res. Board, Spec. Rept. 40.
Wyckoff, R. D., and Botset, H. G. (1936). *Physics* **7**, 325–345.
Wyckoff, R. D., Botset, H. G., Muskat, M., and Reed, D. W. (1933). *Rev. Sci. Instr.* **4**, 394–405.
Young, A., and Low, P. F. (1965). *Bull. Am. Assoc. Petrol. Geologists* **49**, 1004–1007.
Young, A., Low, P. F., and McLatchie, A. S. (1964). *J. Geophys. Res.* **69**, 4237–4245.
Youngs, E. G. (1957). *Soil Sci.* **84**, 283–290.
Youngs, E. G. (1960). *J. Geophys. Res.* **65**, 4025–4030.
Youngs, E. G. (1964). *Soil Sci.* **97**, 307–311.
Zaslavsky, D. (1964a). *Soil Sci.* **97**, 400–410.
Zaslavsky, D. (1964b). *Soil Sci.* **98**, 317–321.
Zemansky, M. W. (1957). "Heat and Thermodynamics," pp. 30–31, McGraw-Hill, New York.

7

Transient Free-Surface Flow in Porous Media

Daniel Dicker

College of Engineering
State University of New York at Stony Brook
Stony Brook, New York

1. Introduction

Free-surface flows in porous media is a branch of free-surface flow in fluid mechanics. The general problem has as its governing partial differential equation the Navier–Stokes equations and the complexity of this problem arises from the nonlinearity of both the governing differential equations and the boundary conditions. Even when the simplification is made that the viscosity is zero, the differential equation remains nonlinear. Cavity flow, a particular free-surface flow, has been of interest in hydrodynamics and engineering applications since Helmholtz formulated the basic steady-state problem about 100 years ago. In the steady-state case, since the free surface is a streamline and the speed, as can be shown, remains constant on the free surface, solutions can be effected by conformal mapping, at least for the two-dimensional case. In the transient problem, the situation is much more complicated and, at this time, there are no general solutions available. A formulation for a perturbation approach is given by Wang and Wu (1965), and Dicker (1966) has shown that if the free streamlines are perturbed about a steady-state value, the solution is given to order ϵ by the steady-state solution. For large changes, finite differences must be employed together with a digital computer. An example of this, the problem of flow under a sluice, was computed by Harlow *et al.* (1965) and Harlow and Welch (1965).

In porous-media flow, the Navier–Stokes equations are replaced by a linear partial differential equation, based, in part, on Darcy's experimental result. The overall boundary value problem, however, remains nonlinear due to the nonlinear boundary condition. Fortunately, however, in applications the flow is frequently essentially horizontal or essentially vertical so that suitable approximations are possible. It remains, however, to prove the validity of the approximation in each instance. In this chapter, after the basic problem has been formulated, several problems recently studied in these two classes will be presented. In Section 2 it will be shown that the classical derivation contains an heretofore undiscovered anomaly, and in Section 4, dealing with horizontal flows, a revised theory is presented which deals more realistically with the phenomenon. Also in Section 4 the Dupuit–Forchheimer equation will be rederived without the assumption that the velocity remains uniform, which is essential for the new theory. Before the equation for vertical flow is derived it will be shown that solutions to the one-dimensional boundary value problem exist and are unique under most circumstances. This digression is necessary since the resulting problem remains nonlinear and it is not easily established, a priori, that solutions are in fact unique. No such proof is necessary for the case of horizontal flow since the theory developed leads to a linear boundary value problem for which existence and uniqueness have previously been established and are well known.

2. Formulation of Boundary Value Problem of Free-Surface Flow

By use of the kinematical free-surface equation, the equation of continuity for an incompressible fluid, and the result known as Darcy's law, the differential equation describing the one-dimensional free surface will be derived.

The experimental result formulated by Henri Darcy (1856), *Darcy's law*, expresses a proportionality relationship between the filtration velocity v (also called the specific discharge when denoted by q) and the change in head $\partial h/\partial x_i$ in the direction of the velocity component. h is defined by $h = (P/\gamma) + z$ where γ is the specific weight of water and z is the vertical distance from some datum to the point in question. Thus,

$$v_{x_i} = K(\partial h/\partial x_i) \tag{1}$$

Equation (1) serves to define the constant of proportionality K, called the hydraulic conductivity. This constant is, of course, a function of fluid properties as well as those of the medium.

Following Muskat (1946) and Polubarinova-Kochina (1962), it is assumed that Darcy's law can be generalized to three dimensions, viz.,

$$u = -K(\partial h/\partial x), \qquad v = -K(\partial h/\partial y), \qquad w = -K(\partial h/\partial z) \tag{2}$$

for an isotropic medium. In recent works this assumption has been questioned, since, as may be shown, it implies that the permeability matrix $\{K_{ij}\}$ admits of a diagonal form. Although there appears to be no theoretical justification for this assumption, experimental evidence seems to indicate the existence of orthogonal principal axes for all samples which have been tested (Collins, 1961).

However, the experimental results represented by Eq. (2) are incomplete, inasmuch as they do not include the effects of inertia. This effect becomes more pronounced when large granules of a porous medium are considered, because the large pores enable the fluid passing through them to come under the influence of inertial forces. A separate study indicates that inertial terms are significant only for a very short time, of the order of a second, so that in the following analysis inertial effects are excluded. It is to be noted, however, that time effects enter via the moving boundary.

The equation of continuity for an incompressible fluid may be written in the form

$$\nabla \cdot \bar{q} = 0 \tag{3}$$

where

$$\bar{q} = u\mathbf{i} + v\mathbf{j} + w\mathbf{k} \qquad \text{and} \qquad \nabla = \mathbf{i}(\partial/\partial x) + \mathbf{j}(\partial/\partial y) + \mathbf{k}(\partial/\partial z)$$

Expressing Eq. (2) in the form

$$\bar{q} = K \operatorname{grad} h = \operatorname{grad} Kh$$

and substituting into (3) gives

$$\nabla \cdot (\operatorname{grad} Kh) = 0$$

Introducing the "potential" $\varphi \equiv Kh$, gives

$$(\partial^2\varphi/\partial x^2) + (\partial^2\varphi/\partial y^2) + (\partial^2\varphi/\partial z^2) = 0 \tag{4}$$

Hence the quantity Kh satisfies Laplace's equation.

Let Eq. (4) be satisfied in a region, shown in Fig. 1, whose boundaries

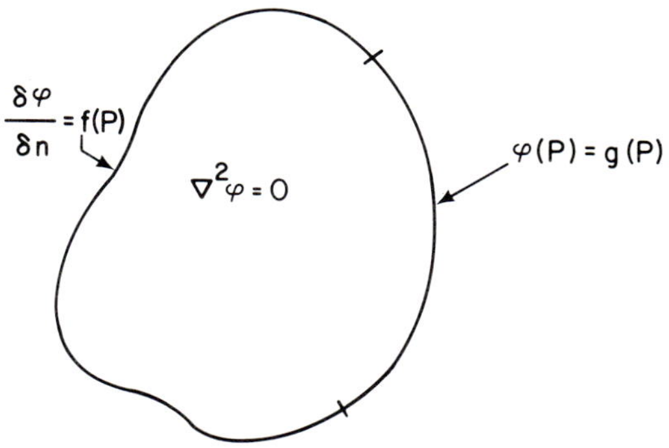

FIG. 1.

are fixed. Also, on one boundary segment let $\varphi(P) = g(P)$, and on the other segment let $\partial\varphi/\partial n = f(P)$, where P denotes any point on the boundary. This represents a basic problem in potential theory and is a well-posed boundary value problem, i.e., a solution exists and in unique. If one of the boundary segments is a free surface, however, the condition on it is of the form

$$G(\varphi, \varphi_x, \varphi_y) = 0 \tag{5}$$

with the condition on the other segment being either $\varphi(P) = g(P)$ or $\partial\varphi/\partial n = f(P)$. Condition (5) may not be sufficient to guarantee uniqueness for all such problems.

In the application of the above theory to flow through embankments, a major discrepancy in the analysis has apparently remained undiscovered up to the present time. To illustrate the error, consider the flow through a vertical embankment (Fig. 2) where the free-surface intercepts the seepage face some distance above the water surface in the reservoir. This problem is formulated and solved according to classical theory in Muskat's book (1946, pp. 303–318). If one investigates the boundary conditions normally taken for the problem on the seepage face, which is the face with which we are concerned, it is seen that the potential above the free liquid level is given by $\varphi = Kz$ and below the liquid surface the potential on the seepage face is given by φ = constant = KH.

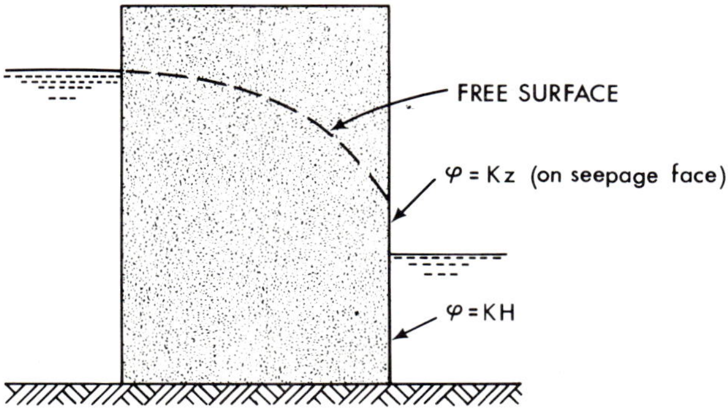

FIG. 2.

The remainder of the boundary conditions are irrelevant to the present discussion. Muskat has solved this problem and plotted various quantities such as a flow velocity on the seepage face. One of these curves will be used to substantiate the point about to be made. Consider the derivative of these potentials given on the seepage surface with respect to the vertical dimension, i.e., $\partial\varphi/\partial z$. It is seen immediately that the $\partial\varphi/\partial z$ above the liquid surface is a nonzero constant; on the other hand, $\partial\varphi/\partial z$ below the liquid surface is zero. But $\partial\varphi/\partial z$ is equal to the vertical velocity. Hence, there must be a discontinuity in the vertical velocity at the liquid surface, according to the boundary conditions as given. Of greater importance perhaps, is the result obtained by this boundary condition and given in Fig. 103 of Muskat (1946, p. 314), namely that the horizontal velocity over the seepage face exhibits a cusp at this liquid surface due to the singularity introduced in the boundary conditions. That is, we see that the horizontal velocity itself exhibits a similar singularity but in its derivative. This is more than a mathematical detail for it immediately raises the question of what is the relevance of this mathematical model to the real problem, inasmuch as the velocity distributions obtained therefrom must be dismissed on physical grounds. Is the model, in any sense, an approximation ? And why has not this inconsistency been uncovered before this time ? We shall comment only upon this last point. It is to be noted that in problems of flow through embankments, the quantity almost always sought is not the velocity or indeed the free-surface location, but simply the total flow, and the flow is obtained by integrating over the vertical dimension. Now the singularity itself is an integrable singularity so that a finite quantity results from the integration of the velocity over the seepage face. Moreover, it has been demonstrated by the Dupuit theory and other approximate theories [see, for example,

Muskat (1946, pp. 377–389)] that the discharge is given to a fair degree of accuracy by any one of a number of formulas, each upon a different theory. This is not surprising, for in engineering we frequently encounter similar examples. Consider, for instance, the bending of a beam. Even without knowing the exact stress distribution one can get quite accurate formulas for the deflection, since the latter depend upon integrated quantities which are related more to the average stress than to the exact distribution. A more pertinent example perhaps, concerns the shear stress distribution caused by the boundary layer flow over a flat plate. It is well known that according to classical theory, the local shear stress at the leading edge of a flat plate is infinite (Goldstein, 1965). More refined theories which tend to eliminate the singular behavior of the shear stress at the origin have been obtained, and one such theory is given by Boley and Friedman (1959), in which it is shown that even when a more exact theory is employed, whereby the local shear stress at the leading edge remains finite, the integrated shear stress differs but little from the more elementary theory which contains the singularity. There is, however, at least one class of problems in porous-media flow in which the free-surface location and velocity determination are important and the total flow relatively unimportant: stability of embankments. In Section 4, dealing with essentially horizontal flows, therefore, a theory will be given which, though approximate, does not possess the singular behavior of the current theory.

Existence and Uniqueness of Solutions

Before proceeding with solutions of various problems involving vertical flow, it will be shown that for the general one-dimensional boundary value problem, solution exist and are unique. In Section 3, the formulas developed previously will be specialized to the case of vertical flow. In this section, the resulting differential equation and the initial and boundary conditions will simply be listed. The only independent spatial variable is z, the vertical coordinate, and the time is given by t. The dependent variables are φ, the velocity potential ($\varphi = K \cdot h$ where h is equal to $p/\gamma + z$), and $\zeta(t)$, the distance to the free surface. In terms of these, the boundary value problem is given by (4) together with

$$\text{at} \quad z = 0, \quad \varphi = KH(t) \tag{6}$$

$$\text{at} \quad z = \zeta, \quad \varphi = K\zeta \tag{7}$$

$$\text{at} \quad z = \zeta, \quad w = n\dot{\zeta} \tag{8}$$

$$\zeta(t) \geqslant z \geqslant 0, \qquad t > 0$$

Assume that there can exist two solutions to the above problem where a solution comprises a function φ together with its associated $\zeta(t)$. One such solution will be designated as φ_1 and ζ_1 and the second solution as φ_2 and ζ_2. For the one-dimensional case, Laplace's equation becomes simply $d^2\varphi/dz^2 = 0$, the general solution of which, for the first case, is $\varphi_1 = A_1(t)\, z + B_1(t)$, and for the second case is $\varphi_2 = A_2(t)\, z + B_2(t)$. From the boundary condition (6) it follows that $B_1(t) = B_2(t) = KH(t)$. Next, if we differentiate each φ with respect to z and utilize boundary condition (8) together with $\partial\varphi/\partial z = w$, it is seen that $A_1(t) = n\dot{\zeta}_1$ and $A_2(t) = n\dot{\zeta}_2$. This, together with the remaining boundary condition (7), gives as the defining equations for the free surface $\epsilon\dot{\zeta}_1\zeta_1 + KH(t) = K\zeta_1$ and $\epsilon\dot{\zeta}_2\zeta_2 + KH(t) = K\zeta_2$. At this point, it is clear that any difference in the two φ's is the direct result of the difference of the two ζ's. Hence the existence and uniqueness will directly follow if it can be shown that the ζ's as given by these differential equations are themselves unique and that a solution exists. Since the two equations are identical except for the subscripts, we need concern ourselves only with the solution of either one of them. We will, therefore, drop the subscripts now and write the result as

$$n(d\zeta/dt) = K - [KH(t)/\zeta] \tag{9}$$

By direct integration we achieve alternatively the integral equation

$$n(\zeta(t) - \zeta(t_0)) = Kt + \int_{t_0}^{t} H(t')\, dt'/\zeta(t') \tag{10}$$

and by means of Picard's theorem, it can be proved that the solution to Eq. (10) exists, is unique and, in fact, continuous, provided that $H(t)$ is continuous and $\zeta(t_0) > 0$ (Wilf, 1962). It remains to account for the case where $H(t)$ is discontinuous, which will also allow for the inclusion of the initial condition $\zeta(t_0) = 0$. Consider the finite discontinuity in $H(t)$ shown in Fig. 3(a). $H(t)$ is presumed to be continuous to the left of the discontinuity as well as to the right. It is sufficient to consider constant values for $H(t)$ for short intervals of time on either side of the discontinuity; see Fig. 3(b). Rewriting the governing differential equation (9) for the two separate integrals gives

$$\zeta\, d\zeta/(K\zeta + H^*) = dt/n, \qquad t_1^* > t \geqslant t_0^* \tag{11}$$

$$\zeta\, d\zeta/(K\zeta + H^{**}) = dt/n, \qquad t^* \geqslant t > t_1^* \tag{12}$$

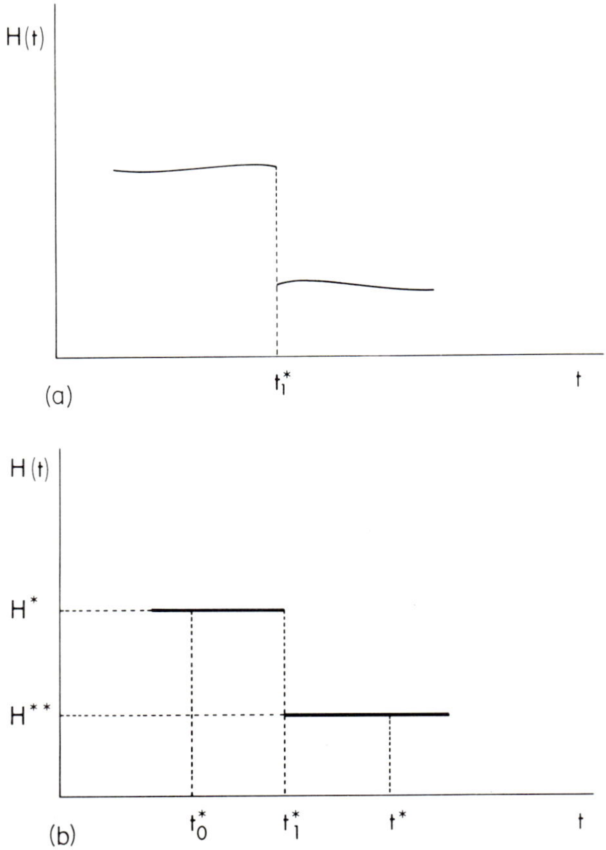

FIG. 3.

where the constants H^* and H^{**} are defined in Fig. 3(b). These equations can be solved exactly so that integrating Eq. (11) gives

$$\zeta + (H^*/K) \ln[(K\zeta + H^*)/(K\zeta_0 + H^*)] = (K/n)(t - t_0^*) \qquad (13)$$

where $\zeta_0 \equiv \zeta(t_0^*)$ and the solution for the second differential equation would be of similar form with H^* replaced by H^{**} and an appropriate change in the limits of integration.

It is easily established that ζ as given by Eq. (13) is an increasing function of t. Designate the solution of the first equation at $t = t_1^*$ as ζ^*. This then is the initial condition for the second integral and it follows that the solution for $t > t_1^*$ is unique, monotonic increasing, and obviously continuous everywhere, from which it is concluded that the problem presented in Eqs. (4), (6), (7), (8) is properly posed.

3. Essentially Vertical Flow†

In this section, certain one-dimensional transient problems are studied for which exact implicit solutions are obtained. Some of these have previously been obtained (Polubarinova-Kochina, 1946, p. 448). However, these have never been exploited for the analysis of seepage problems, the main concern of this section. Moreover, it turns out that the one-dimensional solutions obtained are limiting cases of two- and three-dimensional transient problems, so that there are immediate practical applications of this research, namely in the design of storm-water seepage basins and nuclear waste disposal. With regard to seepage-basin design, it is necessary that the basin be of sufficient dimensions to accommodate the largest anticipated storm runoff without overflowing. To this end, county design criteria, e.g., Suffolk and Nassau Counties on Long Island, New York specify a minimum ratio of basin volume to drainage area. This implies a steady-state condition which is actually almost never attained. Many such basins never contain more than a foot or two of water, although the depths provided are 12–20 ft. Evidently, they are grossly oversized, perhaps by an order of magnitude. A more realistic procedure would be one in which the transient nature of the phenomenon is included, for which the following analysis is pertinent (Section 3.1). With regard to the disposal of low-level wastes at nuclear power plants, it is important that the seepage paths and velocities of radioactive effluent be predictable. The initial motion of such effluent is primarily one dimensional since gravity is the generating force. Hence, the analysis presented herein should be of value in this area also.

3.1. The Equation of One-Dimensional Transient Flow

Consider a semi-infinite porous medium. The standard datum is taken at $z = 0$, z being measured positively downward, and the surface $z = 0$ is taken to be horizontal. The applied head $H(t)$ is measured positively upward from this surface, shown in Fig. 4.

The free surface $F(x, z, t) = 0$, which defines the fluid–gas interface, can be written in the form

$$F(x, z, t) = z - \zeta(x, t) = 0 \tag{14}$$

What is sought is a one-dimensional solution for which the free surface would be horizontal, i.e., $\zeta = \zeta(t)$, so that $F = z - \zeta(t) = 0$.

† Dicker and Sevi (1965).

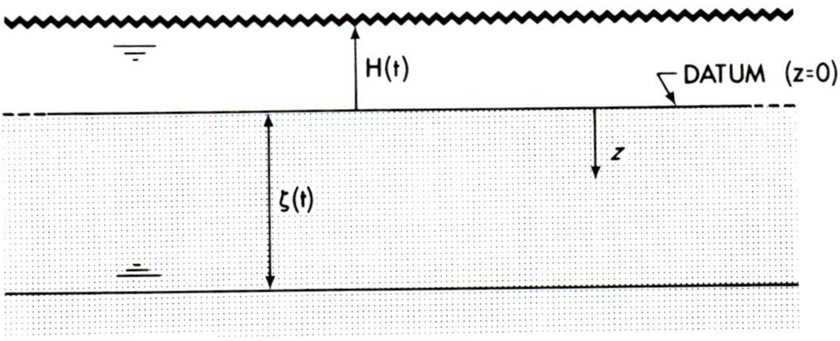

FIG. 4.

The kinematical free surface condition is

$$DF/Dt = 0 \tag{15}$$

where $D/Dt = (\partial/\partial t) + u(\partial/\partial x) + v(\partial/\partial y) + w(\partial/\partial z)$.

Application of (15) to (14) yields the boundary condition in which the subscript f denotes the free surface,

$$[\partial F(z, t)/\partial t] + w_f[\partial F(z, t)/\partial z] = 0 \tag{16}$$

at $z = \zeta$. Laplace's equation becomes, for $t > 0$ and $0 \leqslant z \leqslant \zeta(t)$,

$$\partial^2\varphi/\partial z^2 = 0 \tag{17}$$

the general solution of which is

$$\varphi(z) = Az + B \tag{18}$$

The head $h(z, t)$ is given by†

$$h = (P/\gamma) - z \tag{19}$$

and at $z = 0$,

$$h = P/\gamma \tag{20}$$

In proceeding to derive the transient differential equation, the applied head $h(0, t)$ is represented as $H(t)$, a function of time. From the equation of continuity it is seen that because $\bar{q} = w\mathbf{k}$, $w = w(z, t)$ reduces to $w = w(t)$.

Now the solution given by Eq. (18) is applicable and boundary condition (20) becomes

$$\varphi(0) = KH(t) \tag{21}$$

† The minus sign in (19) is consistent with the positive direction for z being downward.

Since $\varphi(0)$ is the value of the coefficient B, (18) becomes

$$\varphi(z) = Az + KH(t) \tag{22}$$

The filtration velocity may be defined in terms of a total discharge Q over a cross-sectional area A. Thus,

$$w = Q/A \tag{23}$$

A relationship between the filtration velocity, given above, and the free surface velocity, given in Eq. (16), will now be developed.

The porosity n of a given porous medium is defined as the ratio of open space, or voids V' through which the fluid passes, to the total volume of porous media V containing the voids. Thus,

$$n = V'/V = A'/A \tag{24}$$

where A is the total unit cross-sectional area and A' is the area of pores in this total area.

The free-surface velocity is, in terms of the total discharge Q and the pore area A',

$$\dot{\zeta} = Q/A' \tag{25}$$

In view of (24) and (23), the desired relationship between the filtration and free-surface velocities is

$$w = n\dot{\zeta} \tag{26}$$

Alternatively, substitution of (14) and (26) into (16) gives

$$(\partial/\partial y)[z - \zeta(t)] + (w/n)(\partial/\partial z)[z - \zeta(t)] = 0$$

or $w = n\dot{\zeta}$. Since w is independent of z, the free-surface velocity is equal to a constant multiplied by the velocity in the interior.

From (19), since at $z = \zeta$, $P = 0$ (using gage pressure),

$$h = -z = -\zeta$$

on the free surface. Hence at $z = \zeta$, (22) becomes

$$-K\zeta = A\zeta + KH(t) \qquad \text{and} \qquad A = \{K[\zeta + H(t)]\}/\zeta$$

Substitution into (22) gives, for the velocity potential,

$$\varphi = -\{K[\zeta + H(t)]/\zeta\}z + KH(t) \tag{27}$$

But $\partial\varphi/\partial z = -w = -n\dot{\zeta}$. Equation (27) becomes, upon differentiation with respect to z,

$$\partial\varphi/\partial z = -\{K[\zeta + H(t)]\}/\zeta$$

Elimination of $\partial\varphi/\partial z$ from these expressions yields

$$\zeta\dot{\zeta} - (K/n)\,\zeta = (K/n)\,H(t)$$

Equation (28) is the one-dimensional differential equation for transient flow through porous media. Rearranging gives

$$\zeta\, d\zeta/[\zeta + H(t)] = (K/n)\,dt \tag{28}$$

The physical significance of this one-dimensional approach will now be shown to be a limiting case of a two-dimensional problem.

Consider a rectangular channel as illustrated in Fig. 5, along with

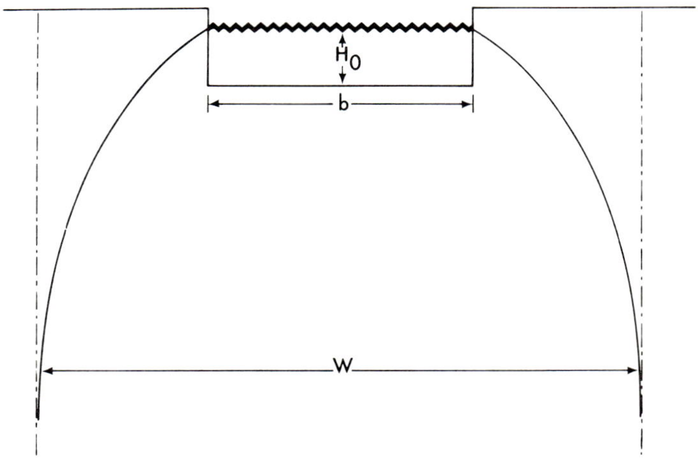

FIG. 5.

its steady-state free surface. Water is contained in this rectangular channel, with fixed dimensions, at a specified height H_0. The free surface is of necessity a function of the two spatial variables and approaches vertical asymptotes.

In the two-dimensional transient problem, the free surface is, of course, time dependent. The one-dimensional model has significance since the free streamlines for the two-dimensional steady state extend downward to a vertical asymptote. Hence W, the horizontal distance between the free streamlines at an infinite depth, is finite. Also, the difference $W - b$ is a function of H_0/b, and as $b \to \infty$ ($H_0/b \to 0$), $W \to b$ (Morel-Seytoux, 1961). Therefore, the one-dimensional problem can be considered an approximation to a two-dimensional problem having a large parameter, i.e., channel width to depth ratio. It is noted, however, that the problem of flow in a half-plane $z \geqslant 0$ where

$$H = H(t), \qquad |x| \leqslant b/2; \qquad H = 0, \qquad |x| > b/2$$

is not a one-dimensional problem. Again, the one-dimensional approximation is valid for large b/H, and becomes exact as $H/b \to 0$.

3.2. The Flow Balance Equation

An equation can be found relating the total inflow, the head $H(t)$, and the penetration depth $\zeta(t)$. Such an equation enables a specification of q; hence, an expression for $H(t)$, which appears in (28), may be found.

The flow balance equation can be stated as:

Total inflow = *Total fluid remaining above the level* $z = 0$
+ *Total fluid penetrated to the depth* $\zeta(t)$

This is, of course, the equation of continuity in integrated form. Alternatively, this may be written as

$$\int Q\, dt = AH(t) + A'\zeta(t) \tag{29}$$

Now by (23), the quantity Q/A is the filtration velocity. Another term sometimes used for this quantity is the specific discharge, defined by

$$q = Q/A \tag{30}$$

which is the inflow in cubic feet per second per unit of horizontal area of basin. Insertion of (30) and (24) into (29) gives the result

$$\int q\, dt = H(t) + n\zeta(t) \tag{31}$$

Equation (31) is the flow balance equation.

3.3. Solutions of the Transient Flow Equation

For the problems to be considered in Sections 3.3.1–3.3.4, the initial condition $\zeta(0) = 0$ will be assumed.

3.3.1. *Constant Head* $[H(t) = C]$

Under this assumption, (28) becomes

$$\zeta\, d\zeta/(\zeta + C) = (K/n)\, dt \tag{32}$$

the solution being

$$\zeta - C \log[(\zeta + C)/C] = (K/n)\, t \tag{33}$$

Equation (33) is plotted in Fig. 6 for a stated range of conductivities,

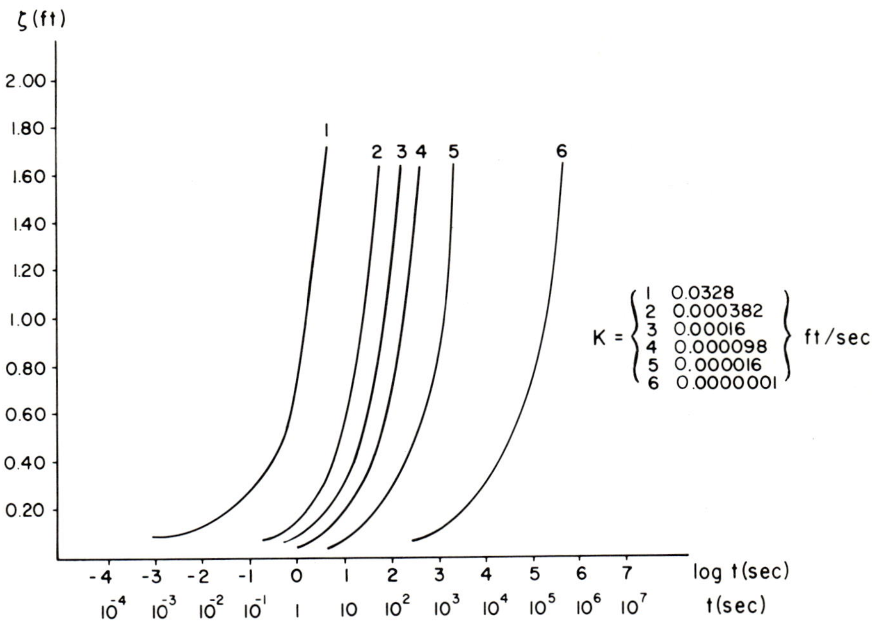

FIG. 6.

a porosity of $n = 0.4$, and constant head $C = 5$ ft. The conductivities listed along with Fig. 6 are for soils that range from coarse sands to tightly packed fine clays.

3.3.2. *Constant Specific Discharge* ($q =$ const)

Under this assumption, (31) becomes

$$H(t) = qt - n\zeta(t) \tag{34}$$

Substitution of (34) into (28) yields

$$\zeta\, d\zeta/[(1 - n)\, \zeta + qt] = (K/n)\, dt$$

Assuming a solution of the form $\zeta(t) = At$ where A is a constant, the following results

$$\zeta(t) = \tfrac{1}{2}\{(K/n)(1 - n) + \{k^2[(1 - n)/n]^2 + (AK/n)\, q\}^{1/2}\}t \tag{35}$$

ζ is thus a linear function of time. We note that substitution of $\zeta(t) = At$ into (34) shows that the same form of solution (35) holds for $H(t) = \text{const} \times t$ as for $q = \text{const}$.

3.3.3. $H(0) = H_0$, $q = 0$, $t \geqslant 0$, *and a Determination of the Total Time for Drainage*

The given condition on q implies a sudden stopping of flow. This means that an initial head of, say, size H_0, remains. Hence (31), for later times, becomes

$$H_0 = H(t) + n\zeta(t)$$

and substitution into (28) gives

$$\zeta\, d\zeta/[(1 - n)\, \zeta + H_0] = (K/n)\, dt$$

Integration gives the solution

$$\frac{1}{1 - n}\left\{(1 - n)\, \zeta - H_0 \log\left[\frac{H_0 + (1 - n)\, \zeta}{H_0}\right]\right\} = \frac{K}{n}\, t \tag{36}$$

The basin is completely drained when $H(t) = 0$. Hence (31) becomes

$$n\zeta = H_0$$

Substituting for ζ into (36) gives the time for complete drainage. Thus,

$$t = \frac{nH_0}{K(1 - n)^2}\left[\frac{1 - n}{n} + \log n\right] \tag{37}$$

3.3.4. $H(t) = \text{const} \times t = Ct$, $0 \leqslant t \leqslant t_0$; $H(t) = \text{const} = Ct_0$, $t_0 < t < t_1$; $H(t_1) = Ct_0$, $q = 0$, $t > t_1$

An application of this problem will be described in Section 3.4, the discussion of a "design storm." As previously noted the specification $H(t) = Ct$ corresponds to a constant q. Also, in view of (31) and (33), the specification $H(t) = Ct_0$ corresponds to a transcendentally decreasing q. The plot of q versus t is shown in Fig. 7.

For the interval $0 \leqslant t \leqslant t_0$, substitution of $H(t) = Ct$ into (28) yields

$$\zeta\, d\zeta/(\zeta + Ct) = (K/n)\, dt$$

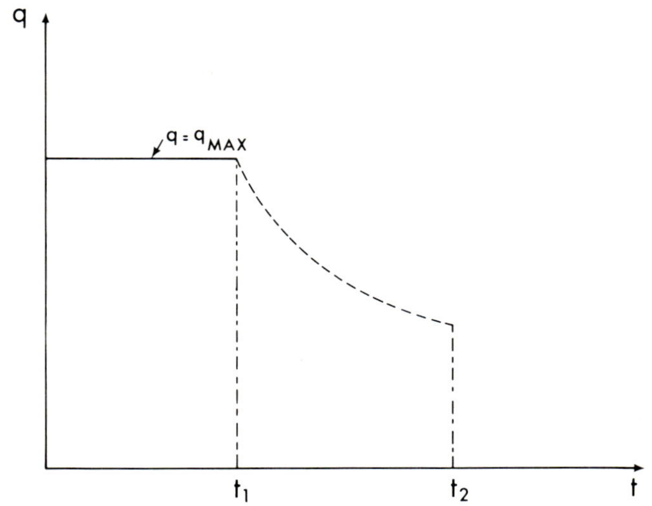

FIG. 7.

The solution being

$$\zeta(t) = \tfrac{1}{2}\{(K/n) + [(K/n)^2 + (4CK/n)]^{1/2}\}t \tag{38}$$

For the interval $t_0 < t \leqslant t_1$, the solution is obtained as before for a constant head, with the initial condition being (38) at $t = t_0$. Thus,

$$\zeta + Ct_0 \log \left\{ \frac{[(a/2) + C]\, t_0}{\zeta + Ct_0} \right\} = \frac{K}{n} t + \frac{b}{2} t_0 \tag{39}$$

where

$$a = \frac{K}{n} + \left[\left(\frac{K}{n}\right)^2 + \frac{4CK}{n}\right]^{1/2}, \qquad b = -\frac{K}{n} + \left[\left(\frac{K}{n}\right)^2 + \frac{4CK}{n}\right]^{1/2}$$

For $t > t_1$, Eq. (31) becomes

$$H(t_1) + n\zeta(t_1) = H(t) + n\zeta(t) \tag{40}$$

where $\zeta(t_1)$ is the initial condition obtained from (39) as an initial condition at the point t_1. Substitution into (28) yields

$$\frac{\zeta \, d\zeta}{(1 - n)\, \zeta + \forall} = \frac{K}{n} \, dt$$

where $\forall = H(t_1) + n\zeta(t_1)$. Integration yields the solution

$$\frac{1}{(1-n)^2}\left\{(1-n)[\zeta - \zeta(t_1)] - \forall \log\left[\frac{\forall + (1-n)\,\zeta}{\forall + (1-n)\,\zeta(t_1)}\right]\right\} = \frac{K}{n}(t - t_1) \tag{41}$$

By means of Eq. (38), (39), and (41) one may easily obtain the overall $\zeta(t)$ as a function of $q(t)$ where the latter is given in Fig. 7.

3.3.5. *Arbitrary Initial Condition on ζ with q = Const*

In the previous section, solutions in the functional form $t = t(\zeta)$ were obtained. For the arbitrary initial condition on the penetration depth, however, both ζ and t are given implicitly. This is shown next.

The arbitrary initial condition is taken to be $\zeta(0) = \zeta_0$, and the initial head to be H_0.

Equation (31) reads, for the given conditions,

$$H(t) + n\zeta(t) = qt + H_0 + n\zeta_0 = qt + \forall_0$$

Substitution into (28) for $H(t)$ gives

$$\frac{\zeta\, d\zeta}{(1-n)\,\zeta + qt + \forall_0} = \frac{K}{n}\,dt$$

Letting $u = t + a$ and setting $a = \forall_0/q$ gives

$$\frac{\zeta\, d\zeta}{(1-n)\,\zeta + qu} = \frac{K}{n}\,du \tag{42}$$

(42) can be made separable by means of change of variables given by $\zeta = xu$, whereby the differential equation becomes

$$\frac{du}{u} = -\frac{x\,dx}{x^2 - (K/n)(1-n)\,x - (K/n)\,q} \tag{43}$$

Integration, use of the initial condition, and substitution back for $\zeta(t)$ provide the result

$$\left(\frac{t+a}{a}\right)\left[\frac{\left(\frac{\zeta}{t+a}\right)^2 - b\,\frac{\zeta}{t+a} - \frac{K}{n}\,q}{\left(\frac{\zeta_0}{a}\right)^2 - b\,\frac{\zeta_0}{a} - \frac{K}{n}\,q}\right]^{1/2} \times \left[\frac{\left(\frac{2\zeta}{t+a} - b - d\right)\left(\frac{2\zeta_0}{a} - b + d\right)}{\left(\frac{2\zeta}{t+a} - b + d\right)\left(\frac{\zeta}{a} - b - d\right)}\right]^{b/(2d)} = 1 \tag{44}$$

where

$$b = (K/n)(1 - n), \qquad d = \{[(1 - n)/n]^2 K^2 + (4K/n) q\}^{1/2}$$

This result for an arbitrary initial condition and constant q may also be used to solve a problem for which an arbitrary $q(t)$ curve is given, by approximating the curve in small intervals by lines of constant q. Note that (41) must be rewritten for a nonzero initial instant, i.e., for $\zeta(t_1) = \zeta_1$. It is more convenient in such cases to work directly with (28) and (31) in finite difference form. Equation (28) becomes

$$\Delta\zeta = \frac{K}{n}\left[\frac{\zeta + H(t)}{\zeta}\right]\Delta t \tag{45}$$

For the given initial conditions, (31) is written as

$$H(t) + n\zeta(t) = H_0 + n\zeta_0 q_i{}'t \tag{46}$$

where q_i' is the average value of q in the subinterval $(t_i, t_i + 1)$ of the $q(t)$ curve. By use of (45) and (46), a finite difference solution may be constructed.

3.3.6. *Dimensionless Variables and the Dimensionless Free-Surface Velocity*

Several of the results previously obtained lend themselves to analysis by means of dimensionless variables. If we let

$$\xi = \zeta/H \tag{47}$$

and

$$\tau = (K/nH)\,t \tag{48}$$

the differential equation (32) becomes, with $H = C$,

$$\xi\, d\xi/(\xi + 1) = d\tau \tag{49}$$

The solution is, with $\xi(0) = 0$,

$$\xi - \log(\xi + 1) = \tau \tag{50}$$

A graph of (50) appears in Fig. 8.

Differentiation of (50) with respect to τ yields the dimensionless free surface velocity

$$d\xi/d\tau = (\xi + 1)/\xi \tag{51}$$

A graph of $d\xi/d\tau$ versus τ, obtained using (51) and (50), is shown in Fig. 9.† From Fig. 9, it is seen that $d\xi/d\tau$ is infinite at the initial instant.

† This discussion follows Polubarinova-Kochina (1962).

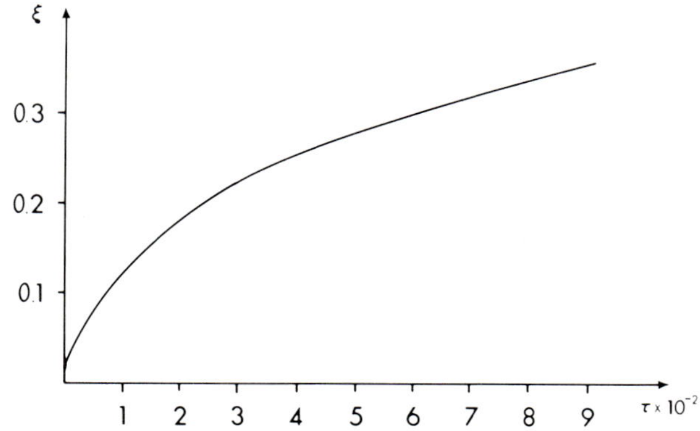

FIG. 8.

Since this is not physically realizable, it must be concluded that either the analysis is incomplete or that the physical model is incorrect. Recent investigations have shown that the singularity at the origin ($\tau = 0$) is removed when inertial terms are included in the differential equation. However, for times other than the initial instant, the inertia effects decay rapidly. The model chosen in this study is, therefore, valid except for the aforementioned minor limitation.

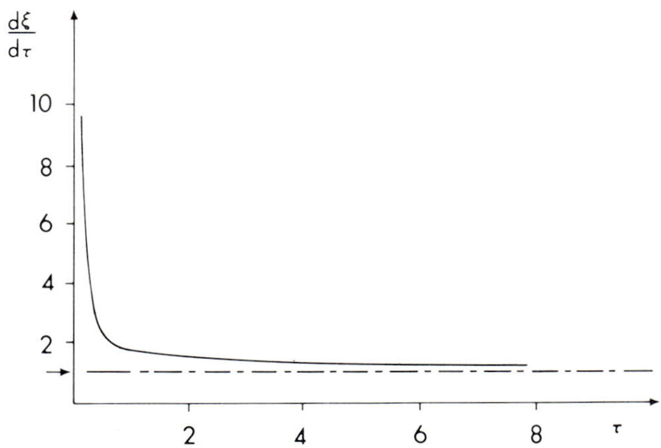

FIG. 9.

3.4. A DESIGN STORM FOR GIVEN PERTINENT PARAMETERS

Up to this point, Eq. (28) has been solved with the aid of (31) for specified $H(t)$ and/or $q(t)$ curves. In this section it will be shown that

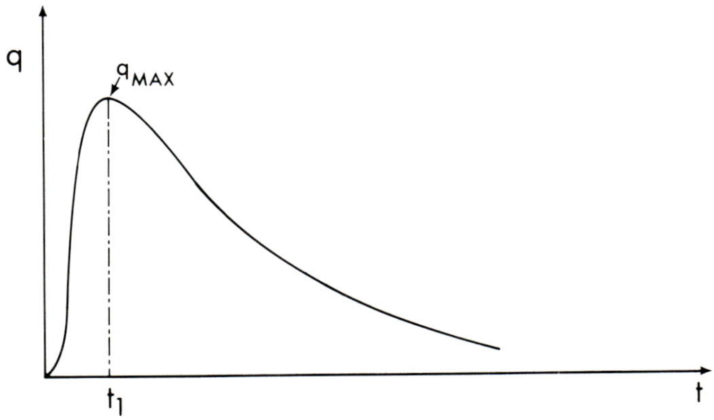

FIG. 10.

the $q(t)$ and $H(t)$ curves can be designed to correspond to the following data supplied for a storm: (a) the maximum intensity, $q_{\max}$; (b) the total inflow, $\int q\,dt \equiv \forall$; and (c) the total duration, t_2 ($q = 0$, $t \geqslant t_2$).

Consider the curve of a typical storm, shown in Fig. 10. After a short time, the intensity reaches a maximum value and then tapers off gradually. This actual $q(t)$ curve may be approximated by a constant q, $q_{\max}$, in the interval $0 \leqslant t \leqslant t_1$, some transcendentally decreasing $q(t)$ [see discussion of Eqs. (38)–(41)] in $t_1 < t \leqslant t_2$, and $q(t) = 0$ for $t > t_2$.

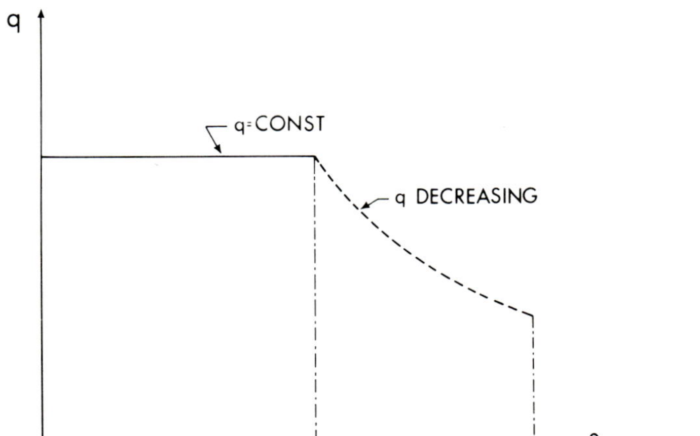

FIG. 11.

This is shown in Fig. 11. It should be noted that, whereas the response would probably not vary much with any of several different approximations, the possibility of analytical computations depends upon being able to provide a specific type of approximation, hence Fig. 11.

In view of the discussion of Eqs. (35) and (38)–(41), the $H(t)$ curve may be constructed as shown in Fig. 12.

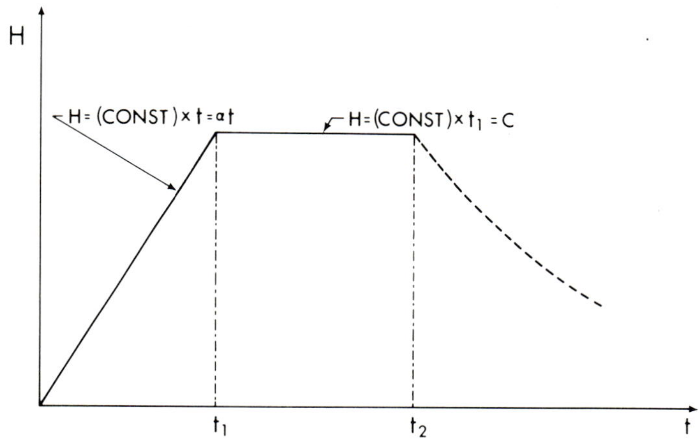

FIG. 12.

The following quantities must be obtained; (1) the depths of penetration in each interval; (2) the expression for the $q(t)$ curve, in the second interval; (3) the complete $H(t)$ curve; and (4) the value of t_1.

By use of (31), Eq. (28) becomes, in each interval:

in $0 \leqslant t \leqslant t_1$,

$$\frac{\zeta\, d\zeta}{(1-n)\,\zeta + q_{\max}t} = \frac{K}{n}\, dt \tag{52}$$

in $t_1 < t \leqslant t_2$,

$$\frac{\zeta\, d\zeta}{\zeta + C} = dt \tag{53}$$

in $t > t_2$,

$$\frac{\zeta\, d\zeta}{(1-n)\,\zeta + \forall} = \frac{K}{n}\, dt \tag{54}$$

where $\forall = H(t_2) + \zeta(t_2) = C + \zeta(t_2)$ with C yet to be determined.

Introduction of (41) and (48) into (52), (53), and (54) gives

$$\frac{\xi\, d\xi}{\xi + (q_{\max}/K)[n/(1+n)]\,\tau} = (1-n)\,d\tau \tag{55}$$

$$\frac{\xi\, d\xi}{\xi + 1} = d\tau \tag{56}$$

$$\frac{\xi\, d\xi}{\xi + (1-n)^{-1}\,[1 + n\xi(\tau_2)]} = (1-n)\,d\tau \tag{57}$$

The solutions are, with the initial condition $\xi(0) = 0$,

$(0 \leqslant \tau \leqslant \tau_1)$:

$$\xi(\tau) = \tfrac{1}{2}\{(1-n) + [(1-n)^2 + (4q_{\max}/K)\,n]^{1/2}\}\,\tau \tag{58}$$

$(\tau_1 < \tau \leqslant \tau_2)$:

$$\xi(\tau) + \log\{(a\tau_1 + 1)/[\xi(\tau) + 1]\} = \tau + \tau_1(a-1) \tag{59}$$

where a is the coefficient of τ in (58),

$(\tau > \tau_2)$:

$$\xi(\tau) - \xi(\tau_2) + (1-n)^{-1}\,[1 + n\xi(\tau_2)]\log\left\{\frac{(1-n)^{-1}\,[1 + \xi(\tau_2)]}{(1-n)^{-1}\,[1+\xi(\tau_2)] + \xi}\right\} = (1-n)(\tau - \tau_2) \tag{60}$$

Equations (58), (59), and (60) specify the $\xi(\tau)$ curve.

In order to solve for $q(t)$ in $t_1 < t \leqslant t_2$, we introduce the dimensionless variables (47) and (48) into the flow balance equation (31) to obtain

$$n\xi(\tau) = n\xi(\tau_1) + \int (n/K)\,q(\tau)\,d\tau$$

Differentiation with respect to τ yields

$$d\xi/d\tau = q(\tau)/K$$

But from (56),

$$d\xi/d\tau = (\xi + 1)/\xi$$

Hence, in $t_1 < t \leqslant t_2$,

$$q(\tau) = K[(\xi + 1)/\xi] \tag{61}$$

Thus, knowing ξ as a function of τ from (59), (τ, as well as C, will subsequently be found) conversion back to dimensional variables will give ζ as a function of t. Substituting (47) into (61) gives $q(t)$ in terms of ζ. Hence, the $q(t)$ curve is obtained.

A flow balance in the interval $0 \leqslant t \leqslant t_1$ gives

$$H(t) + n\zeta(t) = \int q_{\max}\, dt = q_{\max} t \tag{62}$$

Also, in $0 \leqslant t \leqslant t_1$, $H(t) = \alpha t$, and the dimensional solution is $\zeta(t) = At$, A being the coefficient of t in (35), with $q = q_{\max}$. Substituting these results into (62) gives for α,

$$\alpha = q_{\max} - nA \tag{63}$$

or

$$H(t) = (q_{\max} - nA)\, t, \qquad 0 \leqslant t \leqslant t_1 \tag{64}$$

In order to find $H = C$, note that if the total inflow $\forall$ is given, reference to Fig. 11 yields

$$q_{\max} t_1 + \int_{t_1}^{t_2} q(t)\, dt = \forall$$

Substitution of (48) and division by $H(=C)$ gives

$$n(q_{\max}/K)\, \tau_1 + \int_{\tau_1}^{\tau_2} (n/K)\, q(\tau)\, d\tau = \forall/H \tag{65}$$

A flow balance at t_2 gives

$$n\zeta(t_2) + H(t_2) = n\zeta(t_1) + H(t_1) + \int_{t_1}^{t_2} q(t)\, dt$$

Introduction of (47) and (48) into the above yields, since

$$H(t_2) = H(t_1) = C$$

$$n[\xi(\tau_2) - \xi(\tau_1)] = \int_{\tau_1}^{\tau_2} (n/K)\, q(\tau)\, d\tau \tag{66}$$

Substitution of (66) for the integral in (65) gives

$$n[(q_{\max}/K)\, \tau_1 + \xi(\tau_2) - \xi(\tau_1)] = \forall/H$$

However, since $H = C$,

$$H(t) = C = \forall/\{n[(q_{\max}/K)\, \tau_1 + \xi(\tau_2) - \xi(\tau_1)]\} \tag{67}$$

Equation (67) gives $H(t)$ for $t_1 < t \leqslant t_2$. Finally, in $t > t_2$, (31) is

$$n\zeta(t) + H(t) = n\zeta(t_2) + H(t_2)$$

Divide by $H = C = H(t_2)$ to get

$$n\xi(\tau) + H(t)/C = n\xi(\tau_2) + 1$$

or

$$H(t) = C[n(\xi(\tau_2) - \xi(\tau)) + 1] \tag{68}$$

Equations (64), (67), and (68) completely specify the $H(t)$ curve.

The remaining parameter t_1 is found by using (48). Thus,

$$H = C = \alpha t_1 = \alpha[(nH/K)\,\tau_1] = \alpha(nC/K)\,\tau_1 \tag{69}$$

from which

$$C = (\alpha nC/K)\,\tau_1 \qquad \text{or} \qquad \tau_1 = K/(\alpha n) \tag{70}$$

To find the dimensional time t_1, we obtain from (69), $C = \alpha t_1$ or

$$t_1 = C/\alpha \tag{71}$$

Substitution of (67) and (63) into (71) gives

$$t_1 = \forall/\{n(q_{\max} - nA)[(q_{\max}/K)\,\tau_1 + \xi(\tau_2) - \xi(\tau_1)]\} \tag{72}$$

Knowing $\xi(\tau_1)$, $\xi(\tau_2)$, and τ_1, t_1 may be evaluated.

With the expressions for the penetration depth and the specification of the $q(t)$ and $H(t)$ curves, the design problem has been solved.

3.5. Impermeable Barrier at a Finite Depth

All of the preceding discussion dealt with a semi-infinite porous medium. In this section, a differential equation which describes the behavior of the free surface $\zeta(t)$ passing through a medium of bounded depth will be derived, the boundary being impermeable. This physical problem is illustrated in Fig. 13, the finite depth being labeled ζ^*. Examples of such a model may be found abundantly in nature.

Following the development for the semi-infinite medium, it is seen that Laplace's equation is again applicable, the general solution of which is given, in the one-dimensional case, by Eq. (18),

$$\phi = Az + B$$

Boundary conditions are needed to evaluate A and B.

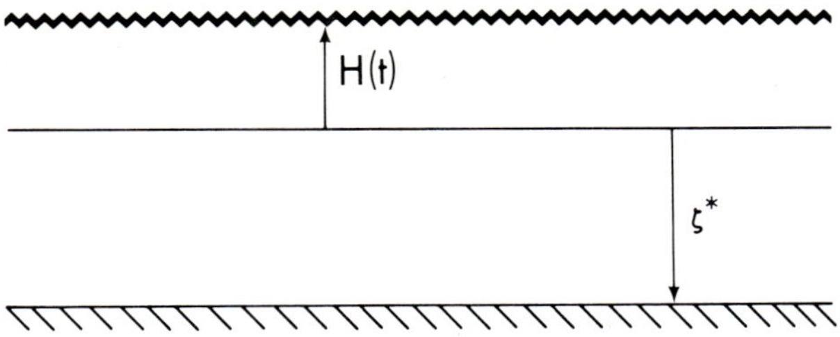

FIG. 13.

Now from (19), at $z = 0^-$ (i.e., approaching $z = 0$ from negative values),

$$h = P(0^-, t)/\gamma$$

Since $\phi = Kh$,

$$\phi(0) = K[P(0^-, t)/\gamma] = B$$

For the ensuing development, the absolute pressure rather than the gage pressure must be used. Thus, at $z = 0^-$,

$$P(0^-, t) = P_{\text{gage}} + P_{\text{at}} = \gamma H(t) + P_{\text{at}} \tag{73}$$

Hence,

$$B = K[H(t) + (P_{\text{at}}/\gamma)] \tag{74}$$

Now at $z = \zeta$,

$$h = [P(\zeta, t)/\gamma] - \zeta$$

Thus, since $\phi = Kh$, use of (18) and (74) at $z = \zeta$ gives

$$K\{[P(\zeta, t)/\gamma] - \zeta\} = A\zeta + K[H(t) + (P_{\text{at}}/\gamma)] \tag{75}$$

To evaluate A it is necessary that $P(\zeta, t)$ be known. Unlike the problems previously studied, the pressure at the interface is not equal to a constant; the thermodynamics of the air-compression process must be taken into consideration.

If it is assumed that the air trapped between the free surface and the impenetrable barrier is compressed in a quasi-steady process, and that the thermodynamic behavior of air may be approximated by an ideal gas, the equation

$$PV = mRT \tag{76}$$

may be used. P is the pressure, V the volume of compressed air, m the number of moles of gas, R the universal gas constant, and T the absolute temperature. Further, if the process is assumed isothermal, the pressure and volume at the initial instant when $\zeta = 0$ may be related to that at an instant when $\zeta \neq 0$, i.e., from (76),

$$P(0^+, t_0)\, V(0^+, t_0) = P(\zeta, t)\, V(\zeta, t) \tag{77}$$

Note that the value of $P(0^+, t_0)$ in (71) is just the pressure of the enclosed mass of air, here taken as atmospheric, whereas in (73), $P(0^-, t)$ represents the pressure exerted by the air and water above the surface $z = 0$. Naturally, for free-surface motion, there may be a discontinuity in pressure at the level $z = 0$, $t < 0$.

If a unit area is considered, $V(0^+, t) = \zeta^*$ and $V(\zeta, t) = \zeta^* - \zeta$. Hence (77) becomes

$$P_{\text{at}}\zeta^* = P(\zeta, t)(\zeta^* - \zeta)$$

Solving for $P(\zeta, t)$ gives

$$P(\zeta, t) = P_{\text{at}}\zeta^*/(\zeta^* - \zeta) \tag{78}$$

Substitution of (78) into (75) yields, upon solution for A,

$$A = \frac{K}{\zeta}\left[\frac{P_{\text{at}}}{\gamma}\frac{\zeta^*}{\zeta^* - \zeta} - \zeta - H(t) - \frac{P_{\text{at}}}{\gamma}\right] \tag{79}$$

Substitution of (79) and (74) into (18) gives the result

$$\phi = \frac{K}{\zeta}\left[\frac{P_{\text{at}}}{\gamma}\frac{\zeta^*}{\zeta^* - \zeta} - \zeta - H(t) - \frac{P_{\text{at}}}{\gamma}\right] z + K\left[H(t) + \frac{P_{\text{at}}}{\gamma}\right] \tag{80}$$

Now by (26),

$$\partial\phi/\partial z = -w = -n\dot{\zeta}$$

Hence, substitution for $\partial\phi/\partial z$, obtained from (80), gives

$$-n\dot{\zeta} = \frac{K}{\zeta}\left[\frac{P_{\text{at}}}{\gamma}\left(\frac{\zeta^*}{\zeta^* - \zeta}\right) - \zeta - H(t) - \frac{P_{\text{at}}}{\gamma}\right]$$

Some rearrangement yields the result

$$\frac{(\zeta - \zeta^*)\,\zeta\, d\zeta}{\zeta^2 - [\zeta^* - H(t) - (P_{\text{at}}/\gamma)]\,\zeta - \zeta^* H(t)} = \frac{K}{n}\, dt \tag{81}$$

Equation (81) will be called the *barrier flow equation.*

Solution of the Barrier Flow Equation with Constant $H(t)$

For $H(t) = \text{const} = C$, Eq. (81) becomes

$$\frac{(\zeta - \zeta^*)\, \zeta\, d\zeta}{\zeta^2 - [\zeta^* - C - (P_{\text{at}}/\gamma)]\, \zeta - C\zeta^*} = \frac{K}{n}\, dt \tag{82}$$

Integration, with $\zeta(0) = 0$, gives

$$\zeta + (r_2 - r_1)^{-1} \{[(a - \zeta^*)\, r_2 + \beta] \log|(\zeta - r_2)/r_2| - [(\alpha - \zeta^*)\, r_1 + \beta] \log|(\zeta - r_1)/r_1|\} = \frac{k}{n} t \tag{83}$$

where

$$\alpha = \zeta^* - C - (P_{\text{at}}/\gamma), \qquad \beta = C\zeta^*, \qquad r_1, r_2 = \tfrac{1}{2}[\alpha \pm (\alpha^2 + 4\beta)^{1/2}]$$

Because $\alpha^2 + 4\beta > 0$, r_1 and r_2 are real.

Equation (83) is the solution to the barrier flow equation for $H(t) = \text{const}$. The maximum penetration depth is given by r_1.

From substitution of the dimensionless variables $\xi = \zeta/\zeta^*$ and $\tau = Kt/n\zeta^*$ into (82) and integration, there results the solution

$$\xi + \frac{1}{R_2 - R_1} \left\{\left[\left(\frac{\alpha}{\zeta^*} - 1\right) R_2 + \frac{\beta}{\zeta^{*2}}\right] \times \log\left|\frac{\xi - R_2}{R_2}\right| - \left[\left(\frac{\alpha}{\zeta^*} - 1\right) R_1 + \frac{\beta}{\zeta^{*2}}\right] \log\left|\frac{\xi - R_1}{R_1}\right|\right\} = \tau \tag{84}$$

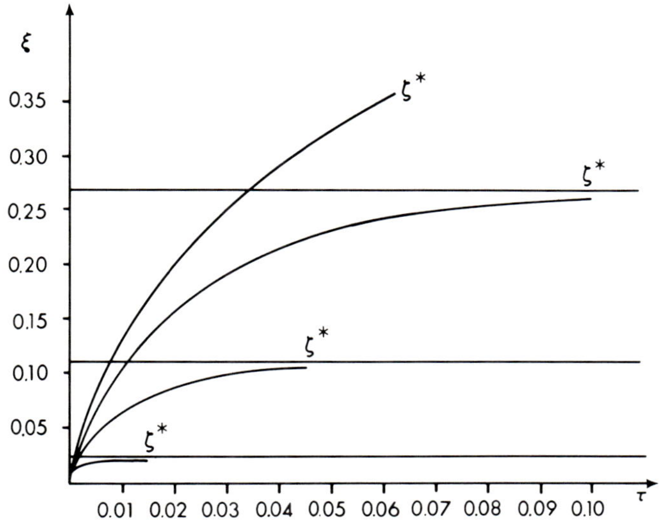

FIG. 14.

where $R_1 = r_1/\zeta^*$ and $R_2 = r_2/\zeta^*$. Here R_1 is the maximum value of the dimensionless penetration depth. Plots of (84) for several values of ζ^* are shown in Fig. 14. For a proper comparison to the ζ/τ plot shown in Fig. 6, the dimensionless variables used are those given by (47) and (48). These are easily obtained from the present dimensionless variables by multiplication by ζ^*/H. In Fig. 14, the asymptotic behavior of ξ ($\tau \to \infty$) is shown, along with the comparison to $\zeta^* = \infty$, i.e., no finite barrier.

4. Essentially Horizontal Flows

Horizontal flows constitute by far a much greater proportion of problems of interest in engineering applications than do vertical flows, and for this reason there have been extensively studied to a greater degree than the latter over the years. Examples of horizontal flows would be flows into wells, flows in embankments, irrigation problems, and flows in leaky acquifers. Here, as in Section 3, the exact theory comprises Laplace's equation, plus an appropriate boundary condition on the free surface, as well as the conditions of known potential or normal gradient on the other boundaries. Because these are multi-dimensional, general solutions of the transient problem are not available for all cases. In fact, they are available only for very exceptional cases, with the result that the transient flow through an embankment, for example, has never been completely solved. De Wiest (1962), utilizing conformal mapping, together with a perturbation procedure, obtained an asymptotic solution for a two-dimensional transient problem. The procedure for extending the range of validity of the solution was presented by Dicker (1962). These methods, however, involve the solution of secular equations of large degree and the solutions are not valid for very early times. The problem, therefore, as studied by most investigators, has been one in which the differential equation and boundary conditions have been recast by suitable approximations into a form whereby the free surface itself becomes the variable, and the resultant equation, usually the "heat" equation, can be solved to yield an approximate solution. This is commonly referred to as the Dupuit–Forchheimer theory. Muskat, however, takes severe issue with this theory (Muskat, 1946, p. 359) because of the nature of the approximations. Nevertheless, the theory has remained a popular one primarily because of the ease of application, the philosophy being apparently that even a poor solution is better than no solution.

In this section, the anomalous boundary condition previously described is replaced with one which does not exhibit the inconsistency. In addition, the Dupuit–Forchheimer differential equation is rederived on a much more rigorous basis in order to overcome the objections raised by Muskat. Utilizing these refinements, a new theory is presented which is then used to explain sloughing of embankments in which the primary cause is the hydrodynamics of the flow in the embankment rather than the idea currently accepted in soil mechanics that failure is due to structural instability.

4.1. Description of Problem

The geophysical problem of sloughing of embankments and landslides, and the engineering design of earth dams are to a large degree related to the manner in which water in the earthen mass flows. Most failures, in fact, occur when an embankment is saturated and the region on one face or the other is void of water. The latter condition can occur in several ways: the water in a reservoir may be suddenly lowered, and because of large friction in the pores of the earthen dam, drainage occurs slowly; or a heavy rainfall may saturate a hill creating a similar situation. The study of the external forces and internal stresses in an earthen mass at failure is within the realm of soil mechanics. However, since failure is closely connected with (or, perhaps, caused by) the hydrodynamic behavior, it is important that the hydrodynamic analysis correctly represent the physical phenomenon. To this end a particular problem is investigated—that of rapid drawdown. Though this problem has been studied by others it is the author's belief that the problem has not yet been properly posed, with the result that the solution presented are of little value in the failure problem described.

4.2. Mathematical Statement of the Problem. Modified Hydraulic Theory

Consider a vertical embankment, and as a convenience, assume that it is founded on an impermeable base and that its face is vertical (Fig. 15). At some time, say $t = 0$, the water level in the reservoir is suddenly lowered to H_2. The problem is to find the location of the free surface for all times $t > 0$. Once this is known the water pressure may be determined anywhere in the embankment from hydrostatics, since the velocity head, being very small, may be neglected.

The basic formula governing flow in porous media is, as given earlier,

$$v_{x_i} = -K(\partial h/\partial x_i) \qquad [1]$$

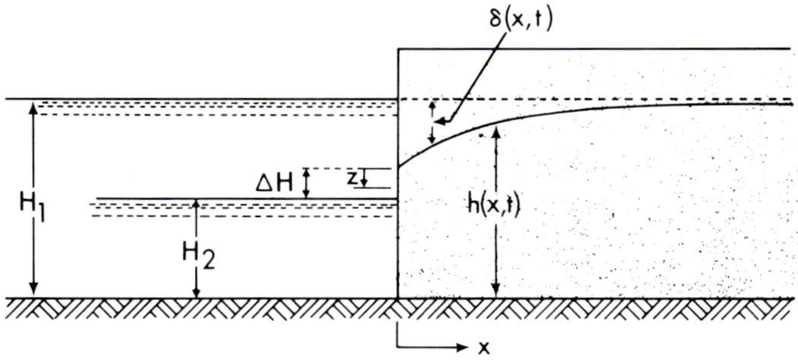

FIG. 15.

From this point onward one may proceed via the "exact," or hydrodynamic theory, or the hydraulic theory. The hydrodynamic theory, as previously discussed, utilizes Darcy's law together with the equation of continuity to yield the Laplace equation as the governing partial differential equation in the flow domain; the boundary conditions along the free surface, whose location is a priori unknown, are time dependent and nonlinear.

The hydraulic theory is suitable where the streamlines are approximately parallel lines with little curvature. The dependent variable is the vertical distance to the free surface. The usual derivation of the partial differential equation utilizes four assumptions: (1) the free surface $h(x, t)$ differs from some known curve, e.g., a horizontal line, by a small amount; (2) the slope is small; (3) Darcy's law at any station is given by $\bar{v} = \bar{K}(\partial h/\partial x)$ where $\bar{v}$ is the average velocity over the cross section and $\bar{K}$ is some mean value of conductivity; and (4) the velocity is essentially horizontal and constant at any cross section. With these assumptions, the differential equation becomes

$$\frac{\partial}{\partial x}\left(\bar{k}h\frac{\partial h}{\partial x}\right) = n\frac{\partial h}{\partial t} \tag{85}$$

For the problem to be solved, it is this latter theory which will be employed. For early times following drawdown, this theory is adequate to describe the behavior, as will be shown, and more important, the essential physical result is not obscured by the complex analysis otherwise required. In the usual formulation for this problem, as given by Polubarinova–Kochina (1962, p. 500) and others, and generally employed with the hydraulic theory, the boundary condition at $x = 0$ is taken as

$$h(0, t) = H_2 \qquad (t > 0) \tag{86}$$

Although for the reverse situation, corresponding to the sudden filling of a reservoir, boundary condition (86) may be an adequate approximation, for the case of drawdown, there is no justification for assuming the level in the embankment at the seepage face will be the same as the reservoir level at all times.

A more nearly correct boundary condition may be obtained as follows: as the reservoir level falls there is an unbalanced head between $x = 0^-$ and $x = 0^+$, and water drains out approximately as it would from a tank whose wall is perforated. The flow is given by a formula similar to the orifice law, the horizontal velocity at a height z being given by

$$u(0, z, t) = f(z, t) \tag{87}$$

Although the exact relationship is unknown, all that will be required is the average horizontal velocity. Therefore, the orifice law will be taken as a first approximation, i.e., $q_x(0, z, t) = C_v(2gz)^{1/2}$, where C_v is the usual velocity coefficient. The average horizontal velocity at the seepage face $\bar{u}(0, t)$ is given by

$$\begin{aligned} \bar{u}(0, t) &= \frac{n}{h(0, t)} \int_0^{h(0,t)} C_V(2gz)^{1/2}\, dz \\ &= \frac{C_V n(2g)^{1/2}}{h(0, t)} \tfrac{2}{3} z^{3/2} \Big|_0^{h(0,t)} \\ &= \tfrac{2}{3} C_V n[2gh(0, t)]^{1/2} \end{aligned} \tag{88}$$

However, the differential equation (85) governing flow in the domain was presumably based on the assumption that the horizontal velocity is uniform in z, thus introducing an apparent inconsistency. While it may be sufficient to argue that the discrepancy merely indicates a transition zone in flow in the vicinity of the seepage face in which the velocity profile changes from a uniform one to a parabolic one at the boundary, it is easily proved that there is, in fact, no anomaly. To do this it is necessary to rederive (85) omitting the fourth assumption. In the course of this development the other major objections listed earlier will also be overcome.

Consider the element shown in Fig. 16. The horizontal velocity at a point z and at a station x is $u(x, z, t)$. The discharge is simply $Q = \int_0^{h(x,t)} u(x, z, t)\, dz$, and the average velocity is thus

$$\bar{u}(x, t) = [h(x, t)]^{-1} \int_0^{h(x,t)} u(x, z, t)\, dz \tag{89}$$

The volume of water contained in the element is approximately

$$\forall = (n/2)[h(x, t) + h(x + \Delta x, t)]\, \Delta x = n[h + \tfrac{1}{2}(\partial h/\partial x)\, \Delta x]\, \Delta x \tag{90}$$

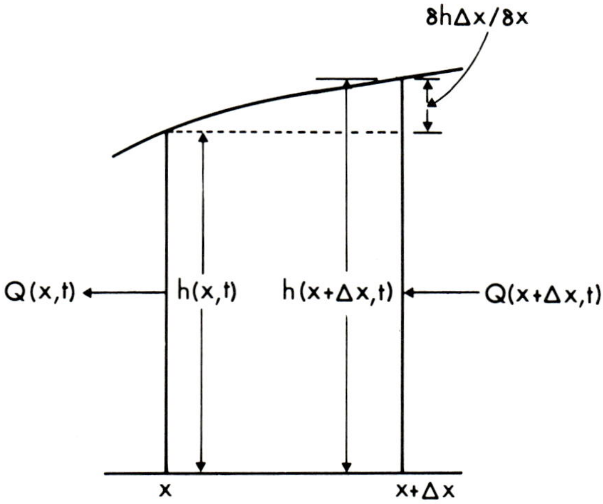

FIG. 16.

where n is the porosity of the domain. From continuity the net flow must equal the time rate of change of the water volume

$$d\forall/dt = Q(x + \Delta x, t) - Q(x, t)$$
$$= \int_0^{h(x+\Delta x,t)} u(x + \Delta x, z, t)\, dz - \int_0^{h(x,t)} u(x, z, t)\, dz \tag{91}$$

But

$$h(x + \Delta x, t) = h(x, t) + \frac{\partial h(x, t)}{\partial x} \Delta x$$

and

$$u(x + \Delta x, z, t) = u(x, z, t) + \frac{\partial u(x, z, t)}{\partial x} \Delta x$$

so the right-hand side of (91) becomes

$$= \int_0^{h+\partial h \Delta x/\partial x} \left[u(x, z, t) + \frac{\partial u(x, z, t)}{\partial x} \Delta x\right] dz - \int_0^h u(x, z, t)\, dz \tag{92}$$

The first integral may be written

$$\int_0^{h+\partial h \Delta x/\partial x} u(x, z, t)\, dz + \int_0^{h+\partial h \Delta x/\partial x} \frac{\partial u(x, z, t)\, \Delta x\, dz}{\partial x}$$
$$= \int_0^h u(x, z, t)\, dz + \int_h^{\partial h \Delta x/\partial x} u(x, z, t)\, dz$$
$$+ \int_0^{h+\partial h \Delta x/\partial x} \frac{\partial u(x, z, t)\, \Delta x\, dz}{\partial x} \tag{93}$$

Substituting in (92) gives

$$\frac{d\forall}{dt} = \int_h^{\partial h \Delta x/\partial x} u(x, z, t)\,dz + \int_0^h \frac{\partial u(x, z, t)\,\Delta x\,dz}{\partial x} + \int_h^{\partial h \Delta x/\partial x} \frac{\partial u(x, z, t)\,\Delta x\,dz}{\partial x} \tag{94}$$

From the rule for differentiation of an integral one obtains

$$\int_0^h \frac{\partial u(x, z, t)}{\partial x}\,\Delta x\,dz = \Delta x \frac{\partial}{\partial x} \int_0^h u(x, z, t)\,dz - u(x, h, t)\frac{\partial h}{\partial x}\,\Delta x \tag{95}$$

and from the mean-value theorem

$$\int_h^{\partial h \Delta x/\partial x} u(x, z, t)\,dz = u(x, h + \theta_1, t)\frac{\partial h}{\partial x}\,\Delta x \tag{96}$$

and

$$\int_h^{\partial h \Delta x/\partial x} \Delta x \frac{\partial u(x, z, t)\,dz}{\partial x} = \frac{\partial u(x, h + \theta_2, t)}{\partial x}\frac{\partial h}{\partial x}(\Delta x)^2 \tag{97}$$

where $\theta_{1,2}$ are some numbers such that $\partial h\,\Delta x/\partial x \geqslant \theta_{1,2} \geqslant 0$.

Substituting (95), (96) and (97) into (94) there results

$$\frac{d\forall}{dt} = u(x, h + \theta_1, t)\frac{\partial h}{\partial x}\,\Delta x + \Delta x \frac{\partial}{\partial x}\int_0^h u(x, z, t)\,dz - u(x, h, t)\frac{\partial h}{\partial x}\,\Delta x + \frac{\partial u(x, h + \theta_2, t)}{\partial x}\frac{\partial h}{\partial x}(\Delta x)^2 \tag{98}$$

The first and third terms cancel after dividing by Δx and taking the limit as $\Delta x \to 0$. It is noted that $\lim_{\Delta x \to 0} \theta_1, \theta_2 = 0$. Because $\partial u/\partial x$ and $\partial h/\partial x$ are bounded functions, the fourth term is of higher order than the others and drops out of the equation in the limit. Differentiating (90) and dividing by Δx gives

$$\frac{1}{\Delta x}\frac{d\forall}{dt} = n\frac{\partial h}{\partial t} + \frac{n}{z}\frac{\partial h}{\partial t}\,\Delta x \tag{99}$$

Taking the limit as $\Delta x \to 0$ and combining with (98) gives, finally,

$$n\frac{\partial h}{\partial t} = \frac{\partial}{\partial x}\int_0^h u(x, z, t)\,dz \tag{100}$$

While this result seems to be an obvious restatement of the equation of continuity, what has been rigorously established is that the time rate of change of h is dependent only on the discharge and that no statement of the detailed velocity distribution (e.g., uniform velocity) is necessary.

It is further seen that the validity of (100) depends on no assumptions regarding h or $\partial h/\partial x$ other than that the latter is bounded.

Moreover, it is easily shown (Harr, 1962) that Darcy's law may be written

$$\bar{u} = \bar{K}(\partial h/\partial x) \tag{101}$$

where $\bar{K} = K + 0(w_f)$ and w_f is the vertical velocity at the free surface. Combining (101) and (100) yields the Dupuit–Forchheimer formula (85) as the governing partial differential equation. The appropriate boundary condition at $x = 0$ (with h as the dependent variable) follows by eliminating $\bar{u}$ in (88) by the use of (101)

$$\bar{K}\frac{\partial h(0, t)}{\partial x} = \tfrac{2}{3}C_V n[2gh(0, t)]^{1/2} \tag{102}$$

4.3. Solution of Rapid Drawdown Problem

The problem solved will be that of complete instantaneous drawdown,† Fig. 15, in an infinite embankment, the complete formulation for which is

$$\frac{\partial}{\partial x}\left(Kh\frac{\partial h}{\partial x}\right) = n\frac{\partial h}{\partial t}, \qquad x \geqslant 0, \quad t > 0 \tag{103}$$

$$h(x, 0) = H_1, \qquad x \geqslant 0 \tag{104}$$

$$K\frac{\partial h(0, t)}{\partial x} = \tfrac{2}{3}C_V n(2g)^{1/2}[h(0, t)]^{1/2} \tag{105}$$

where, based on the assymption that w_f is everywhere small, $\bar{K}$ has been taken equal to K. Since it is expected that initially h will not differ greatly from H_1, we assume for h a perturbation about H_1, i.e.,

$$h(x, t) = H_1 + \delta(x, t) \tag{106}$$

and then linearize the boundary value problem, (103), (104), (105) accordingly. Substituting (106) into the above and neglecting product terms in the dependent variable and/or its derivatives yields

$$\frac{\partial}{\partial x}\left(KH_1\frac{\partial \delta}{\partial x}\right) = n\frac{\partial \delta}{\partial t}, \qquad x \geqslant 0, \quad t > 0 \tag{107}$$

$$\delta(x, 0) = 0 \tag{108}$$

$$K\frac{\partial \delta(0, t)}{\partial x} = \tfrac{2}{3}C_V n(2g)^{1/2}\left(\sqrt{H_1} + \frac{\delta(0, t)}{2\sqrt{H_1}}\right) \tag{109}$$

† Where the drawdown is not complete a modification is necessary. This is subsequently described.

where the binomial expansion has been used to linearize the right-hand side of (109). The solution to (18) is given by Carslaw and Jaeger (1959)

$$\delta(x, t) = -\frac{\varphi_0}{\alpha K} - \frac{2\varphi_0}{\sqrt{\pi}K}\left\{\frac{\sqrt{\pi}}{2}\operatorname{erf}\frac{x}{2\sqrt{\kappa t}} - \frac{e^{\alpha^2\kappa t+\alpha x}}{\alpha}\left[\frac{\sqrt{\pi}}{2} - \frac{\sqrt{\pi}}{2}\operatorname{erf}\frac{x+2\alpha\kappa t}{2\sqrt{\kappa t}}\right]\right\} \tag{110}$$

where

$$\varphi_0 = \tfrac{2}{3}C_V n(2gH_1)^{1/2}, \qquad \kappa = KH_1/n, \qquad \alpha = [C_V(2g)^{1/2}]/(3K\sqrt{H_1})$$

and $\operatorname{erf}\beta = (2/\sqrt{\pi})\int_0^\beta e^{-\sigma^2}\,d\sigma$. At the seepage face $x = 0$, (110) becomes

$$\delta(0, t) = -(\varphi_0/\alpha K) + (\varphi_0/\alpha K)\,e^{\alpha^2\kappa t}[1 - \operatorname{erf}\alpha(\kappa t)^{1/2}] \tag{111}$$

thus completing the solution for the problem posed. It is evident that according to (111), $\delta(0, t)$ is zero initially, in agreement with the *a priori* assumption.

4.4. Steady-State Flow through a Finite Embankment

We next consider the case of incomplete drawdown ($H_2 \neq 0$) for a finite embankment. For the problem of steady flow, a solution is available based on the "exact" theory by the use of conformal mapping. However, the formulation for the "exact" theory contains the anomaly which gives rise to a singular velocity field, as previously discussed.

Proceeding with the hydraulic theory, the differential equation remains as before (85), but with $\partial h/\partial t = 0$. The boundary conditions must be modified to account for the nonzero tailwater depth. Consider Fig. 17.

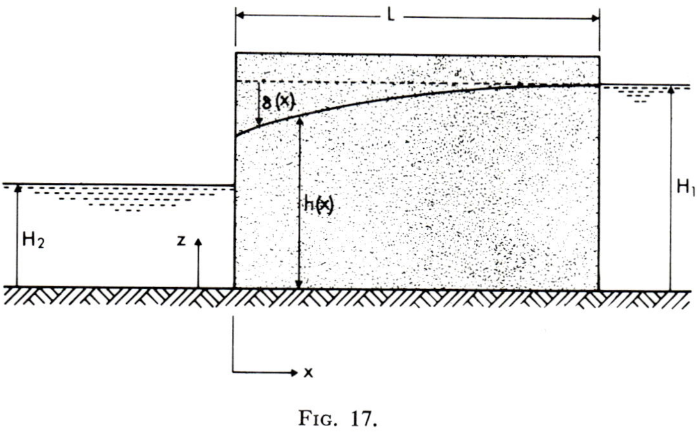

FIG. 17.

Based on the previous analysis, the efflux velocity (at $x = 0$) is given by the following, for $h \geqslant z \geqslant H_2$,

$$u(0, z) = C_V\{2g[h(0) - z]\}^{1/2} \tag{112}$$

For $H_2 \geqslant z \geqslant 0$, the difference in head does not vary with z so that the velocity is constant

$$u(0, z) = C_V\{2g[h(0) - H_2]\}^{1/2} \tag{113}$$

The average velocity is given by

$$\bar{u} = \frac{n}{h(0)} \left(\int_0^{H_2} C_V\{2g[h(0) - H_2]\}^{1/2}\, dz + \int_{H_2}^{h(0)} C_V\{2g[h(0) - z]\}^{1/2}\, dz \right)$$

$$= \frac{2h(0) + H_2}{h(0)} C_V\{2g[h(0) - H_2]\}^{1/2} \tag{114}$$

Combining with (101), the boundary condition is, at $x = 0$,

$$Kh\frac{\partial h}{\partial x} = n(2h + H_2)\, C_V[2g(h - H_2)]^{1/2} \tag{115}$$

The steady-state counterpart of (107) is simply

$$(\partial/\partial x)[Kh(\partial h/\partial x)] = 0, \qquad L \geqslant x \geqslant 0 \tag{116}$$

The solution for which is

$$h^2 = (2c_1/K)x + c_2 \tag{117}$$

For the finite embankment the boundary condition at $x = L$ is

$$h(L) = H_1 \tag{118}$$

Therefore,

$$H_1^2 = (2c_1/K)L + c_2 \tag{119}$$

Substituting the solution (117) into the boundary condition (115) yields the second equation for the coefficients c_1 and c_2.

$$c_1 = (2\sqrt{c_2} + H_2)\, nC_V[2g(\sqrt{c_2} - H_2)]^{1/2} \tag{120}$$

Combining (119) and (120) to eliminate c_1 gives

$$(K/2L)(H_1^2 - c_2) = (2\sqrt{c_2} + H_2)\, nC_v[2g(\sqrt{c_2} - H_2)^{1/2}] \tag{121}$$

Once (121) is solved for $h(0)$, i.e., c_2, the solution for the problem follows directly from (119) and (117).

In comparing the result with Muskat's it is immediately obvious that the solution presented herein contains a parameter—n, the porosity—which plays no role in the "exact" theory. The porosity enters the present theory via the boundary condition on the seepage face. As in all boundary value problems, the law governing the boundary conditions is independent of the law applicable throughout the interior of the domain; in the present case the law itself need not be derivable from the assumptions of porous flow theory since its domain is actually external to the porous medium. This latter argument is tacitly employed, in fact, in the "exact" theory for the domain below H_2 and external to the medium, where, assuming quiescent conditions, the head is constant and it is this constant head which is applied to the boundary as a boundary condition.

If we carry this last idea one step further and allow for nonzero velocities in the liquid external to the medium, it is clear that Bernoulli's law should apply between any two points on this boundary. From the assumption that the horizontal velocity is zero at the upper end of the seepage face, and that the vertical velocity is negligible everywhere on the boundary, one is led directly to the orifice formula introduced earlier.

4.5. Conclusions

It has been shown that the Dupuit–Forchheimer differential equation is valid, regardless of the internal velocity distribution, and with the new boundary condition postulated, the free surface is obtained for both a transient and steady problem.

Previous solutions for the transient case were based on a uniform velocity distribution and assumed further that the downstream end of the free surface followed the downstream water surface. It has been shown that neither of these is strictly correct. The latter point is of great importance in soil mechanics.

Furthermore, the new boundary conditions, derived from physical considerations, leads to a continuous velocity field over the seepage face and thus, more nearly represents the actual phenomenon than that formerly employed in the "exact" theory, since the latter yields an anomaly in the resultant velocity profile which is not explicable on physical grounds.

REFERENCES

Boley, B. A., and Friedman, M. B. (1959). On the viscous flow around the leading edge of a flat plate, *J. Aerospace Sci.* **26**, 453.

Carslaw, H. S., and Jaeger, J. C. (1959). "Conduction of Heat in Solids," 2nd ed., pp. 70, 73. Oxford Univ. Press, London and New York.

Collins, R. E. (1961). "Flow of Fluids Through Porous Materials." Reinhold, New York.

Darcy, H. (1856). "Les fontaines publiques de la ville de Dijon." Victor Dalmont, Paris.

DeWiest, R. J. M. (1962). Free surface flow in homogeneous porous mediums, *Trans. Am. Soc. Civil Engrs*, **127**, 1045–1084.

Dicker, D. (1962). Discussion of free surface flow in homogeneous porous mediums, *Trans. Am. Soc. Civil Engrs.* **127**, 1085–89. 1962.

Dicker, D. (1966). Aerodynamic stability of *H* sections, *J. Eng. Mech. Div.*, ASCE, No. 4764, pp. 1–23.

Dicker, D., and Sevian, W. A. (1965). Transient flow through porous mediums, *J. Geophys. Res.* **70**, 5043–54.

Goldstein, S. (1965). "Modern Developments in Fluid Dynamics," Vol. 1, p. 136. Dover, New York.

Harlow, F. H., and Welch, J. E. (1965). Numerical calculation of time-dependent viscous incompressible flow of fluid with free surace, *Phys. Fluids* **8**, 2182–2189.

Harlow, F. H., Shannon, J. P., and Welch, J. E. (1965). Liquid waves by computer, *Science* **149**, 1092–1093.

Harr, M. E. (1962). "Groundwater and Seepage," p. 40. McGraw-Hill, New York.

Morel-Seytoux, H. J. (1961). Effects of boundary shape on channel seepage, Stanford Univ. Tech. Rept. 7.

Muskat, M. (1946). "The Flow of Homogeneous Fluids Through Porous Media." Edwards, Ann Arbor, Michigan.

Polubarinova-Kochina, P. (1962). "Theory of Ground Water Movement." Princeton Univ. Press, Princeton, New Jersey.

Wang, D. P., and Yoo-Tsu Wu, T. (1965). General formulation of a perturbation theory for unsteady cavity flows, Paper No. 65-FE-18, ASME.

Wilf, H. S. (1962). "Mathematics for the Physical Sciences," p. 145. Wiley, New York.

ACKNOWLEDGMENT

This work was supported by the National Science Foundation. Most of the material in Section 3 was developed by the author together with Mr. Walter A. Sevian of the State University of New York at Stony Brook and appeared in Dicker and Sevian (1965).

8 Elastic Storage of Aquifers

Arnold Verruijt
Department of Civil Engineering
Delft University of Technology
Delft, Netherlands

1. Introduction

The study of unsteady ground water flow, as occurring in compressible soils, connects the theory of ground water movement with soil mechanics. In both branches of science, problems related to unsteady flow of water in a porous medium have received considerable interest.

The first hydrologist to pay attention to the elastic behavior of aquifers seems to have been Meinzer (1928), but the mathematical formulation of a theory was first given by Jacob (1940). In his now classic treatise "The Flow of Homogeneous Fluids through Porous Media," Muskat (1937) considered unsteady flow phenomena due to the compressibility of the fluid and in Jacob's work this was incorporated (Jacob, 1950). Using this theoretical base several practical problems have been solved by Hantush and Jacob (1955), Hantush (1956), and De Wiest (1963), among others.

On the other hand, in soil mechanics it was Terzaghi (1923) who realized the effect of pore water dissipation upon the deformations of a saturated soil. This process is called *consolidation*. In the original development of his theory, in which deformation and ground water flow occur in one direction only, Terzaghi noted the resemblance of his differential equation with the one-dimensional heat diffusion equation. Since the heat diffusion equation has been studied extensively in literature, problems in one-dimensional consolidation were solved analogous to the corresponding heat diffusion problems. The extension towards three-dimensional consolidation was first made by Rendulic (1936), who assumed that the total stresses in a material under constant external loading are independent of time, thus arriving at an equation analogous to the three-dimensional heat diffusion equation.

However, this extension is more in the nature of a mathematical generalization than a physical one. An extension based on physically consistent assumptions was given by Biot (1941). Biot assumed that the soil skeleton behaves as a perfectly elastic material and arrived at differential equations essentially different from those obtained by Rendulic (see also Terzaghi, 1943). The basic difference in the appearance of the differential equations is that Biot's approach leads to an equation of the fourth order, whereas Rendulic's approach leads to an equation of the second order, this equation being a special case of Biot's equation. From a theoretical viewpoint, the more complicated Biot theory is to be preferred because of its self-consistency. It should be noted, however, that the behavior of real soils is more complicated than the behavior of a perfectly elastic material, so that the theory provides merely an approximative description of the consolidation process.

An essential objective of mathematical physics is to account for phenomena observed experimentally. It has been shown (Gibson *et al.*, 1963; Verruijt, 1965) that experimental results for the consolidation of a solid sphere of saturated clay can be explained satisfactory by Biot's theory. Another example in which the computations based on Biot's theory predict a remarkable behavior of the pore water pressures, this time in a two-storied aquifer, similar to the behavior observed in practice, will be given in Section 6 of this study.

In the theory of ground water movement it is usually assumed that the horizontal deformations can be disregarded with respect to the vertical deformations (Jacob, 1950). It will be shown that under these assumptions Biot's theory reduces to Jacob's theory. However, it may be argued that there is no physical reason for the vanishing of the horizontal deformations. Therefore, in the present study Biot's theory will be used as a starting point.

It will be assumed that the soil is almost completely saturated, so that the total amount of air in the soil is small. The compressibility of the air will be taken into account according to Boyle's law. It will be shown that as a first approximation this results in a simple modification of the coefficient of compressibility of the pore fluid. After the derivation of the general differential equations, based upon Hooke's law for the elastic properties of the soil skeleton, and Darcy's law for the movement of the pore fluid, some examples will be given. The first example deals with the problem of a well in a leaky artesian aquifer of small thickness and infinite radial extent. Hantush's solution will be derived from the general basic equations, assuming that the radial displacement vanishes. It will be shown that a different assumption, namely the assumption of plane total stress, leads to a final solution differing from Hantush's solution only in the interpretation of a coefficient. Which solution is to be preferred depends upon the local situation. The second example deals with a two-storied leaky artesian aquifer, a well operating in the upper layer of the aquifer. It will be assumed that Hantush's solution applies for the upper layer. The corresponding solution for the lower layer, which is assumed to extend towards infinity in radial as well as in vertical direction, will be determined starting from the general three-dimensional basic equations. Numerical results will be compared with results obtained in pumping tests. It will appear that in a certain region of the lower aquifer the pore water pressures initially increase during a pumping test. This phenomenon is known as the *Noordbergum effect* and has been observed several times in practice. Although the theory succeeds in predicting this effect, it must be emphasized that the substantial schematization of the situation in the field makes a quantitave comparison of theoretical and experimental results questionable.

2. Basic Equation of Elastic Storage

In this section a general relation between the elementary volume change and the fluid pressure in a porous material will be established. Therefore, a material consisting of three components, A (grains), B (liquid), and C (gas), is considered. The densities of the components will be denoted by ρ_A, ρ_B, and ρ_C. The specific mass discharges (the discharge of mass per unit area of a coordinate system fixed in space) will be denoted by $\mathbf{n}_A$, $\mathbf{n}_B$, and $\mathbf{n}_C$. The specific mass discharges are related to the specific volume discharges (the volume discharges per unit area of a coordinate system fixed in space) $\mathbf{q}_A$, $\mathbf{q}_B$, and $\mathbf{q}_C$ by

$$\mathbf{n}_A = \rho_A \mathbf{q}_A , \qquad \mathbf{n}_B = \rho_B \mathbf{q}_B , \qquad \mathbf{n}_C = \rho_C \mathbf{q}_C \tag{1}$$

If the porosity of the material is denoted by n, then in an elementary volume V the volume of the pore space is nV and the volume of the solid material is $(1 - n)V$. If the degree of saturation is denoted by S_r, then the volume of the liquid in the volume V is $S_r nV$ and the volume of the gas is $(1 - S_r)nV$. With the help of these expressions for the relative volume of the components of the material, the displacement velocities (the actual velocities of the particles of each component) $\mathbf{v}_A$, $\mathbf{v}_B$, and $\mathbf{v}_C$ may be related to the specific volume discharges. This gives

$$\mathbf{q}_A = (1 - n)\,\mathbf{v}_A\,, \qquad \mathbf{q}_B = S_r n \mathbf{v}_B\,, \qquad \mathbf{q}_C = (1 - S_r)\,n\mathbf{v}_C \tag{2}$$

The relative density ϑ_i of the component i is defined as the mass of the component i per unit total volume of the material. Since the density ρ_i is defined as the mass of the component i per unit volume of that component, the relative densities are given by

$$\vartheta_A = (1 - n)\,\rho_A\,, \qquad \vartheta_B = S_r n \rho_B\,, \qquad \vartheta_C = (1 - S_r)\,n\rho_C \tag{3}$$

The basic principle which the specific mass discharges have to satisfy is the conservation of mass, and this principle should hold for every component of the material. If no chemical reactions occur, conservation of mass requires that

$$\partial\vartheta_A/\partial t + \operatorname{div} \mathbf{n}_A = 0 \tag{4}$$

$$\partial\vartheta_B/\partial t + \operatorname{div} \mathbf{n}_B = 0 \tag{5}$$

$$\partial\vartheta_C/\partial t + \operatorname{div} \mathbf{n}_C = 0 \tag{6}$$

The equation of motion of the liquid is postulated in the following form

$$\mathbf{q} = S_r n(\mathbf{v}_B - \mathbf{v}_A) = -K \operatorname{grad} \varphi \tag{7}$$

where φ is the so-called *force potential*,

$$\varphi = \int_{z_0}^{z} d\zeta + \frac{1}{g}\int_{p_0}^{p} \frac{d\pi}{\rho_B(\pi)} \tag{8}$$

In this expression z is the vertical coordinate, positive in the upward direction and p is the pressure of the liquid. The quantities z_0 and p_0 are arbitrary reference constants and g is the gravity constant. Equation (7) is a generalization of Darcy's law. In this form, Darcy's law is in accordance with experimental evidence indicating that the fluid particles do not move with respect to the grains if grad $\varphi = 0$. The first to realize that it should be the velocity of the water relative to the grain skeleton that enters into Darcy's law was Gersevanov in 1934 [see Florin (1961)]. This relative velocity concept was introduced into the three-dimensional theory of consolidation by Biot (1956). The vector $\mathbf{q}$

is called the specific discharge of the fluid. In the present considerations it is assumed that K, the coefficient of permeability, is a constant, although it is well known that K depends upon the degree of saturation S_r, the kinematic viscosity ν of the liquid, as well as upon the porosity n of the soil. The above definition of a force potential corresponds to the definition introduced by Hubbert (1940). The force potential as given by (8) is a potential in the sense that it generates an irrotational vector field, provided that the fluid density ρ_B depends only upon the pressure p (De Wiest, 1966).

Restriction is made to almost completely saturated soils, for which

$$1 - S_r \ll 1 \tag{9}$$

Then it seems reasonable to postulate that

$$\mathbf{v}_C - \mathbf{v}_B = 0 \tag{10}$$

which means that the relative velocity of the gas with respect to the liquid is zero.

The equations of state of the three components are assumed to be

$$\rho_A = \text{constant} \tag{11}$$

$$\rho_B = \rho_0 \exp(\beta p) \quad (\beta \geqslant 0) \tag{12}$$

$$\rho_C = \rho_g(p/p_g) \tag{13}$$

Equation (11) shows that the grains are incompressible, Eq. (12) shows that the liquid is compressible with compressibility β, and Eq. (13) corresponds to Boyle's law. The densities ρ_0 and ρ_g and the pressure p_g are reference quantities. The formulation of the equations of state in the form given above implies that the solubility of the gas in the liquid is disregarded. Also the effect of surface tension at the gas–liquid interface is disregarded, so that the pressures in the liquid and in the gas may be represented by the same symbol p.

Using these definitions, it is possible to rewrite the equations of conservation of mass, Eqs. (4), (5), and (6), in forms suitable for the introduction of Eqs. (7) and (10). With (1), (2), and (3), Eq. (4) gives

$$\partial[(1 - n)\,\rho_A]/\partial t + \operatorname{div}[(1 - n)\,\rho_A \mathbf{v}_A] = 0$$

and using (11), it follows that

$$\partial n/\partial t + \mathbf{v}_A \cdot \operatorname{grad} n - (1 - n) \operatorname{div} \mathbf{v}_A = 0 \tag{14}$$

With (1)–(3), Eq. (5) gives

$$\partial(S_r n \rho_B)/\partial t + \operatorname{div}(S_r n \rho_B \mathbf{v}_B) = 0$$

and after elaboration with (12), the following result is obtained

$$\frac{1}{S_r}\frac{\partial S_r}{\partial t} + \frac{1}{n}\frac{\partial n}{\partial t} + \beta\frac{\partial p}{\partial t} + \frac{\mathbf{v}_B \cdot \operatorname{grad}(S_r n \rho_B)}{S_r n \rho_B} + \operatorname{div}\mathbf{v}_B = 0 \tag{15}$$

With (1)–(3), Eq. (6) gives

$$\partial[(1 - S_r)\, n\rho_C]/\partial t + \operatorname{div}[(1 - S_r)\, n\rho_C \mathbf{v}_C] = 0$$

and using (10) and (13), this may be written as follows

$$\frac{1}{1 - S_r}\frac{\partial S_r}{\partial t} - \frac{1}{n}\frac{\partial n}{\partial t} - \frac{1}{p}\frac{\partial p}{\partial t} - \frac{\mathbf{v}_B \cdot \operatorname{grad}[(1 - S_r)\, n\rho_C]}{(1 - S_r)\, n\rho_C} - \operatorname{div}\mathbf{v}_B = 0 \tag{16}$$

Elimination of the terms containing $\partial S_r/\partial t$ from (15) and (16) gives, after writing out the gradients,

$$\begin{aligned} &\frac{1}{n}\frac{\partial n}{\partial t} + \frac{1 - S_r + S_r p\beta}{p}\frac{\partial p}{\partial t} \\ &\qquad + \mathbf{v}_B \cdot \left[\frac{1}{n}\operatorname{grad} n + \frac{1 - S_r + S_r p\beta}{p}\operatorname{grad} p\right] + \operatorname{div}\mathbf{v}_B = 0 \end{aligned} \tag{17}$$

Taking the divergence of (7) gives

$$\operatorname{div}\mathbf{q} = (\mathbf{v}_B - \mathbf{v}_A) \cdot \operatorname{grad}(S_r n) + S_r n(\operatorname{div}\mathbf{v}_B - \operatorname{div}\mathbf{v}_A) = -K\nabla^2\varphi \tag{18}$$

The right-hand member of (18) can be expressed in terms of the fluid pressure p by using the definition of φ, Eq. (8), and the equation of state (12),

$$\rho_B g \nabla^2\varphi = \nabla^2 p - \beta \operatorname{grad} p \cdot \operatorname{grad} p$$

so that (18) may be written in the following form

$$\begin{aligned} (K/\rho_B g)\, \nabla^2 p = &-(\mathbf{v}_B - \mathbf{v}_A) \cdot \operatorname{grad}(S_r n) - S_r n(\operatorname{div}\mathbf{v}_B - \operatorname{div}\mathbf{v}_A) \\ &+ (K/\rho_B g)\, \beta \operatorname{grad} p \cdot \operatorname{grad} p \end{aligned} \tag{19}$$

Elimination of div $\mathbf{v}_B$ from (17) and (19) yields

$$\begin{aligned} (K/\rho_B g)\, \nabla^2 p = \mathbf{v}_B &\cdot \{-n \operatorname{grad} S_r + S_r n[(1 - S_r + S_r p\beta)/p] \operatorname{grad} p\} \\ &+ \mathbf{v}_A \cdot \{S_r \operatorname{grad} n + n \operatorname{grad} S_r\} + S_r n \operatorname{div}\mathbf{v}_A \\ &+ S_r(\partial n/\partial t) + S_r n[(1 - S_r + S_r p\beta)/p](\partial p/\partial t) \\ &+ (K/\rho_B g)\, \beta \operatorname{grad} p \cdot \operatorname{grad} p \end{aligned}$$

Using (14), the term $S_r(\partial n/\partial t)$ may be eliminated. This gives

$$\begin{aligned}(K/\rho_B g)\,\nabla^2 p = {} & S_r \operatorname{div} \mathbf{v}_A + S_r n[(1 - S_r + S_r p\beta)/p](\partial p/\partial t) \\ & + n\mathbf{v}_A \cdot \operatorname{grad} S_r + n\mathbf{v}_B \\ & \cdot \{-\operatorname{grad} S_r + S_r[(1 - S_r + S_r p\beta)/p] \operatorname{grad} p\} \\ & + (K/\rho_B g)\,\beta \operatorname{grad} p \cdot \operatorname{grad} p\end{aligned}$$

Since it was assumed that $1 - S_r \ll 1$, this equation can be simplified by writing $1 - S_r = \delta$, where now $\delta \ll 1$. In practical problems of ground water flow it can also be assumed that $p\beta \ll 1$. When terms of order δ and of order $p\beta$ are disregarded compared to terms of order 1 the following result is obtained

$$\begin{aligned}(K/\rho_B g)\,\nabla^2 p = {} & \operatorname{div} \mathbf{v}_A + n[\beta + (1 - S_r)/p](\partial p/\partial t) + n\mathbf{v}_A \cdot \operatorname{grad} S_r \\ & + n\mathbf{v}_B \cdot \{-\operatorname{grad} S_r + [\beta + (1 - S_r)/p] \operatorname{grad} p\} \\ & + (K/\rho_B g)\,\beta \operatorname{grad} p \cdot \operatorname{grad} p \qquad (20)\end{aligned}$$

In the theory of linear consolidation, only small incremental deformations are considered. Then the scalar products in the right-hand member of Eq. (20) may be disregarded because they constitute terms of second order. This gives

$$(K/\rho_B g)\,\nabla^2 p = \operatorname{div} \mathbf{v}_A + n\beta'(\partial p/\partial t) \qquad (21)$$

where

$$\beta' = \beta + (1 - S_r)/p, \qquad 1 - S_r \ll 1 \qquad (22)$$

When small deviations from an initial steady state are considered, the pressure p appearing in the definition of β' can be taken equal to the initial value $p^{(0)}$ of the pressure, thus making β' a constant. In the sequel, ρ_B will be written simply as ρ, since no other densities will appear in the equations.

Equation (21) relates the pressure p to the displacement velocity $\mathbf{v}_A$ of the solid material. Since

$$\mathbf{v}_A = \partial \mathbf{u}_A/\partial t \qquad (23)$$

where $\mathbf{u}_A$ is the displacement vector of the solid material, Eq. (21) can be written as

$$(K/\rho g)\,\nabla^2 p = \partial e/\partial t + n\beta'(\partial p/\partial t) \qquad (24)$$

where e is the volume strain,

$$e = \operatorname{div} \mathbf{u}_A \qquad (25)$$

Equation (24) may be called the *storage equation*. It corresponds to the equation given by Biot (1941) [see also de Josselin de Jong (1963)].

Since Eq. (24) contains two variables, the volume strain e and the fluid pressure p, a second equation is needed to describe the phenomenon of unsteady ground water flow. This second equation has to be obtained from the mechanical behavior of the soil skeleton.

3. Elasticity of the Soil Skeleton

It is known from experiments that soils in general have nonlinear, partly irreversible, anisotropic, and sometimes viscous mechanical properties. However, little is known about the quantitative description of these properties. It will therefore be assumed that the soil skeleton behaves like a perfectly elastic solid, i.e., that the soil has linear, reversible, isotropic, nonretarded mechanical properties. It is clear that, due to this assumption, the theory to be developed will provide only an approximative description of the real behavior of the soil.

In Fig. 1 an elementary parallelepiped is drawn, together with the components of stress acting upon its sides. Outward normal directions

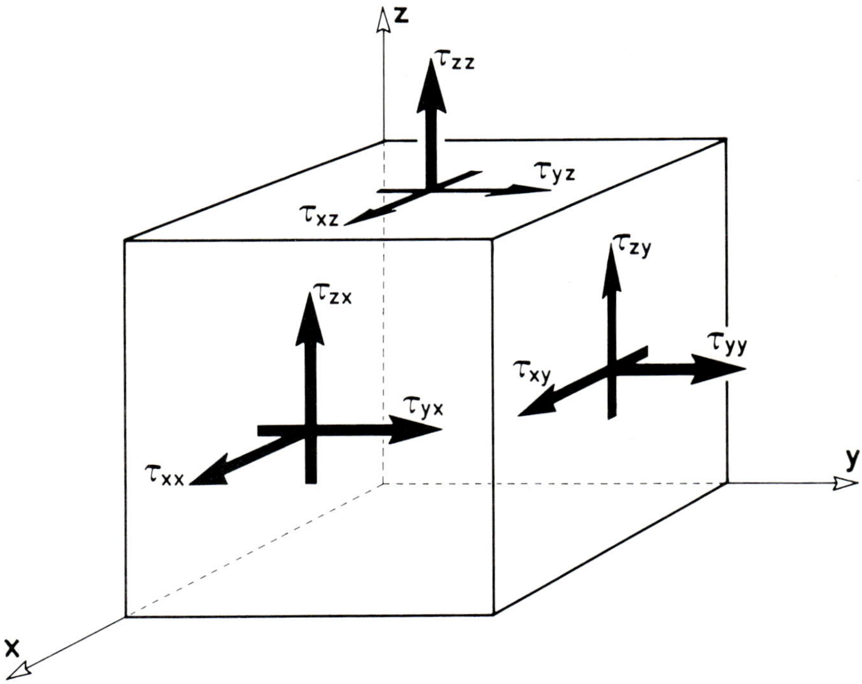

FIG. 1. Elementary parallelepiped with total stresses.

are called positive when pointing in positive coordinate direction and negative when pointing in negative coordinate direction. The components of total stress will be denoted by τ_{ij}, the first index denoting the direction of the stress component and the second index denoting the direction of the normal to the plane, upon which this stress component acts. In accordance with the usual sign convention in the theory of elasticity, τ_{ij} will be called positive when acting in the positive coordinate direction upon a plane having positive normal direction, or when acting in the negative coordinate direction upon a plane having negative normal direction. This implies that unidirectional tension is called positive and that unidirectional compression is called negative.

The deformations of the soil skeleton are caused by the intergranular forces, which, in fact, are concentrated forces. If a plane through the assembly of grains is considered with its normal in the positive j direction, then the stress component τ'_{ij} is defined as the average value per unit area of the components in the positive i direction of the concentrated forces acting from the positive side upon the grains intersected by that plane. The stress components τ'_{ij}, which will be called the *effective stresses*, can be considered to be a measure of the intergranular forces. Hence the deformations of the soil skeleton will be determined by the effective stresses only. The fluid pressure p, which can be considered to be an all-around pressure, acts over the total area of the elementary parallelepiped, since this fluid pressure too causes a pressure of magnitude p in every grain. Now the components of total stress can be decomposed as

$$\begin{aligned} \tau_{xx} &= \tau'_{xx} - p, & \tau_{yz} &= \tau'_{yz}, & \tau_{zy} &= \tau'_{zy} \\ \tau_{yy} &= \tau'_{yy} - p, & \tau_{zx} &= \tau'_{zx}, & \tau_{xz} &= \tau'_{xz} \\ \tau_{zz} &= \tau'_{zz} - p, & \tau_{xy} &= \tau'_{xy}, & \tau_{yx} &= \tau'_{yx} \end{aligned} \tag{26}$$

The minus sign in these equations appears because the fluid pressure p is taken positive for compression, in accordance with the usual sign convention in the theory of flow through porous media.

The total stresses have to satisfy the conditions of equilibrium. If the effects of inertia are disregarded these conditions can be written as

$$\begin{aligned} &\frac{\partial \tau_{xx}}{\partial x} + \frac{\partial \tau_{xy}}{\partial y} + \frac{\partial \tau_{xz}}{\partial z} + P_x = 0, & \tau_{yz} &= \tau_{zy} \\ &\frac{\partial \tau_{yx}}{\partial x} + \frac{\partial \tau_{yy}}{\partial y} + \frac{\partial \tau_{yz}}{\partial z} + P_y = 0, & \tau_{zx} &= \tau_{xz} \\ &\frac{\partial \tau_{zx}}{\partial x} + \frac{\partial \tau_{zy}}{\partial y} + \frac{\partial \tau_{zz}}{\partial z} + P_z = 0, & \tau_{xy} &= \tau_{yx} \end{aligned} \tag{27}$$

where P_x, P_y, and P_z are the components of the body force per unit volume. The three equations in the left part of (27) are obtained from the conditions of equilibrium in the x, y, and z directions, respectively, and the three equations in the right part of (27) are obtained from the conditions of moment equilibrium in the three coordinate directions. With the help of Eqs. (26), the equations of equilibrium can be written in terms of effective stresses and fluid pressure. The equations of equilibrium in the x, y, and z directions are

$$\begin{aligned}
\frac{\partial \tau'_{xx}}{\partial x} + \frac{\partial \tau'_{xy}}{\partial y} + \frac{\partial \tau'_{xz}}{\partial z} + P_x &= \frac{\partial p}{\partial x} \\
\frac{\partial \tau'_{yx}}{\partial x} + \frac{\partial \tau'_{yy}}{\partial y} + \frac{\partial \tau'_{yz}}{\partial z} + P_y &= \frac{\partial p}{\partial y} \qquad (28) \\
\frac{\partial \tau'_{zx}}{\partial x} + \frac{\partial \tau'_{zy}}{\partial y} + \frac{\partial \tau'_{zz}}{\partial z} + P_z &= \frac{\partial p}{\partial z}
\end{aligned}$$

In Section 2 it has been already stated that the considerations are restricted to small departures from an initial steady state. The stresses in this initial state have to satisfy the equations of equilibrium (28). When these initial stresses are characterized by a superscript (0), the effective stress may be written as follows

$$\begin{aligned}
\tau'_{xx} &= \tau'^{(0)}_{xx} + \sigma'_{xx}, \qquad & \tau'_{yz} &= \tau'^{(0)}_{yz} + \sigma'_{yz} \\
\tau'_{yy} &= \tau'^{(0)}_{yy} + \sigma'_{yy}, \qquad & \tau'_{zx} &= \tau'^{(0)}_{zx} + \sigma'_{zx} \qquad (29) \\
\tau'_{zz} &= \tau'^{(0)}_{zz} + \sigma'_{zz}, \qquad & \tau'_{xy} &= \tau'^{(0)}_{xy} + \sigma'_{xy}
\end{aligned}$$

where σ'_{ij} denotes incremental effective stress. The fluid pressure may be written in the form

$$p = p^{(0)} + \sigma \qquad (30)$$

where σ is the incremental fluid pressure.

When the body forces are assumed to be independent of time (as will be the case for body forces resulting from gravity), insertion of the above relations into (28) gives (making use of the fact that the initial state is a state of equilibrium itself),

$$\begin{aligned}
\frac{\partial \sigma'_{xx}}{\partial x} + \frac{\partial \sigma'_{xy}}{\partial y} + \frac{\partial \sigma'_{xz}}{\partial z} &= \frac{\partial \sigma}{\partial x} \\
\frac{\partial \sigma'_{yx}}{\partial x} + \frac{\partial \sigma'_{yy}}{\partial y} + \frac{\partial \sigma'_{yz}}{\partial z} &= \frac{\partial \sigma}{\partial y} \qquad (31) \\
\frac{\partial \sigma'_{zx}}{\partial x} + \frac{\partial \sigma'_{zy}}{\partial y} + \frac{\partial \sigma'_{zz}}{\partial z} &= \frac{\partial \sigma}{\partial z}
\end{aligned}$$

It is assumed, as a first approximation, that the deformations of the soil are related to the effective stresses by Hooke's law, which in its most familiar form states that

$$\begin{aligned} \epsilon_{xx} &= E^{-1}[\sigma'_{xx} - \nu(\sigma'_{yy} + \sigma'_{zz})], & \epsilon_{yz} &= \sigma'_{yz}/2\mu \\ \epsilon_{yy} &= E^{-1}[\sigma'_{yy} - \nu(\sigma'_{zz} + \sigma'_{xx})], & \epsilon_{zx} &= \sigma'_{zx}/2\mu \\ \epsilon_{zz} &= E^{-1}[\sigma'_{zz} - \nu(\sigma'_{xx} + \sigma'_{yy})], & \epsilon_{xy} &= \sigma'_{xy}/2\mu \end{aligned} \tag{32}$$

where the symbol ϵ_{ij} denotes the incremental strain, E is Young's modulus, ν is Poisson's ratio, and μ is the shear modulus related to E and ν by

$$\mu = E/[2(1 + \nu)] \tag{33}$$

It is not necessarily assumed that E and ν are physical constants, since only small departures from an initial state are considered. This means that E and ν may vary with the initial state of stress. In the present study, the incremental elastic coefficients are regarded as given quantities, constant throughout the field. This means that the soil is assumed to behave as a homogeneous material. By solving (32) for the stresses, the following alternative form of Hooke's law is obtained

$$\begin{aligned} \sigma'_{xx} &= 2\mu\epsilon_{xx} + \lambda\epsilon, & \sigma'_{yz} &= 2\mu\epsilon_{yz} \\ \sigma'_{yy} &= 2\mu\epsilon_{yy} + \lambda\epsilon, & \sigma'_{zx} &= 2\mu\epsilon_{zx} \\ \sigma'_{zz} &= 2\mu\epsilon_{zz} + \lambda\epsilon, & \sigma'_{xy} &= 2\mu\epsilon_{xy} \end{aligned} \tag{34}$$

where λ is given by

$$\lambda = \nu E/[(1 - 2\nu)(1 + \nu)] \tag{35}$$

In the theory of elasticity, λ and μ are called Lamé's constants. In Eqs. (34) the incremental volume strain ϵ is introduced according to the relation

$$\epsilon = \epsilon_{xx} + \epsilon_{yy} + \epsilon_{zz} \tag{36}$$

The strain components ϵ_{ij} have to satisfy the conditions of compatibility, which require that the displacement components u_x, u_y, u_z throughout the body be single-valued continuous functions of the coordinates x, y, z. These conditions are most easily satisfied by deducing the strains from the displacements and then simply postulating that the displacement components are single-valued and continuous.

In the case of small deformations, the strains are related to the displacements by the expressions

$$\begin{aligned} \epsilon_{xx} &= \partial u_x/\partial x, & \epsilon_{yz} &= \tfrac{1}{2}(\partial u_y/\partial z + \partial u_z/\partial y) \\ \epsilon_{yy} &= \partial u_y/\partial y, & \epsilon_{zx} &= \tfrac{1}{2}(\partial u_z/\partial x + \partial u_x/\partial z) \\ \epsilon_{zz} &= \partial u_z/\partial z, & \epsilon_{xy} &= \tfrac{1}{2}(\partial u_x/\partial y + \partial u_y/\partial x) \end{aligned} \tag{37}$$

With (36) it follows that

$$\epsilon = \partial u_x/\partial x + \partial u_y/\partial y + \partial u_z/\partial z = \operatorname{div} \mathbf{u} \tag{38}$$

which corresponds to the definition of e given in Eq. (25). It is to be noted that in (25), $\mathbf{u}_A$ denotes the displacement from the nonstressed-state, whereas here $\mathbf{u}$ denotes the incremental displacement vector. The volume strain e and the incremental volume strain ϵ are related by

$$e = e^{(0)} + \epsilon$$

Substitution of (37) into (34) gives the stresses expressed in the displacements,

$$\begin{aligned} \sigma'_{xx} &= 2\mu(\partial u_x/\partial x) + \lambda\epsilon, & \sigma'_{yz} &= \mu(\partial u_y/\partial z + \partial u_z/\partial y) \\ \sigma'_{yy} &= 2\mu(\partial u_y/\partial y) + \lambda\epsilon, & \sigma'_{zx} &= \mu(\partial u_z/\partial x + \partial u_x/\partial z) \\ \sigma'_{zz} &= 2\mu(\partial u_z/\partial z) + \lambda\epsilon, & \sigma'_{xy} &= \mu(\partial u_x/\partial y + \partial u_y/\partial x) \end{aligned} \tag{39}$$

With the help of these relations, the equations of equilibrium (31) can be written in terms of the displacements. This gives

$$\begin{aligned} \mu \nabla^2 u_x + (\lambda + \mu)(\partial\epsilon/\partial x) &= \partial\sigma/\partial x \\ \mu \nabla^2 u_y + (\lambda + \mu)(\partial\epsilon/\partial y) &= \partial\sigma/\partial y \\ \mu \nabla^2 u_z + (\lambda + \mu)(\partial\epsilon/\partial z) &= \partial\sigma/\partial z \end{aligned} \tag{40}$$

where as before,

$$\nabla^2 = \partial^2/\partial x^2 + \partial^2/\partial y^2 + \partial^2/\partial z^2$$

Together with Eq. (24), which in terms of incremental stresses and strains can be written as follows,

$$(K/\rho g)\, \nabla^2\sigma = \partial\epsilon/\partial t + n\beta'(\partial\sigma/\partial t) \tag{41}$$

the Eqs. (38) and (40) form a system of five equations in the five unknown variables: σ, ϵ, u_x, u_y, u_z. In several types of problems it is useful to reduce the number of equations to two. This can be done by differentiating the first of Eqs. (40) with respect to x, the second with respect to y, the third with respect to z, and then adding the results. This gives

$$(\lambda + 2\mu)\, \nabla^2 \epsilon = \nabla^2 \sigma \tag{42}$$

Equations (41) and (42) together form a system of two equations in the two independent variables ϵ and σ. Integration of (42) gives

$$(\lambda + 2\mu)\, \epsilon = \sigma + f \tag{43}$$

where f is a function of x, y, z, and t which has to satisfy Laplace's equation $\nabla^2 f = 0$, for every value of the time t. Only in the case that f is identically zero, Eqs. (41) and (43) may be combined to give

$$(\alpha + n\beta')(\partial\sigma/\partial t) = (K/\rho g)\, \nabla^2 \sigma \tag{44}$$

This equation appears in many places in the literature, sometimes apart from the term due to the compressibility of the liquid–gas mixture. In the one-dimensional case, when fluid flow as well as deformation occur in one direction only, this equation is correct. It was originally derived by Terzaghi (1923). In a later publication, Terzaghi (1943) gave the equation (44) as a mathematical generalization of his one-dimensional equation. The correct mathematical derivation of the differential equations (41) and (42), which are based upon the physical assumptions of the validity of Darcy's law and Hooke's law, was first given by Biot (1941). In the theory of ground water flow many authors, [Jacob (1940, 1950); Hantush (1964); De Wiest (1965)] arrived at equations similar to (44) by assuming that displacements occur in one direction only and that in that same direction, the total stresses do not change. If that direction is taken to be the z direction, it follows from (38) that $\partial u_z/\partial z = \epsilon$ since u_x and u_y are assumed to vanish. Then it follows from the last of (39) that

$$\sigma'_{zz} = (\lambda + 2\mu)\, \epsilon$$

and the condition $\sigma_{zz} = 0$ then implies $\sigma'_{zz} = \sigma$ so that it is found that

$$(\lambda + 2\mu)\, \epsilon = \sigma$$

Indeed, under these simplifying assumptions it appears that the function f in Eq. (43) vanishes, which leads to Eq. (44). However, the assumption that no displacements occur in the x and y direction is subject to doubt. Therefore, some cases of practical interest will be studied in detail in the following sections, starting from the general three-dimensional differential equations derived above. It will appear that in some cases the deviations of the results from the results obtained by solving the simplified equation (44) are negligible, but it will also appear that in some cases, the more complicated theory is the only one that succeeds in explaining experimental results.

For further general information with regard to the function f, the reader is referred to de Josselin de Jong (1963), who showed that a necessary condition for the vanishing of f is the vanishing of the rotations throughout the body. De Leeuw (1965) considered in detail the effect of vertical sand drains upon the fluid pressures in a consolidating soil, assuming axial symmetry and plane strain. For this special case De Leeuw succeeded in deriving necessary and sufficient conditions for the vanishing of the function f.

4. Radial Flow in a Leaky Aquifer of Small Thickness

In this section the problem of a leaky artesian aquifer of small thickness will be considered. For the special case in which the aquifer extends in the radial direction toward infinity, the final solution will be derived.

Cylindrical coordinates r, θ, z will be used,

$$r = (x^2 + y^2)^{1/2}, \qquad \theta = \arctan(y/x)$$

The z axis will be taken positive in the upward direction. The thickness of the aquifer will be denoted by b_1. It is assumed that all boundary conditions possess axial symmetry so that it can be assumed that the solution will also present axial symmetry. In order to be able to realize what simplifications follow from this assumption, the general equations of elasticity for a porous material, given in cartesian coordinates in Section 3, will be rewritten in cylindrical coordinates. As in Section 3, restriction is made to small deviations from an initial state of equilibrium in which steady flow occurs. The incremental stresses will again be denoted by the symbols σ_{ij} for total stress, σ'_{ij} for effective stress, and σ for the pore water pressure. Incremental strains will be denoted by ϵ_{ij}.

The strain components ϵ_{ij} $(i, j = r, \theta, z)$ are related to the displacement components u_r, u_θ, u_z by the following expressions (Sokolnikoff, 1956),

$$\begin{aligned}
\epsilon_{rr} &= \frac{\partial u_r}{\partial r}, & \epsilon_{\theta z} &= \frac{1}{2}\left(\frac{\partial u_\theta}{\partial z} + \frac{1}{r}\frac{\partial u_z}{\partial \theta}\right) \\
\epsilon_{\theta\theta} &= \frac{1}{r}\frac{\partial u_\theta}{\partial \theta} + \frac{u_r}{r}, & \epsilon_{zr} &= \frac{1}{2}\left(\frac{\partial u_z}{\partial r} + \frac{\partial u_r}{\partial z}\right) \\
\epsilon_{zz} &= \frac{\partial u_z}{\partial z}, & \epsilon_{r\theta} &= \frac{1}{2}\left(\frac{1}{r}\frac{\partial u_r}{\partial \theta} + \frac{\partial u_\theta}{\partial r} - \frac{u_\theta}{r}\right)
\end{aligned} \tag{45}$$

The components of effective stress are given by the following relations, which follow from Hooke's law,

$$\begin{aligned}
\sigma'_{rr} &= 2\mu\epsilon_{rr} + \lambda\epsilon, & \sigma'_{\theta z} &= 2\mu\epsilon_{\theta z} \\
\sigma'_{\theta\theta} &= 2\mu\epsilon_{\theta\theta} + \lambda\epsilon, & \sigma'_{zr} &= 2\mu\epsilon_{zr} \\
\sigma'_{zz} &= 2\mu\epsilon_{zz} + \lambda\epsilon, & \sigma'_{r\theta} &= 2\mu\epsilon_{r\theta}
\end{aligned} \tag{46}$$

where $\epsilon = \epsilon_{rr} + \epsilon_{\theta\theta} + \epsilon_{zz}$.

The equations of equilibrium are (Sokolnikoff, 1956)

$$\begin{aligned}
\frac{\partial \sigma_{rr}}{\partial r} + \frac{1}{r}\frac{\partial \sigma_{r\theta}}{\partial \theta} + \frac{\partial \sigma_{rz}}{\partial z} + \frac{\sigma_{rr} - \sigma_{\theta\theta}}{r} &= 0, & \sigma_{\theta z} &= \sigma_{z\theta} \\
\frac{\partial \sigma_{\theta r}}{\partial r} + \frac{1}{r}\frac{\partial \sigma_{\theta\theta}}{\partial \theta} + \frac{\partial \sigma_{\theta z}}{\partial z} + 2\,\frac{\sigma_{\theta r}}{r} &= 0, & \sigma_{zr} &= \sigma_{rz} \\
\frac{\partial \sigma_{zr}}{\partial r} + \frac{1}{r}\frac{\partial \sigma_{z\theta}}{\partial \theta} + \frac{\partial \sigma_{zz}}{\partial z} + \frac{\sigma_{zr}}{r} &= 0, & \sigma_{r\theta} &= \sigma_{\theta r}
\end{aligned} \tag{47}$$

where σ_{ij} $(i, j = r, \theta, z)$ are the components of total stress which can be resolved into the components of effective stress σ'_{ij} and the pore water pressure σ in the following way:

$$\begin{aligned}
\sigma_{rr} &= \sigma'_{rr} - \sigma, & \sigma_{\theta z} &= \sigma'_{\theta z}, & \sigma_{z\theta} &= \sigma'_{z\theta} \\
\sigma_{\theta\theta} &= \sigma'_{\theta\theta} - \sigma, & \sigma_{zr} &= \sigma'_{zr}, & \sigma_{rz} &= \sigma'_{rz} \\
\sigma_{zz} &= \sigma'_{zz} - \sigma, & \sigma_{r\theta} &= \sigma'_{r\theta}, & \sigma_{\theta r} &= \sigma'_{\theta r}
\end{aligned} \tag{48}$$

These equations are equivalent to Eqs. (26). They show that the normal components of stress are carried both by the grain skeleton and the water, whereas the shear stresses have to be carried by the grain skeleton only. Using (48), the equations of equilibrium can be expressed in terms of effective stress and pore water pressure. The three equations in the left part of (47) become

$$\begin{aligned}
\frac{\partial \sigma'_{rr}}{\partial r} + \frac{1}{r}\frac{\partial \sigma'_{r\theta}}{\partial \theta} + \frac{\partial \sigma'_{rz}}{\partial z} + \frac{\sigma'_{rr} - \sigma'_{\theta\theta}}{r} &= \frac{\partial \sigma}{\partial r} \\
\frac{\partial \sigma'_{\theta r}}{\partial r} + \frac{1}{r}\frac{\partial \sigma'_{\theta\theta}}{\partial \theta} + \frac{\partial \sigma'_{\theta r}}{\partial z} + 2\frac{\sigma'_{\theta r}}{r} &= \frac{1}{r}\frac{\partial \sigma}{\partial \theta} \\
\frac{\partial \sigma'_{zr}}{\partial r} + \frac{1}{r}\frac{\partial \sigma'_{z\theta}}{\partial \theta} + \frac{\partial \sigma'_{zz}}{\partial z} + \frac{\sigma'_{zr}}{r} &= \frac{\partial \sigma}{\partial z}
\end{aligned} \tag{49}$$

In the above equations no use has yet been made of the conditions of symmetry. In case of axial symmetry,

$$u_\theta = 0, \qquad \partial u_r/\partial \theta = 0, \qquad \partial u_z/\partial \theta = 0, \qquad \partial \sigma/\partial \theta = 0 \tag{50}$$

Then Eqs. (45) reduce to

$$\begin{aligned}
\epsilon_{rr} &= \partial u_r/\partial r, & \epsilon_{\theta z} &= 0 \\
\epsilon_{\theta\theta} &= u_r/r, & \epsilon_{zr} &= \tfrac{1}{2}(\partial u_z/\partial r + \partial u_r/\partial z) \\
\epsilon_{zz} &= \partial u_z/\partial z, & \epsilon_{r\theta} &= 0
\end{aligned} \tag{51}$$

and substitution of (51) into (46) gives

$$\begin{aligned}
\sigma'_{rr} &= 2\mu(\partial u_r/\partial r) + \lambda\epsilon, & \sigma'_{\theta z} &= 0 \\
\sigma'_{\theta\theta} &= 2\mu(u_r/r) + \lambda\epsilon, & \sigma'_{zr} &= \mu(\partial u_z/\partial r + \partial u_r/\partial z) \\
\sigma'_{zz} &= 2\mu(\partial u_z/\partial z) + \lambda\epsilon, & \sigma'_{r\theta} &= 0
\end{aligned} \tag{52}$$

where now

$$\epsilon = \partial u_r/\partial r + u_r/r + \partial u_z/\partial z \tag{53}$$

The equations of equilibrium, Eqs. (49), reduce to

$$\begin{aligned}
\frac{\partial \sigma'_{rr}}{\partial r} + \frac{\partial \sigma'_{rz}}{\partial z} + \frac{\sigma'_{rr} - \sigma'_{\theta\theta}}{r} &= \frac{\partial \sigma}{\partial r} \\
\frac{\partial \sigma'_{\theta\theta}}{\partial \theta} &= 0 \\
\frac{\partial \sigma'_{zr}}{\partial r} + \frac{\partial \sigma'_{zz}}{\partial z} + \frac{\sigma'_{zr}}{r} &= \frac{\partial \sigma}{\partial z}
\end{aligned} \tag{54}$$

In this stage of considerations, two different approaches can be used for further simplification of the problem. The first approach is to assume that the total vertical stresses do not change during the process of consolidation and that the radial displacements vanish identically. The first assumption, which is reasonable when the aquifer is very thin, leads to

$$\sigma_{zz} = 0$$

since σ_{zz} is the incremental stress component. It then follows from (48) that

$$\sigma'_{zz} = \sigma$$

and with the last equation in the left part of (52), the following result is obtained

$$\sigma = 2\mu(\partial u_z/\partial z) + \lambda\epsilon \tag{55}$$

Furthermore, it follows from (53) and the second assumption, that

$$\epsilon = \partial u_z/\partial z \tag{56}$$

Combination of the results (55) and (56) shows that

$$(\lambda + 2\mu)\,\epsilon = \sigma \tag{57}$$

from which it follows that the storage equation (41) reduces to

$$(\alpha + n\beta')(\partial\sigma/\partial t) = (K/\rho g)\,\nabla^2\sigma \tag{58}$$

where $\alpha = 1/(\lambda + 2\mu)$. This is the simplified equation given by Jacob (1946) and solved by Hantush and Jacob (1955) for the case of an aquifer of small thickness and infinite radial extent.

The second approach for further simplification is to assume that the aquifer deforms under the conditions of plane total stress

$$\sigma_{zz} = 0, \qquad \sigma_{rz} = 0 \tag{59}$$

These assumptions will be reasonable when the aquifer is confined between two soft layers (such as clay layers) which cannot resist shear stresses. The second approach seems more realistic than the first since the vanishing of the radial displacement is possible only if the confining layers are completely rigid in the radial direction and if no slip occurs along the common boundaries of the confining layers and aquifer. However, it will appear that the second approach, which will be worked out in the sequel, leads to a final solution for the aquifer of small thickness and infinite radial extent which differs from the solution given by

Hantush and Jacob (1955) only in the interpretation of one of the parameters appearing in the solution.

From (59), it follows that

$$\sigma'_{zz} = \sigma, \qquad \sigma'_{rz} \doteq 0 \tag{60}$$

so that the equation of equilibrium in the r direction, the first of Eqs. (54), reduces to

$$\frac{\partial \sigma'_{rr}}{\partial r} + \frac{\sigma'_{rr} - \sigma'_{\theta\theta}}{r} = \frac{\partial \sigma}{\partial r} \tag{61}$$

Substitution of the expressions for σ'_{rr} and $\sigma'_{\theta\theta}$ given in Eq. (52) into (61) gives

$$2\mu\left(\frac{\partial^2 u_r}{\partial r^2} + \frac{1}{r}\frac{\partial u_r}{\partial r} - \frac{u_r}{r^2}\right) + \lambda\frac{\partial \epsilon}{\partial r} - \frac{\partial \sigma}{\partial r} = 0$$

which can also be written in the form

$$2\mu\frac{\partial}{\partial r}\left(\frac{\partial u_r}{\partial r} + \frac{u_r}{r}\right) + \lambda\frac{\partial \epsilon}{\partial r} - \frac{\partial \sigma}{\partial r} = 0 \tag{62}$$

Integration of (62) gives

$$2\mu\left(\frac{\partial u_r}{\partial r} + \frac{u_r}{r}\right) + \lambda\epsilon - \sigma = 2F(t, z) \tag{63}$$

where $F(t, z)$ is an arbitrary function of t and z, the factor 2 being added for convenience.

It follows from the first of (60) and the last equation in the left part of (52) that

$$\sigma = 2\mu(\partial u_z/\partial z) + \lambda\epsilon$$

and with (53) it follows that

$$2\mu(\partial u_r/\partial r + u_r/r) = (\lambda + 2\mu)\,\epsilon - \sigma \tag{64}$$

Comparison of (64) and (43) shows that in this special case,

$$f = 2\mu(\partial u_r/\partial r + u_r/r) \tag{65}$$

which shows that in this case the function f, which describes the deviation of the simplified Terzaghi–Jacob theory from the Biot theory, is completely determined by the radial displacement. However, this result will not be needed in the following developments.

Combination of (63) and (64) gives

$$(\lambda + \mu)\,\epsilon = \sigma + F(t, z) \tag{66}$$

Use will be made of the assumption that the thickness of the aquifer is very small, from which it follows that all quantities can be approximated by their average values over the thickness b_1 of the aquifer. The average values will be denoted by underbars. They are defined by

$$\underline{G}(r, t) = (1/b_1)\int_0^{b_1} G(r, t, z)\, dz$$

where G may represent any function of the variables r, t, and z.

Multiplication of the storage equation (41), which in cylindrical coordinates states that

$$\frac{\partial \epsilon}{\partial t} = -n\beta' \frac{\partial \sigma}{\partial t} + \frac{K}{\rho g}\left(\frac{\partial^2 \sigma}{\partial r^2} + \frac{1}{r}\frac{\partial \sigma}{\partial r} + \frac{\partial^2 \sigma}{\partial z^2}\right)$$

by dz, and integration from 0 to b_1 gives, after dividing through by b_1,

$$\frac{\partial \underline{\epsilon}}{\partial t} = -n\beta' \frac{\partial \underline{\sigma}}{\partial t} + \frac{K}{\rho g}\left(\frac{\partial^2 \underline{\sigma}}{\partial r^2} + \frac{1}{r}\frac{\partial \underline{\sigma}}{\partial r}\right) + \frac{K}{\rho g b_1}\left\{\left[\frac{\partial \sigma}{\partial z}\right]_{z=h} - \left[\frac{\partial \sigma}{\partial z}\right]_{z=0}\right\} \tag{67}$$

In this equation, the values of $\partial\sigma/\partial z$ for $z = h$ and $z = 0$ have to be determined from specified boundary conditions along the upper and lower boundaries of the aquifer. From (66), it follows that

$$(\lambda + \mu)\,\underline{\epsilon} = \underline{\sigma} + \underline{F}(t) \tag{68}$$

The average radial displacement is found from (63) or from (64). In its simplest form the expression for the radial displacement is written as

$$\frac{2\mu}{r}\frac{\partial(\underline{u}_r r)}{\partial r} = \underline{F}(t) + \mu\underline{\epsilon} \tag{69}$$

In the sequel also an expression for the stress component $\underline{\sigma}_{rr}$ will be needed. It follows from the first of (48) and the first of (52) that

$$\underline{\sigma}_{rr} = 2\mu(\partial \underline{u}_r/\partial r) + \lambda\underline{\epsilon} - \underline{\sigma}$$

Using (68) and (69), this equation can be rewritten in the following form

$$\underline{\sigma}_{rr} = -(2\mu/r)\,\underline{u}_r + 2\underline{F}(t) \tag{70}$$

For a leaky aquifer extending towards infinity (Fig. 2), the boundary conditions are:

$$\begin{aligned} z = 0: &\quad \partial\sigma/\partial z = 0 \\ z = h: &\quad \partial\sigma/\partial z = -\underline{\sigma}/(b'K_1/K') \end{aligned} \tag{71}$$

$$\begin{aligned} r \to \infty: &\quad \underline{\sigma} \to 0 \\ r \to \infty: &\quad \underline{\sigma}_{rr} \to 0 \\ r = r_0: &\quad \underline{u}_r = 0 \\ r = r_0: &\quad \partial\underline{\sigma}/\partial r = \begin{cases} 0, & t < 0 \\ \dfrac{Q_0}{(K_1/\rho g)\, 2\pi r b_1}, & t > 0 \end{cases} \end{aligned} \tag{72}$$

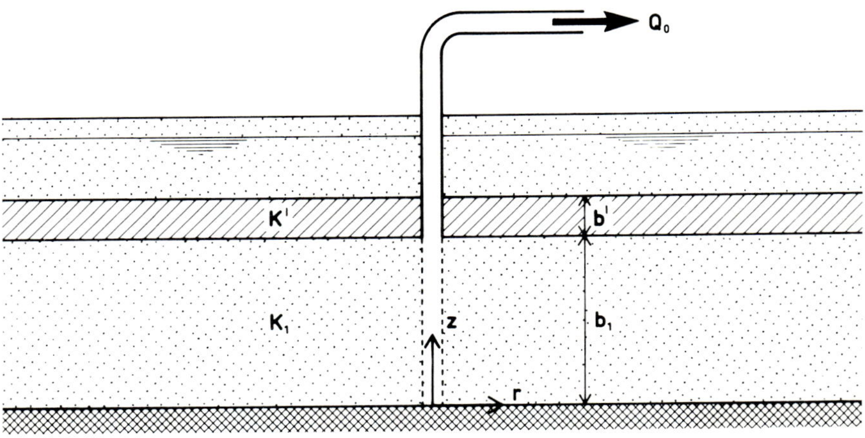

FIG. 2. Well in a leaky artesian aquifer.

The first of conditions (71) shows that the bottom layer is impermeable and the second condition shows that through the confining layer above the aquifer, with thickness b' and coefficient of permeability K', water enters the aquifer. Actually, this condition is the condition of continuity along the boundary $z = h$. The vertical component of the specific discharge is given by

$$q_z = -K_1[\partial\sigma/\partial z]_{z=h} = K'(\underline{\sigma} - \sigma_0)/b'$$

where σ_0 is the incremental pore water pressure in the layer above the confining layer. By putting $\sigma_0 = 0$ (which is justified if the head in the upper layer is constant), the second equation of (71) is obtained. The first two of the conditions (72) express that the stresses at infinity are supposed to vanish. The last two of the conditions (72) express that the aquifer is bounded internally by a rigid cylinder and that, starting at time $t = 0$, a constant amount of water Q_0 is extracted per unit time from the aquifer.

With (71), the storage equation (67) becomes

$$\frac{\partial \underline{\epsilon}}{\partial t} = -n\beta' \frac{\partial \underline{\sigma}}{\partial t} + \frac{K_1}{\rho g}\left(\frac{\partial^2 \underline{\sigma}}{\partial r^2} + \frac{1}{r}\frac{\partial \underline{\sigma}}{\partial r} - \frac{\underline{\sigma}}{B^2}\right) \tag{73}$$

where $B^2 = b_1 b' K_1/K'$.

A solution will be sought by means of operational mathematics. In the present case the Laplace transform (Churchill, 1958) is particularly suitable. The Laplace transform of a quantity, say the volume strain $\underline{\epsilon}$, is defined by

$$\bar{\underline{\epsilon}} = \int_0^\infty \exp(-st)\,\underline{\epsilon}\,dt \tag{74}$$

and will be denoted by overbars. It can be shown (Churchill, 1958) that the Laplace transform of $\partial\underline{\epsilon}/\partial t$ is given by

$$\int_0^\infty \exp(-st)(\partial\underline{\epsilon}/\partial t)\,dt = s\bar{\underline{\epsilon}} - \lim_{t\downarrow 0} \underline{\epsilon} \tag{75}$$

Under the conditions (72), the external loading of the aquifer does not present a discontinuity at time $t = 0$, which means that stresses and strains are continuous functions of time at $t = 0$. Hence, using the fact that for $t < 0$ all quantities are zero, it follows that the limit in the right-hand member of (75) is zero. The essential property of the Laplace transform follows from Eq. (75). This equation shows that the operation $\partial/\partial t$ is transformed into the much simpler operation of multiplication by s. The Laplace transform of Eq. (73) in total is

$$s\bar{\underline{\epsilon}} = -n\beta' s\bar{\underline{\sigma}} + \frac{K_1}{\rho g}\left(\frac{d^2\bar{\underline{\sigma}}}{dr^2} + \frac{1}{r}\frac{d\bar{\underline{\sigma}}}{dr} - \frac{\bar{\underline{\sigma}}}{B^2}\right) \tag{76}$$

Since the Laplace transform does not affect the differentiation with respect to r, the transforms of Eqs. (68)–(70) are

$$(\lambda + \mu)\,\bar{\underline{\epsilon}} = \bar{\underline{\sigma}} + C_3 \tag{77}$$

$$\frac{2\mu}{r}\frac{d(\bar{\underline{u}}_r r)}{dr} = C_3 + \mu\bar{\underline{\epsilon}} \tag{78}$$

$$\bar{\underline{\sigma}}_{rr} = -(2\mu/r)\,\bar{\underline{u}}_r + 2C_3 \tag{79}$$

where C_3 is the Laplace transform of $\underline{F}(t)$.

First the general solution of the transformed problem will be determined. From (76) and (77), the following differential equation in $\bar{\underline{\sigma}}$ is obtained,

$$\frac{d^2\bar{\underline{\sigma}}}{dr^2} + \frac{1}{r}\frac{d\bar{\underline{\sigma}}}{dr} - \bar{\underline{\sigma}}\left[\frac{1}{B^2} + \frac{\rho g(\alpha' + n\beta')\,s}{K_1}\right] = \frac{\rho g\alpha' s}{K_1}\,C_3 \tag{80}$$

where

$$\alpha' = 1/(\lambda + \mu) \tag{81}$$

The solution of the ordinary differential equation (80) is

$$\bar{\underline{\sigma}} = C_1 I_0(\omega r) + C_2 K_0(\omega r) - (\rho g\alpha'/K_1)(s/\omega^2)\,C_3 \tag{82}$$

where $\omega^2 = B^{-2} + [\rho g(\alpha' + n\beta')s/K_1]$ and where $I_0(x)$ and $K_0(x)$ are modified Bessel functions of order zero and of the first and second kind, respectively. The integration constants C_1 and C_2 as well as C_3 may in general depend upon s, which in the differential Eq. (80) can be regarded as a parameter.

The volume strain $\bar{\underline{\epsilon}}$ is obtained from (77) and (82). This gives

$$\bar{\underline{\epsilon}} = C_1\alpha' I_0(\omega r) + C_2\alpha' K_0(\omega r) + \alpha'[1 - (\rho g\alpha'/K_1)(s/\omega^2)]\,C_3 \tag{83}$$

Substitution of this result into (78) gives

$$\frac{2\mu}{r}\frac{d(\bar{\underline{u}}_r r)}{dr} = C_1\mu\alpha' I_0(\omega r) + C_2\mu\alpha' K_0(\omega r) + [1 + \mu\alpha' - \mu\alpha'(\rho g\alpha'/K_1)(s/\omega^2)]\,C_3$$

which can be written as

$$\frac{d(\bar{\underline{u}}_r r)}{dr} = \tfrac{1}{2}C_1\alpha' r I_0(\omega r) + \tfrac{1}{2}C_2\alpha' r K_0(\omega r) + \tfrac{1}{2}[\mu^{-1} + \alpha' - \alpha'(\rho g\alpha'/K_1)(s/\omega^2)]\,rC_3$$

Integration gives, when the result is divided through by r,

$$\bar{\underline{u}}_r = (C_1\alpha'/2\omega)\,I_1(\omega r) - (C_2\alpha'/2\omega)\,K_1(\omega r) + \tfrac{1}{4}[\mu^{-1} + \alpha' - \alpha'(\rho g\alpha'/K_1)(s/\omega^2)]\,rC_3 + C_4/r \tag{84}$$

Equation (79) now gives

$$\bar{\underline{\sigma}}_{rr} = -(C_1\alpha'\mu/\omega)\,I_1(\omega r) + (C_2\alpha'\mu/\omega)\,K_1(\omega r) + \tfrac{1}{2}[3 - \mu\alpha' + \mu\alpha'(\rho g\alpha'/K_1)(s/\omega^2)]\,C_3 - 2\mu C_4/r^2 \tag{85}$$

Equations (82)–(85) together constitute the general solution of the transformed problem of a leaky artesian aquifer of small thickness under the conditions of axial symmetry and plane total stress. In the general solution, four constants C_1, C_2, C_3, and C_4 appear. The values of these constants have to be determined from the boundary conditions. Since there are two boundaries and on each boundary a condition for the pore pressure, as well as a condition for the loading of the grain skeleton can be imposed, there are four boundary conditions to satisfy. From these four boundary conditions the four constants can be determined.

In the special case of an aquifer of infinite radial extent, the boundary conditions are given by Eqs. (72). Since the general solution is in the form of Laplace transforms, the boundary conditions have to be transformed first. It follows from the definition (74) that the Laplace transform of a function which is zero for all values of t is zero itself, and it can be shown (Churchill, 1958) that the Laplace transform of a function which is zero for $t < 0$ and takes a finite value, say 1, for $t > 0$ (this is usually called Heaviside's unit-step function $H(t)$) equals s^{-1}, in formula:

$$\text{when} \quad H(t) = \begin{cases} 0, & t < 0 \\ 1, & t > 0 \end{cases} \quad \text{then} \quad \int_0^\infty \exp(-st)\, H(t) = s^{-1} \tag{86}$$

Using the above results the Laplace transforms of the boundary conditions are:

$$\begin{aligned} r \to \infty: &\quad \bar{\underline{\sigma}} \to 0, \qquad r \to \infty: \quad \bar{\underline{\sigma}}_{rr} \to 0 \\ r = r_0: &\quad \bar{\underline{u}}_r = 0 \\ r = r_0: &\quad d\bar{\underline{\sigma}}\; dr = Q_0/[(K_1/\rho g)\, 2\pi r b_1 s] \end{aligned} \tag{87}$$

When r approaches infinity the Bessel functions $K_0(\omega r)$ and $K_1(\omega r)$ tend to zero, but the Bessel functions $I_0(\omega r)$ and $I_1(\omega r)$ increase indefinitely. It now follows from (82) and (85) that the first two conditions of (87) can be satisfied if and only if

$$C_1 = C_3 = 0 \tag{88}$$

The result $C_3 = 0$ is very interesting since this means that (77) can be written as $(\lambda + \mu)\bar{\underline{\epsilon}} = \bar{\underline{\sigma}}$ so that $(\lambda + \mu)\underline{\epsilon} = \underline{\sigma}$. Using this result, Eq. (73) gives

$$(\alpha' + n\beta') \frac{\partial \underline{\sigma}}{\partial t} = \frac{K_1}{\rho g} \left(\frac{\partial^2 \underline{\sigma}}{\partial r^2} + \frac{1}{r} \frac{\partial \underline{\sigma}}{\partial r} - \frac{\underline{\sigma}}{B^2} \right)$$

and this is, apart from the difference between α and α', exactly the same equation as solved by Hantush and Jacob (1955). Hence, the only difference between the two approaches of the problem for an aquifer of infinite radial extent is the different interpretation of the constant α, which describes the elastic behaviour of the soil skeleton. Although the solution was extensively discussed (Hantush, 1956), its complete deriviation will be given here, because some results are needed for the considerations in Section 6.

With $C_1 = C_3 = 0$, the expression (82) reduces to

$$\bar{\underline{\sigma}} = C_2 K_0(\omega r) \tag{89}$$

from which it follows that

$$d\bar{\underline{\sigma}}/dr = -C_2 \omega K_1(\omega r)$$

The constant C_2 can be determined from the last of the conditions (87). This gives

$$C_2 = -Q_0 \rho g / [2\pi K_1 b_1 s \omega r_0 K_1(\omega r_0)] \tag{90}$$

and substitution of this result into (89) yields

$$\bar{\underline{\sigma}} = -\frac{Q_0 \rho g}{2\pi K_1 b_1 \omega r_0} \frac{1}{s} \frac{K_0(\omega r)}{K_1(\omega r_0)} \tag{91}$$

When ωr_0 approaches zero, which will be the case if the radius r_0 of the well is very small, one may write $\omega r_0 K_1(\omega r_0) = 1$ and (91) reduces to

$$\bar{\underline{\sigma}} = -\frac{Q_0 \rho g}{2\pi K_1 b_1} \frac{K_0(\omega r)}{s} \tag{92}$$

where

$$\omega = [B^{-2} + \rho g(\alpha' + n\beta')\, s / K_1]^{1/2}$$

The problem will be completely solved if the inverse transform of this expression can be obtained. Although a general method of determining the inverse transform exists (actually this method will be used in Section 6), the simplest way of obtaining the inverse transform of (92) is by using a table of Laplace transforms. In such a table (e.g., Churchill, 1958) the following formula may be found

$$K_0(ks^{1/2}) = \int_0^\infty \exp(-st)[(2t)^{-1} \exp(-k^2/4t)]\, dt$$

Substitution of $s + a$ for the variable s in this formula gives

$$K_0[k(s + a)^{1/2}] = \int_0^\infty \exp(-st)[(2t)^{-1} \exp(-k^2/4t) \exp(-at)]\, dt$$

Finally, using the theorem that multiplication by s^{-1} of the Laplace transform of a function corresponds to integration with respect to the time of the original function, it is found that

$$s^{-1}K_0[k(s + a)^{1/2}] = \int_0^\infty \exp(-st) \left\{\int_0^t (2\tau)^{-1} \exp(-k^2/4\tau) \exp(-a\tau)\, d\tau\right\} dt \tag{93}$$

Since in the expression (92),

$$\begin{aligned} \omega r &= [B^{-2} + \rho g(\alpha' + n\beta')\, s/K_1]^{1/2}\, r \\ &= [r^2 \rho g(\alpha' + n\beta')/K_1]^{1/2} \{s + K_1/[\rho g(\alpha' + n\beta')\, B^2]\}^{1/2} \end{aligned}$$

the formula (93) can be used for the determination of the inverse transform of (92). This gives

$$\underline{\sigma} = -\frac{Q_0 \rho g}{4\pi K_1 b_1} \int_0^t \frac{1}{\tau} \exp\left[-\frac{r^2 \rho g(\alpha' + n\beta')}{4K_1 \tau} - \frac{K_1 \tau}{\rho g(\alpha' + n\beta')\, B^2}\right] d\tau$$

which by putting $y = r^2 \rho g(\alpha' + n\beta')/(4K_1\tau)$, may alternatively be written in the following form

$$\underline{\sigma}/\rho g = -[Q_0/(4\pi K_1 b_1)] \int_{r^2 \rho g(\alpha' + n\beta')/(4K_1 t)}^{\infty} y^{-1} \exp[-y - r^2/(4B^2 y)]\, dy \tag{94}$$

Apart from the difference between α and α' this formula corresponds to the result given by Hantush and Jacob (1955). The integral in the right-hand member of (94) was tabulated by Hantush (1956). Since in many cases the outer boundary of the aquifer can be considered to lie at infinity, Hantush's formula is of wide applicability, and many measurements have confirmed its validity. As an example, in Fig. 3 some experimental results are compared with the theoretical values. The experimental data were obtained by the Netherlands Government Institute for Water Supply during a pumping test in the polder Vierbannen on the island of Schouwen-Duiveland in the south western part of the Netherlands. This pumping test has been described by Van Eyden *et al.* (1963). For comparison of theoretical and experimental

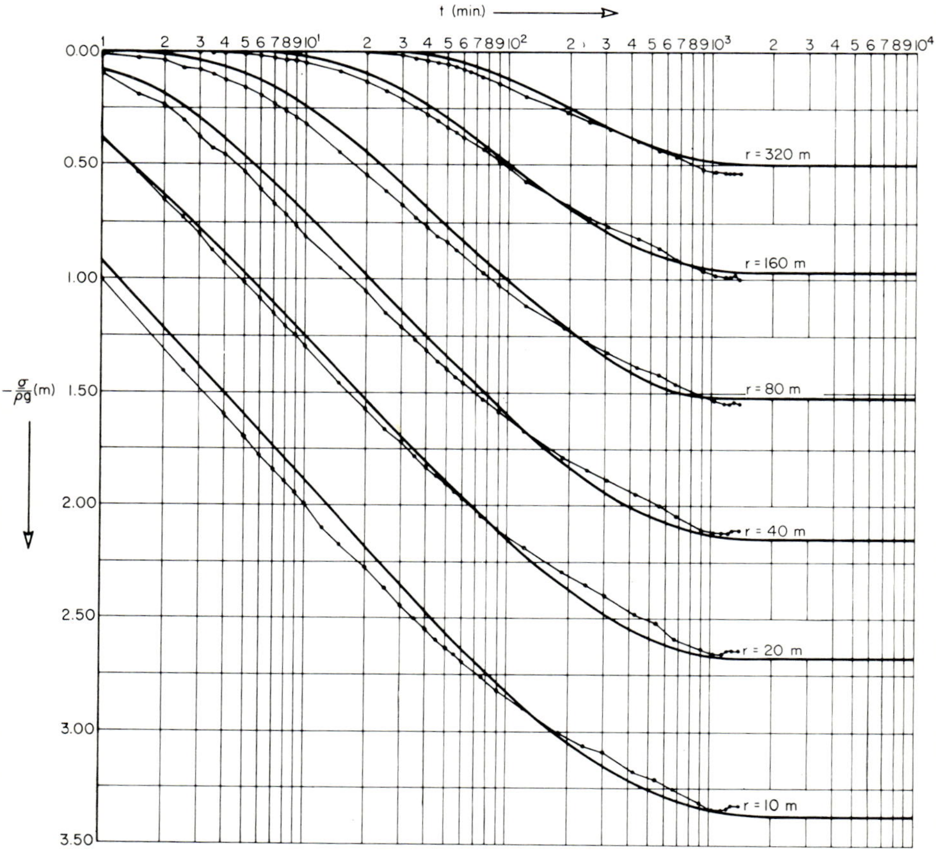

FIG. 3. Theoretical and experimental results for the drawdown in a leaky artesian aquifer during a pumping test.

results the following numerical values were introduced into Hantush's formula,

$$Q_0 = 1.74 \times 10^{-2} \text{ m}^3/\text{sec}, \qquad K_1 b_1 = 0.30 \times 10^{-2} \text{ m}^2/\text{sec}$$

$$B = 400 \text{ m}, \qquad \rho g b(\alpha' + n\beta') = 5.6 \times 10^{-4}$$

Various values for r/B were used, in accordance with the location of the observation wells in the field. As can be seen from Fig. 3, experimental and theoretical results rather closely fit each other despite the fact that the proper situation at Vierbannen is more complicated than the situation for which the theory was established. Actually, a more realistic description of the situation at Vierbannen would involve an aquifer consisting of two layers, the two layers being separated by a semi-impermeable layer. In Section 6, such an aquifer will be studied in detail.

It should be noted that the reduction of the solution of the problem considered in this section to the well-known solution of the simplified differential equation is entirely due to the fact that the outer boundary of the aquifer was supposed to lie at infinity. When aquifers of finite radial extent are considered, the solution of the general differential equations of Biot, under the conditions of axial symmetry and plane total stress, may differ substantially from the solution obtained by assuming that the radial displacements are zero.

5. Axially Symmetric Problems for the Half-Space

An important class of problems is concerned with solids of revolution in which the boundary conditions for the stresses, the displacements, and the pore water pressure are symmetric with respect to the axis of revolution. Then, provided that the soil properties are isotropic, the solution will also present axial symmetry. In this section, the basic equations which govern the axially symmetric problems will be given. The generalization with respect to the considerations in Section 4 is that it will no longer be assumed that the thickness of the aquifer is very small. For the special case of a half-space bounded by a plane perpendicular to the axis of symmetry, a general method of solution, due to McNamee and Gibson (1960b), will be described.

Again the basic equations are the general differential equations (41) and (42). In order to simplify the analysis, restriction will be made to completely saturated soils ($S_r = 1, \beta' = \beta$) and it will be assumed that the compressibility of the pore fluid itself can be disregarded, $\beta = 0$. With $\beta' = 0$, the basic equations (41) and (42) reduce to

$$\partial\epsilon/\partial t = (K/\rho g)\,\nabla^2\sigma \tag{95}$$

$$(\lambda + 2\mu)\,\nabla^2\epsilon = \nabla^2\sigma \tag{96}$$

In cylindrical coordinates r, θ, z, the operator ∇^2 is, in case of axial symmetry

$$\nabla^2 = \frac{\partial^2}{\partial r^2} + \frac{1}{r}\frac{\partial}{\partial r} + \frac{\partial^2}{\partial z^2} \tag{97}$$

Since the boundary conditions in general are expressed in terms of stresses and displacements, and not in terms of the volume strain ϵ, the relations between stresses, displacements, and volume strain have to be known. The effective stresses are related to the displacement components by the following equations, which were derived from Hooke's law in Section 4, Eq. (52),

$$
\begin{aligned}
\sigma'_{rr} &= 2\mu(\partial u_r/\partial r) + \lambda\epsilon, & \sigma'_{\theta z} &= 0 \\
\sigma'_{\theta\theta} &= 2\mu(u_r/r) + \lambda\epsilon, & \sigma'_{zr} &= \mu(\partial u_z/\partial r + \partial u_r/\partial z) \\
\sigma'_{zz} &= 2\mu(\partial u_z/\partial z) + \lambda\epsilon, & \sigma'_{r\theta} &= 0
\end{aligned}
\tag{98}
$$

where u_r and u_z are the displacement components in the r and z directions and where the volume strain ϵ is related to u_r and u_z by

$$\epsilon = \partial u_r/\partial r + u_r/r + \partial u_z/\partial z \tag{99}$$

In order to obtain equations in which the displacement components are related to ϵ and σ, use has to be made of the equations of equilibrium in the r and z directions, which are given by the first and the last of Eqs. (54),

$$
\begin{aligned}
\frac{\partial\sigma'_{rr}}{\partial r} + \frac{\sigma'_{rr} - \sigma'_{\theta\theta}}{r} + \frac{\partial\sigma'_{rz}}{\partial z} &= \frac{\partial\sigma}{\partial r} \\
\frac{\partial\sigma'_{zz}}{\partial z} + \frac{\partial\sigma'_{zr}}{\partial r} + \frac{\sigma'_{zr}}{r} &= \frac{\partial\sigma}{\partial z}
\end{aligned}
\tag{100}
$$

Substitution of (98) into (100) yields the equations of equilibrium expressed in terms of displacements

$$
\begin{aligned}
\mu\,\nabla^2 u_r - \mu u_r/r^2 + (\lambda + \mu)(\partial\epsilon/\partial r) &= \partial\sigma/\partial r \\
\mu\,\nabla^2 u_z + (\lambda + \mu)(\partial\epsilon/\partial z) &= \partial\sigma/\partial z
\end{aligned}
\tag{101}
$$

The first of these equations relates the radial displacement component u_r to the quantities ϵ and σ, and the second relates the vertical displacement component u_z to these quantities.

One of the main difficulties of the problem is that it is not sufficient to solve the basic differential equations (95) and (96), since the boundary conditions for the elastic material will give conditions which the stresses or the displacements have to satisfy. Therefore, it is necessary to solve Eqs. (101) once that the general solutions of (95) and (96) have been found. Only when general expressions for the displacements have been obtained, can the boundary conditions be introduced to select the appropriate solution of the problem from the general solution. Since Eqs. (101) are differential equations of the second order, this procedure of solving particular problems is rather complicated. A method which simplifies this procedure was given by McNamee and Gibson (1960a). This method will be used in the sequel, the only differences from the original publications of McNamee and Gibson being differences in notations and sign convention.

A function $\mathscr{S}$ is introduced such that

$$\sigma = (\lambda + 2\mu)\, \epsilon + 2\mu(\partial \mathscr{S}/\partial z) \tag{102}$$

In order to satisfy Eq. (96), it is necessary that $\nabla^2(\partial \mathscr{S}/\partial z) = 0$, which surely will be satisfied by putting

$$\nabla^2 \mathscr{S} = 0 \tag{103}$$

A second function, $\mathscr{E}$, is introduced such that

$$\begin{aligned} u_r &= \partial \mathscr{E}/\partial r + z\, \partial \mathscr{S}/\partial r \\ u_z &= \partial \mathscr{E}/\partial z + z\, \partial \mathscr{S}/\partial z - \mathscr{S} \\ \epsilon &= \nabla^2 \mathscr{E} \end{aligned} \tag{104}$$

Substitution of these expressions into (101) shows that the equations of equilibrium are satisfied identically, provided that (103) is valid. Substitution into (95) shows that $\mathscr{E}$ has to satisfy the following differential equation

$$\nabla^2(\partial \mathscr{E}/\partial t) = C\, \nabla^2\, \nabla^2 \mathscr{E} \tag{105}$$

where $c = K(\lambda + 2\mu)/\rho g$, the so-called *consolidation coefficient.*

The differential equations (103) and (105) together form the new system of equations to be solved. The principal advantage of the introduction of the functions $\mathscr{E}$ and $\mathscr{S}$ is that the displacements can directly be obtained from them by differentiation, as is seen from (104). The functions $\mathscr{E}$ and $\mathscr{S}$ are therefore called *displacement functions.* Since the effective stresses can be obtained from the displacements with the help of Eqs. (98), they can be expressed in terms of the displacement functions by substituting (104) into (98). The pore water pressure is given by (102) and the total stresses can be obtained from the effective stresses and the pore water pressure by means of relations of the form (48). The quantities which will be used in the sequel are:

$$\begin{aligned} \sigma &= (\lambda + 2\mu)\, \nabla^2 \mathscr{E} + 2\mu\, \partial \mathscr{S}/\partial z \\ \partial \sigma/\partial z &= (\lambda + 2\mu)\, \nabla^2(\partial \mathscr{E}/\partial z) + 2\mu\, \partial^2 \mathscr{S}/\partial z^2 \\ \sigma_{zz}/2\mu &= \partial^2 \mathscr{E}/\partial z^2 - \nabla^2 \mathscr{E} + z\, \partial^2 \mathscr{S}/\partial z^2 - \partial \mathscr{S}/\partial z \\ \sigma_{rz}/2\mu &= \partial^2 \mathscr{E}/\partial r\, \partial z + z \partial^2 \mathscr{S}/\partial r\, \partial z \end{aligned} \tag{106}$$

The last two quantities of (106) represent the components of total stress acting upon a plane with normal in the z direction, i.e., a horizontal plane.

A general solution of the partial differential equations can be obtained by using operational mathematics. Therefore, two integral transforms are introduced. For general information regarding integral transforms, the reader is referred to Sneddon (1951). The Hankel transform is defined by

$$F_h(\xi, z, t) = \int_0^\infty r J_0(r\xi) F(r, z, t)\, dr \tag{107}$$

with the inverse transform

$$F(r, z, t) = \int_0^\infty \xi J_0(r\xi) F_h(\xi, z, t)\, d\xi \tag{108}$$

Furthermore, the Laplace transform is defined by

$$\bar{F}_h(\xi, z, s) = \int_0^\infty \exp(-st) F_h(\xi, z, t)\, dt \tag{109}$$

with the inverse transform

$$F_h(\xi, z, t) = \frac{1}{2\pi i} \int_{-i\infty}^{\gamma+i\infty} \exp(st)\, \bar{F}_h(\xi, z, s)\, ds \tag{110}$$

where γ is a real number which should be taken larger than the real part of the singularity with the largest real part of all singularities of $\bar{F}_h(\xi, z, s)$. In the sequel, Hankel transforms will be indicated by the suffix h and Laplace transforms will be indicated by an overbar. The utility of the transforms appears when the differential equations are transformed. This can be done by multiplying Eqs. (103) and (105) by

$$r J_0(r\xi) \exp(-st)\, dr\, dt$$

and then integrating over r and t from zero to infinity. The Hankel transform has the property that the operator $\partial^2/\partial r^2 + r^{-1}\partial/\partial r$ is transformed into multiplication by $-\xi^2$ and the Laplace transform has the property that the operator $\partial/\partial t$ is transformed into multiplication by s and substraction of the initial value of the transformed quantity. The result of the successive transforms is a pair of ordinary differential equations in which only derivatives with respect to z appear and in which ξ and s can be considered as parameters. Thus the differential equations (103) and (105) are transformed into the following ordinary differential equations

$$\left(\frac{d^2}{dz^2} - \xi^2 - \frac{s}{c}\right)\left(\frac{d^2}{dz^2} - \xi^2\right) \bar{\mathscr{E}}_h = 0, \qquad \left(\frac{d^2}{dz^2} - \xi^2\right) \bar{\mathscr{S}}_h = 0 \tag{111}$$

where use has been made of the fact that the initial value of the volume strain is zero in saturated soils. The general solutions of these equations are

$$\begin{aligned}\bar{\mathscr{E}}_h &= A_1 \exp(+z\xi) + A_2 \exp[+z(\xi^2 + s/c)^{1/2}] \\ &\quad + A_3 \exp(-z\xi) + A_4 \exp[-z(\xi^2 + s/c)^{1/2}] \\ \bar{\mathscr{S}}_h &= B_1 \exp(+z\xi) + B_2 \exp(-z\xi)\end{aligned} \tag{112}$$

In the present considerations, restriction is made to solutions for the half-space $z \leqslant 0$. It is assumed that the stresses at infinity vanish. In regard with (106) and (112), this assumption leads to the following simplifications:

$$A_3 = A_4 = B_2 = 0 \tag{113}$$

Inversion of the general solution (112) of the transformed problem gives, with (108) and (110) as the general solution of the original problem,

$$\mathscr{E} = \frac{1}{2\pi i}\int_0^\infty \xi J_0(r\xi)\, d\xi \int_{\gamma-i\infty}^{\gamma+i\infty} \{A_1 \exp(z\xi) + A_2 \exp[z(\xi^2 + s/c)^{1/2}]\} \exp(st)\, ds \tag{114}$$

$$\mathscr{S} = \frac{1}{2\pi i}\int_0^\infty \xi J_0(r\xi)\, d\xi \int_{\gamma-i\infty}^{\gamma+i\infty} B_1 \exp(z\xi) \exp(st)\, ds \tag{115}$$

Using the fact that $J_0(r\xi)$ satisfies Bessel's equation,

$$\frac{\partial^2 J_0(r\xi)}{\partial r^2} + \frac{1}{r}\frac{\partial J_0(r\xi)}{\partial r} + \xi^2 J_0(r\xi) = 0$$

the expressions (106) give, with (114) and (115),

$$\begin{aligned}\sigma = \frac{1}{2\pi i}\int_0^\infty \xi J_0(r\xi)\, d\xi \int_{\gamma-i\infty}^{\gamma+i\infty} &\{(\lambda + 2\mu) A_2(s/c) \exp[z(\xi^2 + s/c)^{1/2}] \\ &+ 2\mu B_1 \xi \exp(z\xi)\} \exp(st)\, ds\end{aligned} \tag{116}$$

$$\begin{aligned}\partial\sigma/\partial z = \frac{1}{2\pi i}\int_0^\infty \xi J_0(r\xi)\, d\xi \int_{\gamma-i\infty}^{\gamma+i\infty} &\{(\lambda + 2\mu) A_2(s/c)(\xi^2 + s/c)^{1/2} \\ &\times \exp[z(\xi^2 + s/c)^{1/2}] + 2\mu B_1 \xi^2 \exp(z\xi)\} \exp(st)\, ds\end{aligned} \tag{117}$$

$$\begin{aligned}\sigma_{zz}/2\mu = \frac{1}{2\pi i}\int_0^\infty \xi^2 J_0(r\xi)\, d\xi \int_{\gamma-i\infty}^{\gamma+i\infty} &\{A_1 \xi \exp(z\xi) + A_2 \xi \exp[z(\xi^2 + s/c)^{1/2}] \\ &+ B_1(z\xi - 1) \exp(z\xi)\} \exp(st)\, ds\end{aligned} \tag{118}$$

$$\begin{aligned}\sigma_{rz}/2\mu = -\frac{1}{2\pi i}\int_0^\infty \xi^2 J_1(r\xi)\, d\xi \int_{\gamma-i\infty}^{\gamma+i\infty} &\{A_1 \xi \exp(z\xi) + A_2(\xi^2 + s/c)^{1/2} \\ &\times \exp[z(\xi^2 + s/c)^{1/2}] + B_1 z\xi \exp(z\xi)\} \exp(st)\, ds\end{aligned} \tag{119}$$

Similar expressions may be obtained for other quantities, such as the displacement components.

The integration constants A_1, A_2, and B_1, which in general depend upon the parameters ξ and s, have to be determined from the boundary conditions along the surface $z = 0$. When these constants have been found, the solution of the problem is complete. However, mathematical difficulties may arise in the evaluation of the integrals in the s and ξ planes, depending upon the nature of the boundary conditions. Only in cases with very simple boundary conditions can the integrals be evaluated in closed form. In Section 6 an application of the method described above will be discussed.

6. The Noordbergum Effect

As an example, the approximate solution of the problem of a pumping test in the upper layer of a two-storied leaky artesian aquifer of infinite radial extent will be determined (Fig. 4). It is assumed that the lower layer of the aquifer tends towards infinity in negative z direction. Storage in the semi-impermeable layer separating the two stories of the aquifer will be disregarded, which means that this layer is assumed to be incompressible. Although a correct formulation of the problem has to account for the coupling of the flow phenomena in the two layers, it will be assumed that the pore water pressures in the upper layer satisfy Hantush's equation. This assumption seems reasonable if the replenishment from the lower layer is small. The above assumptions make it possible to use the method of Section 5 for the solution of the problem in the lower layer, by using the solution for the upper layer as a boundary condition. Furthermore, it will be assumed that the total vertical load acting upon the lower layer does not change, so that the incremental stress σ_{zz} vanishes along the surface $z = 0$. It is also assumed that along the surface $z = 0$, the shear stress σ_{rz} vanishes. The coefficient of permeability of the upper layer of the aquifer will be denoted by K_1 and its thickness by b_1. The coefficient of permeability of the lower layer of the aquifer will be denoted by K_2. The coefficients of permeability of the two semi-impermeable layers will be denoted by K' for the upper one and K'' for the lower one. Their respective thicknesses will be denoted by b' and b''. For the pore water pressure in the upper layer, the symbol $\underline{\sigma}_1$ will be used. The symbols σ and σ_{ij} refer to the pore water pressure and the components of total stress in the lower aquifer.

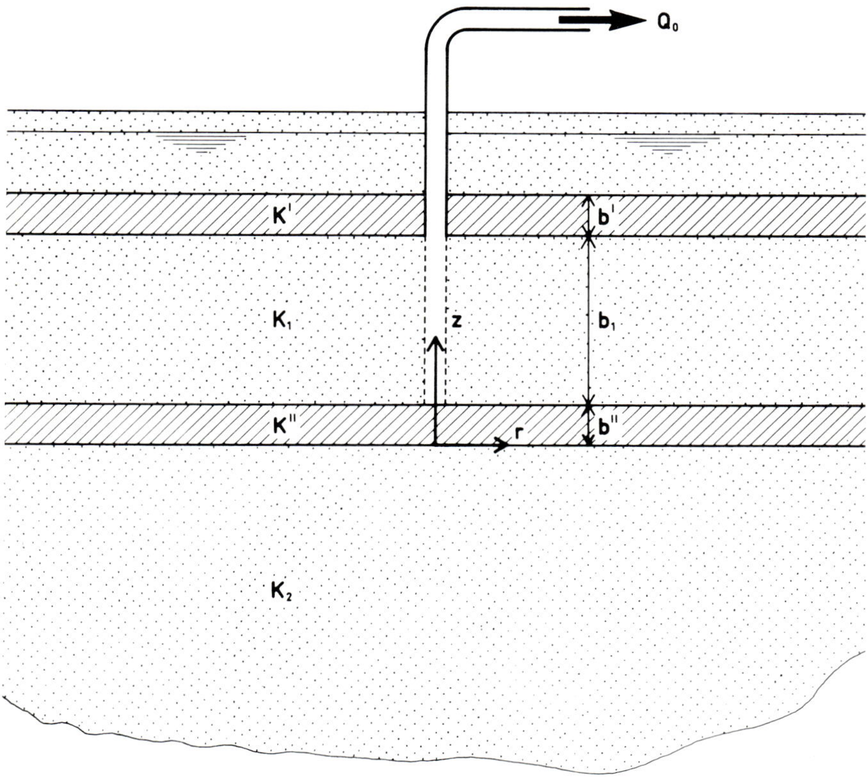

FIG. 4. Well in the upper layer of a two-storied leaky artesian aquifer.

The boundary conditions for the lower layer are:

$$z = 0: \qquad \sigma_{zz} = 0 \tag{120}$$

$$z = 0: \qquad \sigma_{rz} = 0 \tag{121}$$

$$z = 0: \qquad -(K_2/\rho g)(\partial\sigma/\partial z) = (K''/\rho g)(\sigma - \underline{\sigma}_1)/b'' \tag{122}$$

The condition (122) follows from the continuity of the vertical component of the specific discharge along the surface $z = 0$. Using the notation $L = b''K_2/K''$, this condition can be written as

$$z = 0: \qquad \partial\sigma/\partial z = -(\sigma - \underline{\sigma}_1)/L \tag{123}$$

where $\underline{\sigma}_1$ is assumed to be given by Hantush's formula, Eq. (94).

In view of the general expressions (118) and (119) for the stresses

σ_{zz} and σ_{rz}, it immediately follows from the conditions (120) and (121) that

$$\xi A_1 + \xi A_2 - B_1 = 0 \tag{124}$$

$$\xi A_1 + (\xi^2 + s/c)^{1/2} A_2 = 0 \tag{125}$$

In order to find the condition for the constants A_1, A_2, and B_1, resulting from the boundary condition (123), it is necessary to have an integral representation of $\underline{\sigma}_1$, because σ and $\partial\sigma/\partial z$ are given as integral representations. The Laplace transform of $\underline{\sigma}_1$ is given by (92),

$$\bar{\underline{\sigma}}_1 = -(P/s)\, K_0(\omega r) \tag{126}$$

where for simplicity the notation $P = Q_0\rho g/(2\pi K_1 b_1)$ is introduced and where

$$\omega^2 = B^{-2} + \rho g(\alpha' + n\beta')\, s/K_1 = (s + c'/B^2)/c' \tag{127}$$

The constant $c' = K_1/[\rho g(\alpha' + n\beta')]$ is called the *consolidation coefficient* of the upper layer of the aquifer. The Hankel transform of (126) is (Bateman, 1954),

$$\begin{aligned}\bar{\underline{\sigma}}_{1-h} &= -(P/s)\int_0^\infty r J_0(r\xi)\, K_0(\omega r)\, dr = -P/[s(\omega^2 + \xi^2)] \\ &= -Pc'/[s(s + c'\xi^2 + c'B^{-2})]\end{aligned}$$

Hence, one may write

$$\underline{\sigma}_1 = (2\pi i)^{-1}\int_0^\infty \xi J_0(r\xi)\, d\xi \int_{\gamma-i\infty}^{\gamma+i\infty} \{-Pc'/[s(s + c'\xi^2 + c'B^{-2})]\}\exp(st)\, ds \tag{128}$$

The boundary condition (123) now gives, with (116), (117), and (128),

$$\begin{aligned}&(\lambda + 2\mu)(s/c)(\xi^2 + s/c)^{1/2} A_2 + 2\mu\xi^2 B_1 \\ &\quad = -L^{-1}\{(\lambda + 2\mu)(s/c)\, A_2 + 2\mu\xi B_1 + Pc'/[s(s + c'\xi^2 + c'B^{-2})]\}\end{aligned} \tag{129}$$

The equations (124), (125), and (129) together form a system of three equations with three unknowns. Solving for A_1, A_2, and B_1 gives

$$\begin{aligned}A_1 &= \frac{Pc'}{2\mu\xi sL}\frac{(\xi^2 + s/c)^{1/2}}{N(s, \xi)}, \qquad A_2 = -\frac{Pc'}{2\mu sL}\frac{1}{N(s, \xi)} \\ B_1 &= \frac{Pc'}{2\mu sL}\frac{(\xi^2 + s/c)^{1/2} - \xi}{N(s, \xi)}\end{aligned} \tag{130}$$

where

$$N(s, \xi) = (s + c'\xi^2 + c'B^{-2})\{(2ms/c)[(\xi^2 + s/c)^{1/2} + L^{-1}] - \xi(\xi + L^{-1})[(\xi^2 + s/c)^{1/2} - \xi]\}$$

and where $m = (\lambda + 2\mu)/4\mu$. Substitution of the values for the constants A_1, A_2, and B_1 given in (130) into Eq. (116) yields the following expression for the pore water pressure

$$\sigma = -\frac{Pc'}{2\pi i L}\int_0^\infty \xi J_0(r\xi)\, d\xi \times \int_{\gamma-i\infty}^{\gamma+i\infty} \frac{\left[\begin{aligned}&\{(2ms/c)\exp[z(\xi^2 + s/c)^{1/2}]\\ &\quad - \xi[(\xi^2 + s/c)^{1/2} - \xi]\exp(z\xi)\}\exp(st)\end{aligned}\right]}{\left[\begin{aligned}&s(s + c'\xi^2 + c'B^{-2})\{(2ms/c)[(\xi^2 + s/c)^{1/2} + L^{-1}]\\ &\quad - \xi(\xi + L^{-1})[(\xi^2 + s/c)^{1/2} - \xi]\}\end{aligned}\right]}\, ds \tag{131}$$

The problem is now reduced to the evaluation of the integrals appearing in Eq. (131). Since it seems impossible to obtain a general expression in terms of elementary functions, some special cases will be considered in the sequel.

Steady State Solution

When $t \to \infty$, the system will tend towards a steady state. The pore water pressures in this steady state are most easily obtained by using a theorem from the theory of the Laplace transformation (Carslaw and Jaeger, 1947, p. 256) which states that

$$\lim_{t\to\infty} \sigma = \lim_{s\to 0} s\bar{\sigma} \tag{132}$$

where $\bar{\sigma}$ is the Laplace transform of σ. The theorem holds under the condition that the limit in the right-hand member exists. Evaluation of this limit with (131) gives

$$\lim_{t\to\infty} \sigma = -\frac{Pc'}{L}\int_0^\infty \xi J_0(r\xi)\, d\xi \times \lim_{s\to 0} \frac{\left[\begin{aligned}&\{(2ms/c)\exp[z(\xi^2 + s/c)^{1/2}]\\ &\quad - \xi[(\xi^2 + s/c)^{1/2} - \xi]\exp(z\xi)\}\end{aligned}\right]}{\left[\begin{aligned}&(s + c'\xi^2 + c'B^{-2})\{(2ms/c)[(\xi^2 + s/c)^{1/2} + L^{-1}]\\ &\quad - \xi(\xi + L^{-1})[(\xi^2 + s/c)^{1/2} - \xi]\}\end{aligned}\right]}$$

or

$$\sigma_\infty = \lim_{t\to\infty} \sigma = -\frac{P}{L}\int_0^\infty \frac{\xi J_0(r\xi)\exp(z\xi)}{(\xi^2 + B^{-2})(\xi + L^{-1})}\, d\xi \tag{133}$$

No explicit expression for this integral is known to the author. Numerical evaluation is possible by writing $\xi = \rho B^{-1}$. Then the formula (133) can be written in the following dimensionless form:

$$\frac{\sigma_\infty}{P} = -\frac{B}{L}\int_0^\infty \frac{\rho J_0(\rho r/B)\exp(\rho z/B)}{(\rho^2+1)(\rho+B/L)}\,d\rho \tag{134}$$

In Fig. 5, numerical values are represented graphically for the case $B/L = 0.35$.

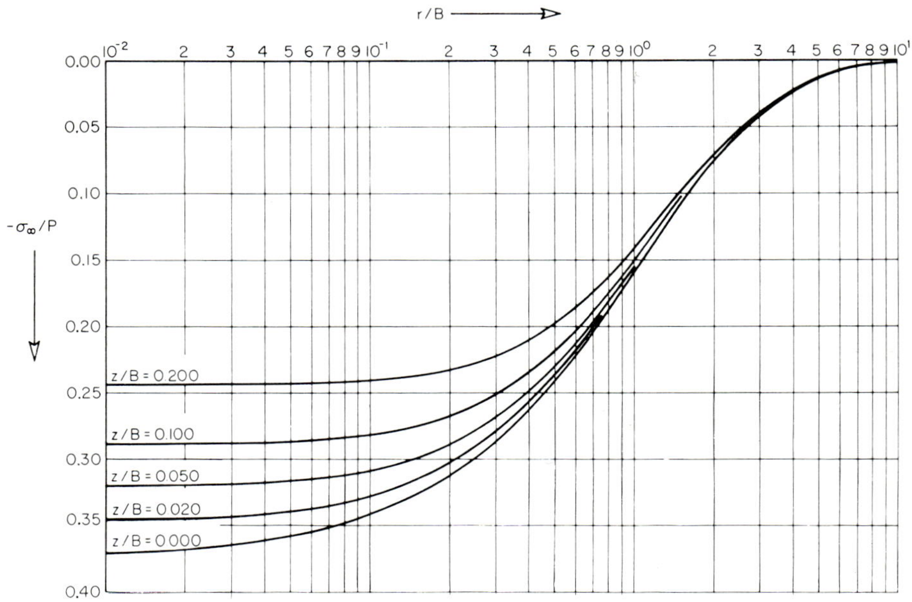

FIG. 5. Final values of the drawdown in the lower layer of a two-storied leaky artesian aquifer under the influence of a well in the upper layer.

Solution for Small Values of the Time

From the definition of the Laplace transformation,

$$\bar{\sigma} = \int_0^\infty \exp(-st)\,\sigma\,dt$$

it follows that when s is very large, the integral will mainly consist of contributions for small values of the time because the factor $\exp(-st)$ will tend to zero very rapidly. Therefore, an approximation suitable for small values of the time is obtained by taking s large in the transformed function.

When s is very large, the exponential form $\exp[z(\xi^2 + s/c)^{1/2}]$ in Eq. (131) will vanish because $z < 0$. Hence (131) reduces to

$$\sigma = [Pc'/(2\pi iL)] \int_0^\infty \xi^2 J_0(r\xi)\, d\xi \int_{\gamma-i\infty}^{\gamma+i\infty} (s/c)^{1/2}[(1 + \xi^2 c/s)^{1/2} - \xi(c/s)^{1/2}]$$
$$\times \exp(z\xi) \exp(st)\, s^{-1}(s + c'\xi^2 + c'B^{-2})^{-1} \{(2ms/c)(s/c)^{1/2} [(1 + \xi^2 c/s)^{1/2}$$
$$+ L^{-1}(c/s)^{1/2}] - \xi(\xi + L^{-1})(s/c)^{1/2} [(1 + \xi^2 c/s)^{1/2} - \xi(c/s)^{1/2}]\}^{-1}\, ds$$

Expansion in decreasing powers of s gives

$$\sigma = [Pc'c/(4\pi imL)] \int_0^\infty \xi^2 J_0(r\xi) \exp(z\xi)\, d\xi$$
$$\times \int_{\gamma-i\infty}^{\gamma+i\infty} s^{-3}\{1 - (\xi + L^{-1})(c/s)^{1/2} + \cdots\} \exp(st)\, ds$$

The integrals in the complex s plane are well-known inverse Laplace transforms (Bateman, 1954). Hence,

$$\sigma = [Pc'c/(2mL)] \int_0^\infty \xi^2 J_0(r\xi) \exp(z\xi)$$
$$\times \{\tfrac{1}{2}t^2 - (8/15)\, \pi^{-1/2}(\xi + L^{-1})\, c^{1/2}t^{5/2} + \cdots\}\, d\xi$$

and since $z < 0$, the above integrals are well-known convergent integrals which happen to be of the Laplace transform type. Evaluation of these integrals gives

$$\frac{\sigma}{P} = \frac{c'c}{2mL} \left\{ \frac{t^2}{2} \frac{2z^2 - r^2}{(z^2 + r^2)^{5/2}} + \frac{8t^2}{15} \left(\frac{ct}{\pi}\right)^{1/2} \frac{6z^3 - 9r^2 z}{(z^2 + r^2)^{7/2}} \right.$$
$$\left. - \frac{8t^2}{15} \left(\frac{ct}{\pi}\right)^{1/2} \frac{2z^2 - r}{L(z^2 + r^2)^{5/2}} + \cdots \right\}$$

Using the notation $R = (z^2 + r^2)^{1/2}$, this result can be written in the following form,

$$\frac{\sigma}{P} = \frac{1}{4m} \frac{c'ct^2}{R^4} \left\{ \frac{2z^2 - r^2}{RL} \right.$$
$$\left. - \frac{16}{15\pi^{1/2}} \left(\frac{ct}{R^2}\right)^{1/2} \frac{-3zL(2z^2 - 3r^2) + R^2(2z^2 - r^2)}{R^2L^2} + \cdots \right\} \tag{135}$$

It appears that a series of increasing powers of t is obtained. This series is useful for small values of the time. Since the first term is positive when $2z^2 - r^2 > 0$, it follows that the pore water pressure initially increases in the region for which $2z^2 - r^2 > 0$. This region is the interior of the cone $r = -z\sqrt{2}$, $(z < 0)$.

The result that in a certain region the pore water pressures initially increase during a pumping test may be surprising at a first glance. However, this phenomenon may be explained from the elastic behavior of the aquifer as follows. Due to the decreasing pore water pressures in the upper layer, this layer will attract water from the upper part of the lower layer. The loss of water involves a decrease in volume of the upper part of the lower layer. All elements of a circular ring around the origin will decrease in volume, thus producing a radial displacement towards the center to keep the ring closed. For the lower parts of the layer, this radial displacement presents a loading due to which a tendency for compression will be present. Since the pore water opposes volume compression of the soil, this results in an increase of the pore water pressure.

Similar effects have been obtained in several types of problems in three-dimensional consolidation (Mandel, 1950, 1953; Cryer, 1963; De Leeuw, 1964; Schiffman, 1965). Moreover, experimental evidence in support of the theoretical results obtained by Cryer for a solid sphere of saturated soil has been reported by Gibson *et al.* (1963) and Verruijt (1965).

The present computations show increasing pore pressures in a pumping test in a two-storied aquifer. In several pumping tests this initial increase was, in fact, observed first at the village of Noordbergum in the Netherlands. Theoretical and experimental results will be compared below.

Numerical Solution for $m = \frac{1}{2}$ *and* $c' = c$

Although an analytical evaluation of the integral expression (131) seems impossible, numerical results may be obtained by means of a computer. However, numerical evalutation of the integral in the complex s plane presents the difficulties that along the path of integration, the integrand is an oscillating function because of the factor $\exp(st)$. Furthermore the convergence is rather slow. Therefore, it is reasonable to try to convert the path of integration using the theorems of the theory of functions. For simplicity it will be assumed that $c' = c$ and that $m = \frac{1}{2}$ (which means that Poisson's ratio is assumed to be zero).

With $c' = c$ and $m = \frac{1}{2}$, Eq. (131) reduces to

$$\frac{\sigma}{P} = -\frac{1}{2\pi i}\frac{c}{L}\int_0^\infty \xi J_0(r\xi)\, d\xi$$

$$\times \int_{\gamma-i\infty}^{\gamma+i\infty} \frac{\left[\begin{array}{l}\{(s/c)\exp[z(\xi^2+s/c)^{1/2}] \\ \qquad -\xi[(\xi^2+s/c)^{1/2}-\xi]\exp(z\xi)\}\exp(st)\end{array}\right]}{\left[\begin{array}{l}s(s+c\xi^2+cB^{-2})\{(s/c)[(\xi^2+s/c)^{1/2}+L^{-1}] \\ \qquad -\xi(\xi+L^{-1})[(\xi^2+s/c)^{1/2}-\xi]\}\end{array}\right]}\, ds \tag{136}$$

Introducing the dimensionless variables

$$\rho = \xi B, \quad y = sB^2/c, \quad \eta = r/B, \quad \zeta = -z/B, \quad \psi = B/L, \quad \tau = ct/B^2$$

Eq. (136) can be written in the following form

$$\frac{\sigma}{P} = -\frac{\psi}{2\pi i}\int_0^\infty \rho J_0(\eta\rho)\, d\rho \int_{\gamma-i\infty}^{\gamma+i\infty} \frac{\begin{bmatrix}\{y\exp[-\zeta(\rho^2+y)^{1/2}] \\ -\rho[(\rho^2+y)^{1/2}-\rho]\exp(-\zeta\rho)\}\exp(\tau y)\end{bmatrix}}{\begin{bmatrix}y(y+\rho^2+1)\{y[(\rho^2+y)^{1/2}+\psi \\ -\rho(\rho+\psi)][(\rho^2+y)^{1/2}-\rho]\}\end{bmatrix}}\, dy$$

By putting $x = y + \rho^2$, it appears that the numerator and the denominater of the integrand have a common factor $x^{1/2} - \rho$, and the following simplified result is obtained

$$\frac{\sigma}{P} = -\frac{\psi}{2\pi i}\int_0^\infty \rho J_0(\eta\rho)$$

$$\times \exp(-\tau\rho^2)\, d\rho \int_{\gamma-i\infty}^{\gamma+i\infty} \frac{\{(x^{1/2}+\rho)\exp(-\zeta x^{1/2}) - \rho\exp(-\zeta\rho)\}\exp(\tau x)}{(x-\rho^2)(x+1)[x+(\psi+\rho)\, x^{1/2}-\rho^2]}\, dx \tag{137}$$

First the integral in the complex x plane will be considered. The singularities of the integrand are a branch point in the origin ($x = 0$) and poles of order one in the points $x = \rho^2$, $x = -1$, $x = x_1$, and $x = x_2$, where x_1 and x_2 are the zeros of $x + (\psi + \rho)x^{1/2} - \rho^2$. The branch point $x = 0$ makes it necessary to introduce a branch cut in the x plane from $x = 0$ to infinity. In the plane with the branch cut, the integrand of (137) is single valued so that the theory of analytic functions may be applied in this plane. As branch cut, the negative part of the real axis is chosen. This means that the argument of x is restricted to the interval

$$-\pi < \arg(x) \leqslant \pi \tag{138}$$

The zeros of $x + (\psi + \rho)x^{1/2} - \rho^2$ are $x = x_1$ and $x = x_2$ where

$$\begin{aligned}(x_1)^{1/2} &= \tfrac{1}{2}\{[(\psi+\rho)^2 + 4\rho^2]^{1/2} - (\psi+\rho)\} \\ (x_2)^{1/2} &= -\tfrac{1}{2}\{[(\psi+\rho)^2 + 4\rho^2]^{1/2} + (\psi+\rho)\}\end{aligned} \tag{139}$$

and since $0 \leqslant \rho < \infty$ and $0 \leqslant \psi < \infty$, it follows that

$$0 \leqslant (x_1)^{1/2} \leqslant \rho, \qquad -\infty < (x_2)^{1/2} \leqslant 0$$

This means that $(x_2)^{1/2} = a \exp(i\pi)$ with $0 \leqslant a < \infty$ so that $x_2 = a^2 \exp(2i\pi)$. The argument of x_2 appears to be 2π and since restriction is made to values of $\arg(x)$ in the interval $(-\pi, +\pi)$, it follows that x_2 is not a pole in the relevant plane. The singularity $x = x_1$, however, lies on the positive part of the real axis, $\arg(x_1) = 0$, hence $x = x_1$ is indeed a pole to be taken into account.

Now consider the integral

$$I_{AB} = \frac{1}{2\pi i} \int_{\gamma - i\infty}^{\gamma + i\infty} \frac{\{(x^{1/2} + \rho) \exp(-\zeta x^{1/2}) - \rho \exp(-\zeta\rho)\} \exp(\tau x)}{(x - \rho^2)(x + 1)[x + (\psi + \rho)\, x^{1/2} - \rho^2]}\, dx \tag{140}$$

In Fig. 6 a contour in the complex x plane is shown. It is supposed

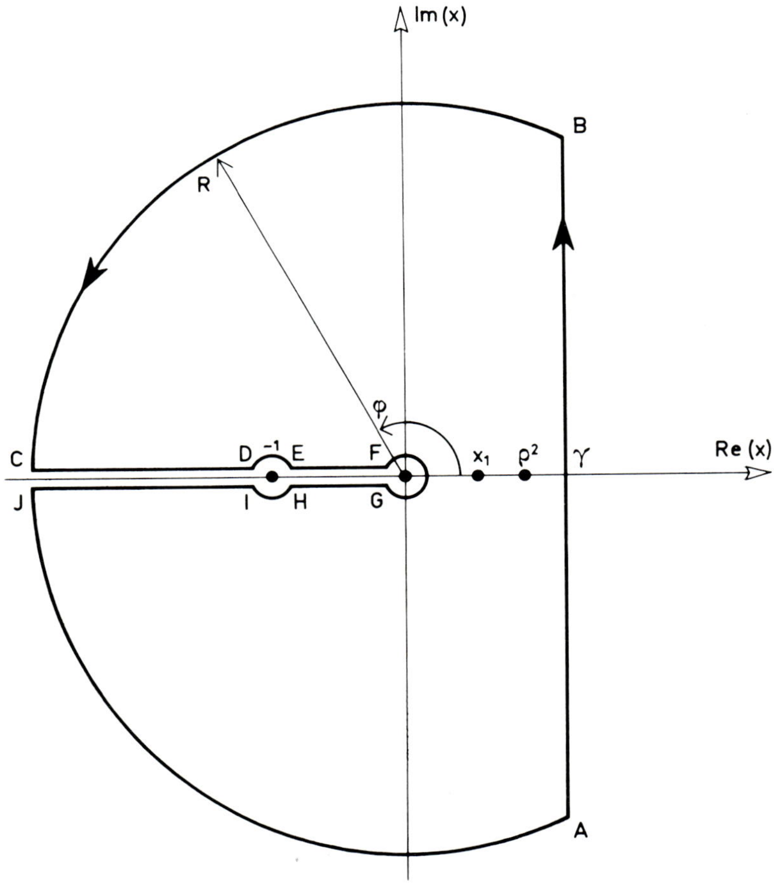

FIG. 6. Contour in the complex x plane.

that the radius R of the circle segments is very large. The closed contour is an extension of the integration path AB, hence

$$I_{AB} = I - I_{BC} - I_{CD} - I_{DE} - I_{EF} - I_{FG} - I_{GH} - I_{HI} - I_{IJ} - I_{JA} \qquad (141)$$

where I denotes the contour integral. Since the contour does not cross the branch cut and the integrand is a single-valued analytic function inside the contour except in the simple poles $x = x_1$ and $x = \rho^2$, it now follows from the residue theorem (Titchmarsh, 1939) that

$$I = \sum \text{ residues in } \quad x = x_1 \quad \text{and} \quad x = \rho^2$$

Evaluation of these residues gives

$$I = \frac{\exp(-\zeta\rho)\exp(\tau\rho^2)}{(\rho^2+1)(\rho+\psi)} - \frac{2x_1^{1/2}\{(x_1^{1/2}+\rho)\exp(-\zeta x_1^{1/2}) - \rho\exp(-\zeta\rho)\}\exp(\tau x_1)}{(\rho^2 - x_1)(1+x_1)(2x_1^{1/2}+\psi+\rho)} \qquad (142)$$

The integral along AB will be known if the integrals along the parts $BCDEFGHIJA$ are known. It can be shown that

$$\lim_{R\to\infty} \{I_{BC} + I_{JA}\} = 0 \qquad (143)$$

The proof of the vanishing of this integral, which will not be given in detail, makes use of the fact that along the parts BC and JA of the contour

$$|\exp(-\zeta x^{1/2})| \leqslant 1 \qquad \text{and} \qquad |\exp(\tau x)| \leqslant \exp(\tau y)$$

Hence the numerator tends to infinity of order $R^{1/2}$ when $R \to \infty$. Since the denominator tends to infinity of order R^3 and the length of the path of integration is of order R, it follows that the integral tends to zero of order $R^{-3/2}$ when $R \to \infty$. The contribution of the integral along the small circle FG vanishes because the integrand is bounded in $x = 0$ and the length of the integration path tends to zero,

$$I_{FG} = 0 \qquad (144)$$

The integrals along DE and IH can be evaluated by putting

$$x = \exp(i\pi) + a\exp(i\varphi) \qquad (0 < \varphi < \pi)$$

and

$$x = \exp(-i\pi) + a\exp(i\varphi) \qquad (-\pi < \varphi < 0)$$

respectively, and then passing into the limit $a \to 0$. This gives

$$I_{DE} + I_{HI}$$
$$= -\frac{\{(\rho^3 - \psi)\cos\zeta + (2\rho^2 + \rho\psi + 1)\sin\zeta - \rho(1 + \rho^2)\exp(-\zeta\rho)\}\exp(-\tau)}{(1 + \rho^2)[(1 + \rho^2)^2 + (\psi + \rho)^2]} \tag{145}$$

Finally, the integrals along $CD + EF$ and $GH + IJ$ are obtained by putting $x = w\exp(i\pi)$ and $x = w\exp(-i\pi)$, respectively. This gives

$$I_{CD+EF+GH+IJ} = -\pi^{-1} \oint_0^\infty -[(2\rho^2 + \rho\psi + w)\, w^{1/2}\cos(\zeta w^{1/2})$$
$$+ (\rho^3 - w\psi)\sin(\zeta w^{1/2}) + \rho(\psi + \rho)\, w^{1/2}\exp(-\zeta\rho)]$$
$$\times \exp(-\tau w)\{(w + \rho^2)(1 - w)[(w + \rho^2)^2 + w(\psi + \rho)^2]\}^{-1}\, dw \tag{146}$$

where the $\oint$ indicates that the integral is a Cauchy principal value. With (142)–(146), Eq. (141) now gives

$$I_{AB} = \exp(-\zeta\rho)\exp(\tau\rho^2)[(\rho^2 + 1)(\psi + \rho)]^{-1}$$
$$- 2x_1^{1/2}\{(x_1^{1/2} + \rho)\exp(-\zeta x_1^{1/2}) - \rho\exp(-\zeta\rho)\}$$
$$\times \exp(\tau x_1)\{(\rho^2 - x_1)(1 + x_1)(2x_1^{1/2} + \psi + \rho)\}^{-1}$$
$$+ \{(\rho^3 - \psi)\cos\zeta + (2\rho^2 + \rho\psi + 1)\sin\zeta - \rho(1 + \rho^2)\exp(-\zeta\rho)\}$$
$$\times \exp(-\tau)\{(\rho^2 + 1)[(\rho^2 + 1)^2 + (\psi + \rho)^2]\}^{-1}$$
$$+ \pi^{-1}\oint_0^\infty [-(2\rho^2 + \rho\psi + w)\, w^{1/2}\cos(\zeta w^{1/2})$$
$$+ (\rho^3 - w\psi)\sin(\zeta w^{1/2}) + \rho(\psi + \rho)\, w^{1/2}\exp(-\zeta\rho)]\exp(-\tau w)$$
$$\times \{(w + \rho^2)(1 - w)[(w + \rho^2)^2 + w(\psi + \rho)^2]\}^{-1}\, dw \tag{147}$$

where $x_1^{1/2} = \frac{1}{2}\{[(\psi + \rho)^2 + 4\rho^2]^{1/2} - (\psi + \rho)\}$.

No explicit solution of the integral in Eq. (147) is known to the author. Comparison of the two equivalent expressions (140) and (147) for I_{AB} shows that in both expressions a complicated integral appears. However, the integral in (140) is slowly convergent, whereas the integral in (147) is rapidly convergent, especially for large values of the time parameter τ, because of the factor $\exp(-\tau w)$. Hence, for numerical evaluation, (147) is to be preferred. Actually, in this case the above procedure of converting the path of integration is justified only because it yields an integral more suitable for numerical evaluation.

The formula for the pore water pressure is, with (137) and (140),

$$\sigma/P = -\psi \int_0^\infty \rho J_0(\eta\rho) \exp(-\tau\rho^2) I_{AB} \, d\rho \tag{148}$$

This integral is also rapidly convergent, especially for large values of τ. In Fig. 7 numerical results are shown as a function of time for the special

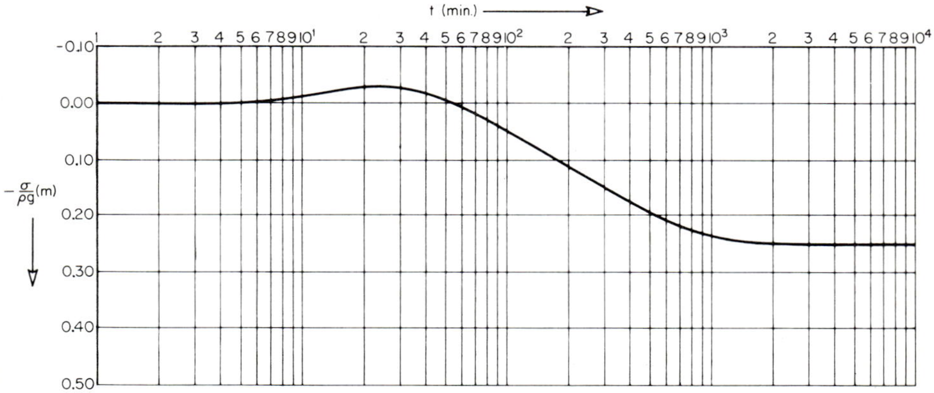

FIG. 7. Theoretical results for the drawdown in the lower layer of a two-storied leaky artesian aquifer during a pumping test in the upper layer.

case $\psi = B/L = 0.35$, $\zeta = -z/B = 0.200$, $\rho = r/B = 0.025$. Numerical values were obtained from Eq. (135) for small values of the time parameter and from Eqs. (147) and (148) for large values of the time parameter. In order to plot the drawdown $-\sigma/\rho g$ as a function of time, the definitions of P and τ

$$P = Q_0\rho g/(2\pi K_1 b_1), \qquad \tau = ct/B^2$$

were used. Since the consolidation coefficient c of the lower layer of the aquifer was set equal to the consolidation coefficient c' of the higher layer, its value is given by

$$c = c' = K_1 b_1/[\rho g b_1(\alpha' + n\beta')]$$

Numerical values for $K_1 b_1$, Q_0, B, and $\rho g b_1(\alpha' + n\beta')$ were chosen in accordance with the values used for Fig. 3, i.e.,

$$Q_0 = 1.74 \times 10^{-2} \text{ m}^3/\text{sec}, \qquad K_1 b_1 = 0.30 \times 10^{-2} \text{ m}^2/\text{sec}$$

$$B = 400 \text{ m}, \qquad \rho g b_1(\alpha' + n\beta') = 5.6 \times 10^{-4}$$

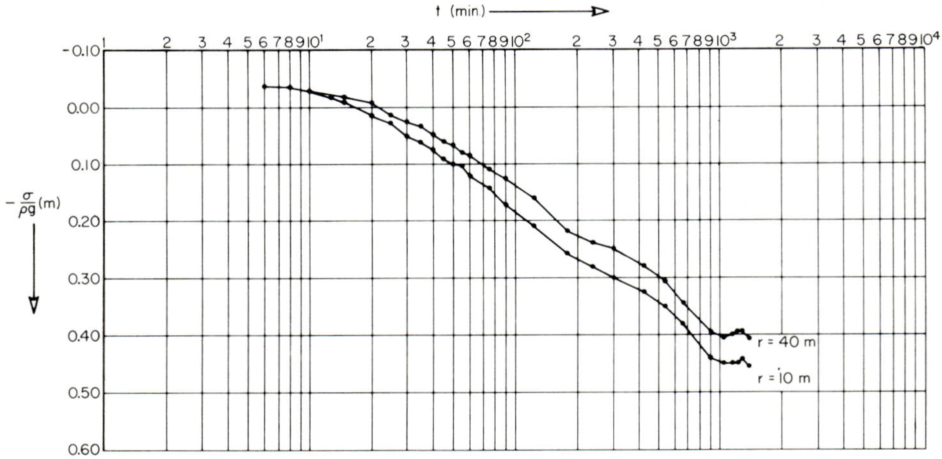

FIG. 8. Experimental results for the drawdown in the lower layer of a two-storied leaky artesian aquifer during a pumping test in the upper layer.

In Fig. 8, some experimental results obtained during the pumping test in the polder Vierbannen, already mentioned in Section 4, are represented graphically. It appears that the theoretical curves of Fig. 7 and the experimental data of Fig. 8 show similar tendencies. A closer fit might be obtained by choosing different values for the soil constants. However, the purpose of the present study is to give a qualitative explanation of the fact that the pore water pressures in the lower layer of a two-storied aquifer initially may increase during a pumping test in the higher layer, and not to describe a method for the determination of the soil constants. Indeed, both the theoretical and experimental curves show this effect, and the orders of magnitude correspond to each other.

7. Conclusion

The reader looking for the definition of the storage coefficient in this study, which according to its title deals with elastic storage of aquifers, may be disappointed not to find it. Although the author is aware that for many practical purposes the simple differential equation, in connection to which this storage coefficient usually appears, may give satisfactory results, he wants to underscore that on theoretical grounds the more complicated Biot theory based on the theory of elasticity has to be preferred. Therefore, in the present work the use of the words "storage coefficient" has been avoided carefully, even in cases where the Biot theory reduces to the simpler Jacob theory.

To support the theory advocated in this paper, experimental evidence was produced indicating that a remarkable effect occurring in two-storied aquifers (or more generally, in nonhomogeneous aquifers), called the *Noordbergum effect*, can be explained qualitatively by this theory.

Acknowledgments

The work described in this paper was carried out in close cooperation with Professor Dr. Ir. G. de Josselin de Jong, to whom the author is much indebted. The concept of a relative velocity to be used in Darcy's law was brought to the author's attention by Professor Dr. R. E. Gibson of King's College, London University, during a lecture at Delft in 1962.

Many thanks are due to the Netherlands Governmental Institute for Water Supply for permission to publish some results of the pumping test in the polder Vierbannen, and to Mr. P. ten Kortenaar of the Stevin Laboratory, who took care of the calculation of the numerical values used for the construction of Figs. 5 and 7.

Finally, the author would like to mention the discussions during the 1965 Summer Institute sponsored by the National Science Foundation, held at Princeton, New Jersey under the direction of Professor Dr. R. J. M. de Wiest. These discussions have resulted in many refinements and improvements in the present work.

REFERENCES

Bateman, H. (1954). "Tables of Integral Transforms," Vols. 1 and 2. McGraw-Hill, New York.

Biot, M. A. (1941). General theory of three-dimensional consolidation, *J. Appl. Phys.* **12**, 155–164.

Biot, M. A. (1956). Theory of deformation of a porous viscoelastic anisotropic solid, *J. Appl. Phys.* **27**, 459–467.

Carslaw, H. S., and Jaeger, J. C. (1947). "Operational Methods in Applied Mathematics." Oxford Univ. Press, London and New York.

Churchill, R. V. (1958). "Operational Mathematics," 2nd ed. McGraw-Hill, New York.

Cryer, C. W. (1963). A comparison of the three-dimensional consolidation theories of Biot and Terzaghi, *Quart. J. Mech. Appl. Math.* **16**, 401–412.

De Josselin de Jong, G. (1963). Consolidatie in drie dimensies, *L.G.M.-Mededelingen* **7**, 57–73.

De Leeuw, E. H. (1964). Consolidatie in drie dimensies (vervolg), *L.G.M.-Mededelingen* **9**, 17–48.

De Leeuw, E. H. (1965). The theory of three-dimensional consolidation applied to cylindrical bodies. *Proc. 6th Intern. Conf. Soil Mech. and Found. Eng., Montreal, 1965*, Vol. 1, pp. 287–290.

De Wiest, R. J. M. (1963). Flow to an eccentric well in a leaky circular aquifer with varied lateral replenishment, *Geofis. Pura Appl.* **54**, 87–102.

De Wiest, R. J. M. (1965). "Geohydrology." Wiley, New York.

De Wiest, R. J. M. (1966). On the storage coefficient and the equations of ground-water flow, *J. Geophys. Res.* **71**, 1117–1122, 5016–5017.

Florin, V. A. (1961). "Fundamentals of Soil Mechanics," Vol. 2 (in Russian). State Publ. House, Leningrad and Moscow.

Gibson, R. E., Knight, K., and Taylor, P. W. (1963). A critical experiment to examine theories of three-dimensional consolidation, *Proc. European Conf. Soil Mech. and Found. Eng., Wiesbaden, 1963*, Vol. 1, pp. 69–76.

Hantush, M. S. (1956). Analysis of data from pumping tests in leaky aquifers, *Trans. Am. Geophys. Union* **37**, 702–714.

Hantush, M. S. (1964). Hydraulics of wells, *Advan. Hydrosci.* **1**, 281–432.

Hantush, M. S., and Jacob, C. E. (1955). Non-steady radial flow in an infinite leaky aquifer, *Trans. Am. Geophys. Union* **36**, 95–100.

Hubbert, M. K. (1940). The theory of ground-water motion, *J. Geol.* **48**, 785–944.

Jacob, C. E. (1940). The flow of water in an elastic artesian aquifer, *Trans. Am. Geophys. Union* **21**, 574–586.

Jacob, C. E. (1946). Radial flow in a leaky artesian aquifer, *Trans. Am. Geophys. Union* **27**, 198–205.

Jacob, C. E. (1950). Flow of ground water, *in* "Engineering Hydraulics" (H. Rouse, ed.), pp. 321–386. Wiley, New York.

McNamee, J., and Gibson, R. E. (1960a). Displacement functions and linear transforms applied to diffusion through porous elastic media, *Quart. J. Mech. Appl. Math.* **13**, 98–111.

McNamee, J., and Gibson, R. E. (1960b). Plane strain and axially symmetric problems of the consolidation of a semi-infinite clay medium, *Quart. J. Mech. Appl. Math.* **13**, 210–227.

Mandel, J. (1950). Étude mathématique de la consolidation des sols, *Actes Colloq. Intern. Mech. (Poitiers)* **4**, 9–19.

Mandel, J. (1953). Consolidation des sols, *Geotechnique* **3**, 287–299.

Meinzer, O. E. (1928). Compressibility and elasticity of artesian aquifers, *Econ. Geol.* **23**, 263–291.

Muskat, M. (1937). "The Flow of Homogeneous Fluids through Porous Media." McGraw-Hill, New York.

Rendulic, L. (1936). Porenziffer und Porenwasserdruck in Tonen, *Der Bauingenieur* **17**, 559–564.

Schiffman, R. L. (1965). Discussion, *Proc. 6th Intern. Conf. Soil Mech. and Found. Eng., Montreal, 1965*, Vol. 3, pp. 394–397.

Sneddon, I. N. (1951). "Fourier Transforms." McGraw-Hill, New York.

Sokolnikoff, I. S. (1956). "Mathematical Theory of Elasticity," 2nd ed. McGraw-Hill, New York.

Terzaghi, K. (1923). Die Berechnung des Durchlässigkeitsziffer des Tones aus dem Verlauf der hydrodynamischen Spannungserscheinungen, *Sitz. Akad. Wiss. Wien*, Abt. IIa, **132**, 125–138.

Terzaghi, K. (1943). "Theoretical Soil Mechanics." Chapman and Hall, London.

Titchmarsh, E. C. (1939). "The Theory of Functions," 2nd ed. Oxford Univ. Press, London.

Van Eyden, W. A. A., Kuper, H. W., and Santema, P. W. (1963). Some methods used in the geohydrological survey of the south-western deltaic area in the Netherlands, *Proc. Gen. Assembly Berkeley I.U.G.G.*, Publ. No. 63, I.A.S.H., pp. 528–557.

Verruijt, A. (1965). Discussion, *Proc. 6th Intern. Conf. Soil Mech. and Found. Eng., Montreal, 1965*, Vol. 3, pp. 401–402.

9

Generating Functions in the Theory of Flow through Porous Media

G. de Josselin de Jong
Civil Engineering Department
Delft University of Technology
Delft, the Netherlands

1. Introduction

In this study the flow of fluids through porous media is considered for the case in which fluid and porous medium are inhomogeneous. The properties that may differ from place to place in the field are the density and viscosity of the fluid and the intrinsic permeability of the medium. These inhomogeneities are responsible for rotations in the flow pattern. Since the Darcian flow of fluids through porous media is usually irrotational, it was considered instructive in this presentation to elaborate especially the character of these rotations.

For the determination of the distribution of the pressure and the specific discharge vector it is convenient to make use of potential and streamfunctions. These functions may be called generating functions, because their gradients generate the specific discharge vector. For incompressible homogeneous fluids these functions obey Laplace equations of the type

$$\nabla^2 \mathscr{F} = 0$$

where $\mathscr{F}$ is the relevant function. The solution of a problem is obtained by selecting from the functions that obey this basic equation those that satisfy the boundary conditions.

In this study, functions will be determined that in a similar way generate the specific discharge vector by differentiation if inhomogeneities are present. The basic equations obeyed by these generating functions will turn out to be Poisson equations. These are equations of the type

$$\nabla^2\mathscr{F} = \mathscr{G}$$

where $\mathscr{G}$ is a function of place.

If the value of $\mathscr{G}$ is known everywhere, it is possible to solve $\mathscr{F}$ in a region enclosed by boundaries along which either the value of $\mathscr{F}$ or its derivative normal to the boundary is known. The solution of the Poisson equation will be given in the form of an integral which has to be extended over all points where $\mathscr{G}$ has a nonvanishing value. This method of solving may be called the *solution by singularities.* In addition, solutions of the Laplace equation have to be applied to satisfy the remaining boundary conditions.

If $\mathscr{G}$ is an unknown function (for instance, if $\mathscr{G}$ depends on $\mathscr{F}$), it is not possible to use the solution by means of the singularities mentioned. It may be possible that other kinds of singularities are able to cope with Poisson equations where $\mathscr{G}$ is a linear function of $\mathscr{F}$. This possibility is not investigated here and only functions obeying a Poisson equation with known values for $\mathscr{G}$ are considered.

The generating functions appropriate for flow of fluids with varying properties may be called *multiple fluid functions.* It will be required that the multiple fluid functions are single valued in the entire field irrespective of how many fluids are involved. The multiple fluid functions will be chosen such that they resemble as closely as possible the potential and streamfunctions for the single fluid case.

The multiple fluid functions developed here are essentially the same as those described in a previous article (de Josselin de Jong, 1960). In that paper Poisson's equations were established for the streamfunction Ψ and a multiple fluid potential, Θ. Yih (1961) arrived at the same result with respect to Ψ. Knudsen (1962) showed that the pressure obeys a Poisson equation of a kind very similar to that for Ψ.

Lusczynski (1961) suggested several types of heads for multiple fluid flow. De Wiest (1964, 1965) reviewed this work and selected H_{ip}, the type called "point water head" by Lusczynski, as the most suitable physical quantity to use. It will be shown here that Lusczynski's "point water head" H_{ip} is directly related to Θ. By elaborating the second-order differential equation governing H_{ip} and Θ, it turns out, however, that H_{ip} obeys a Poisson equation that is less convenient to solve than the relation governing Θ.

The name chosen by de Josselin de Jong (1960) to indicate Θ may be confusing because Θ is not a potential in the sense that grad Θ is proportional to the specific discharge vector. Because there is a rotation in the fluid with varying density, the specific discharge cannot be derived from a potential. Until a better name is suggested, the term *multiple fluid potential* will be used here, with the stipulation that it is not a velocity potential in the usual sense.

In Sections 2–8, the case of a two-dimensional aquifer is treated. In Section 9, formulas are given for the three-dimensional case.

The density is considered variable throughout the entire chapter. The viscosity of the fluids and the permeability of the porous medium are taken constant, except in Section 8. In that section, an electric analogy is described which provides a means of treating variability of these other properties.

All fluids will be considered to be incompressible. This means that fluid elements will preserve their density during displacement. The variability of density considered is the difference in density between fluid elements at different places in the field. No surface tension will be assumed to exist between fluids of different properties. The porous medium is isotropic.

The multiple fluid functions generate the fluid displacement at a given instant. It is assumed in this analysis that the distribution of the fluids and their properties is known at that moment. From the multiple fluid functions it is then possible to determine the displacement of fluid particles during a small time interval. By taking into account the manner in which the properties of the fluids are convected by this displacement, the distribution of the properties after that time interval can be determined. This creates a new situation which requires new multiple fluid functions. It is not the objective of this study to pursue what is happening at successive time intervals, but only to describe the instantaneous flow situation.

2. Basic Properties of the Flow

Since the fluids are assumed to be incompressible, conservation of mass requires that the specific discharge vector $\mathbf{q}$ obeys the continuity equation

$$\operatorname{div} \mathbf{q} = \frac{\partial q_x}{\partial x} + \frac{\partial q_z}{\partial z} = 0 \tag{1}$$

The second requirement is Darcy's law which can be written in terms of an equilibrium of forces. When the forces acting on the fluid per unit volume of the fluid are equated, the equilibrium equation results:

$$(\mu/k)\,\mathbf{q} = -\text{grad}\,p - \rho\,\text{grad}\,F \tag{2}$$

The left side of (2) expresses the force per unit volume of fluid necessary to drive a specific discharge $\mathbf{q}$ through the porous medium according to Darcy's law. In (2) μ is the dynamic viscosity, k is the intrinsic permeability, and ρ is the density of the fluid. The two terms on the right of (2) represent the two forces that can work on the fluid and produce the driving action. The first term is obtained from the gradient of the pressure p. The second term represents a body force which can be derived from a potential F acting per unit mass of the fluid.

If only gravity works, F is equal to gz, in which z is measured vertically upwards. The equilibrium equation becomes in this case

$$(\mu/k)\,\mathbf{q} = -\text{grad}\,p - \rho g\,\text{grad}\,z \tag{3}$$

Equations (1) and (2) are the basic relationships sufficient for the establishment of the multiple fluid functions. They are relations describing the fluid behavior in a point. Because (3) describes a point relationship it is irrelevant for the last term whether the density is a constant or a variable in the field.

Since this point was questioned, it may be helpful to derive this term by integrating the forces acting on the fluid in an infinitesimally small circular disc with radius r_1 (Fig. 1).

Let ξ be a coordinate running in the direction of grad ρ and let ζ be a coordinate perpendicular to ξ. Further, let the subscript zero in the expressions ρ_0, $(\partial\rho/\partial\xi)_0$,..., etc., indicate the value of these quantities in the center of the circle. Then the density in the circle is approximated by its Taylor series:

$$\rho = \rho_0 + \xi(\partial\rho/\partial\xi)_0 + \tfrac{1}{2}\xi^2(\partial^2\rho/\partial\xi^2)_0 + \xi\zeta(\partial^2\rho/\partial\xi\,\partial\zeta)_0 + \tfrac{1}{2}\zeta^2(\partial^2\rho/\partial\zeta^2)_0 + \cdots \tag{4}$$

The force acting on a unit of mass in the gravity field is $-g$ grad z. In order to find the total force on the fluid element, the mass of that fluid amount has to be evaluated by integration over the volume of the disc. If f is the thickness of the disc and n the porosity, this gives

$$M = \int_0^{r_1}\int_0^{2\pi} \rho \cdot nf \cdot dr \cdot r\,d\theta \tag{5}$$

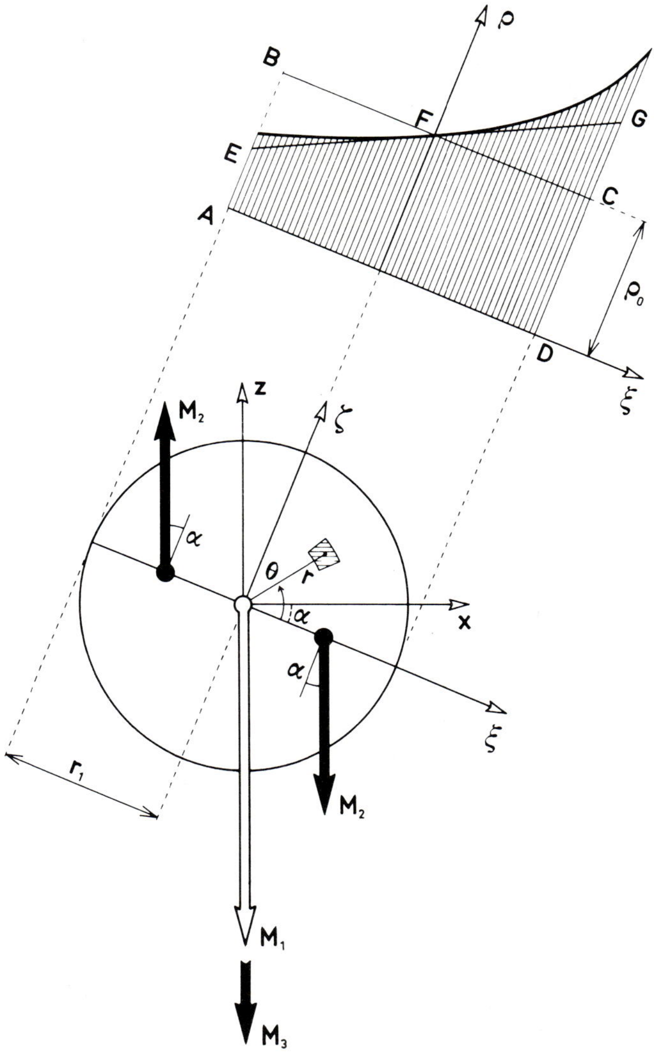

FIG. 1.

where ρ can be replaced by the series in (4). Let the series be split in three parts, giving for the integral three contributions M_1, M_2, M_3, respectively, such that M_1 is obtained from the first term in the series, M_2 from the second term, and M_3 from the third, fourth, and fifth terms together.

The result is

$$M_1 = \rho_0 nf \cdot \pi r_1^2, \qquad M_2 = 0$$
$$M_3 = [(\partial^2\rho/\partial\xi^2)_0 + (\partial^2\rho/\partial\zeta^2)_0]\, nf \cdot \tfrac{1}{4}\pi r_1^4 \tag{6}$$

Since M_3 is of the order r_1^4, its value may be disregarded with respect to M_1 which is of order r_1^2, because in the limit r_1 tends to zero.

To obtain the force per unit volume of the fluid, (6) has to be multiplied by g grad z and divided by $nf \cdot \pi r_1^2$, the volume of the fluid in the disc. Thus, the force on the infinitesimal element finally becomes $-\rho_0 g$ grad z as indicated in (3). The second term from the series expression for ρ does not contribute to a force on the element because M_2 is zero.

In the diagram for ρ in Fig. 1, $(\partial\rho/\partial\xi)_0$ is the inclination of the line *EFG*, the tangent to the curve for ρ. The second term, $\xi(\partial\rho/\partial\xi)_0$, covers the triangles *BFE* and *CFG*, which being similar and opposite give cancelling forces. However, they contribute a clockwise torque which can be written as

$$T = g \cos\alpha \int_0^{r_1} \int_0^{2\pi} \xi \cdot \xi(\partial\rho/\partial\xi)_0\, nf\, dr \cdot r\, d\theta$$

where α is the angle between ξ and x. By evaluating this integral it is found that

$$T = g \cos\alpha(\partial\rho/\partial\xi)_0 \cdot \tfrac{1}{4}\pi r_1^4 \cdot nf$$

Since $\cos\alpha = (\partial\xi/\partial x)$, the torque can be written as

$$T = \tfrac{1}{4}\pi r_1^4 g(\partial\rho/\partial x)_0\, nf \tag{7}$$

This torque is responsible for the rotational properties of the flow fields, as will be shown in Section 6.

The expressions (1) and (2) are therefore point relationships which only require ρ_0, the value of ρ at the point considered. The variability of ρ with location only contributes if higher order differential equations are established. This actually happens in the subsequent treatment.

3. Single-Valuedness of the Functions

For the multiple fluid functions to be applicable in the analysis it is of importance that they be single-valued because otherwise they have no physical meaning and may become misleading in computations.

The single-valuedness of a function Φ is established if according to Young's theorem,

$$\partial^2\Phi/\partial x\, \partial z = \partial^2\Phi/\partial z\, \partial x \tag{8}$$

This theorem can be proved by starting from the contour integral expression

$$\int_C (\partial\Phi/\partial s)\, ds = 0 \tag{9}$$

which indicates that for every closed contour C, the value of Φ will return to its original value. After changing to x, z coordinates, and introducing $M = \partial\Phi/\partial x$, $N = \partial\Phi/\partial z$, this contour integral can be converted into a surface integral by use of Green's theorem

$$\begin{aligned}\int_C (\partial\Phi/\partial s)\, ds &= \int_C [(\partial\Phi/\partial x)\, dx + (\partial\Phi/\partial z)\, dz] \\ &= \int_C (M\, dx + N\, dz) \\ &= \iint_A [(\partial N/\partial x) - (\partial M/\partial z)]\, dx\, dz \\ &= \iint_A [(\partial^2\Phi/\partial z\, \partial x) - (\partial^2\Phi/\partial x\, \partial z)]\, dA = 0 \end{aligned}\tag{10}$$

Since (9) applies for every contour C in the field, (10) applies for every area A enclosed by that contour. By reducing the contour to as small a circle as one wishes, the last integral of (10) indicates that (8) applies to every point in the field.

Wherever expression (8) will occur in the sequel, it will be concluded that single-valuedness of the relevant function is established.

4. Multiple Fluid Functions for μ and k Constant, ρ Variable

Of the multiple fluid functions, the stream function Ψ resembles most the conventional stream function. It will be defined in two-dimensional flow as

$$q_x = -(\partial\Psi/\partial z), \qquad q_z = +(\partial\Psi/\partial x) \tag{11}$$

This means that the specific discharge vector has the magnitude of the gradient of Ψ and runs perpendicular to grad Ψ such that the angle $\frac{1}{2}\pi$ from $\mathbf{q}$ to grad Ψ is traced clockwise (Fig. 2).

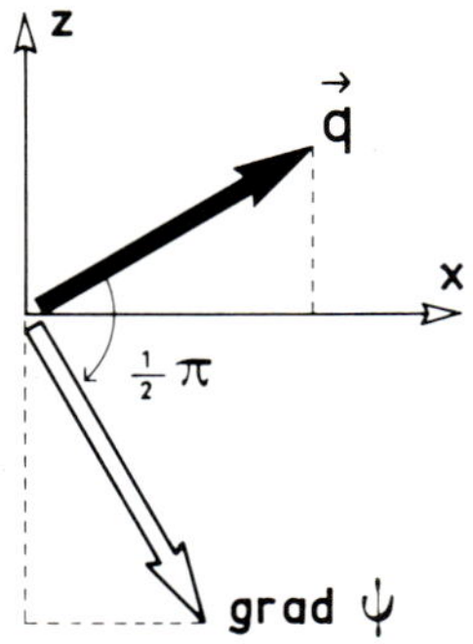

FIG. 2.

Application of the continuity Eq. (1) gives

$$-(\partial^2\Psi/\partial z\,\partial x) + (\partial^2\Psi/\partial x\,\partial z) = 0 \tag{12}$$

which shows that Ψ is a single-valued function irrespective of the properties of the fluids involved.

More information about Ψ may be obtained with the equilibrium Eq. (2). This equation written in its x, z components gives

$$(\mu/k)\,q_x = -(\partial p/\partial x), \qquad (\mu/k)\,q_z = -(\partial p/\partial z) - \rho g \tag{13}$$

From the physical viewpoint the pressure p must be single-valued because it is a physical scalar quantity which has one value in every point of the field. Therefore, the relation

$$-(\partial^2 p/\partial x\,\partial z) + (\partial^2 p/\partial z\,\partial x) = 0 \tag{14}$$

is satisfied everywhere in the field.

The identity (14) can be used to eliminate p from (13). Since (μ/k) is a constant, differentiating Eqs. (13) with respect to z and x, respectively, and subtracting gives

$$\begin{aligned}(\mu/k)[(\partial q_x/\partial z) - (\partial q_z/\partial x)] &= -(\partial^2 p/\partial x\,\partial z) + (\partial^2 p/\partial z\,\partial x) + [\partial(\rho g)/\partial x]\\ &= \partial(\rho g)/\partial x\end{aligned} \tag{15}$$

Applying expressions (11) to this result gives

$$-(\partial^2\Psi/\partial z^2) - (\partial^2\Psi/\partial x^2) = (k/\mu)[\partial(\rho g)/\partial x]$$

or

$$\boxed{\nabla^2\Psi = -(k/\mu)[\partial(\rho g)/\partial x]} \tag{16}$$

This relation, obtained previously (de Josselin de Jong, 1960), permits computation of Ψ with the aid of singularities of rotational type if the distribution of ρ is known. This will be shown in Section 5.

Applying the continuity Eq. (1) to (13) gives

$$(\mu/k)\ \mathrm{div}\ \mathbf{q} = -(\partial^2 p/\partial x^2) - (\partial^2 p/\partial z^2) - [\partial(\rho g)/\partial z] = 0$$

or

$$\boxed{\nabla^2 p = -\partial(\rho g)/\partial z} \tag{17}$$

This relation for p was obtained by Knudsen (1961). The two Poisson Eqs. (16) and (17) are very similar because they contain the derivatives of the density in the x and z directions, respectively. For p, solutions of boundary value problems can be obtained by use of singularities in a similar way as used for Ψ.

In a previous article (de Josselin de Jong, 1960), a function was introduced which can be considered the equivalent of the potential function for the flow of homogeneous fluids. This function Θ was called the *multiple fluid potential*, although it is not a potential in the sense that grad Θ is proportional to the discharge vector. This function Θ is related to pressure p and elevation z by the expression

$$(\mu/k)\ \Theta = p + \rho g z \tag{18}$$

Here ρ is the density of the fluid at the point of consideration. Since p, ρ, and z are single valued, Θ is also single valued. From (18) it follows that the relation between Θ and H_{ip}, the *point water head* as defined by Lusczynski (1961), is

$$(\mu/k)\ \Theta = \rho_i g H_{ip} \tag{19}$$

Taking the gradient of (18) gives, since (μ/k) is constant,

$$(\mu/k)\ \mathrm{grad}\ \Theta = \mathrm{grad}\ p + z\ \mathrm{grad}(\rho g) + (\rho g)\ \mathrm{grad}\ z \tag{20}$$

A comparison with (3) shows that this expression contains one term more than the expression for $(\mu/k)\mathbf{q}$. Therefore, Θ is not a regular potential.

By use of the equilibrium Eq. (3), the pressure can be eliminated to give

$$(\mu/k)\ \mathbf{q} = -(\mu/k)\ \mathrm{grad}\ \Theta + z\ \mathrm{grad}(\rho g) \tag{21}$$

Taking the divergence of the expression in order to establish continuity gives

$$(\mu/k)\ \text{div}\ \mathbf{q} = 0 = -(\mu/k)\ \nabla^2\Theta + \text{div}[z\ \text{grad}(\rho g)] \tag{22}$$

This finally gives for Θ the Poisson equation

$$\boxed{\nabla^2\Theta = +(k/\mu)\ \text{div}[z\ \text{grad}(\rho g)]} \tag{23}$$

The Poisson equation (23) shows that Θ can be solved by use of singularities.

Since H_{ip}, the "point water head," is so closely related to Θ it might be expected that H_{ip} also obeys some convenient Poisson equation. Elaboration along the same lines as above gives

$$\nabla^2 H_{ip} = (\rho g)^{-1}\{\text{div}[z\ \text{grad}(\rho g)] - 2\ \text{grad}\ H_{ip}\cdot\text{grad}(\rho g) - H_{ip}\nabla^2(\rho g)\}$$

Since the right-hand side contains H_{ip}, this equation is inconvenient in the analysis. H_{ip} is therefore less suitable then Θ.

The results of this section are summarized in Table 1.

TABLE 1

Multiple Fluid Functions

Multiple fluid functions:	Ψ	p	$\Theta = (k/\mu)(p + \rho g z)$
$\frac{\mu}{k} q_x =$	$-\frac{\mu}{k}\frac{\partial\Psi}{\partial z} =$	$-\frac{\partial p}{\partial x} =$	$-\frac{\mu}{k}\frac{\partial\Theta}{\partial x} + z\frac{\partial(\rho g)}{\partial x}$
$\frac{\mu}{k} q_z =$	$+\frac{\mu}{k}\frac{\partial\Psi}{\partial x} =$	$-\frac{\partial p}{\partial z} - \rho g =$	$-\frac{\mu}{k}\frac{\partial\Theta}{\partial z} + z\frac{\partial(\rho g)}{\partial z}$
Single-valuedness established by:	div $\mathbf{q} = 0$	Physical necessity	Physical necessity
Poisson equation established by:	Elimination of p	div $\mathbf{q} = 0$	div $\mathbf{q} = 0$
Gives:	$\nabla^2\Psi = -\frac{k}{\mu}\frac{\partial(\rho g)}{\partial x}$	$\nabla^2 p = -\frac{\partial(\rho g)}{\partial z}$	$\nabla^2\Theta = +\frac{k}{\mu}\text{div}[z\ \text{grad}(\rho g)]$
Type of singularity	Vortex	Source or sink	Source–sink dipole

5. Velocity of the Flow Field

The Poisson equation (16) indicates that the flow contains rotation in those points where the gradient of the fluid density has a horizontal component. Although reasons can be found in textbooks why this equation represents vorticity, it is helpful for the subsequent treatment to show this here.

Consider the case that in a circular area A_2 with radius r_2 (Fig. 3), the following relation holds

$$\nabla^2 \Psi = 2\omega, \qquad 0 < r < r_2 \tag{24}$$

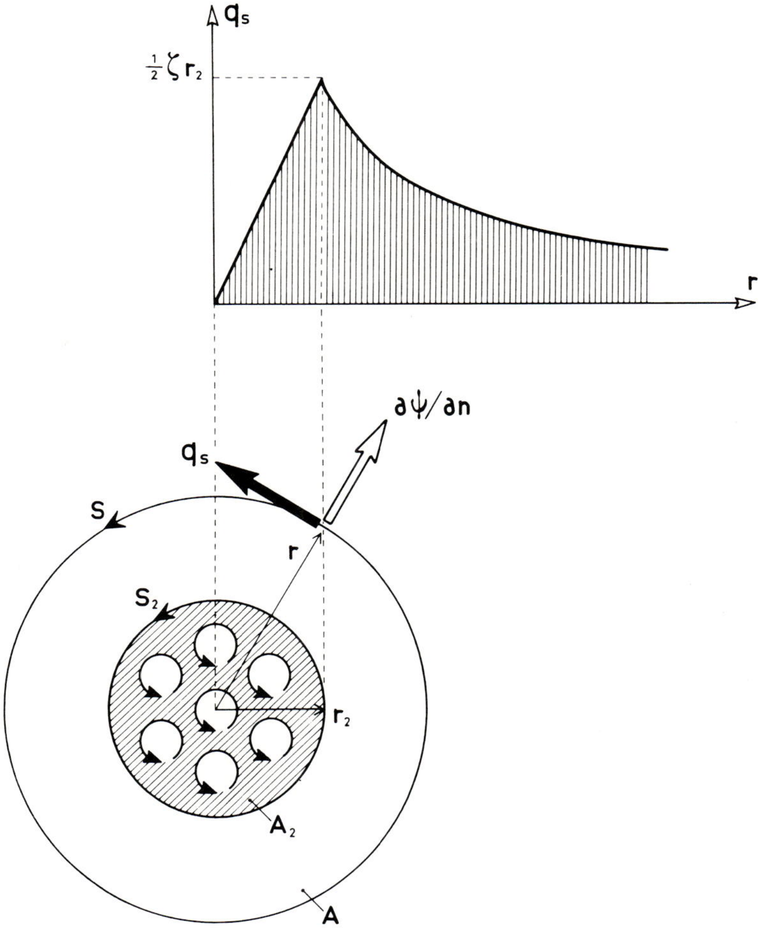

FIG. 3.

with ω = constant, and that outside A_2 it is

$$\nabla^2\Psi = 0, \qquad r_2 < r < \infty \tag{25}$$

Let S be a circular contour, concentric with A_2 and with a radius r that can be either larger or smaller than r_2.

Consider first the case where r is smaller than r_2, Then only (24) is relevant. Integration of (24) over the area A gives

$$\iint_A \nabla^2\Psi \, dA = \iint_A 2\omega \, dA = 2\omega \iint_A dA = 2\omega\pi r^2 \tag{26}$$

By Gauss' theorem the first integral can be written

$$\iint_A \nabla^2\Psi \, dA = \int_S (\partial\Psi/\partial n) \, dS \tag{27}$$

In this expression n is the outer normal to the circle S, which is related to the radius r in such a manner that $(\partial\Psi/\partial n) = (\partial\Psi/\partial r)$. Because the system is axial symmetric, $(\partial\Psi/\partial r)$ is constant over the circle S, and (27) becomes

$$\iint_A \nabla^2\Psi \, dA = (\partial\Psi/\partial r) \int_S dS = (\partial\Psi/\partial r) \, 2\pi r \tag{28}$$

Combination of (26) and (28) gives

$$(\partial\Psi/\partial r) = \omega r, \qquad 0 < r < r_2 \tag{29}$$

From Fig. 2 it is seen that $(\partial\Psi/\partial r)$ represents a discharge q_S tangential to the circle S. Hence, (29) can be written

$$q_S = \omega r, \qquad 0 < r < r_2 \tag{30}$$

and this indicates that ω in the region A creates a tangential flow circulating around this region with a discharge velocity proportional to the distance to the center. ω is called the *vorticity*.

The second case to consider is r larger than r_2. Since according to (25), the value of $\nabla^2\Psi$ is zero outside A_2, integration of $\nabla^2\Psi$ over the area A now only consists of a contribution over the area A_2. This gives

$$\iint_A \nabla^2\Psi \, dA = \iint_{A_2} \nabla^2\Psi \, dA = 2\omega\pi {r_2}^2 \tag{31}$$

Again, the first integral of (31) can be elaborated by the Gauss theorem as (27) to yield a result (28) with r now the radius larger than r_2. By combining (28) and (31), there results

$$\partial\Psi/\partial r = \omega(r_2{}^2/r)$$

which shows that the discharge circulating around A_2 is

$$q_S = \omega(r_2{}^2/r), \qquad r_2 < r < \infty \tag{32}$$

This expression indicates that the discharge outside the region A_2, which contains the vorticity, circulates around that region with a velocity which is inversely proportional to the distance from the center.

The distribution of the specific discharge as given by the Eqs. (30) and (32) is represented in Fig. 3. In that figure it is seen that on the boundary of the region A_2 containing the vorticity, the circulating discharge q_s is continuous. This follows from (30) and (32) which give the same value for q_s if r is made equal to r_2. The continuity of q_s is a direct consequence of the above analysis, which shows that a discontinuity can only exist in a place where ω is infinite.

From (16) and (24) it follows that the vorticity is equal to

$$\omega = -\tfrac{1}{2}(k/\mu)[\partial(\rho g)/\partial x] \tag{33}$$

This expression shows that ω is only infinite if $\partial\rho/\partial x$ is infinite. This is the case if ρ changes abruptly, as is encountered at a sharp interface between two different fluids. Since this aspect was treated extensively by de Josselin de Jong (1960), its treatment will be omitted here. Only gradually changing densities (grad ρ = finite) will be taken into account in the sequel.

A further conclusion from (33) is that every small region, where the density varies in the horizontal x direction, contributes a vorticity similar to the region A_2 in the example. Around every elementary region ΔA_2 containing density variation, a circulating discharge Δq_S is created which obeys Eq. (32). The influence of the density variations in the entire field is thus obtained by adding vectorially all elementary discharges Δq_S created by all elementary areas ΔA_2. By choosing the elementary areas ΔA_2 infinitesimally small, the discharge Δq_S in the area itself obeying Eq. (30) can be disregarded.

If this reasoning is correct it is also necessary that the discharge (30) be obtained in this way by integrating discharges of the character (32) as generated by infinitesimally small vortices over the region A_2.

This may seem contradictory because it requires that the result obtained for the region outside A_2 (which differs from the solution inside), when applied to the region inside, actually gives that different inside solution. A lengthy but straightforward evaluation of the integral involved shows, however, that the solution (30) is indeed obtained. The verification is an entertaining exercise in elliptic integrals.

It may seem to be a difficulty that the discharge according to (32) becomes infinite for $r = 0$. This means that the vortex at the point of consideration contributes an infinitely large discharge. However, this difficulty disappears since the magnitude of the vortex at a point, being proportional to the area, becomes infinitely small of the same order. Therefore, the integral remains finite. This exorbitant behavior at a point containing vorticity is the reason we apply the word *singularity*. Using the solution (32) for the discharges around points containing vorticity and adding those in an integral is called the *integral solution with singularities*.

Instead of adding vectorially elementary discharges, it is mathematically more convenient to add the elementary streamfunctions which generate them. This gives an integral expression for Ψ which is explained in Section 6.

6. Solutions with Singularities

If a small region of the size $dx_0\, dz_0$ around the point x_0, z_0 contains vorticity $\omega(x_0, z_0)$, the discharge is rotating around that point according to (32) with a magnitude

$$q_S = 2\omega(x_0, z_0)\, dx_0\, dz_0/2\pi r \tag{34}$$

where r is the distance from the point of vorticity. Written in terms of the streamfunction Ψ, this is

$$(\partial\Psi/\partial r) = 2\omega(x_0, z_0)\, dx_0\, dz_0/2\pi r$$

which by integration gives

$$\Psi = (1/\pi)\, \omega(x_0, z_0) \ln r\, dx_0\, dz_0 + \text{constant} \tag{35}$$

The constant is not a function of θ, because $q_r = 0$ since the discharge only rotates around the point of vorticity. Thus, $\partial\Psi/\partial\theta = 0$. Since we are only interested in the derivative of Ψ, the constant is irrelevant and can be taken as zero without loss of generality.

Because all the regions containing vorticity contribute to Ψ in this manner, the total expression for Ψ as created by all the vortices is the integral

$$\Psi(x, z) = (1/\pi) \iint \omega(x_0, z_0) \ln r \, dx_0 \, dz_0 \tag{36}$$

where

$$r = [(x - x_0)^2 + (z - z_0)^2]^{1/2} \tag{37}$$

In the multiple fluid case where density variations create vorticity according to (33), the expression for Ψ becomes

$$\Psi_{\text{I}}(x, z) = (-k/2\pi\mu) \iint [\partial(\rho g)/\partial x]_0 \ln r \, dx_0 \, dz_0 \tag{38}$$

The surface integral for Ψ_{I} covers the entire region between the boundaries containing fluids. This expression for Ψ_{I} forms the vortex part of the streamfunction.

In general, the values of Ψ_{I} on the boundaries differ from the values required by the boundary conditions. These differences form a new boundary value problem to be satisfied by a harmonic streamfunction Ψ_{II} obeying the Laplace equation $\nabla^2\Psi_{\text{II}} = 0$. The function Ψ_{II} forms the second part of the solution. The final solution to the boundary value problem is $\Psi = \Psi_{\text{I}} + \Psi_{\text{II}}$.

This way of solving the problem is known in potential theory as the solution of Poisson's equation by singularities. In this case the Poisson equation for Ψ_{I} is (16) which has (38) as its solution.

In a similar way, a solution for the pressure p or the multiple fluid potential Θ can be given because they obey similar Poisson equations i.e., (17) and (23). This gives

$$p_{\text{I}} = (-1/2\pi) \iint [\partial(\rho g)/\partial z]_0 \ln r \, dx_0 \, dz_0 \tag{39}$$

$$\Theta_{\text{I}} = (+k/2\pi\mu) \iint [z \, \nabla^2(\rho g) + \partial(\rho g)/\partial z]_0 \ln r \, dx_0 \, dz_0 \tag{40}$$

Also in these cases, additional functions p_{II} and Θ_{II} obeying Laplace's equation must be introduced to complete the solution and to satisfy all boundary conditions.

An alternative way of decomposing the solution of Poisson's equation for boundary conditions is pointed out by De Wiest (1969). Instead of $\ln r$ which can be considered as a Green's function for the infinite domain, Greens functions can be introduced which have zero conditions along the boundaries of the domain considered in each specific problem. (The function or its normal derivative is zero along the boundary.) The differences with the required boundary values are then those boundary values themselves, creating a boundary value problem with a solution satisfying Laplace's equation.

Since, however, a boundary value problem has to be solved anyway, the method proposed here seems less laborious because it circumvents the construction of a Green's function, which in contradistinction to $\ln r$ is different for every boundary geometry and for every point in the field.

We are now ready to compare the torque, caused by density variation in a circular area as determined in Section 4, to the driving force necessary for the circulating discharge caused by the vorticity. If, however, the vorticity is limited to the circular area, the available torque is not sufficient to support the discharge outside the area. This outside discharge requires a torque of infinite magnitude.

This discrepancy is a consequence of the vorticity, being caused by density variations. In other instances vortices exist which can be limited to isolated regions. In this case it is impossible to visualize an isolated region containing a density gradient. Isolation means that it is surrounded by a field which, up to infinity, has the same density. The gradient actually means that the field contains fluids of different density. If this difference exists up to infinity, this entails also a torque which is infinitely large.

In order to circumvent this difficulty and to obtain a realistic situation, the case of a circular aquifer of finite dimensions and surrounded by an impermeable boundary will be considered (see Fig. 4). The radius is r_3, the thickness f, and the porosity n.

For the analysis to remain simple the density will be taken to vary linearly over the field according to

$$\rho = \rho_0 + \xi \rho_0' \tag{41}$$

where ρ_0' is a constant.

For this situation, $(\partial\rho/\partial x)$ is constant over the field and equal to

$$\frac{\partial\rho}{\partial x} = \frac{\partial\rho}{\partial\xi} \cdot \frac{\partial\xi}{\partial x} = \rho_0' \cos\alpha \tag{42}$$

From (33), the vorticity is

$$\omega = -\tfrac{1}{2}(k/\mu)[\partial(\rho g)/\partial x], \qquad 0 < r < r_3$$

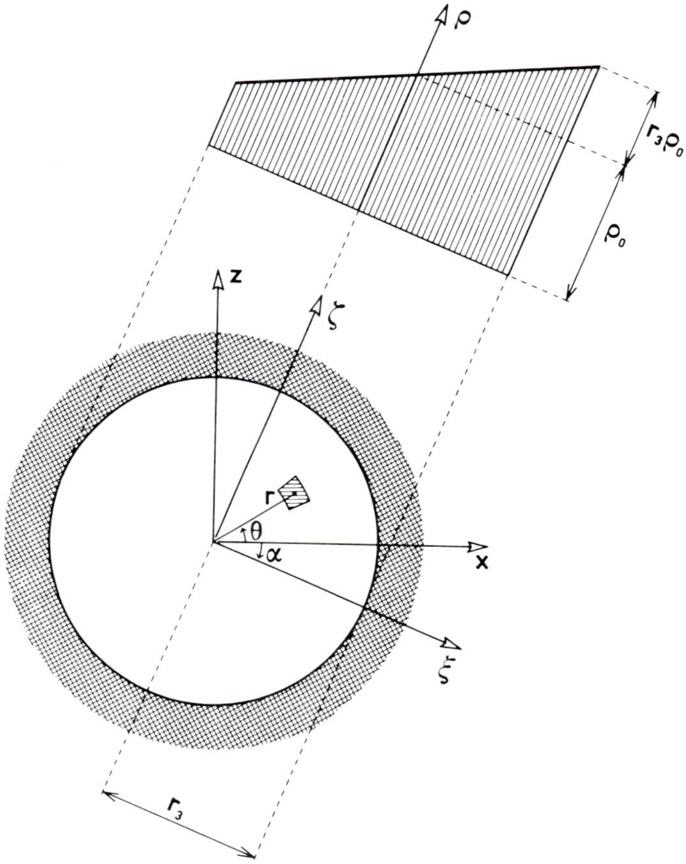

FIG. 4.

which is also constant. As explained above, an integration of the influence according to (32) of all the vortices in the entire region gives the solution similar to (30)

$$q_S = \omega r = -\tfrac{1}{2}(k/\mu)[\partial(\rho g)/\partial x]\, r \tag{43}$$

Since this gives directly a discharge perpendicular to the radius, this solution satisfies the boundary conditions which have actually been chosen in this example in such a way that the discharge obeys the simple formula (43). Therefore, the second part of the solution is zero.

Since in Fig. 3, q_S is positive for counter clockwise rotation, (43) gives a clockwise rotation for positive ρ_0'. In order to show that the clockwise torque (7) is apt to create this circulating flow, the torque will be computed here by considering the required forces.

The force necessary to drive fluid through the ring between r and $r + dr$ is

$$(\mu/k)\, q_S n 2\pi r\, dr\, f$$

Hence the total torque clockwise in the circular region is

$$T = \int_0^{r_3} r \left(\frac{\mu}{k}\right) \frac{1}{2} \left(\frac{k}{\mu}\right) \frac{\partial(\rho g)}{\partial x}\, rn2\pi rf\, dr$$

$$= \tfrac{1}{4}\pi r_3{}^4 \left[\frac{\partial(\rho g)}{\partial x}\right] nf \tag{44}$$

A comparison with (7) shows that if $r = r_3$, the two torques are equal.

From this analysis it follows that the rotational flow within the circular area is possible because a torque created by density variations acts on the fluid of this region. This is comparable with the general concept that forces of the boundaries cannot create rotations in the interior of the fluid. Vorticity can only occur in those places where body forces act in the form of torques on the fluid.

The geometry of Fig. 4 gives a flow pattern, which by its simplicity permits us to predict the rotation of the fluid at all subsequent times. The discharge is such that the fluid rotates as a rigid body. Therefore, regardless of dispersion and diffusion, the fluid conserves its density distribution (41) if the coordinate ξ is allowed to rotate clockwise with the fluid. This means that the angular velocity is equal to

$$\frac{d\alpha}{dt} = \frac{-q_S}{nr} = \frac{1}{2}\left(\frac{k}{\mu}\right)\left[\frac{\rho_0' g \cos\alpha}{n}\right]$$

Integration gives for $\alpha = 0$, $t = 0$,

$$\tan(\tfrac{1}{4}\pi + \tfrac{1}{2}\alpha) = \exp[\tfrac{1}{2}(k/\mu)(\rho_0' g/n)\, t] \tag{45}$$

7. Multiple Fluid Potential for Variable μ, ρ, and k

If also the fluid viscosity and the intrinsic permeability of the porous medium are variable, only the multiple fluid potential Θ will give a Poisson equation which permits solutions by singularities. This is seen as follows.

Write (21) as

$$\mathbf{q} = -\text{grad}\,\Theta + (kz/\mu)\,\text{grad}(\rho g) \tag{46}$$

Then the continuity equation gives

$$\text{div}\ \mathbf{q} = 0 = -\text{div grad}\ \Theta + \text{div}(kz/\mu)\ \text{grad}(\rho g)$$

or, in the form of a Poisson equation,

$$\nabla^2\Theta = +\text{div}[(kz/\mu)\ \text{grad}(\rho g)] \tag{47}$$

If μ, ρ, k have known values in the entire field, the right side of (47) can be computed. This then permits us to evaluate the integral expression

$$\Theta_{\text{I}} = (1/2\pi) \iint \text{div}[(kz/\mu)\ \text{grad}(\rho g)]_0 \ln r\ dx_0\ dz_0 \tag{48}$$

which represents the rotational part of the solution.

For Ψ and p, the equations are

$$\nabla^2\Psi = -\frac{k}{\mu}\frac{\partial(\rho g)}{\partial x} - \frac{k}{\mu}\ \text{grad}\left(\frac{\mu}{k}\right) \cdot \text{grad}\ \Psi \tag{49}$$

$$\nabla^2 p = -\frac{\partial(\rho g)}{\partial z} - \text{grad}\left(\frac{\mu}{k}\right) \cdot [\text{grad}\ p + \rho g\ \text{grad}\ z] \tag{50}$$

Both these expressions contain Ψ and p, respectively, on the right sides. Therefore, the solution cannot be evaluated with the integral expression because the integrand still contains unknown terms.

A possible way to obtain a solution is by electric analogy. [See Malavard (1956).] Although this manner of solving a multiple fluid problem may only be of academic value, it will be given here for the sake of completeness.

8. Electric Analogy for Variable μ, ρ, and k

8.1. Potential Lines

If the objective is to determine the pressure in the fluid, the electric potential can be identified with the multiple fluid potential multiplied by (μ/k). This slightly revised function will be called Θ^*.

The analogy between fluid and electric quantities is

$$\Theta^* = (p + \rho g z) \leftrightarrow E \tag{51}$$

$$(\mu/k)\ \mathbf{q} - z\ \text{grad}(\rho g) \leftrightarrow \rho_e \mathbf{i}_e \tag{52}$$

$$(\mu/k) \leftrightarrow \rho_e \tag{53}$$

in which the electric quantities are: E, the electric potential, ρ_e, the specific electric resistance, and $\mathbf{i}_e$, the specific electric current vector.

Because p, z, μ, ρ, and k are single-valued functions, Θ^* is single-valued and therefore, the single-valuedness of E is assured. In the electrical model the current will flow according to Ohm's law, which can be written as

$$\rho_e \mathbf{i}_e = -\text{grad } E \tag{54}$$

which holds also if ρ_e is variable in the field. For the fluids this means that the following relation is supposed to hold

$$(\mu/k)\,\mathbf{q} - z \text{ grad}(\rho g) = -\text{grad } p - z \text{ grad}(\rho g) - (\rho g) \text{ grad } z \tag{55}$$

A comparison with (3) shows that this is indeed required.

Application of the continuity equation for the fluids can be effected after dividing (52) by (μ/k) at the left side and by ρ_e at the right side. Taking the divergence then gives

$$\text{div } \mathbf{q} - \text{div}(kz/\mu) \text{ grad}(\rho g) \leftrightarrow \text{div } \mathbf{i}_e \tag{56}$$

Requirement (1) indicates that div $\mathbf{i}_e$ must be made equal to the electric analogy of $-\text{div}(kz/\mu) \text{ grad}(\rho g)$. Since div $\mathbf{i}_e$ is the amount of electric current that has to be injected into the electric flow field, it means that in those places where ρ or k or μ and ρ vary, current has to be fed into the model.

In an electric model, only electric potential can be observed or for that matter, equal potential lines may be drawn. From these lines the values of the pressure can be found by means of (51).

8.2. Streamlines

If the objective is to determine the streamlines in the fluid, the electric model must be used in an inverse way such that electric potential lines are related to the streamlines of the fluid. The inversion of streamlines and potential lines is obtained by inverting the boundary conditions. But besides that, in the interior, the inversion also has its consequences because the analogy of the resistance must be made reciprocal.

The analogy between fluid and electric properties will now be taken as

$$q_x \leftrightarrow -(\partial E/\partial z), \qquad q_z \leftrightarrow +(\partial E/\partial x) \tag{57}$$

$$\mu/k \leftrightarrow 1/\rho_e \tag{58}$$

Continuity of the fluids requires that

$$-(\partial^2 E/\partial z\,\partial x) + (\partial^2 E/\partial x\,\partial z) = 0$$

which indicates that E is single-valued.

Ohm's law applied to the electric system gives with (57), (58), and (13) the following relation between the specific electric current and the specific discharge of the fluids

$$\begin{aligned} i_{e_x} &= -\frac{1}{\rho_e}\frac{\partial E}{\partial x} \leftrightarrow -\frac{\mu}{k} q_z = +\frac{\partial p}{\partial z} + \rho g \\ i_{e_z} &= -\frac{1}{\rho_e}\frac{\partial E}{\partial z} \leftrightarrow +\frac{\mu}{k} q_x = -\frac{\partial p}{\partial x} \end{aligned} \tag{59}$$

Taking the divergence of $\mathbf{i}_e$ and using the single-valuedness of p then gives

$$\text{div } \mathbf{i}_e \leftrightarrow [\partial(\rho g)/\partial x] \tag{60}$$

From (60) it follows that also in this case an injection in the electric flow field is necessary with a magnitude of the electric current equivalent to $[\partial(\rho g)/\partial x]$. Because of the inversion, injection in this case has the effect of vorticity.

From (57) it follows directly that lines of equal electric potential will trace streamlines.

9. Variable Density Flow in Three Dimensions

For three dimensions, Poisson equations similar to the two dimensional case can be derived. For the pressure p, this was done by Knudsen (1962). Writing the equilibrium equation (3) in vector notation gives

$$(\mu/k)\,\mathbf{q} = -\nabla p - \rho g\mathbf{k} \tag{61}$$

where the vector $\mathbf{k}$ indicates the unit vector in the z direction, vertically upward. Continuity applied to this expression if μ and k are constant, leads to the following result in view of (1),

$$(\mu/k)\,\nabla\cdot\mathbf{q} = -\nabla^2 p - \nabla\rho\cdot g\mathbf{k} = 0$$

or

$$\nabla^2 p = -g(\partial\rho/\partial z) \tag{62}$$

For Θ, the Poisson equation is already stated in general coordinates by (23) giving

$$\nabla^2\Theta = +(k/\mu)\,\nabla\cdot[z\,\nabla\rho] \tag{63}$$

For the streamfunction it is more complicated since the transition to three dimensions requires the introduction of two streamfunctions. How this can be done was described by Yih (1957), who showed that the discharge can be written as

$$\mathbf{q} = (\nabla\Psi) \times (\nabla\chi) \tag{64}$$

Since the establishment of the appropriate Poisson equations for Ψ and χ was too difficult to achieve, it will be shown here that the discharge itself obeys a Poisson's equation which permits a direct computation with the solution by singularities.

Application of the curl operation to (61) gives, if μ and k are constant,

$$(\mu/k)\,\nabla \times \mathbf{q} = -\nabla \times \nabla p - \nabla \times \rho g\mathbf{k} \tag{65}$$

Expansion of the first term to the right side gives

$$\begin{aligned}\nabla \times \nabla p = {} & \mathbf{i}[(\partial^2 p/\partial y\,\partial z) - (\partial^2 p/\partial z\,\partial y)] \\ & + \mathbf{j}[(\partial^2 p/\partial z\,\partial x) - (\partial^2 p/\partial x\,\partial z)] \\ & + \mathbf{k}[(\partial^2 p/\partial x\,\partial y) - (\partial^2 p/\partial y\,\partial x)]\end{aligned}$$

For physical reasons p is single-valued, hence by use of (10), all three vectors are zero and the term vanishes. Thus, (65) becomes

$$(\mu/k)\,\nabla \times \mathbf{q} = -\nabla \times \rho g\mathbf{k} = g[\mathbf{i}(\partial\rho/\partial y) - \mathbf{j}(\partial\rho/\partial x)] \tag{66}$$

Since $\nabla \times \mathbf{q}$ represents a vortex, this expression shows that the axis of the vortex lies in the horizontal plane and has the direction of lines for ρ constant in that plane. Further, since $\mathbf{i}(\partial\rho/\partial x) + \mathbf{j}(\partial\rho/\partial y)$ is the horizontal component of the density gradient, the expression (66) shows that the vortex strength is proportional to this horizontal component of grad ρ.

The curl operation on (66) gives

$$(\mu/k)\,\nabla \times (\nabla \times \mathbf{q}) = -\nabla \times (\nabla \times \rho g\mathbf{k})$$

The first term can be written as

$$\nabla \times (\nabla \times \mathbf{q}) = \nabla(\nabla\cdot\mathbf{q}) - \nabla^2\mathbf{q}$$

Continuity requires that $\nabla\cdot\mathbf{q}$ be zero so there remains

$$\nabla^2\mathbf{q} = (k/\mu)\,\nabla \times (\nabla \times \rho g\mathbf{k}) \tag{67}$$

This is Poisson's equation for $\mathbf{q}$. Solutions of this equation can be given in the form of the integral

$$\mathbf{q}_{\mathrm{I}} = +(k/4\pi\mu) \int_{\forall} \nabla \times (\nabla \times \rho g \mathbf{k})\, r^{-1}\, d\forall \tag{68}$$

This integral describes the part of the discharge vector that is created by the vortices caused by density variations. A second part, $\mathbf{q}_{\mathrm{II}}$, that is irrotational has to be added to satisfy the boundary conditions. The integrand of (68) contains the density variations because the development of the curls gives

$$\nabla \times (\nabla \times \rho g \mathbf{k}) = -(\partial^2\rho/\partial x\, \partial z)\, \mathbf{i} - (\partial^2\rho/\partial y\, \partial z)\, \mathbf{j} + [(\partial^2\rho/\partial x^2) + (\partial^2\rho/\partial y^2)]\, \mathbf{k}$$

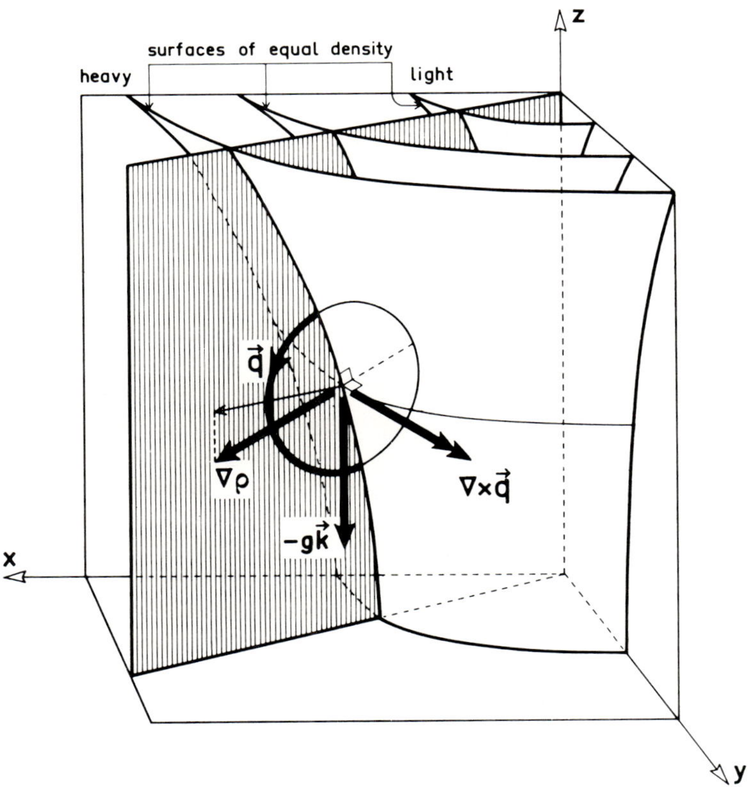

FIG. 5.

The behavior of the discharge is represented in Fig. 5 where a three-dimensional picture is given of surfaces of equal density. In the region where the density varies continuously, many of these sheets can be visualized as lying perpendicular to the density gradient. According to (66), the discharge vortex created by the density gradient has its axis in the direction of horizontal lines on these sheets.

The vortex axis, which is horizontal, is perpendicular to the vertical direction of gravity; it lies in the isopicnic (of equal density) sheet and is perpendicular to the gradient of density. Therefore, the vortex axis is perpendicular to the plane through the vertical and the density gradient (shaded area in Fig. 5).

The discharge from the vortex rotates in a plane perpendicular to the vortex axis. Hence, finally, the discharge vortex in a point is rotating in the plane through the vertical and the density gradient in that point. The strength of the vortex is proportional to the horizontal component of the density gradient.

REFERENCES

de Josselin de Jong, G. (1960). Singularity distributions for the analysis of multiple fluid flow through porous media, *J. Geophys. Res.* **65**, 3739–3758.

De Wiest, R. J. M. (1964). Dispersion and Salt water Intrusion, *Groundwater, J. of NWWA* **2**, 39–40.

De Wiest, R. J. M. (1965). "Geohydrology," pp. 305–313. Wiley, New York.

De Wiest, R. J. M. (1969). Green's functions in the flow through porous media, this volume, Chapter X.

Knudsen, W. C. (1962). Equations of fluid flow through porous media—incompressible fluids of varying density, *J. Geophys. Res.* **67**, 733–737.

Lusczynski, N. J. (1961). Head and Flow of Groundwater of variable Density, *J. Geophys. Res.* **66**, 4247–4256.

Malavard, L. C. (1956). The use of rheoelectrical analogies in aerodynamics, *AGARDograph* 18, NATO Advisory Group of Aeronautical Res. and Devel., Paris.

Yih, C. S. (1957). Fonctions de courant dans les écoulements à trois dimensions, *La Houille Blanche* **12**, 439-444.

Yih, C. S. (1961). Flow of a nonhomogeneous fluid in a porous medium, *J. Fluid Mech.* **10**, pp. 133–141.

10

Green's Functions in the Flow through Porous Media

Roger J. M. De Wiest
Department of Civil and Geological Engineering
Princeton University
Princeton, New Jersey

1. Green's Identities

The construction of the Green's function to solve a partial differential equation of mathematical physics for given boundary (and initial) conditions is accomplished in the classical approach (Kellogg, 1929) with the help of Green's identities, which are briefly reviewed. The underlying basic theorem is the divergence theorem, sometimes known as Gauss' theorem or even Green's theorem. The equations are given for two dimensions but they can easily be extended to three dimensions.

1.1. Divergence Theorem

$$\iint_A \operatorname{div} \mathbf{F}\, dA = \oint_C F_n\, ds \tag{1}$$

or

$$\iint_A \left(\frac{\partial X}{\partial x} + \frac{\partial Y}{\partial y}\right) dA = \oint_C (Xl + Ym)\, ds \tag{2}$$

in which X, Y are the components of the vector field $\mathbf{F}$, and l and m are the direction cosines of the normal n to s, directed outward from C, where C is the contour of the simply connected domain A described counter clockwise (Fig. 1). With the sign convention of Fig. 1, it follows that the orthogonality requirement for the tangent along s counter clockwise and the outward normal is

$$\frac{dx}{dn} = +\frac{dy}{ds}, \qquad \frac{dy}{dn} = -\frac{dx}{ds} \tag{3}$$

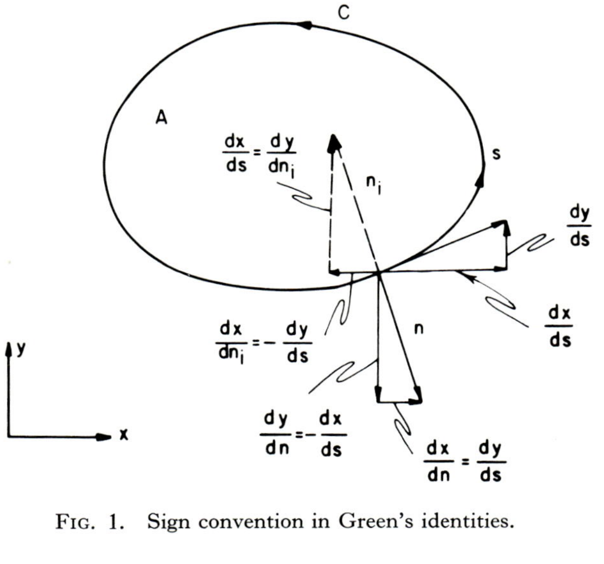

Fig. 1. Sign convention in Green's identities.

For the inward normal n_i, equations similar to (3) would hold, but with the signs reversed. In the sequel, the outward normal n is always used without subscript. Substitution of (3) into the divergence theorem (2) leads to a frequently used form of Green's theorem

$$\iint_A \left(\frac{\partial X}{\partial x} + \frac{\partial Y}{\partial y}\right) dA = \oint_C X\,dy - Y\,dx \tag{4}$$

1.2. Green's First Identity

Let u and v be two functions defined in A, continuous in A and with continuous partial derivatives of the second order† in A. Let the vector field $\mathbf{F}$ in (2) be represented by

$$X = v\frac{\partial u}{\partial x}, \qquad Y = v\frac{\partial u}{\partial y}$$

so that (2) becomes

$$\iint_A v\,\nabla^2 u\,dA + \iint_A \nabla u \cdot \nabla v\,dA = \oint_C v\frac{\partial u}{\partial n}\,ds \tag{5}$$

Equation (5) is known as *Green's first identity.*

1.3. Green's Second Identity

If the roles of u and v in (5) are interchanged and if (5) is subtracted from the resulting equation, *Green's second identity* is obtained as

$$\iint_A (u\,\nabla^2 v - v\,\nabla^2 u)\,dA = \oint_C \left(u\frac{\partial v}{\partial n} - v\frac{\partial u}{\partial n}\right) ds \tag{6}$$

1.4. Green's Third Identity

Let $v = \log r$ in (6), in which r is the distance from any point $P(x, y)$ to a point Q with running coordinates (ξ, η), or variables of integration in (6). Since v is not defined in A when Q covers P, (6) can only be applied to the domain A' which results from A when P is excluded from it by a small circle with radius ϵ which in the limit shrinks to zero. The contour of integration now consists of the original C, described counter clockwise, plus the circle C_1, described clockwise, and linked to C by two closely spaced lines described in opposite direction (Fig. 2).

† For the first identity, only continuity of the partial derivative of the first order is required for v.

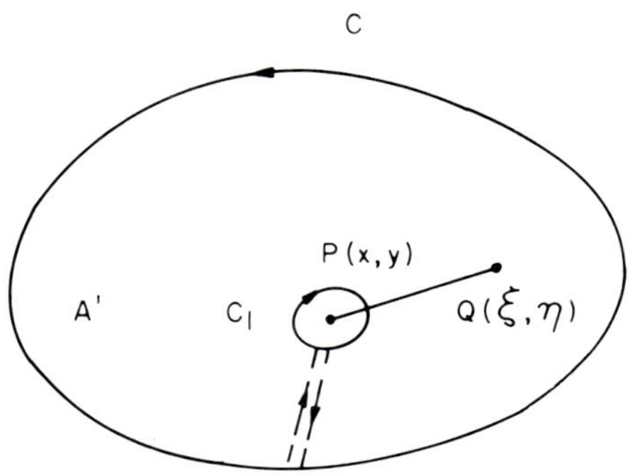

FIG. 2. Derivation of Green's third identity.

Application of (6) to A' leads to

$$-\iint_{A'} \log r\, \nabla^2 u\, dA = \oint_C \left(u \frac{\partial v}{\partial n} - v \frac{\partial u}{\partial n}\right) ds + \oint_{C_1} \left(u \frac{\partial v}{\partial n} - v \frac{\partial u}{\partial n}\right) ds \quad (7)$$

Since on C_1, n has the opposite direction from r, the second contour integral can be written as

$$\lim_{\epsilon \to 0} \int_0^{2\pi} \left[-u \frac{1}{\epsilon} + \log \epsilon \left.\frac{\partial u}{\partial r}\right|_{r=\epsilon}\right] \epsilon\, d\theta = -2\pi u(P)$$

in which $u(P) = u(x, y)$.

Also in the limit, for $\epsilon \to 0$, $A' \to A$, and (7) becomes

$$u(x, y) = \frac{1}{2\pi} \iint_A \log r\, \nabla^2 u\, dA + \frac{1}{2\pi} \oint_C \left(u \frac{\partial}{\partial n} \log r - \log r \frac{\partial u}{\partial n}\right) ds \quad (8)$$

Equation (8) is known as *Green's third identity.*

2. Green's Functions for Self-Adjoint Differential Equations

For steady state conditions of flow, the equations which govern the variation of head or drawdown in aquifers are of the self-adjoint type, namely,

Laplace's equation: $\nabla^2 u = 0$

Poisson's equation: $\nabla^2 u = f(x, y)$

and the elliptic equation in canonical form

$$\nabla^2 u - k^2 u = f(x, y)$$

Laplace's equation, for example, holds for the head in a confined aquifer. Poisson's equation is the approximate equation for the variation of head in an unconfined aquifer with replenishment through percolation. $\nabla^2 u - k^2 u = f(x, y)$ is the approximate equation for the variation of the head in an aquifer replenished from above through percolation and receiving or losing water from below through leakage. In the last equation, the constant k^2 stands for $1/B^2$ where B is the leakage factor. As may be seen from (41) for the case of the nonself-adjoint differential equation, the Laplacian operator ∇^2 in (8) may be replaced by the self-adjoint operator $\mathscr{L} = \nabla^2 - k^2$ without invalidating (8).

The general procedure to solve $\mathscr{L}u = f(x, y)$ or $\mathscr{L}u = 0$ (for boundary conditions as stipulated hereafter) by means of the Green's function of the problem is to consider a solution $v(x, y; \xi, \eta)$ of $\mathscr{L}v = 0$ which is composed of two parts:

(a) a part which has a singularity inside the domain A at the source point or pole $P(x, y)$, say $\log r(x, y; \xi, \eta)$, called the principal or fundamental solution;

(b) a part $w(x, y; \xi, \eta)$ which is regular or smooth in the entire domain A, and which will be constructed so as to make $v(x, y; \xi, \eta)$ satisfy certain boundary conditions.

The resulting function $v = \log r + w$ is called the *Green's function G* of the problem. As pointed out in (73), this statement is equivalent to saying that $G(x, y; \xi, \eta)$ satisfies the equation

$$\mathscr{L}G = \delta(\xi - x)\,\delta(\eta - y) \tag{9}$$

in which $\delta(\xi - x)$ and $\delta(\eta - y)$ are Dirac's delta functions.

If $v = \log r + w$ is inserted in (6), in which ∇^2 is replaced by $\mathscr{L}$, considerations similar to those used in the derivation of (7) at once lead to

$$u(x, y) = \frac{1}{2\pi}\iint_A (\log r + w)\,\mathscr{L}u\,dA + \frac{1}{2\pi}\oint_C \left[u\frac{\partial}{\partial n}(\log r + w) - (\log r + w)\frac{\partial u}{\partial n}\right] ds \tag{10}$$

or

$$u(x, y) = \frac{1}{2\pi}\iint_A G\mathscr{L}u\,dA + \frac{1}{2\pi}\oint_C \left(u\frac{\partial G}{\partial n} - G\frac{\partial u}{\partial n}\right) ds \tag{11}$$

2.1. Boundary Value Problems

The boundary conditions associated with the equations $\mathscr{L}u = 0$ or $\mathscr{L}u = f$ are commonly distinguished from each other as giving rise to three different boundary value problems, namely:

The first boundary value problem, in which a solution of equations $\mathscr{L}u = 0$ or $\mathscr{L}u = f$ is sought in a domain A when the value of u is prescribed along the boundary C of A. This problem is called the *Dirichlet problem*.

The second boundary value problem, in which the value of $\partial u/\partial n$ is prescribed along the boundary C of A, also called the *von Neumann problem*.

The third boundary value problem, in which a linear combination of u and $\partial u/\partial n$ is prescribed along the boundary C of A.

Sometimes the Dirichlet and von Neumann problems are combined into a mixed boundary value problem (Muskat, 1935) in which the value of u is prescribed along part of the boundary and the value of $\partial u/\partial n$ is given along the remainder of the boundary.

2.2. Dirichlet Problem

Since only u is known along C and no information about $\partial u/\partial n$ is available along C, (11) will render a solution if G is so constructed that it vanishes along C, so that (11) becomes, with $u = u(s)$,

$$u(x, y) = \frac{1}{2\pi} \iint_A G\mathscr{L}u \, dA + \frac{1}{2\pi} \oint_C u \frac{\partial G}{\partial n} \, ds \tag{12}$$

2.3. Method of Images

The construction of G becomes particularly easy when the domain A can be treated by a simple reflection method in which singularities are reflected across the boundaries of the domain. The reflection process is simple for certain configurations of A such as the infinite half-plane, the infinite quadrant, the infinite and semi-infinite strip, the circle and the rectangle. In some cases, a finite number of singularities will suffice to construct the Green's function, while in other cases an infinite number of singularities is necessary.

Example 1. Find the head distribution $h(x, y)$ in a semi-infinite confined aquifer completely intersected by a rectilinear canal with variable stage $h(\xi)$ (Fig. 3). The problem is to solve $\nabla^2 h = 0$ in the domain A, given $h(\xi)$ along the boundary of the domain.

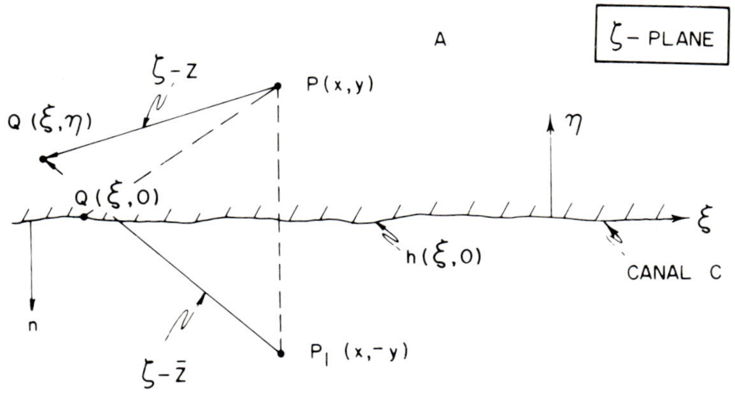

FIG. 3. Solution of $\nabla^2 h = 0$ for aquifer bounded by canal with variable $h(\xi)$.

Here the pole $P(x, y)$ is reflected into the ξ axis which coincides with the canal and the image of P is $P_1(x, -y)$. If we take $w = -\log r_{P_1Q}$, then

$$G(x, y; \xi, \eta) = \log(r_{PQ}/r_{P_1Q}) \tag{13}$$

is indeed zero when Q is on the contour C. Furthermore, w is a regular harmonic function because Q cannot cover P_1, Hence, we have found the Green's function for the problem. Considering that $n = -\eta$, $h(s) = h(\xi, 0)$, $ds = d\xi$, we find from (12),

$$h(x, y) = -\frac{1}{2\pi}\int_{-\infty}^{\infty} h(\xi, 0)\left\{\frac{\partial}{\partial\eta}\log\left[\frac{(x-\xi)^2+(y-\eta)^2}{(x-\xi)^2+(y+\eta)^2}\right]^{1/2}\right\}_{\eta=0} d\xi$$

or

$$h(x, y) = \frac{y}{\pi}\int_{-\infty}^{\infty}\frac{h(\xi, 0)\,d\xi}{(x-\xi)^2+y^2} \tag{14}$$

If the aquifer were leaky, the equation would be $\nabla^2 s - (s/B^2) = 0$, in which s represents the drawdown and B is the leakage factor and the problem would be to solve this equation, given $s(\xi, 0)$. Now

$$G(x, y; \xi, \eta) = -K_0\left[\frac{1}{B}[(x-\xi)^2+(y-\eta)^2]^{1/2}\right]$$
$$+ K_0\left[\frac{1}{B}[(x-\xi)^2+(y+\eta)^2]^{1/2}\right] \tag{15}$$

and we find from (12)

$$s(x, y) = \frac{1}{2\pi}\int_{-\infty}^{\infty} s(\xi, 0)\left\{\frac{\partial}{\partial\eta}\left[K_0\left(\frac{1}{B}r_{PQ}\right) - K_0\left(\frac{1}{B}r_{P_1Q}\right)\right]\right\}_{\eta=0} d\xi$$

or

$$s(x, y) = \frac{y}{B\pi} \int_{-\infty}^{\infty} \frac{s(\xi, 0)\, K_1\{(1/B)[(x - \xi)^2 + y^2]^{1/2}\}}{[(x - \xi)^2 + y^2]^{1/2}}\, d\xi \tag{16}$$

If we let $B \to \infty$ (i.e., no leakage) in (16), we replace

$$K_1 \left[\frac{1}{B} [(x - \xi)^2 + y^2]^{1/2} \right]$$

by its asymptotic value for small arguments, which is simply the inverse of its argument and we find

$$s(x, y) = \frac{y}{\pi} \int_{-\infty}^{\infty} \frac{s(\xi, 0)\, d\xi}{(x - \xi)^2 + y^2} \tag{17}$$

a formula which is compatible with (14).

It should be noted that in (15) we have used $-K_0(r/B)$ for the function with the singularity at $r = 0$ because for small values of its argument, K_0 approaches asymptotically the negative value of the logarithm of its argument. Thus we are compatible with the construction of G as resulting from $v = \log r + w$, a concept used in the derivation of (11).

Example 2. Find the drawdown distribution $s(x, y)$ in an infinite strip of leaky aquifer bounded by rectilinear water bodies of variable head (Fig. 4). The problem is to solve $\nabla^2 s - (s/B^2) = 0$ in the domain A, given $s(\xi, 0)$ and $s(\xi, 2a)$ along boundaries of the domain.

The construction of the Green's function by means of the reflection process (method of images) is indicated in Fig. 4. To obtain a vanishing G along the boundaries $\eta = 0$ and $\eta = 2a$ for the principal solution $-K_0(r/B)$, it is necessary to add to it

$$w = K_0 \left\{ \frac{1}{B} [(\xi - x)^2 + (\eta + y - 4a)^2]^{1/2} \right\} + K_0 \left\{ \frac{1}{B} [(\xi - x)^2 + (\eta + y)^2]^{1/2} \right\} \tag{18}$$

These functions may be thought of as having singularities in the images of the source point P with respect to $\eta = 2a$ and $\eta = 0$, respectively. (In the domain A, however, they are regular.) For this reason it has become customary and convenient to introduce the "method of images," in which the source point P [associated with a minus sign here because of the principal solution $-K_0(r/B)$] is reflected in the boundaries to give rise to images of opposite sign (for Dirichlet conditions). Each of the K_0 functions of (18) combined with the principal solution would give a vanishing Green's function along one of the boundaries, but

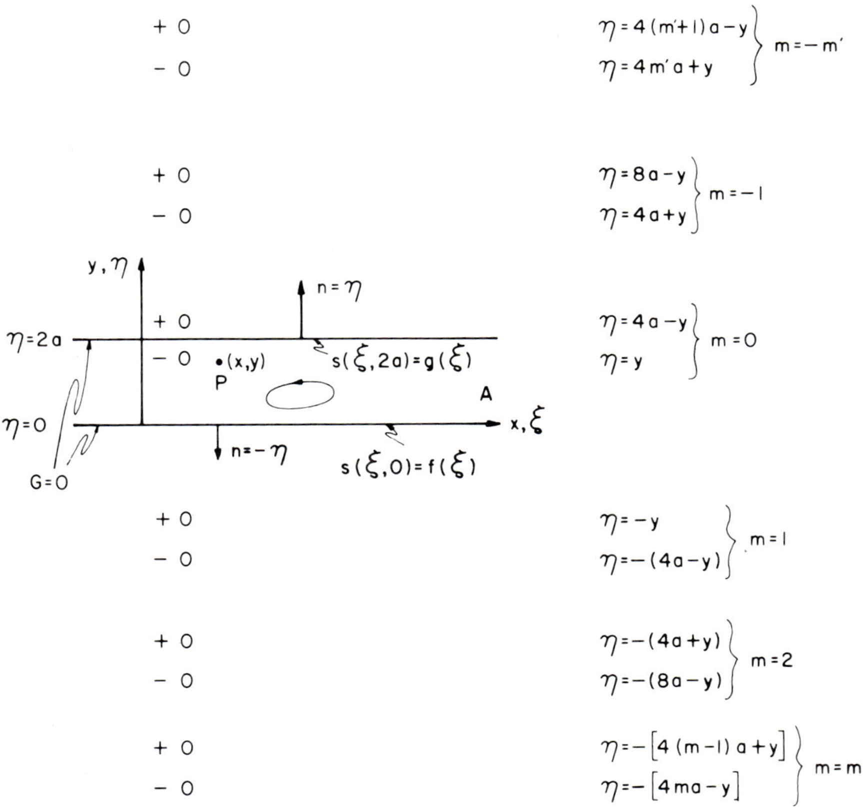

FIG. 4. Method of images for infinite strip of leaky aquifer between two waterbodies of variable head.

the addition of its partner in (18) disturbs this condition of vanishing along that particular boundary. The expression of w in (18) is therefore not complete as new contributions will arise due to the reflection of the two positive singularities into the boundaries. It is easy to see that this reflection process leads to infinite series of K_0 functions, one for the singularities associated with a negative sign and one for the singularities associated with a positive sign. From Fig. 4 it follows that

$$G(x, y; \xi, \eta) = -\sum_{m=-\infty}^{\infty} K_0 \left\{ \frac{1}{B} [(\eta - y - 4ma)^2 + (\xi - x)^2]^{1/2} \right\}$$

$$+ \sum_{m=-\infty}^{\infty} K_0 \left\{ \frac{1}{B} [(\eta + y - 4a - 4ma)^2 + (\xi - x)^2]^{1/2} \right\} \tag{19}$$

Hantush and Jacob (1954) have shown that these series may be simplified for computation by the use of properties of the Fourier cosine transform. This procedure however is laborious and a simpler method of constructing the Green's function of this problem is given in Section 4.3.1. The result, as derived in that section, is

$$G(x, y; \xi, \eta) = \frac{1}{\pi} \sum_{m-1}^{\infty} \frac{\exp\left\{-\frac{\pi}{2a} | x - \xi | \left[m^2 + \left(\frac{2a}{\pi B}\right)^2\right]^{1/2}\right\}}{\left[m^2 + \left(\frac{2a}{\pi B}\right)^2\right]^{1/2}} \sin \frac{m\pi y}{2a} \sin \frac{m\pi\eta}{2a} \tag{20}$$

and the solution of the problem is

$$\begin{aligned} s(x, y) = & -\frac{1}{2\pi} \int_{-\infty}^{\infty} s(\xi, 0) \\ & \times \left[\sum_{m=1}^{\infty} \frac{m}{2a} \frac{\exp\left\{-\frac{\pi}{2a} | x - \xi | \left[m^2 + \left(\frac{2a}{\pi B}\right)^2\right]^{1/2}\right\}}{\left[m^2 + \left(\frac{2a}{\pi B}\right)^2\right]^{1/2}} \sin \frac{m\pi y}{2a} \right] d\xi \\ & + \frac{1}{2\pi} \int_{+\infty}^{-\infty} s(\xi, 2a) \\ & \times \left[\sum_{m=1}^{\infty} \frac{m}{2a} \frac{\exp\left\{-\frac{\pi}{2a} | x - \xi | \left[m^2 + \left(\frac{2a}{\pi B}\right)^2\right]^{1/2}\right\}}{\left[m^2 + \left(\frac{2a}{\pi B}\right)^2\right]^{1/2}} \sin \frac{m\pi y}{2a} \cos m\pi \right] d\xi \end{aligned} \tag{21}$$

2.4. Von Neumann Problem

Since only $\partial u/\partial n$ is known along C and no information about u is available along C, (11) will render a solution if G is so constructed that $\partial G/\partial n$ vanishes along C, so that (11) becomes

$$u(x, y) = \frac{1}{2\pi} \iint_A G\mathscr{L}u \, dA - \frac{1}{2\pi} \oint_C G \frac{\partial u}{\partial n} ds \tag{22}$$

The Green's function for the von Neumann problem can also be constructed by means of the method of images. A given singularity is now reflected across the boundary into an image singularity of the same kind (i.e., of the same sign). Its construction becomes evident from the examination of an example of a mixed boundary value problem.

2.5. Mixed Boundary Value Problem

The problem is to solve $\mathscr{L}u = f$ in A when u and $\partial u/\partial n$ are preassigned arbitrarily over parts of the boundary. A "mixed" Green's function is defined such that it vanishes over part C_1 of the boundary and its normal derivative vanishes over the remainder C_2. The solution of the problem is expressed, from (11), as

$$u(x, y) = \frac{1}{2\pi} \oint_{C_1} u \frac{\partial G}{\partial n} ds - \frac{1}{2\pi} \oint_{C_2} G \frac{\partial u}{\partial n} ds + \frac{1}{2\pi} \iint_A G\mathscr{L}u \, dA \tag{23}$$

Muskat (1935) showed how the solutions could be obtained for more geometrically complicated regions once a solution was constructed for a simple prototype region such as the infinite quadrant. It is sufficient to map these regions onto the infinite quadrant and to transfer the boundary conditions properly.

2.5.1. *Infinite Quadrant*

The problem is to solve $\nabla^2 u = 0$ in the upper right quadrant of the ξ, η plane, given $u(\xi, 0)$ and $(\partial u/\partial \xi)(0, \eta)$. The Green's function is constructed by means of the method of images as indicated in Fig. 5 and it may be verified that

$$G = \tfrac{1}{2} \log \left(\frac{r_{PQ} r_{P_2Q}}{r_{P_1Q} r_{P_3Q}} \right)^2 \tag{24}$$

satisfies the Dirichlet conditions along the ξ axis and the von Neumann conditions along the η axis.

Insertion of (24) into (23) leads to

$$\begin{aligned} u(x, y) = {} & \frac{1}{2\pi} \int_0^\infty f(\xi) \\ & \times \left\{ \left(-\frac{\partial}{\partial \eta} \right) \frac{1}{2} \log \frac{[(\xi - x)^2 + (\eta - y)^2][(\xi + x)^2 + (\eta - y)^2]}{[(\xi + x)^2 + (\eta + y)^2][(\xi - x)^2 + (\eta + y)^2]} \right\} d\xi \\ & \qquad \text{at} \quad \eta = 0 \\ & - \frac{1}{2\pi} \int_0^\infty \frac{1}{2} \left\{ \log \frac{[(\xi - x)^2 + (\eta - y)^2][(\xi + x)^2 + (\eta - y)^2]}{[(\xi + x)^2 + (\eta + y)^2][(\xi - x)^2 + (\eta + y)^2]} \right\} \left(-\frac{\partial u}{\partial \xi} \right) d\eta \\ & \qquad \text{at} \quad \xi = 0 \end{aligned} \tag{25}$$

or

$$\begin{aligned} u(x, y) = {} & \frac{y}{\pi} \int_0^\infty f(\xi) \left\{ \frac{1}{(\xi - x)^2 + y^2} + \frac{1}{(\xi + x)^2 + y^2} \right\} d\xi \\ & - \frac{1}{2\pi} \int_0^\infty F(\eta) \log \frac{x^2 + (\eta - y)^2}{x^2 + (\eta + y)^2} d\eta \end{aligned} \tag{26}$$

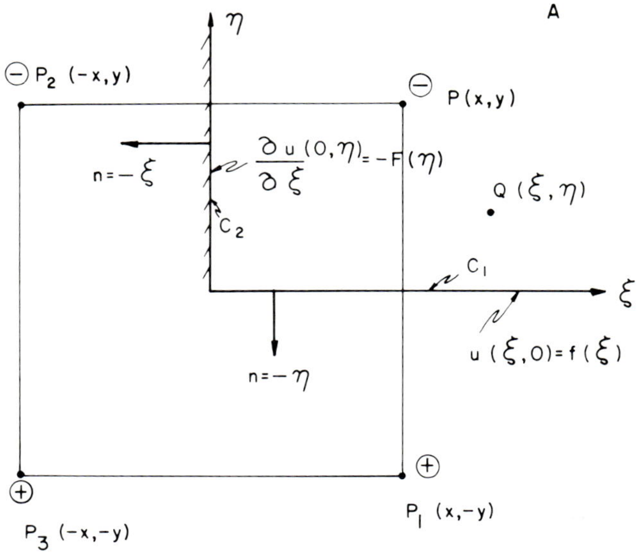

FIG. 5. Solution of $\nabla^2 u = 0$, mixed boundary value problem, for infinite quadrant.

2.5.2. *Infinite Quadrant with u or $\partial u/\partial n$ Given over Parts of Both Axes*

Through the function

$$z^2 = z'^2 - x_0'^2 \tag{27}$$

the z' plane is mapped onto the z plane so that the boundary along which $\partial u/\partial n$ is given is mapped onto the η axis and the boundary along which u is given is mapped onto the ξ axis. Hence, the problem is reduced to the prototype infinite quadrant, but it remains to express the boundary conditions in the z plane, given those in the z' plane. This is equivalent to finding the values of $f(\xi)$ and $F(\eta)$ in (26). After $u(x, y)$ is found by means of (26), it is possible to express $u(x', y')$ through the mapping function (27). From (27) it follows that

$$x^2 - y^2 = x'^2 - y'^2 - x_0'^2 \tag{28a}$$

$$2xy = 2x'y' \tag{28b}$$

and also

$$\xi^2 - \eta^2 = x'^2 - y'^2 - x_0'^2 \tag{29a}$$

$$2\xi\eta = 2x'y' \tag{29b}$$

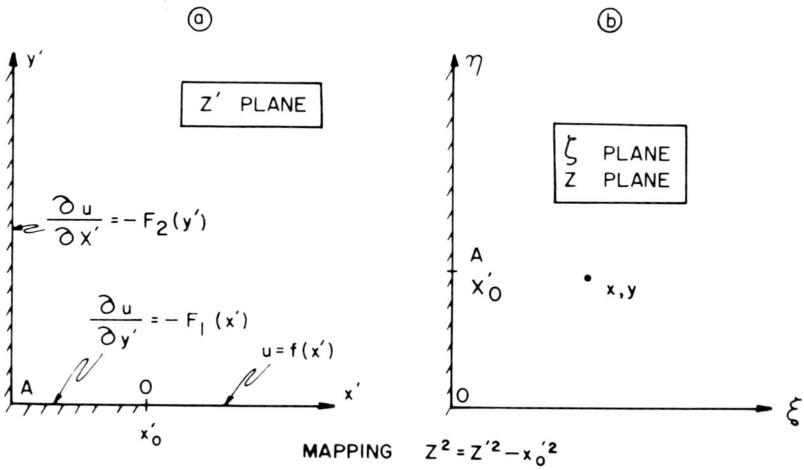

FIG. 6. Solution of $\nabla^2 u = 0$, mixed boundary value problem, for infinite quadrant with u or $\partial u/\partial n$ given over parts of both axes.

Let us first evaluate $F(\eta)$ of (26) or its equivalent $(-\partial u/\partial \xi)_{\xi=0}$ of (25). We therefore write

$$-\frac{\partial u}{\partial \xi}\bigg|_{\xi=0} = -\left[\frac{\partial u}{\partial y'}\bigg|_{\xi=0} \cdot \frac{\partial y'}{\partial \xi}\bigg|_{\xi=0} + \frac{\partial u}{\partial x'}\bigg|_{\xi=0} \bigg| \frac{\partial x'}{\partial \xi}\bigg|_{\xi=0}\right] \tag{30}$$

Now $(\partial u/\partial y')|_{\xi=0} = -F_1(x')|_{\xi=0}$. If $\xi = 0$, then from (29b), since $x' \neq 0$ here, we must have $y' = 0$, and from (29a) we find that $x' = (x_0'^2 - \eta^2)^{1/2}$. Therefore,

$$\frac{\partial u}{\partial y'}\bigg|_{\xi=0} = -F_1\{(x_0'^2 - \eta^2)^{1/2}\}$$

Also from (29b),

$$\frac{\partial y'}{\partial \xi} = \frac{\eta}{x'}$$

and at $\xi = 0$, we find

$$\frac{\partial y'}{\partial \xi}\bigg|_{\xi=0} = \frac{\eta}{(x_0'^2 - \eta^2)^{1/2}}$$

The first part of the right-hand side of (30), which is the expression of $F(\eta)$ for $0 < \eta < x_0'$, becomes

$$-\frac{\partial u}{\partial y'}\bigg|_{\xi=0} \cdot \frac{\partial y'}{\partial \xi}\bigg|_{\xi=0} = \frac{\eta}{(x_0'^2 - \eta^2)^{1/2}} \cdot F_1\{(x_0'^2 - \eta^2)^{1/2}\}$$

In a similar way we find the second part of $F(\eta)$. Now

$$\left.\frac{\partial u}{\partial x'}\right|_{\xi=0} = -F_2(y')\Big|_{\xi=0}$$

If $\xi = 0$, then from (29b), since $y' \neq 0$ here, we must have $x' = 0$, and from (29a) we find that $y' = (\eta^2 - x_0'^2)^{1/2}$. Therefore,

$$\left.\frac{\partial u}{\partial x'}\right|_{\xi=0} = -F_2\{(\eta^2 - x_0'^2)^{1/2}\}$$

Also from (29b),

$$\frac{\partial x'}{\partial \xi} = \frac{\eta}{y'}$$

and at $\xi = 0$, we find

$$\left.\frac{\partial x'}{\partial \xi}\right|_{\xi=0} = \frac{\eta}{(\eta^2 - x_0'^2)^{1/2}}$$

The expression for $F(\eta)$ in the interval $x_0' < \eta < \infty$ becomes

$$-\left.\frac{\partial u}{\partial x'}\right|_{\xi=0} \cdot \left.\frac{\partial x'}{\partial \xi}\right|_{\xi=0} = \frac{\eta}{(\eta^2 - x_0'^2)^{1/2}} \cdot F_2\{(\eta^2 - x_0'^2)^{1/2}\}$$

and we have the complete expression of $F(\eta)$ as follows:

$$F(\eta) = \underbrace{\frac{\eta}{(x_0'^2 - \eta^2)^{1/2}} F_1\{(x_0'^2 - \eta^2)^{1/2}\}}_{0 < \eta < x_0'} + \underbrace{\frac{\eta}{(\eta^2 - x_0'^2)^{1/2}} F_2\{(\eta^2 - x_0'^2)^{1/2}\}}_{x_0' < \eta < \infty}$$

Let us now evaluate $f(\xi)$ in (26). Since $y' = 0$ and $\eta = 0$, from (29a) it follows that $x' = (\xi^2 + x_0'^2)^{1/2}$ and that $f(x') = f\{(\xi^2 + x_0'^2)^{1/2}\}$. Therefore, the transferred Dirichlet condition is simply $u(\xi, 0) = f\{(\xi^2 + x_0'^2)^{1/2}\}$. Finally, the application of (26) gives the result:

$$\begin{aligned} u(x, y) = {} & \frac{y}{\pi} \int_0^\infty f\{(\xi^2 + x_0'^2)^{1/2}\} \left\{\frac{1}{(\xi - x)^2 + y^2} + \frac{1}{(\xi + x)^2 + y^2}\right\} d\xi \\ & - \frac{1}{2\pi} \int_0^{x_0'} \frac{\eta}{(x_0'^2 - \eta^2)^{1/2}} F_1\{(x_0'^2 - \eta^2)^{1/2}\} \log \frac{x^2 + (\eta - y)^2}{x^2 + (\eta + y)^2} d\eta \\ & - \frac{1}{2\pi} \int_{x_0'}^\infty \frac{\eta}{(\eta^2 - x_0'^2)^{1/2}} F_2\{(\eta^2 - x_0'^2)^{1/2}\} \log \frac{x^2 + (\eta - y)^2}{x^2 + (\eta + y)^2} d\eta \end{aligned} \tag{31}$$

Once (31) is computed, $u(x', y')$ can be derived by means of (27). It may be noticed that (31) remains valid for the special case of $x_0' = 0$, which leads at once to (26) as it should.

For the case of mixed boundary value problems of the infinite strip and semi-infinite strip and circle, the reader is referred to the original paper by Muskat (1935).

2.6. Green's Function and Conformal Mapping

The upper half-plane of Fig. 3 can be mapped in a one-to-one conformal way onto the interior of the unit circle of the w plane of Fig. 7 so that the point z and a given direction through z correspond

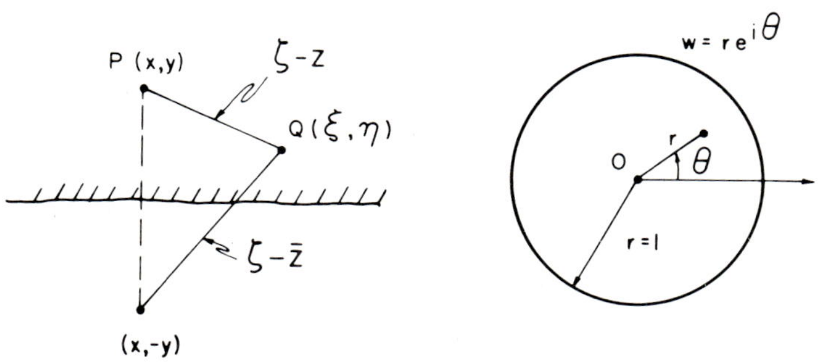

FIG. 7. Mapping of upper half-plane onto unit circle: $w = [(\zeta - z)/(\zeta - \bar{z})]\, e^{i\varphi}$.

to the center of the circle and the positive real axis by the one and only function

$$w = (\zeta - z)/(\zeta - \bar{z}) \tag{32}$$

(i.e., we assume $e^{i\varphi} = 1$ in the mapping of Fig. 7). We find that $\log |w|$ is nothing but the Green's function (13) for the upper half-plane.

Actually, this way of constructing the Green's function is not limited to the half-plane. It can easily be proved (Kellogg, 1929) that if $w = f(\zeta)$ maps the simply connected domain T of the ζ plane on the interior of the unit circle in the w plane in a one-to-one conformal manner, then $\log |f(\zeta)|$ is Green's function for T, the pole being the point of the ζ plane corresponding to $w = 0$. Conversely, if Green's function for T is known, the mapping function can be determined in a straightforward way.

The problem of determining Green's function for T and the problem of mapping T by an analytic function in a one-to-one manner on the interior of the unit circle are equivalent. This fact is illustrated by Riemann's fundamental theorem on mapping:

The interior T of any simply connected region whose boundary contains more than one point, can be mapped in a one-to-one conformal manner on the interior of the unit circle. The theorem is equivalent to asserting the existence of Green's function for T and therefore, the existence of the solution u of a certain Dirichlet problem.

3. Green's Functions for Not Self-Adjoint Differential Equations

The equations considered thus far happened to be formally self-adjoint. However, the equation of unsteady flow in a leaky aquifer is not formally self-adjoint as will become evident from the following analysis. The construction of the Green's function for the not formally self-adjoint equation is given in the case of the Dirichlet problem and the von Neumann problem and it is shown that Eqs. (12) and (22) remain valid.

3.1. Formal Adjoint

The problem is to find the formal adjoint of a given linear partial differential equation. Let the operator

$$\mathscr{L} = A \frac{\partial^2}{\partial x^2} + 2B \frac{\partial^2}{\partial x \, \partial y} + C \frac{\partial^2}{\partial y^2} + D \frac{\partial}{\partial x} + E \frac{\partial}{\partial y} + F \tag{33}$$

operate on $u(x, y)$, $A, B, \ldots, F$ being given functions of x, y and having sufficiently many derivatives.

By definition, the formal adjoint operator $\mathscr{M}$ operating on $v(x, y)$ is determined by the equation

$$v\mathscr{L}(u) - u\mathscr{M}(v) = \frac{\partial X}{\partial x} + \frac{\partial Y}{\partial y} \tag{34}$$

in which X and Y may be thought of as being the components of the abstract vector field of Eq. (1). In other words, to find the formal adjoint $\mathscr{M}$, we write the product $v\mathscr{L}(u)$ and then transform the expression into the sum of $u\mathscr{M}(v)$ and a number of divergence expressions. For example, the term $vA(\partial^2 u/\partial x^2)$ is transformed as follows:

$$\begin{aligned} vA\frac{\partial^2 u}{\partial x^2} &= \frac{\partial}{\partial x}\left(vA\frac{\partial u}{\partial x}\right) - \frac{\partial u}{\partial x}\frac{\partial}{\partial x}(vA) \\ &= \frac{\partial}{\partial x}\left(vA\frac{\partial u}{\partial x}\right) - \left[\frac{\partial}{\partial x}\left(u\frac{\partial vA}{\partial x}\right) - u\frac{\partial^2 vA}{\partial x^2}\right] \\ &= u\frac{\partial^2 vA}{\partial x^2} + \frac{\partial}{\partial x}\left(vA\frac{\partial u}{\partial x}\right) - \frac{\partial}{\partial x}\left(u\frac{\partial vA}{\partial x}\right) \end{aligned}$$

If all terms are treated in the same way, the result is

$$\begin{aligned} \mathscr{M}(v) = {} & A\frac{\partial^2 v}{\partial x^2} + 2B\frac{\partial^2 v}{\partial x\,\partial y} + C\frac{\partial^2 v}{\partial y^2} \\ & + \frac{\partial v}{\partial x}\left(2\frac{\partial A}{\partial x} - D + 2\frac{\partial B}{\partial y}\right) + \frac{\partial v}{\partial y}\left(2\frac{\partial C}{\partial y} - E + 2\frac{\partial B}{\partial x}\right) + Fv \\ & + v\left(\frac{\partial^2 A}{\partial x^2} + 2\frac{\partial^2 B}{\partial y\,\partial x} + \frac{\partial^2 C}{\partial y^2} - \frac{\partial D}{\partial x} - \frac{\partial E}{\partial y}\right) \end{aligned} \tag{35}$$

Also,

$$\begin{aligned} X &= A\left(v\frac{\partial u}{\partial x} - u\frac{\partial v}{\partial x}\right) + B\left(v\frac{\partial u}{\partial y} - u\frac{\partial v}{\partial y}\right) + \left(D - \frac{\partial A}{\partial x} - \frac{\partial B}{\partial y}\right)uv \\ Y &= B\left(v\frac{\partial u}{\partial x} - u\frac{\partial v}{\partial x}\right) + C\left(v\frac{\partial u}{\partial y} - u\frac{\partial v}{\partial y}\right) + \left(E - \frac{\partial B}{\partial x} - \frac{\partial C}{\partial y}\right)uv \end{aligned} \tag{36}$$

The operator $\mathscr{M}$ defined by (35) is the formal adjoint of $\mathscr{L}$. The conditions for formal self-adjointness, i.e., which make $\mathscr{M} = \mathscr{L}$, are

$$D = \frac{\partial A}{\partial x} + \frac{\partial B}{\partial y}, \qquad E = \frac{\partial B}{\partial x} + \frac{\partial C}{\partial y} \tag{37}$$

Moreover, if the boundary conditions imposed on $\mathscr{L}(u) = 0$ and $\mathscr{M}(v) = 0$ are the same, $\mathscr{L}$ and $\mathscr{M}$ are self-adjoint under conditions (37).

3.2. Elliptic Partial Differential Equations

3.2.1. *Green's Second Identity*

Consider first the elliptic partial differential equation with constant coefficients, written in its canonical form, i.e., $A = C = 1$ and $B = 0$ in (33).

Then

$$\mathscr{L}(u) = \frac{\partial^2 u}{\partial x^2} + \frac{\partial^2 u}{\partial y^2} + D\frac{\partial u}{\partial x} + E\frac{\partial u}{\partial y} + Fu = 0 \tag{38}$$

Apply now the divergence theorem to (34) in the domain A of Fig. 1. In view of the conventions for the outer normal (3) and for Green's theorem (4), we find

$$\iint_A [v\mathscr{L}(u) - u\mathscr{M}(v)]\, dA = \oint_C X\, dy - Y\, dx \tag{39}$$

With

$$X = \left(v\frac{\partial u}{\partial x} - u\frac{\partial v}{\partial x}\right) + Duv, \qquad Y = \left(v\frac{\partial u}{\partial y} - u\frac{\partial v}{\partial y}\right) + Euv$$

from (36), transformation of (39) leads to Green's second identity for not self-adjoint equations:

$$\iint_A [u\mathscr{M}(v) - v\mathscr{L}(u)]\, dA = \oint_C \left(u\frac{\partial v}{\partial n} - v\frac{\partial u}{\partial n}\right) ds$$
$$- \oint_C [D\cos(n, x) + E\cos(n, y)]\, uv\, ds \tag{40}$$

3.2.2. *Green's Functions*

The construction of the Green's function proceeds in a manner similar to that for the self-adjoint equation. Now a solution v of the formally adjoint equation $\mathscr{M}(v) = 0$ is considered. As before, v is composed of two parts, one having a logarithmic singularity inside the domain at the source point $P(x, y)$, the other being smooth in the entire domain A. Insertion of the first part, $v = \log r$, into (40) gives

$$2\pi u(x, y) = \iint_A \log r\mathscr{L}(u)\, dA + \oint_C \left(u\frac{\partial \log r}{\partial n} - \log r\frac{\partial u}{\partial n}\right) ds$$
$$- \oint_C [D\cos(n, x) + E\cos(n, y)]\, u\log r\, ds \tag{41}$$

Insertion of the regular part $v = w$ into (40) leads to

$$-\iint_A w\mathscr{L}(u)\, dA = \oint_C \left(u\frac{\partial w}{\partial n} - w\frac{\partial u}{\partial n}\right) ds$$
$$- \oint_C [D\cos(n, x) + E\cos(n, y)]\, uw\, ds \tag{42}$$

Addition of (41) and (42) gives

$$u(x, y) = \frac{1}{2\pi} \iint_A (\log r + w)\, \mathscr{L}(u)\, dA$$

$$+ \frac{1}{2\pi} \oint_C \left[u \frac{\partial}{\partial n} (\log r + w) - (\log r + w) \frac{\partial u}{\partial n}\right] ds$$

$$- \frac{1}{2\pi} \oint_C u(\log r + w)[D \cos(n, x) + E \cos(n, y)]\, ds \qquad (43)$$

Let $G(x, y; \xi, \eta) = \log r + w$, then for the *Dirichlet problem* we have the following properties:

(a) $\mathscr{M}(G) = 0$;
(b) G vanishes on the boundary so that (43) reduces to (12);
(c) $G(x, y; \xi, \eta) = H(\xi, \eta; x, y)$.

Here H is Green's function for the adjoint equation to $\mathscr{M}(v) = 0$, and hence, it satisfies the equation $\mathscr{L}(H) = 0$. Proof of this property is given in Appendix E.

For the *von Neumann problem* we have:

(a) $\mathscr{M}(G) = 0$;
(b) on the boundary

$$\frac{\partial G}{\partial n} - G[D \cos(n, x) + E \cos(n, y)] = 0$$

so that (43) reduces to (22);

(c) $G(x, y; \xi, \eta) = H(\xi, \eta; x, y)$.

3.3. Parabolic Partial Differential Equations

The most common equation of this type occurring in ground-water flow is that for the leaky aquifer,

$$\mathscr{L}(s) = \frac{\partial^2 s}{\partial x^2} + \frac{\partial^2 s}{\partial y^2} - \frac{1}{\nu} \frac{\partial s}{\partial t} - \frac{s}{B^2} = 0 \qquad (44)$$

in which s is the drawdown, x and y are the coordinates of the horizontal plane, $\nu = T/S$ is the ratio of the transmissivity and storage coefficient and is sometimes called the hydraulic diffusivity, B is the leakage factor, and t is the time.

The construction of the Green's function for (44) and the solution of (44) for a given problem is very similar to the treatment of heat conduction problems (Sommerfeld, 1964). The construction of the Green's function requires the construction of a principal solution of the equation $\mathscr{M}(v) = 0$, formally adjoint to (44). Subsequent application of Green's second identity (40) generalized to include t as an independent variable and use of the method of images for bounded regions will lead to a solution of (44) for given boundary and initial conditions.

3.3.1. *Construction of Principal Solution*

To construct a principal solution of $\mathscr{M}(v) = 0$, let us first construct a principal solution of (44) for the case of an aquifer of infinite extent. Let the drawdown $s(x, y)$ be given at $t = 0$

$$s(x, y) = f(x, y); \qquad t = 0, \qquad -\infty < x, y < +\infty \tag{45}$$

and consider its representation by a Fourier integral (see Appendix A)

$$f(x, y) = [(2\pi)^2]^{-1} \int_{-\infty}^{+\infty} d\alpha \int_{-\infty}^{+\infty} d\alpha' \int_{-\infty}^{+\infty} \int_{-\infty}^{+\infty} f(\xi, \eta) \times \exp[i\alpha(x - \xi) + i\alpha'(y - \eta)]\, d\xi\, d\eta \tag{46}$$

Let

$$t' = \nu t \tag{47}$$

and require that

$$s(x, y, t') = \phi(t') \exp[i\alpha(x - \xi) + i\alpha'(y - \eta)] \tag{48}$$

with

$$\phi(0) = 1 \tag{49}$$

be a solution of (44). Insertion of (48) into (44) leads to

$$[-(\alpha^2 + \alpha'^2) - (B^2)^{-1}]\, \phi = d\phi/dt' \tag{50}$$

or

$$\phi = \exp\{-[\alpha^2 + \alpha'^2 + (B^2)^{-1}]\, t'\} \tag{51}$$

after integration of (50) and considering (49). The general solution of (44) for the problem defined by (45) can therefore be expressed by replacing in (46)

$$\exp[i\alpha(x - \xi)] \qquad \text{by} \qquad \exp[i\alpha(x - \xi) - \alpha^2 t']$$

and

$$\exp[i\alpha'(y - \eta)] \qquad \text{by} \qquad \exp[i\alpha'(y - \eta) - \alpha'^2 t']$$

and by introducing the multiplier $\exp(-t'/B^2)$. Therefore,

$$s(x, y, t) = [(2\pi)^2]^{-1} \int_{-\infty}^{+\infty} \int_{-\infty}^{+\infty} e^{(-\nu/B^2)t} f(\xi, \eta)\, d\xi\, d\eta$$
$$\times \int_{-\infty}^{+\infty} \exp[i\alpha(x - \xi) - \alpha^2 \nu t]\, d\alpha \int_{-\infty}^{\infty} \exp[i\alpha'(y - \eta) - \alpha'^2 \nu t]\, d\alpha' \tag{52}$$

or (see Appendix B)

$$s(x, y, t) = \int_{-\infty}^{+\infty} \int_{-\infty}^{+\infty} f(\xi, \eta)\, d\xi\, d\eta\, e^{(-\nu/B^2)t} \frac{\exp\{-[(x - \xi)^2 + (y - \eta)^2]/(4\nu t)\}}{4\pi\nu t} \tag{53}$$

Let

$$U = e^{(-\nu/B^2)t} \frac{\exp\{-[(x - \xi)^2 + (y - \eta)^2]/(4\nu t)\}}{4\pi\nu t} \tag{54}$$

then (53) becomes

$$s(x, y, t) = \int_{-\infty}^{+\infty} \int_{-\infty}^{+\infty} f(\xi, \eta)\, U\, d\xi\, d\eta \tag{55}$$

For $t \to 0$, we have $s(x, y, t) \to f(x, y)$ and (55) becomes

$$f(x, y) = \int_{-\infty}^{+\infty} \int_{-\infty}^{+\infty} f(\xi, \eta)\, U\, d\xi\, d\eta \tag{56}$$

This shows that at $t = 0$, U behaves as a δ function. U is the principal solution of $\mathscr{L}(s) = 0$ (44).

Physically, U may be compared with the function describing the space–time behavior of a unit heat source or of a heat pole of unit strength, multiplied by the factor $e^{(-\nu/B^2)t}$. This multiplier arises from the leakage into the aquifer. It is independent of the coordinates, showing that leakage takes place under uniform conditions of the geological formations. Because of this multiplier, $s \to 0$ for $t \to \infty$ and also $s \to 0$ for $B \to 0$. The first asymptotic condition means that leakage eventually offsets any drawdown, given sufficient time, while the second asymptotic condition means that it is impossible to maintain a drawdown when the semiconfining structure disappears, which is obvious.

The behavior of U as a δ function at $t = 0$ means that U vanishes in the limit $t \to 0$ for all values of $(x, y) \neq (\xi, \eta)$ and becomes infinite when the action point $Q(\xi, \eta)$ covers the source point $P(x, y)$ and, because of (56), in such a way that

$$\int_{-\infty}^{+\infty} \int_{-\infty}^{+\infty} U\, d\xi\, d\eta = 1 \tag{57}$$

Hantush and Jacob (1955b) showed that U may be obtained from the solution of the flow to a well in a leaky aquifer. For a well pumped at a constant flow rate Q, Hantush and Jacob (1955a) showed that the drawdown s is given by

$$s = (Q/4\pi T)\int_u^\infty \exp[-m - (r^2/4B^2m)](dm/m) \tag{58}$$

in which m is an integration variable, $u = r^2S/4Tt$, and the other symbols are as defined before. The integral of (58) is known as $W[u, (r/B)]$, the well function for leaky aquifers. It is easy to prove that

$$\frac{S}{4\pi T}\frac{\partial}{\partial t}W(t) = U \tag{59}$$

To go from the principal solution U to the Green's function, and to Green's second identity, it is convenient to replace t in (54) by the integration variable τ and then to introduce the difference $\tau - t$ as follows

$$V = \exp\left[-\frac{\nu}{B^2}(\tau - t)\right]\frac{\exp\{-[(x-\xi)^2 + (y-\eta)^2]/4\nu(\tau - t)\}}{4\pi\nu(\tau - t)} \tag{60}$$

In subsequent derivations, t will be associated with the source point (x, y) and τ with the action point (ξ, η). Sometimes $\tau = t$ is called the general initial time, but this is a misleading definition as $\tau = 0$ is still the initial time.

The formally adjoint equation of (44) can be expressed as

$$\mathscr{M}(v) = \frac{\partial^2 v}{\partial x^2} + \frac{\partial^2 v}{\partial y^2} + \frac{1}{\nu}\frac{\partial v}{\partial t} - \frac{v}{B^2} = 0 \tag{61}$$

The adjoint operator $\mathscr{M}$ is determined as explained in Section 3.1 by the property that

$$v\mathscr{L}(s) - s\mathscr{M}(v) = \frac{\partial X}{\partial x} + \frac{\partial Y}{\partial y} + \frac{\partial Z}{\partial t} \tag{62}$$

in which

$$X = v\frac{\partial s}{\partial x} - s\frac{\partial v}{\partial x}, \qquad Y = v\frac{\partial s}{\partial y} - s\frac{\partial v}{\partial y}, \qquad Z = -\frac{S}{T}sv \tag{63}$$

3.3.2. *Construction of Green's Function*

First we wish to construct the principal solution V of (61). A comparison of (61) with (44) shows that this can be done by reversing the sign affecting the time components of (60), i.e.,

$$V = \exp\left[-\frac{\nu}{B^2}(t - \tau)\right]\frac{\exp\{-[(x-\xi)^2 + (y-\eta)^2]/4\nu(t - \tau)\}}{4\pi\nu(t - \tau)} \tag{64}$$

The singularity hereby remains situated at the point $x, y = \xi, \eta$ and $t = \tau$. The properties of V are:

(1) V behaves as a δ function for $\tau = t$.

(2) V is defined only for $\tau < t$, which is the past of t, in contrast with the principal solution U which is regular only for the future of t, i.e., for $\tau > t$.

V is the Green's function for the infinite plane. If the domain A is finite, as in Fig. 8, the construction of the Green's function $G(x, y, t;$

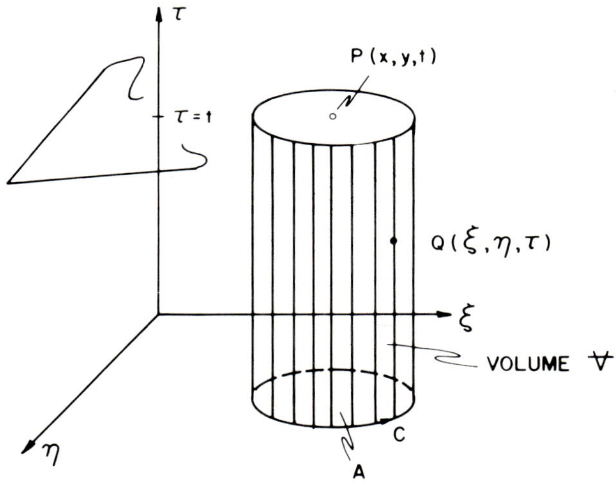

FIG. 8. Solution of Eq. (44) for leaky aquifer.

$\xi, \eta, \tau)$ is achieved by a reflection process (method of images) as explained earlier. In the simplest case the construction is such that G or its normal derivative vanishes along the boundary, depending on the knowledge of s and $\partial s/\partial n$ along the boundary.

3.3.3. *Solution for Finite Domain*

The problem now is to find a solution of (44) for the domain A of Fig. 8, given the initial drawdown distribution $s(x, y, 0)$ over the domain A, the value of s or $\partial s/\partial n$ along the contour C, and the Green's function $G(x, y, t; \xi, \eta, \tau)$ for A.

To find this solution, consider the space–time prism $\forall$ of Fig. 8 and integrate (62) over this volume

$$\iiint [v\mathscr{L}(s) - s\mathscr{M}(v)]\, d\forall = \iiint \left(\frac{\partial X}{\partial \xi} + \frac{\partial Y}{\partial \eta} + \frac{\partial Z}{\partial \tau}\right) d\forall \qquad (65)$$

or apply the divergence theorem to the right-hand side of (65)

$$\iiint [v\mathscr{L}(s) - s\mathscr{M}(v)]\, d\forall = \iint [X\cos(n, \xi) + Y\cos(n, \eta) + Z\cos(n, \tau)]\, d\sigma \tag{66}$$

in which σ represents the surface area of $\forall$. Since we are looking for a solution of $\mathscr{L}(s) = 0$ and since $v = G$ satisfies $\mathscr{M}(G) = 0$, the left hand side of (66) vanishes, so that

$$\iint [X\cos(n, \xi) + Y\cos(n, \eta) + Z\cos(n, \tau)]\, d\sigma = 0 \tag{67}$$

In (67), we replace X, Y, Z by their expression (63) in which we make $v = G$. The result is

$$\iint \left\{\left[G\frac{\partial s}{\partial \xi} - s\frac{\partial G}{\partial \xi}\right]\cos(n, \xi) + \left[G\frac{\partial s}{\partial \eta} - s\frac{\partial G}{\partial \eta}\right]\cos(n, \eta) - \frac{S}{T}Gs\cos(n, \tau)\right\} d\sigma = 0 \tag{68}$$

To compute (68), the surface integral applied to the prism of Fig. 8 is broken down as follows:

(1) *Integral over base of prism*, i.e., for $\tau = 0$. Here $\cos(n, \xi) = \cos(n, \eta) = 0$, $\cos(n, \tau) = -1$, and $d\sigma = d\xi\, d\eta = dA$. Therefore, the contribution of (68) becomes

$$(S/T)\iint_A G(x, y, t; \xi, \eta, 0)\, s(\xi, \eta, 0)\, d\xi\, d\eta$$

(2) *Integral over top of prism* where $\tau = t$. Here $\cos(n, \xi) = \cos(n, \eta) = 0$, $\cos(n, \tau) = 1$, and $d\sigma = d\xi\, d\eta = dA$. Therefore, the contribution of (68) becomes

$$-(S/T)\iint_A G(x, y, t; \xi, \eta, t)\, s(\xi, \eta, t)\, d\xi\, d\eta$$

But the Green's function behaves as a δ function for $\tau = t$ and therefore, the last integral reduces to

$$-(S/T)\, s(x, y, t)$$

(3) *Integral over lateral surface of prism* of Fig. 8 for $0 < \tau < t$. Over this surface $\cos(n, \tau) = 0$ and $d\sigma = dC\, d\tau$, in which dC is the

elemental length of the contour C so that the contribution of (68) becomes

$$\int_0^t \oint_C \left(G \frac{\partial s}{\partial n} - s \frac{\partial G}{\partial n}\right)_b dC\, d\tau$$

in which the subscript b indicates that the values inside the parentheses are evaluated at the lateral surface.

The final result is obtained as

$$s(x, y, t) = \iint_A G(x, y, t; \xi, \eta, 0)\, s(\xi, \eta, 0)\, d\xi\, d\eta + \frac{T}{S} \int_0^t \oint_C \left(G \frac{\partial s}{\partial n} - s \frac{\partial G}{\partial n}\right)_b dC\, d\tau \tag{69}$$

4. Green's Functions and Dirac's Delta Function

A more elegant as well as simpler approach to the solution of differential equations by means of Green's functions than has been discussed thus far is based upon the spectral representation of the differential operators. In the case of a partial differential operator, this approach becomes particularly useful when the operator is the sum of two or more commutative operators. A simple example of such operators is the Laplacian

$$\mathscr{L}u = u_{xx} + u_{yy} = \mathscr{L}_1 u + \mathscr{L}_2 u \tag{70}$$

in which $\mathscr{L}_1$ and $\mathscr{L}_2$ are commutative because $\mathscr{L}_1\mathscr{L}_2 u = \mathscr{L}_2\mathscr{L}_1 u$. In this case, either $\mathscr{L}_1$ or $\mathscr{L}_2$ may be considered as a constant in finding the inverse of $\mathscr{L} = \mathscr{L}_1 + \mathscr{L}_2$. The problem of finding the inverse operator $\mathscr{L}^{-1}$ is therefore reduced to that of finding the inverse of an ordinary differential operator.

The inverse of a differential operator is an integral operator whose kernel is the Green's function of the differential operator. Let $\mathscr{L}$ be a linear ordinary differential operator acting on a space of functions $u(x)$. Let its inverse $\mathscr{L}^{-1}$ be an integral operator with kernel $G(x, \xi)$ so that

$$\mathscr{L}^{-1}u = \int G(x, \xi)\, u(\xi)\, d\xi \tag{71}$$

Then

$$u(x) = \int \mathscr{L}G(x, \xi)\, u(\xi)\, d\xi \tag{72}$$

because $\mathscr{L}\mathscr{L}^{-1} = I$, the identity operator, provided that the limiting operations are interchangeable (Friedman, 1956). A comparison of (72) and (56) leads us to write

$$\mathscr{L}G(x, \xi) = \delta(\xi - x) \tag{73}$$

which is compatible with (9) and in which $\delta(\xi, x)$ is Dirac's delta function. Equation (72) becomes

$$u(x) = \int \delta(\xi - x)\, u(\xi)\, d\xi \tag{74}$$

This result is in agreement with Dirac's definition: the δ function is zero for every value of ξ except $\xi = x$ where it is infinite in such a way that

$$\int_{-\infty}^{\infty} \delta(\xi - x)\, d\xi = 1 \tag{75}$$

In the sequel, the operator $\mathscr{L}$, in the case of ordinary differential equations as well as partial differential equations, will include only first-order and second-order operators.

The methods used in Section 4 are covered extensively by Friedman (1956).

4.1. Green's Functions for Ordinary Differential Equations

The Green's function for a second-order differential equation in one dimension differs from that in more than one dimension by the fact that the discontinuity has passed from the function to its derivative. The function itself is continuous at the source. For example, the potential equation in one dimension becomes $d^2u/dx^2 = 0$, which renders $du/dx =$ constant. The constant assumes different values on the right and left side of the source, respectively, and in order to have a unit source, the difference of these values is 1.

If the operator is simple, Green's function may sometimes be found by straightforward integration of (73). In more general cases, methods are available to construct Green's function (Friedman, 1956). In any case, $G(x, \xi)$ is a continuous, piecewise differentiable function and if $f(x)$ is any integrable function which vanishes outside a finite interval, then

$$u(x) = \int G(x, \xi) f(\xi)\, d\xi \tag{76}$$

satisfies the equation

$$\mathscr{L}u = f(x) \tag{77}$$

Indeed, if $\mathscr{L}$ operates on both sides of (76), the result is

$$\mathscr{L}u = \mathscr{L} \int G(x, \xi) f(\xi)\, d\xi = \int \mathscr{L}Gf\, d\xi = \int \delta(\xi - x) f(\xi)\, d\xi = f(x)$$

in view of (73) and (74).

Also, $G(x, \xi)$ as a function of x satisfies the same conditions as those satisfied by the solutions of the differential Eq. (77). These conditions on u determine the domain $\mathscr{D}$ of the operator $\mathscr{L}$, which in ground-water flow is only operating on such functions u that $\mathscr{L}u$ exists and belongs to the space $\mathscr{S}$ of all real-valued and integrable functions u.

Let us first consider the operator

$$\mathscr{L}_0 = \frac{-d}{dx}\left(p \frac{d}{dx}\right)$$

with some homogeneous boundary conditions in its finite domain $\mathscr{D}$. Here $p(x)$ is a continuous function different from zero in the domain $\mathscr{D}$. Let $G_0(x, \xi)$ be the solution of (73) for $\mathscr{L} = \mathscr{L}_0$,

$$-\frac{d}{dx}\left(p \frac{dG_0}{dx}\right) = \delta(\xi - x) \tag{78}$$

Equation (78) can be integrated immediately with $H(x - \xi)$ as the Heaviside function,

$$p \frac{dG_0}{dx} = -H(x - \xi) + \alpha(\xi) \tag{79}$$

and with $\alpha(\xi)$ as the integration constant. A second integration renders

$$G_0(x, \xi) = -H(x - \xi) \int_\xi^x \frac{dk}{p(k)} + \alpha(\xi) \int_0^x \frac{dk}{p(k)} + \beta(\xi) \tag{80}$$

where $\beta(\xi)$ is another integration constant. This result shows that $G_0(x, \xi)$ is a continuous function of x and that its derivative has a jump of magnitude $-1/p(\xi)$ at $x = \xi$.

For a formally self-adjoint operator $\mathscr{L}$ of the second order and for homogeneous boundary conditions, Green's function $G(x, \xi)$ has the same properties as $G_0(x, \xi)$. The definition of the operator $\mathscr{L}$ may be extended to include nonhomogeneous boundary conditions as shown in the following example of flow into a trench.

4.1.1. *Flow into a Trench*

Figure 9 represents idealized field conditions for flow into a trench parallel to a river and cutting through an unconfined aquifer of average

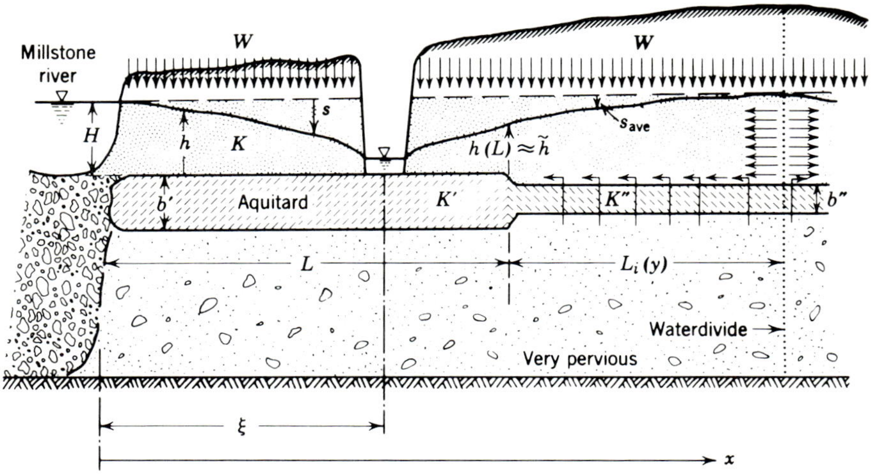

FIG. 9. Idealized field conditions for flow into trench. (After De Wiest 1963b).

transmissivity $T = K\tilde{h}$ (De Wiest, 1963b). The equation for the drawdown s is

$$(d^2s/dx^2) - (s/B^2) = W/T \tag{81}$$

in which W is the areal recharge and B is the leakage factor. Equation (81) is made homogeneous by a change of variables, i.e., $s = s_1 - B^2W/T$,

$$(d^2s_1/dx^2) - (s_1/B^2) = 0 \tag{82}$$

with corresponding boundary conditions

$$s_1(0) = \alpha \tag{83}$$

$$s_1(L) = \beta s_1'(L) + \epsilon \tag{84}$$

in which α, β, and ϵ are physical constants and L is the length of aquitard (mathematically, the interval in which the function is defined).

In order to determine the drawdown for these boundary conditions and for a constant withdrawal q (dimensions L^2/T) per unit length of trench located at a distance ξ from the river, Eq. (85) must first be solved:

$$\frac{d^2G(x, \xi)}{dx^2} - \frac{G(x, \xi)}{B^2} = -\frac{q}{T}\,\delta(x - \xi) \tag{85}$$

for the homogeneous boundary conditions

$$G(0, \xi) = 0 \tag{86}$$

$$G(L, \xi) = \beta G_x(L, \xi) \tag{87}$$

in which $G(x, \xi)$ is Green's function for the operator

$$\mathscr{L} = (d^2/dx^2) - (B^2)^{-1} \tag{88}$$

with the conditions expressed by (86) and (87). This renders the drawdown for a system in which the actual trench is replaced by a mathematical line-sink of strength q (De Wiest, 1963a). To this drawdown, the drawdown due to the inhomogeneity of the boundary conditions must be added.

The Green's function is constructed as follows (Friedman, 1956). Let

$$w_1(x) = \sinh(x/B) + \lambda_1 \cosh(x/B) \tag{89}$$

be a linear combination of two independent solutions of (85) made homogeneous ($q = 0$) and satisfying (86). This requires $\lambda_1 = 0$. Let

$$w_2(x) = \lambda_2 \sinh(x/B) + \cosh(x/B) \tag{90}$$

be a similar combination that satisfies (87). This requires

$$\lambda_2 = \frac{(\beta/B)\sinh(L/B) - \cosh(L/B)}{\sinh(L/B) - (\beta/B)\cosh(L/B)} \tag{91}$$

Consider now the products

$$\begin{aligned} P(x, \xi) &= w_1(x)\, w_2(\xi) \qquad \text{for} \quad x < \xi \\ &= w_2(x)\, w_1(\xi) \qquad \text{for} \quad x > \xi \end{aligned} \tag{92}$$

and compute the jump J in the derivative

$$J = \partial P(x, \xi)/\partial x \qquad \text{at} \quad x = \xi$$

Then

$$G(x, \xi) = -(q/TJ)[w_1(x)\, w_2(\xi)\, H(\xi - x) + w_2(x)\, w_1(\xi)\, H(x - \xi)] \tag{93}$$

is the desired Green's function. The result of the computations here renders

$$G(x, \xi) = (qB/T)\sinh(x/B)\,\varphi(\xi)\, H(\xi - x) + (qB/T)\sinh(\xi/B)\,\varphi(x)\, H(x - \xi) \tag{94}$$

in which

$$\varphi(t) = \frac{\sinh[(L - t)/B] - (\beta/B)\cosh[(L - t)/B]}{\sinh(L/B) - (\beta/B)\cosh(L/B)} \tag{95}$$

To determine the drawdown due to the boundary conditions of (83) and (84), as if there were no trench, (82) is solved straightforwardly to find

$$s_1 = -\alpha \frac{(\beta/B)\cosh[(L - x)/B] - \sinh[(L - x)/B]}{\sinh(L/B) - (\beta/B)\cosh(L/B)} + \epsilon \frac{\sinh(x/B)}{\sinh(L/B) - (\beta/B)\cosh(L/B)} \tag{96}$$

Equation (96) is obtained by merely writing the solution

$$s_1 = C_1 \sinh(x/B) + C_2 \cosh(x/B)$$

and by determining the constants C_1 and C_2 by means of the boundary conditions (83) and (84). If these conditions were more complicated, then it might be advantageous to extend the definition of the operator $\mathscr{L}$ by the use of some properties of the delta function, as outlined in Appendix D.

4.2. Green's Function and Spectral Representation of the δ Function†

The spectral representation of the δ function in terms of the eigenfunctions of a given differential operator for specific boundary conditions provides a powerful tool in the solution of linear differential equations. The method is based upon the spectral theory of operators. Since we shall only be interested in partial differential operators, which are sums of commutative ordinary differential operators, the expansion of the δ function in one dimension suffices for the present purpose.

A function $u(x)$ is an eigenfunction of the operator $\mathscr{L}$ defined over the domain $\mathscr{D}$ if it satisfies the equation

$$\mathscr{L}u = \lambda u \tag{97}$$

In particular, the eigenfunction $u_n(x)$ corresponds to the eigenvalue λ_n so that

$$\mathscr{L}u_n = \lambda_n u_n \tag{98}$$

† Friedman (1956).

For a simple operator $\mathscr{L}$, the eigenfunctions are complete and every square integrable function $u(x)$ may be expanded uniquely in terms of these eigenfunctions as follows:

$$u(x) = \sum \alpha_n u_n(x) \tag{99}$$

where $\alpha_n = \langle v_n, u \rangle$, $v_n(x)$ being the eigenfunctions of the adjoint $\mathscr{M}$. It is well known that eigenfunctions of $\mathscr{L}$, corresponding to λ, are orthogonal to eigenfunctions of $\mathscr{M}$ corresponding to $\lambda' \neq \lambda$, and that if λ is an eigenvalue of $\mathscr{L}$, then it is also an eigenvalue of $\mathscr{M}$. If $\mathscr{L}$ is self-adjoint, the eigenfunctions form an orthogonal set, that is,

$$\langle u_l, u_m \rangle = \delta_{lm}$$

A real self-adjoint differential operator has only real eigenvalues if it acts on a domain which contains $\overline{u(x)}$ whenever it contains $u(x)$. In particular, if $\mathscr{L}$ is positive-definite, i.e., $\langle u, \mathscr{L}u \rangle > 0$ for all nonzero u in $\mathscr{D}$, the eigenvalues of $\mathscr{L}$ are positive.

If we operate with $\mathscr{L}$ on u of (99) and use the property of (98), we find

$$\mathscr{L}u = \sum \alpha_n \mathscr{L} u_n = \sum \alpha_n \lambda_n u_n \tag{100}$$

It is easy to see that because of the linearity of the operator, $\mathscr{L}$ may be replaced by a polynomial in $\mathscr{L}$. This property can be generalized to include any continuous function $f(\mathscr{L})$, so that (100) can be extended as

$$f(\mathscr{L})\, u(x) = \sum f(\lambda_n)\, \alpha_n u_n(x) \tag{101}$$

In particular, consider $f(\mathscr{L}) = 1/(\mathscr{L} - \lambda)$; then

$$\frac{1}{\mathscr{L} - \lambda} u(x) = -\sum \frac{\alpha_n u_n(x)}{\lambda - \lambda_n} \tag{102}$$

We can find an equivalent expression of the left side of (102) if we know the Green's function $G(x, \xi; \lambda)$ of the operator $\mathscr{L} - \lambda$. Indeed, the solution of $u(x) = (\mathscr{L} - \lambda)\, m(x)$ according to (76) and (77) may be expressed as

$$m(x) = \int G(x, \xi, \lambda)\, u(\xi)\, d\xi \tag{103}$$

which is therefore the desired equivalent expression. On the other hand, integration of (102) over a large circle in the complex λ plane leads to

$$\lim_{R\to\infty} \frac{1}{2\pi i} \oint \frac{d\lambda}{\mathscr{L} - \lambda} u(x) = -\sum \alpha_n u_n(x) = -u(x) \tag{104}$$

If we replace the left side of (102) by its equivalent expression (103) and perform the same integration, we find, using (104),

$$(2\pi i)^{-1} \oint d\lambda \oint G(x, \xi, \lambda)\, u(\xi)\, d\xi = \int [(2\pi i)^{-1} \oint G\, d\lambda]\, u(\xi)\, d\xi$$
$$= -u(x) \qquad (105)$$

Equation (105) shows that the bracketed quantity has the property of a δ function and therefore,

$$\lim_{R\to\infty} (2\pi i)^{-1} \int G(x, \xi, \lambda)\, d\lambda = -\delta(x - \xi) \qquad (106)$$

Green's function $G(x, \xi, \lambda)$ is an analytic function except for pole and branch-point singularities. The integral of (106) therefore reduces to a sum of residues at the poles plus integrals along the branch cuts. Branch cuts occur only when the operator is defined in an infinite domain. For finite domains, the sum of residues corresponds to the discrete spectrum of the operator. In this case we have

$$\delta(x - \xi) = \sum \varphi_n(x)\, \varphi_n(\xi) \qquad (106a)$$

where $\varphi_n(x)$ are the complete and orthonormal eigenfunctions of $\mathscr{L}$ in the domain for which the operator is defined.

The expansion of any continuous function $u(x)$ can be obtained by multiplying the expansion of $\delta(x - \xi)$ by $u(\xi)$ and integrating with respect to ξ.

The spectral expansion of the δ function is especially useful in the solution of partial differential equations by means of Green's functions as will be shown in Section 4.3 and its applications.

4.3. Green's Functions for Partial Differential Equations

In the preceding section it was shown that the spectral representation of the δ function for an ordinary differential operator $\mathscr{L}$ could be obtained by integrating the Green's function for $\mathscr{L} - \lambda$ around a large circle in the λ plane. The same procedure remains valid for the case of a partial differential operator $\mathscr{L}$ which is the sum of two or more commutative operators. If $\mathscr{L}_1$ and $\mathscr{L}_2$ commute, the inverse of $\mathscr{L}_1 + \mathscr{L}_2$ may be found by considering $\mathscr{L}_2$ (or $\mathscr{L}_1$) as a constant. The result is a function of $\mathscr{L}_2$ (or $\mathscr{L}_1$) and should be interpreted by using the spectral representation of the δ function for $\mathscr{L}_2$ (or $\mathscr{L}_1$). For the proof of this statement, the reader is referred to Friedman (1956). We shall illustrate the technique by several examples.

4.3.1. *Well in an Infinite Aquifer Strip*

As a preliminary step to this problem let us compute the Green's function of Example 2, Section 2.3, Fig. 4. The boundary value problem is to solve the equation

$$u_{xx} + u_{yy} - (u/B^2) = 0 \tag{107}$$

for the boundary conditions

$$\begin{aligned} u(x, 0) &= u(x, +2a) = 0 \\ u(x, y) &\quad \text{bounded for} \quad x = \pm\infty \end{aligned} \tag{108}$$

If we consider $\mathscr{L}$ as the positive-definite operator

$$\mathscr{L} = -\nabla^2 + (B^2)^{-1} \tag{109}$$

then the Green's function is found by solving the equation

$$-(u_{xx} + u_{yy}) + (u/B^2) = \delta(x - \xi)\,\delta(y - \eta) \tag{110}$$

for the boundary conditions (108). In (110), define the positive-definite operator $-d^2/dy^2$ as λ by the equation

$$-u_{yy} = \lambda u \tag{111}$$

for the boundary condition $u(0) = u(+2a) = 0$ and treat λ as a constant in the equation derived from (110) as

$$u_{xx} - [\lambda + (B^2)^{-1}]\, u = -\delta(x - \xi)\,\delta(y - \eta) \tag{112}$$

with boundary conditions $u(x)$ bounded for $x = \pm\infty$. This means that (112) will be treated as an ordinary differential equation in x, which may depend also on a parameter y. The Green's function of (112) is constructed as explained in Section 4.1.1. Let $\lambda + 1/B^2 = p^2$, then

$$\begin{aligned} g(x, \xi) &= e^{px}e^{-p\xi} \qquad \text{for} \quad -\infty < x < \xi \\ &= e^{-px}e^{p\xi} \qquad \text{for} \quad \xi < x < +\infty \end{aligned}$$

The jump condition in dg/dx at $x = \xi$ gives $J = -2p$. The jump should be -1; therefore, the solution of (112) is

$$u(x, y) = (e^{-p|x-\xi|}/2p)\,\delta(y - \eta) \tag{113}$$

The eigenvalue λ follows from the solution of (111) with the boundary condition $u(+2a) = 0$ as

$$\lambda = (m^2\pi^2)/(2a)^2 \qquad \text{with} \quad m = 0, 1, 2, \ldots \tag{114}$$

Since the operator $-d^2/dx^2 + p^2$ of (112) is positive-definite, the eigenvalue p is positive and therefore,

$$p = \frac{\pi}{2a}\left[m^2 + \left(\frac{2a}{\pi B}\right)^2\right]^{1/2} \tag{115}$$

Finally,

$$\delta(y - \eta) = \frac{1}{a}\sum_{m=1}^{\infty} \sin\frac{m\pi y}{2a} \sin\frac{m\pi\eta}{2a} \tag{116}$$

as will be shown further in the derivation of (126). The combination of (113), (115), and (116) leads at once to (20).

Let us now consider a well located in the point x_0, y_0 of the aquifer of Fig. 4 and let it be pumped at a flow rate Q of dimensions L^3/time. Moreover let us assume a leaky boundary condition along $y = 2a$, i.e., a condition similar to Newton's law of cooling in heat conduction, in which the temperature is proportional to its normal derivative along the boundary. The problem is to solve the equation

$$\nabla^2 s - (s/B^2) = -(Q/T)\,\delta(x - x_0)\,\delta(y - y_0) \tag{117}$$

with the boundary conditions

$$s(x, 0) = 0, \qquad s(x, 2a) = \beta(\partial s/\partial y)(x, 2a) \tag{118}$$

and

$$s(\pm\infty, y) \quad \text{is finite} \tag{119}$$

From the preliminary step, we can at once write an expression equivalent to (113),

$$s(x, y) = \frac{Q}{T}\frac{\exp(-p \mid x - x_0 \mid)}{2p}\,\delta(y - y_0) \tag{120}$$

The eigenvalue λ follows from the solution of (111) now with the second boundary condition of (118) as

$$\sin(\sqrt{\lambda}2a) - \beta\sqrt{\lambda}\cos(\sqrt{\lambda}2a) = 0$$

or

$$\sqrt{\lambda_n} = \beta^{-1} tg(2a\sqrt{\lambda_n}) \tag{121}$$

in which $n = 1, 2, \ldots,$ indicates the succession of the roots. With $2a\sqrt{\lambda} = z$, it follows that the eigenvalues can be found by graphically solving the equation $tg z = (\beta/2a)z$. Their number is infinite.

To find the expansion of $\delta(y - y_0)$, the Green's function is constructed for

$$d^2s/dy^2 = -\lambda s$$

with the boundary conditions $s(0) = 0$, $s(2a) = \beta s'(2a)$. The derivation proceeds as in Section 4.1.1. The result is

$$\begin{aligned} g(y, y_0, \lambda) = & \\ & \frac{\sin(y\sqrt{\lambda})}{\sqrt{\lambda}} \left\{ \frac{-\beta\sqrt{\lambda}\cos[(2a - y_0)\sqrt{\lambda}] + \sin[(2a - y_0)\sqrt{\lambda}]}{\sin(2a\sqrt{\lambda}) - \beta\sqrt{\lambda}\cos(2a\sqrt{\lambda})} \right\} \\ & \times H(y_0 - y) + \frac{\sin(y_0\sqrt{\lambda})}{\sqrt{\lambda}} \\ & \times \left\{ \frac{-\beta\sqrt{\lambda}\cos[(2a - y)\sqrt{\lambda}] + \sin[(2a - y)\sqrt{\lambda}]}{\sin(2a\sqrt{\lambda}) - \beta\sqrt{\lambda}\cos(2a\sqrt{\lambda})} \right\} \\ & \times H(y - y_0) \end{aligned} \tag{122}$$

It is easy to prove that $\lambda = 0$ is neither a pole nor a branch point for this function. The singularities of $g(y, y_0, \lambda)$ are simple poles at the zeros of

$$\sin(2a\sqrt{\lambda}) - \beta\sqrt{\lambda}\cos(2a\sqrt{\lambda}) = 0$$

or

$$\sqrt{\lambda_n} = \beta^{-1} tg(2a\sqrt{\lambda_n})$$

as expressed by (121). The residue at each pole is

$$\frac{2\sin(y\sqrt{\lambda_n})\sin(y_0\sqrt{\lambda_n})}{\beta\cos^2(2a\sqrt{\lambda_n}) - 2a}$$

and the expansion of the δ function is

$$\delta(y - y_0) = 2\sum_{n=1}^{\infty} \frac{\sin(y\sqrt{\lambda_n})\sin(y_0\sqrt{\lambda_n})}{-\beta\cos^2(2a\sqrt{\lambda_n}) + 2a} \tag{123}$$

The final solution follows from (120) in which p and $\delta(y - y_0)$ are replaced by their computed values

$$s(x, y) = \frac{Q}{T}\sum_{n=1}^{\infty} \frac{\exp\{(-|x - x_0|)[\lambda_n + (B^2)^{-1}]^{1/2}\}}{[\lambda_n + (B^2)^{-1}]^{1/2}} \frac{\sin(y\sqrt{\lambda_n})\sin(y_0\sqrt{\lambda_n})}{[2a - \beta\cos^2(2a\sqrt{\lambda_n})]} \tag{124}$$

Special Cases

(1) $\beta = 0$. This case was examined by Hantush and Jacob (1954) who used the theory of images to construct the Green's functions. The same results can be found here as a limit case. The Green's function follows from (122) for $\beta = 0$. This function now has simple poles for

$$\lambda_n = n^2\pi^2/4a^2, \qquad n = 1,2,\ldots \tag{125}$$

It is easy to prove that $\lambda_n = 0$ is not a pole. The expansion of the δ function follows from (123) for $\beta = 0$ and for λ_n replaced by its value (125) so that

$$\delta(y - y_0) = \frac{1}{a}\sum_{n=1}^{\infty} \sin\frac{n\pi y}{2a} \sin\frac{n\pi y_0}{2a} \tag{126}$$

and the value of the drawdown s becomes

$$s(x, y) = \frac{Q}{2aT}\sum_{n=1}^{\infty} \frac{\exp(-|x - x_0|)[(n^2\pi^2/4a^2) + (B^2)^{-1}]^{1/2}}{[(n^2\pi^2/4a^2) + (B^2)^{-1}]^{1/2}} \sin\frac{n\pi y}{2a} \sin\frac{n\pi y_0}{2a}$$

(2) $\beta = \infty$. The Green's function follows again from (122) in the limit for $\beta \to \infty$. The expansion of the δ function follows from the integration of this new Green's function in the λ plane. The result is

$$\delta(y - y_0) = a^{-1}\sum_{n=1}^{\infty} \sin[(n + \tfrac{1}{2})\,\pi(y/2a)] \sin[(n + \tfrac{1}{2})\,\pi(y_0/2a)] \tag{127}$$

and

$$\lambda_n = [(2n + 1)^2\,\pi^2]/(4a^2) \tag{128}$$

This is the case of the infinite strip with boundary conditions $s(x, 0) = 0$ and $\partial s/\partial y(x, 2a) = 0$. It is easy to show that the analogous problem with boundary conditions $\partial s/\partial y(x, 0) = 0$ and $s(x, 2a) = 0$ leads to the same eigenvalues (128) and the following expansion of the δ function

$$\delta(y - y_0) = a^{-1}\sum_{n=0}^{\infty} \cos[(n + \tfrac{1}{2})\,\pi(y/2a)] \cos[(n + \tfrac{1}{2})\,\pi(y_0/2a)] \tag{129}$$

In both cases the drawdown follows from (120).

4.3.2. *Eccentric Well in Leaky Aquifer with Lateral Inflow*

Flow to an eccentric well in a leaky circular aquifer with varied lateral replenishment was treated by De Wiest (1963a) (Figs. 10 and 11).

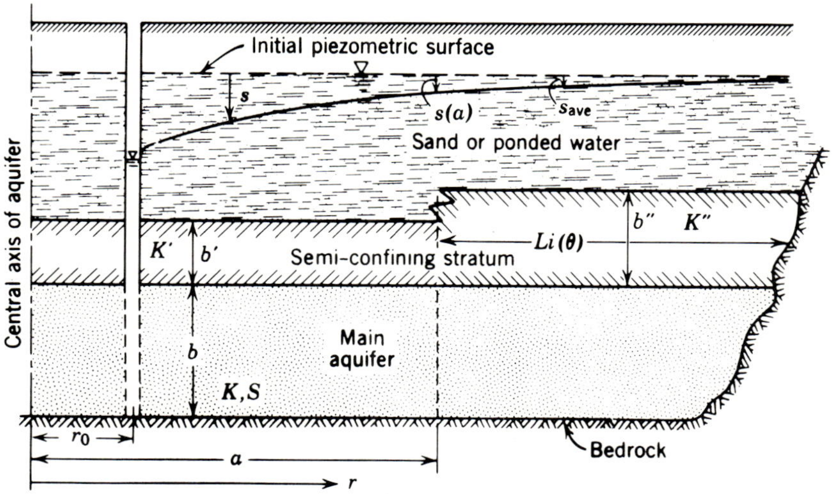

FIG. 10. Aquifer with vertical leakage and varied lateral replenishment. (After De Wiest, 1963a).

The differential equation for the drawdown $s(r, \theta, t)$ may be written as

$$\frac{\partial^2 s}{\partial r^2} + \frac{1}{r}\frac{\partial s}{\partial r} + \frac{1}{r^2}\frac{\partial^2 s}{\partial \theta^2} - \frac{s}{B^2} = \frac{S}{T}\frac{\partial s}{\partial t} \tag{130}$$

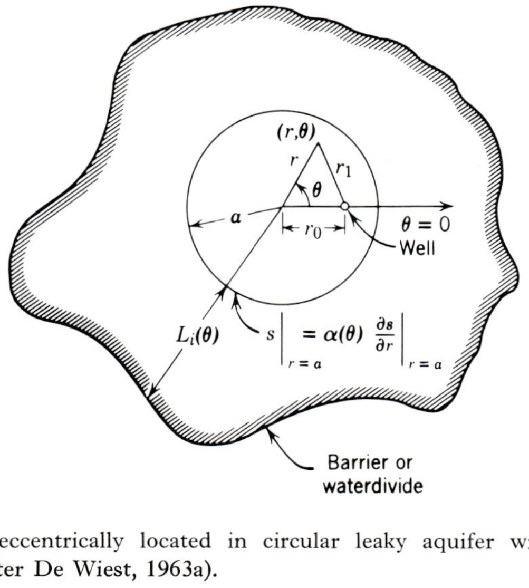

FIG. 11. Well eccentrically located in circular leaky aquifer with varied lateral replenishment. (After De Wiest, 1963a).

The boundary condition expressing the varied lateral replenishment along the perimeter of the circular aquifer with radius a is

$$s(a) = \alpha(\theta) \frac{\partial s}{\partial r}\bigg|_{r=a} \tag{131}$$

Here $\alpha(\theta)$ is always negative as is β of (84) and (118).

Steady State: Well located at (r_0, θ_0) pumped at constant discharge Q. To solve this problem one puts a mathematical sink of strength ϵ in point (r_0, θ_0) and finds the Green's function $g(r, \theta)$ by integrating

$$\frac{\partial^2 g}{\partial r^2} + \frac{1}{r}\frac{\partial g}{\partial r} + \frac{1}{r^2}\frac{\partial^2 g}{\partial \theta^2} - \frac{g}{B^2} = \frac{\epsilon}{r}\,\delta(r - r_0)\,\delta(\theta - \theta_0) \tag{132}$$

for the boundary conditions

$$g(0) \text{ is finite,} \qquad g(a, \theta) = \alpha(\partial g/\partial r)|_{r=a} \tag{133}$$

Here, $(1/r)\,\delta(r - r_0)\,\delta(\theta - \theta_0)$ is Dirac's delta function in polar coordinates.

The Green's function $g(r, \theta; r_0, \theta_0)$ thus obtained is valid for the mathematical sink. To relate it to the well of discharge Q, an additional condition is imposed

$$\lim_{r_1 \to 0} r_1 \frac{\partial s}{\partial r_1} = -\frac{Q}{2\pi T} \tag{134}$$

This determines the strength ϵ and leads at once to the final form of the drawdown $s(r, \theta)$.

To integrate (132), use is made of the fact that the differential operators in r and θ are commutative. Therefore, as in (111), we write

$$-\partial^2 g/\partial\theta^2 = \lambda g \tag{135}$$

and consider λ as an operator, with eigenvalues and eigenfunctions that can be found by solving (135) with the periodic boundary conditions

$$g(0) = g(2\pi), \qquad \partial g(0)/\partial\theta = \partial g(2\pi)/\partial \tag{136}$$

The spectral representation of $\delta(\theta - \theta_0)$ in these eigenfunctions can be found by integrating the Green's function for (135) and (136) in the complex λ plane. The eigenvalues are

$$\lambda_n = n^2, \qquad n = 1, 2, \ldots \tag{137}$$

and

$$\delta(\theta - \theta_0) = (2\pi)^{-1} + \pi^{-1} \sum_{n=1}^{\infty} \cos n(\theta - \theta_0) \tag{138}$$

With the eigenvalues $\lambda = n^2$ known, (132) is transformed into

$$\frac{d^2g}{dr^2} + \frac{1}{r}\frac{dg}{dr} + \frac{n^2}{r^2}g - \frac{g}{B^2} = \frac{\epsilon}{r}\,\delta(r - r_0)\,\delta(\theta - \theta_0) \tag{139}$$

and we determine the Green's function of this ordinary differential equation in which θ is treated as a parameter. To construct this Green's function, two independent solutions, say $I_n(r/B)$ and $K_n(r/B)$, are taken for the homogeneous equation resulting from (139) if the left side is made equal to zero. Next, two linear combinations of these solutions, say $q_1(r/B)$ and $q_2(r/B)$, are chosen so as to satisfy the boundary conditions (133). It follows that

$$q_1(r/B) = I_n(r/B), \qquad q_2(r/B) = I_n(r/B) + C_n K_n(r/B)$$

where

$$C_n = \frac{(\alpha/B)[I_{n+1}(a/B) + I_{n-1}(a/B)] - 2I_n(a/B)}{(\alpha/B)[K_{n+1}(a/B) + K_{n-1}(a/B)] + 2K_n(a/B)} \tag{140}$$

The Green's function can now be found by forming the products

$$\begin{aligned} q(r/B, r_0/B) &= q_1(r/B)\, q_2(r_0/B) \qquad \text{for} \quad r < r_0 \\ &= q_1(r_0/B)\, q_2(r/B) \qquad \text{for} \quad r > r_0 \end{aligned}$$

and by satisfying the condition that the derivative of q at $r = r_0$ has a jump $1/r = 1/r_0$.

With $\epsilon = -Q/T$ from (134) as further shown, and with $\delta(\theta - \theta_0)$ from (138), the solution of the steady state problem becomes:

$$\begin{aligned} s(r, \theta; r_0, \theta_0) &= \frac{Q}{2\pi T}\left\{\frac{I_0(r/B)}{C_0}[I_0(r_0/B) + C_0K_0(r_0/B)]\right. \\ &\quad \left. + \sum_{n=1}^{\infty} \frac{2}{C_n} I_n(r/B)[I_n(r_0/B) + C_nK_n(r_0/B)] \cos n(\theta - \theta_0)\right\} \\ &\qquad\qquad \text{for} \quad 0 \leqslant r \leqslant r_0 \qquad (141) \\ &= \frac{Q}{2\pi T}\left\{\frac{I_0(r_0/B)}{C_0}[I_0(r/B) + C_0K_0(r/B)]\right. \\ &\quad \left. + \sum_{n=1}^{\infty} \frac{2}{C_n} I_n(r_0/B)[I_n(r/B) + C_nK_n(r/B)] \cos n(\theta - \theta_0)\right\} \\ &\qquad\qquad \text{for} \quad r_0 \leqslant r \leqslant a \end{aligned}$$

with C_n given by (140).

Special Case: *Concentric well*, $r_0 = 0$. Due to the radial symmetry, the terms in $n \neq 0$ do not exist. To check the value of ϵ, the second part of the Green's function for $r_0 \leqslant r \leqslant a$ is taken. Since $I_0(0) = 1$, this part becomes

$$g(r, 0) = -(\epsilon/2\pi C_0)[I_0(r/B) + C_0 K_0(r/B)]$$

For $\alpha = 0$, the value of C_0 is

$$C_0 = -I_0(a/B)/K_0(a/B)$$

and

$$g(r, 0) = -(\epsilon/2\pi)[K_0(r/B) - K_0(a/B)\, I_0(r/B)/I_0(a/B)]$$

According to (134),

$$\lim_{r \to 0} r(\partial g/\partial r) = -Q/2\pi T$$

or

$$\lim_{r \to 0} r(\epsilon/2\pi B)\, K_1(r/B) = -Q/2\pi T \qquad \text{and} \qquad \epsilon = -Q/T$$

Therefore, for the concentric well and $\alpha = 0$,

$$s_0 = (Q/2\pi T)[K_0(r/B) - K_0(a/B)\, I_0(r/B)/I_0(a/B)]$$

For $\alpha = \infty$, $C_0 = -I_1(a/B)/K_1(a/B)$ so that

$$s_0 = (Q/2\pi T)[K_0(r/B) + K_1(a/B)\, I_0(r/B)/I_1(a/B)]$$

These equations check with those obtained previously by the method of images (Hantush and Jacob, 1954).

Unsteady State: A problem of practical interest is that in which the discharge Q of a well located at (r_0, θ_0) is suddenly given an increment ΔQ, say at $t = 0$, and in which the influence of ΔQ on the drawdown is the unknown. To find $s(r, \theta, t)$, it is possible to consider s as the superposition of the existing steady state drawdown $s(r, \theta)$ given by (141) and of $s_u(r, \theta, t)$, satisfying (130) and the following conditions:

Boundary conditions

$$s_u(0, 0, t) \text{ is finite}$$

$$s_u(a, \theta, t) = \alpha \frac{\partial s_u(a, \theta, t)}{\partial r} \tag{142}$$

Initial condition

$$s_u(r, \theta, 0) = (\Delta Q/T)\, r^{-1}\, \delta(r - r_0)\, \delta(\theta - \theta_0) \tag{143}$$

The initial condition as formulated by (143) takes care of the strength condition

$$\lim_{r_1 \to 0} r_1 \frac{\partial s_u(r, \theta, t)}{\partial r_1} = -\frac{\Delta Q}{2\pi T}$$

In view of the fact that all differential operators of (130) are commutative, the problem is solved in the way previously shown for the steady state case. Equation (130) is rewritten as

$$\frac{\partial^2 s_u}{\partial r^2} + \frac{1}{r}\frac{\partial s_u}{\partial r} + \frac{1}{r^2}\frac{\partial^2 s_u}{\partial \theta^2} - \frac{s_u}{B^2} = \frac{S}{T}\frac{\partial s_u}{\partial t} \tag{144}$$

Consider first the operator μ in

$$\mu s_u = \frac{S}{T}\frac{\partial s_u}{\partial t} \tag{145}$$

and solve (145) for the initial condition (143), and treat μ as a constant. The result is

$$s_u = (\Delta Q/T)\, e^{T\mu t/S} r^{-1}\, \delta(r - r_0)\, \delta(\theta - \theta_0) \tag{146}$$

Moreover, since the conditions on θ have not changed as compared to the steady state case, $\partial^2 s_u/\partial\theta^2$ is replaced by $(-n^2 s_u)$. Equation (144) now becomes

$$\frac{\partial^2 s_u}{\partial r^2} + \frac{1}{r}\frac{\partial s_u}{\partial r} + \left(-\frac{n^2}{r^2} - \frac{1}{B^2} - \mu\right) s_u = 0 \tag{147}$$

The eigenvalues of μ and the spectral representation of the Dirac function $r^{-1}\,\delta(r - r_0)$ may be found by integrating the Green's function of (147) for the boundary conditions (142) in the complex μ-plane. This Green's function is:

$$\begin{aligned}
g_u(r, r_0) = & -(2\pi)^{-1} C_0^{-1} I_0[r(1/B^2 + \mu)^{1/2}]\{I_0[r_0(1/B^2 + \mu)^{1/2}] \\
& + C_0 K_0[r_0(1/B^2 + \mu)^{1/2}]\} - \sum_{n=1}^{\infty} C_n^{-1}\pi^{-1} I_n[r(1/B^2 + \mu)^{1/2}] \\
& \times \{I_n[r_0(1/B^2 + \mu)^{1/2}] + C_n K_n[r_0(1/B^2 + \mu)^{1/2}]\} \quad \text{for} \quad 0 \leqslant r \leqslant r_0 \\
= & -(2\pi)^{-1} C_0^{-1} I_0[r_0(1/B^2 + \mu)^{1/2}]\,\{I_0[r(1/B^2 + \mu)^{1/2}] \\
& + C_0 K_0[r(1/B^2 + \mu)^{1/2}]\} - \sum_{n=1}^{\infty} C_n^{-1}\pi^{-1} I_n[r_0(1/B^2 + \mu)^{1/2}] \\
& \times \{I_n[r(1/B^2 + \mu)^{1/2}] + C_n K_n[r(1/B^2 + \mu)^{1/2}]\} \quad \text{for} \quad r_0 \leqslant r \leqslant a
\end{aligned} \tag{148}$$

with C_n as in (140) and Bessel functions for argument $a[(B^2)^{-1} + \mu]^{1/2}$.

Before solving the case for general values of α, it is worthwhile to examine what happens for $\alpha = 0$. For this special case, (148) becomes:

$$g_u(r, r_0) = -(1/2\pi)\, I_0[r(1/B^2 + \mu)^{1/2}] \left\{ K_0[r_0(1/B^2 + \mu)^{1/2}] - \frac{K_0[a(1/B^2 + \mu)^{1/2}]\, I_0[r_0(1/B^2 + \mu)^{1/2}]}{I_0[a(1/B^2 + \mu)^{1/2}]} \right\}$$
$$- (\pi)^{-1} \sum_{n=1}^{\infty} I_n[r(1/B^2 + \mu)^{1/2}] \left\{ K_n[r_0(1/B^2 + \mu)^{1/2}] - \frac{K_n[a(1/B^2 + \mu)^{1/2}]\, I_n[r_0(1/B^2 + \mu)^{1/2}]}{I_n[a(1/B^2 + \mu)^{1/2}]} \right\} \quad \text{for } 0 \leqslant r \leqslant r_0 \tag{149}$$

plus a symmetric term with r, r_0 reversed for $r_0 \leqslant r \leqslant a$. The terms for $n = 1, 2, 3, \ldots$ have simple poles for

$$I_n[a(1/B^2 + \mu)^{1/2}] = (i^n)^{-1}\, J_n[ia(1/B^2 + \mu)^{1/2}] = 0 \tag{150}$$

i.e., for the m zeros of $J_n(u) = 0$. Call these roots $\alpha_{n,m}$ so that the eigenvalues of μ follow from

$$ia(1/B^2 + \mu)^{1/2} = \alpha_{n,m} \qquad \text{or} \qquad \mu_{n,m} = -[1/B^2 + \alpha_{n,m}^2/a^2] \tag{151}$$

The residues at the simple poles give the eigenfunctions of the discrete spectrum of Dirac's delta function $r^{-1}\,\delta(r - r_0)$ in (146). The expansion of $\delta(\theta - \theta_0)$ is given in (138). A typical computation goes like this:

$$\frac{N(\mu)}{D(\mu)} = \frac{-I_n[r(1/B^2 + \mu)^{1/2}]\, K_n[a(1/B^2 + \mu)^{1/2}]\, I_n[r_0(1/B^2 + \mu)^{1/2}]}{I_n[a(1/B^2 + \mu)^{1/2}]}$$

The residue is $N(\mu_{n,m})/D'(\mu_{n,m})$. One computes

$$D'(\mu) = \frac{1}{i^n}\frac{d}{d\mu}\, J_n[ia(1/B^2 + \mu)^{1/2}] = -\frac{a^2}{2i^n\alpha_{n,m}}\, J_n'[ia(1/B^2 + \mu)^{1/2}]$$

and

$$\frac{N(\mu_{n,m})}{D'(\mu_{n,m})} = \frac{2\alpha_{n,m}K_n(-i\alpha_{n,m})\, J_n(\alpha_{n,m}r/a)\, J_n(\alpha_{n,m}r_0/a)}{i^n a^2 J_n'(\alpha_{n,m})}$$

A typical term of (146) due to this discrete spectrum is then

$$T_{n,m} = \frac{\Delta Q}{\pi T}\, \frac{2\alpha_{n,m}K_n(-i\alpha_{n,m})\, J_n(\alpha_{n,m})\, r/a)\, J_n(\alpha_{n,m}r_0/a) \cos n(\theta - \theta_0)}{i^n a^2 J_n'(\alpha_{n,m})}$$
$$\times \exp\left[-\frac{Tt}{S}\left(\frac{1}{B^2} + \frac{\alpha_{n,m}^2}{a^2}\right)\right] \tag{152}$$

These terms lead to a double series $\sum_{n=1}^{\infty} \sum_{m=0}^{\infty} T_{n,m}$. The first term of the Green's function gives also a contribution to the discrete spectrum, namely,

$$T^*_{0,m} = \frac{\Delta Q}{\pi T} \frac{\alpha_{0,m} K_0(-i\alpha_{0,m}) \, J_0[\alpha_{0,m}(r/a)] \, J_0[\alpha_{0,m}(r_0/a)]}{a^2 J_0'(\alpha_{0,m})}$$

$$\times \exp \left[-\frac{Tt}{S} \left(\frac{\alpha_{0,m}^2}{a^2} + \frac{1}{B^2} \right) \right] \tag{153}$$

and to a series $\sum_{m=0}^{\infty} T^*_{0,m}$ in the final solution. The terms $T^*_{0,m}$, $T_{n,m}$ may still be simplified by the use of (Watson, 1944)

$$K_n(\zeta e^{i\pi 3/2}) = -(\pi/2)(e^{-in\pi 3/2})[Y_n(\zeta) + i(1 + 2e^{2n\pi i}) \, J_n(\zeta)]$$

It follows that $K_n(-i\alpha_{n,m}) = -(\pi/2) \, i^n Y_n(\alpha_{n,m})$ because $J_n(\alpha_{n,m}) = 0$.

By inspection of (149) as a Green's function, it is easy to see that the terms $I_n K_n$ do vanish at infinity in the μ plane, independently of the order of n. To see this, use is made of the asymptotic formulas

$$I_n(X) \approx e^X/(2\pi X)^{1/2}, \qquad K_n(X) = e^{-X}(\pi/2X)^{1/2}$$

Furthermore, the product $I_n(X) \, K_n(X)$ is finite for $X = 0$ and $n \neq 0$, and because of the product form, $I_n(\sqrt{X}) \, K_n(\sqrt{X})$ has no branch points at $X = 0$ and $n \neq 0$. However, the term for $n = 0$ in (149) requires more study. For $\mu = -1/B^2$,

$$K_0[a(1/B^2 + \mu)^{1/2}] \qquad \text{and} \qquad K_0[r(1/B^2 + \mu)^{1/2}]$$

become infinite and have a branch point. The other branch point is at infinity and to integrate the Green's function properly, a cut has to be made as in Fig. 12. At first glance, it seems that besides the discrete spectrum, a continuous spectrum would arise as a result of integration along the branch cut. This would be unusual because of the finiteness of the domain in which solutions are sought. Therefore, perform the integration along the branch cut. Let

$$1/B^2 + \mu = \nu^2$$

and assume

$$(1/B^2 + \mu)^{1/2} = -\nu \quad \text{along the lower side of the cut}$$

$$= +\nu \quad \text{along the upper side of the cut}$$

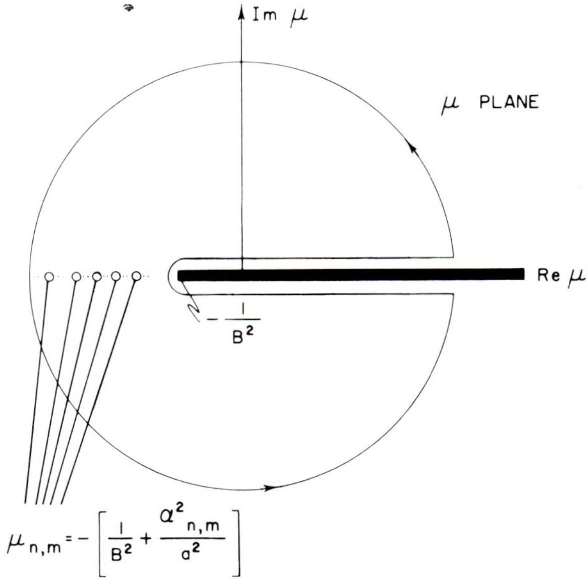

FIG. 12. Poles and branch cut in complex μ plane. (After De Wiest, 1963a).

Then

$$(2\pi i)^{-1} \int_{\text{cut}} I_0[r(1/B^2 + \mu)^{1/2}]\, K_0[r_0(1/B^2 + W)^{1/2}]\, d\mu$$
$$= (2\pi i)^{-1} \left[\int_{\infty}^{0} I_0(-\nu r)\, K_0(-\nu r_0)\, 2\nu\, d\nu + \int_0^{\infty} I_0(\nu r)\, K_0(\nu r_0)\, 2\nu\, d\nu\right] = 0$$

because $I_0(x)$ and $K_0(x)$ are even.

In a similar way it can be shown that the second term for $n = 0$ in (149) gives no contribution along the branch cut. The spectrum then is merely discrete and the final solution for $\alpha = 0$ can be written as

$$s_u(r, \theta, t; r_0, \theta_0, 0) = \sum_{m=0}^{\infty} T^*_{0,m} + \sum_{n=1}^{\infty} \sum_{m=0}^{\infty} T_{n,m} \tag{154}$$

in which

$$T^*_{0,m} = -\frac{\Delta Q}{2T} \frac{\alpha_{0,m} Y_0(\alpha_{0,m})\, J_0(\alpha_{0,m} r/a)\, J_0(\alpha_{0,m} r_0/a)}{a^2 J_0'(\alpha_{0,m})}$$
$$\times \exp\left[-\frac{Tt}{S}\left(\frac{\alpha^2_{n,m}}{a^2} + \frac{1}{B^2}\right)\right]$$

$$T_{n,m} = -\frac{\Delta Q}{T}\frac{\alpha_{n,m}Y_n(\alpha_{n,m})\,J_n(\alpha_{n,m}r/a)\,J_n(\alpha_{n,m}r_0/a)}{a^2J_n'((\alpha_{n,m})}$$

$$\times \cos n(\theta - \theta_0) \exp\left[-\frac{Tt}{S}\left(\frac{\alpha_{n,m}^2}{a^2} + \frac{1}{B^2}\right)\right]$$

These formulas are valid for $0 \leqslant r \leqslant a$ because the result of the integration of both expressions of the Green's function (149) is identical.

The same procedure can be repeated for $\alpha = \infty$ and for any value of α. In the general case, the poles of the Green's function follow for the zeros of $C_0 = 0$ and $C_n = 0$, i.e., the following transcendental equations for μ must be solved.

$$-I_0[a(1/B^2 + \mu)^{1/2}] + (\alpha/B)\,I_1[a(1/B^2 + \mu)^{1/2}] = 0$$

$$-I_n[a(1/B^2 + \mu)^{1/2}] + (\alpha/B)\,I_n'[a(1/B^2 + \mu)^{1/2}] = 0$$

Numerical Results: see De Wiest (1963a, 1967).

Appendix A. A Note on the Fourier Integral

Fourier series are useful for representing a wide class of functions in a finite interval, but are of no value if the interval is of infinite extent. An extension of the basic approach to an infinite interval leads to the Fourier integral representation. The following analysis does not constitute a rigorous proof, but indicates the appropriate form of the Fourier integral (Wayland, 1957).

A.1. Basic Fourier Series

Let $f(x)$ be defined arbitrarily within the interval S where $a \leqslant x < a + p$ except for the following restrictions.

(1) $f(x)$ is single-valued in S;

(2) $f(x)$ possesses a finite number of maxima and minima in S;

(3) $f(x)$ is continuous in S except for a finite number of finite discontinuities and a finite number of infinities; and

(4) $\int_a^{a+p} |f(x)|\, dx$ exists.

Then if

$$a_n = \frac{2}{p}\int_a^{a+p} f(\xi)\cos\frac{2n\pi\xi}{p}\,d\xi, \qquad b_n = \frac{2}{p}\int_a^{a+p} f(\xi)\sin\frac{2n\pi\xi}{p}\,d\xi \tag{A.1}$$

the series

$$\frac{a_0}{2} + \sum_{n=1}^{\infty} \left(a_n \cos \frac{2n\pi x}{p} + b_n \sin \frac{2n\pi x}{p} \right) \tag{A.2}$$

converges at every point in which S is not one of the infinities of $f(x)$ and its sum is

$$\tfrac{1}{2}[f(x+0) + f(x-0)] \tag{A.3}$$

If $f(x)$ is continuous at the point in question, the series converges to the value of $f(x)$.

A.2. Passage to the Fourier Integral

Let $a = -L$, $a + p = L$, i.e., $p = 2L$ in the previous formulas. Then

$$a_n = \frac{1}{L} \int_{-L}^{L} f(\xi) \cos \frac{n\pi\xi}{L} \, d\xi; \qquad b_n = \frac{1}{L} \int_{-L}^{L} f(\xi) \sin \frac{n\pi\xi}{L} \, d\xi$$

and $f(x)$ becomes, with a_0, a_n, b_n replaced by their values,

$$\begin{aligned} f(x) &= \frac{1}{2L} \int_{-L}^{L} f(\xi)\, d\xi + \sum_{n=1}^{\infty} \frac{1}{L} \int_{-L}^{L} f(\xi) \left[\cos \frac{n\pi x}{L} \cos \frac{n\pi\xi}{L} \right. \\ &\qquad \left. + \sin \frac{n\pi x}{L} \sin \frac{n\pi\xi}{L} \right] d\xi \\ &= \frac{1}{2L} \int_{-L}^{L} f(\xi)\, d\xi + \sum_{n=1}^{\infty} \frac{1}{L} \int_{-L}^{L} f(\xi) \cos \frac{n\pi}{L} (x - \xi)\, d\xi \end{aligned} \tag{A.4}$$

when $L \to \infty$, the fourth condition underlying the basic Fourier series becomes

$$\int_{-\infty}^{\infty} |f(\xi)|\, d\xi < \infty$$

In this case,

$$\lim_{L\to\infty} (2L)^{-1} \int_{-L}^{L} f(\xi)\, d\xi \to 0$$

and (A.4) becomes

$$f(x) = \lim_{L\to\infty} \sum_{n=1}^{\infty} L^{-1} \int_{-L}^{L} f(\xi) \cos(n\pi/L)(x - \xi)\, d\xi \tag{A.5}$$

Now let $\pi/L = \Delta\alpha$ in (A.5) and consider

$$\int_{-L}^{L} f(\xi) \cos n\,\Delta\alpha(x - \xi)\, d\xi = g(n\,\Delta\alpha)$$

But

$$\lim_{\Delta\alpha \to 0} \sum_{n=1}^{\infty} g(n\,\Delta\alpha)\,\Delta\alpha = \int_0^{\infty} g(\alpha)\, d\alpha$$

so that (A.5) becomes

$$f(x) = \pi^{-1} \int_0^{\infty} d\alpha \int_{-\infty}^{\infty} f(\xi) \cos \alpha(x - \xi)\, d\xi \tag{A.6}$$

if the various limiting processes are valid.

A.3. Exponential Fourier Integral

By means of Euler's formula, (A.6) can be transformed into

$$\begin{aligned} f(x) &= \pi^{-1} \int_0^{\infty} d\alpha \int_{-\infty}^{\infty} \tfrac{1}{2}[e^{i\alpha(x-\xi)} + e^{-i\alpha(x-\xi)}] f(\xi)\, d\xi \\ &= (2\pi)^{-1} \int_0^{\infty} d\alpha \int_{-\infty}^{\infty} e^{i\alpha(x-\xi)} f(\xi)\, d\xi \\ &\quad + (2\pi)^{-1} \int_0^{\infty} d\alpha \int_{-\infty}^{\infty} e^{-i\alpha(x-\xi)} f(\xi)\, d\xi \\ &= \mathscr{T}_1 + \mathscr{T}_2 \end{aligned}$$

In $\mathscr{T}_1$ let $\alpha = -\alpha'$, $d\alpha = -d\alpha'$, so that $f(x)$ becomes

$$\begin{aligned} f(x) &= (2\pi)^{-1} \int_0^{-\infty} -d\alpha' \int_{-\infty}^{\infty} e^{-i\alpha'(x-\xi)} f(\xi)\, d\xi + \mathscr{T}_2 \\ &= (2\pi)^{-1} \int_{-\infty}^{0} d\alpha' \int_{-\infty}^{\infty} e^{-i\alpha'(x-\xi)} f(\xi)\, d\xi + \mathscr{T}_2 \end{aligned}$$

If in this expression we replace the dummy integration variable α' by α, we find

$$f(x) = (2\pi)^{-1} \int_{-\infty}^{\infty} d\alpha \int_{-\infty}^{\infty} e^{-i\alpha(x-\xi)} f(\xi)\, d\xi \tag{A.7}$$

This may be rewritten as

$$f(x) = (2\pi)^{-1} \int_{-\infty}^{\infty} e^{-i\alpha x}\, d\alpha \int_{-\infty}^{\infty} e^{i\alpha\xi} f(\xi)\, d\xi$$

Call

$$\int_{-\infty}^{\infty} e^{i\alpha\xi} f(\xi)\, d\xi = F(\alpha)$$

the Fourier transform then

$$f(x) = (2\pi)^{-1} \int_{-\infty}^{\infty} e^{-i\alpha x} F(\alpha)\, d\alpha$$

is called the inversion formula.

Equation (A.7) conforms to the notation used by Churchill (1958). However, it is easy to see that there is an alternative formula,

$$f(x) = (2\pi)^{-1} \int_{-\infty}^{\infty} d\alpha \int_{-\infty}^{\infty} e^{i\alpha(x-\xi)} f(\xi)\, d\xi \tag{A.8}$$

which stems from the fact that the cosine is an even function so that the sign of the argument in (A.4) may be changed. This leads to (A.8), a notation used by Sommerfeld (1964).

Appendix B. Derivation of Principal Solution for Equation of Heat Conduction

To explain the step from (52) to (53), it is sufficient to examine a linear heat conductor infinite in both directions. This problem is treated almost in all textbooks of mathematical physics (see for example, Sommerfeld, 1964; Tikhonov and Samarskii, 1963). The integral that occurs is

$$(2\pi)^{-1} \int_{-\infty}^{\infty} \exp[i\alpha(x-\xi) - \alpha^2 \nu t]\, d\alpha \tag{B.1}$$

Abbreviate the exponent in (B.1) by $-m\alpha^2 + n\alpha$ and complete the square

$$-m\alpha^2 + n\alpha = -m(\alpha - n/2m)^2 + n^2/4m$$

Let $p = \alpha - n/2m$ so that (B.1) becomes

$$\begin{aligned}(2\pi)^{-1} &\int_{-\infty}^{\infty} \exp[-\alpha^2 \nu t + i\alpha(x-\xi)]\, d\alpha \\ &= (2\pi)^{-1} \exp[-(x-\xi)^2/(4\nu t)] \int_{-\infty}^{+\infty} e^{-mp^2}\, dp \end{aligned} \tag{B.2}$$

The formula for the Laplace integral is

$$\int_{-\infty}^{\infty} e^{-p^2}\, dp = \sqrt{\pi}; \qquad \text{therefore,} \quad \int_{-\infty}^{\infty} e^{-mp^2}\, dp = \sqrt{(\pi/m)}$$

We therefore can express the right-hand side of (B.2) as

$$U = (4\pi\nu t)^{-1/2} \exp[-(x - \xi)^2/(4\nu t)] \tag{B.3}$$

U describes the space–time behavior of a unit heat source or of a heat pole of unit strength. In two dimensions, the expression for U is

$$U = (4\pi\nu t)^{-1} \exp\{-[(x - \xi)^2 + (y - \eta)^2]/(4\nu t)\} \tag{B.4}$$

In the three-dimensional case, U is the product of three factors of the form (B.3)

$$U = (4\pi\nu t)^{-3/2} \exp\{-[(x - \xi)^2 + (y - \eta)^2 + (z - \zeta)^2]/(4\nu t)\} \tag{B.5}$$

Appendix C. Flow to Wells in Aquifers

The flow in the aquifer penetrated by a number of wells and with variable head or variable derivative of the head along the boundaries can be found by superposition. A simple example of steady flow is given in Fig. 13 where a single well penetrates an aquifer shaped as an infinite quadrant. The head along the edges of the quadrant is variable. The well is pumped at a constant flow rate Q, and is located at (x_1, y_1). The problem is to find the head h in $P(x, y)$. This h may be thought of as the sum of the head h_1, which would prevail if the head along the

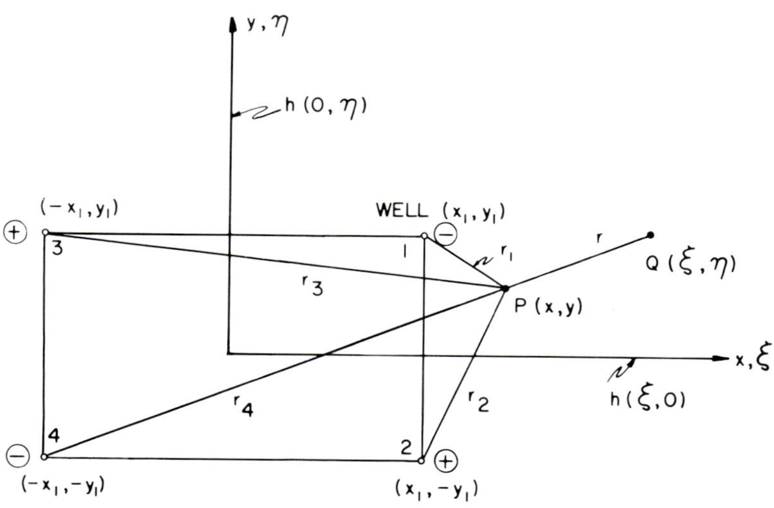

FIG. 13. Flow to well in aquifer.

boundaries were constant and if the well were pumped at the flow rate Q, and of the head h_2, which would prevail if there were no well, and for the actual boundary conditions.

To find h_1, we assume that the head along the boundaries is zero. By inserting image wells in 2, 3 and 4, we find

$$h_1 = \frac{Q}{2\pi T} \log \frac{r_1 r_4}{r_2 r_3} \tag{C.1}$$

To find h_2, we construct Green's function and P becomes a fictitious source point. Q is introduced as an action point and Green's function is constructed again by the method of images. We find that

$$h_2(x, y) = (2\pi)^{-1} \left[\int_0^\infty h(\xi, 0)(-\partial G/\partial\eta)_{\eta=0}\, d\xi + \int_0^\infty h(0, \eta)(-\partial G/\partial\xi)_{\xi=0}\, d\eta \right] \tag{C.2}$$

and $h(x, y) = h_1 + h_2$.

This principle remains applicable for the case of several wells.

Appendix D. Definition of Operator $\mathscr{L}$ Extended for Nonhomogeneous Boundary Conditions

Consider the operator $\mathscr{L}$ defined by (88)

$$\mathscr{L} = d/dx^2 - 1/B^2 \tag{D.1}$$

in the domain (region) $\mathscr{D}$ defined by

$$u(0) = 0, \qquad u(L) = \beta u'(L) \tag{D.2}$$

Find the adjoint operator $\mathscr{M}$ and its domain $\mathscr{D}^*$ by considering

$$\langle \mathscr{M} v, u \rangle = \langle v, \mathscr{L} u \rangle \tag{D.3}$$

in which $\langle\ \rangle$ represents the scalar product in the interval $(0, L)$ so that

$$\begin{aligned} \langle v, \mathscr{L} u \rangle &= \int_0^L v(u'' - u/B^2)\, dx \\ &= [vu' - v'u]_0^L + \int_0^L u(v'' - v/B^2)\, dx \end{aligned} \tag{D.4}$$

Let the conjunct J^* be

$$J\ (v, u) = [vu' - v'u]_0^L \tag{D.5}$$

It follows that $J^* = 0$ if the conditions (D.2) are inserted and when identical conditions on v are required, i.e.,

$$v(0) = 0, \qquad v(L) = \beta v'(L) \tag{D.6}$$

It follows that $\mathscr{D}^* = \mathscr{D}$ and that $\mathscr{M} = \mathscr{L}$, so that $\mathscr{L}$ is self-adjoint or Hermitian. If J^* became zero for conditions on v that were not identical with those on u, $\mathscr{L}$ would only be *formally self-adjoint.*

Now we consider the function $w(x)$ belonging to $\mathscr{S}$, not necessarily to $\mathscr{D}$. We wish to define $\mathscr{L}w$. The result may not be a function, but it will be called a *symbolic function.*

If v is a testing function for $\mathscr{L}$, that is, if v belongs to $\mathscr{D}^*$, the scalar product $\langle \mathscr{M}v, w\rangle$ has a meaning. This scalar product is a continuous linear functional on the space $\mathscr{D}^*$ of testing functions for $\mathscr{L}$. We put

$$\langle \mathscr{M}v, w\rangle = \langle v, \mathscr{L}w\rangle \tag{D.7}$$

where $\mathscr{L}w$ is the symbolic function defined by this equality,

$$\begin{aligned}\langle \mathscr{M}v, w\rangle &= \int_0^L (v'' - v/B^2)\, w\, dx \\ &= \int_0^L v(w'' - w/B^2)\, dx + [wv' - w'v]_0^L \end{aligned} \tag{D.8}$$

Now we make use of the fact that v is in $\mathscr{D}^* = \mathscr{D}$ so that (D.8) becomes

$$\langle v, \mathscr{L}w\rangle = \int_0^L v(w'' - w/B^2)\, dx + v'(L)[w(L) - \beta w'(L)] - w(0)\, v'(0) \tag{D.9}$$

With

$$v'(0) = -\int_0^L v\, \delta'(x)\, dx \quad \text{and} \quad v'(L) = -\int_0^L v\, \delta'(x - L)\, dx$$

it is convenient to rewrite (D.9) as

$$\langle v, \mathscr{L}w\rangle = \int_0^L v\{w'' - w/B^2 + \delta'(x)\, w(0) - \delta'(x - L)[w(L) - \beta w'(L)]\}\, dx \tag{D.10}$$

The definition of the operator $\mathscr{L}$ to take account of nonhomogeneous boundary conditions is therefore

$$\mathscr{L}w = w'' - w/B^2 + \delta'(x)\, w(0) - \delta'(x - L)[w(L) - \beta w'(L)] \tag{D.11}$$

To the solution w_1 of $w_1'' - w_1/B^2 = 0$ for $w_1(0) = 0$ and $w_1(L) - \beta w_1'(L) = 0$, we must add a function w_2 which satisfies $w_2'' - w_2/B^2 = 0$ with conditions $w_2(0) = \alpha$, $w_2(L) - \beta w_2'(L) = \epsilon$. It is therefore required to find a solution of

$$\mathscr{L}w = \alpha\,\delta'(x) - \epsilon\,\delta'(x - L) \tag{D.12}$$

To solve (D.12), it must be remembered that a solution of

$$\mathscr{L}G = \delta(x - \xi) \tag{D.13}$$

is already available if (94) is divided by $(-q/T)$. From (13) it follows that

$$\mathscr{L}\alpha[\partial G(x, 0)/\partial\xi] = -\alpha\,\delta'(x) \tag{D.14}$$

in which

$$\partial G(x, 0)/\partial\xi = \partial G(x, \xi)/\partial\xi\,|_{\xi=0} \tag{D.15}$$

The derivative is readily computed from (94), and after division by $(-q/T)$, we have

$$\begin{aligned}\partial G(x, \xi)/\partial\xi = &-B\sinh(x/B)\,H(\xi - x)(\partial/\partial\xi)\,\varphi(\xi)\\ &-B\sinh(x/B)\,\varphi(\xi)\,\delta(\xi - x)\\ &-\cosh(\xi/B)\,\varphi(x)\,H(x - \xi)\\ &-B\sinh(\xi/B)\,\varphi(x)[-\delta(x - \xi)]\end{aligned} \tag{D.16}$$

in which φ is given by (95).

Only the third of these four terms in the derivative does not vanish for $\xi = 0$. This term, evaluated at $\xi = 0$, renders

$$\frac{\partial G(x, 0)}{\partial\xi} = -\left[\frac{-\dfrac{\beta}{B}\cosh\dfrac{L - x}{B} + \sinh\dfrac{L - x}{B}}{\sinh\dfrac{L}{B} - \dfrac{\beta}{B}\cosh\dfrac{L}{B}}\right] \tag{D.17}$$

A comparison of (D.12) and (D.14) shows that part of the solution of (D.12) is given by

$$-\alpha[\partial G(x, 0)/\partial\xi]$$

or

$$\alpha\left[\frac{-\left(\dfrac{\beta}{B}\right)\cosh\dfrac{L - x}{B} + \sinh\dfrac{L - x}{B}}{\sinh\dfrac{L}{B} - \dfrac{\beta}{B}\cosh\dfrac{L}{B}}\right]$$

which is indeed the first term of the right side of (96). From (D.13) it also follows that

$$\mathscr{L}\epsilon[\partial G(x, \xi)/\partial\xi]|_{\xi=L} = -\epsilon\, \delta'(x - L) \tag{D.18}$$

so that $\epsilon(\partial G/\partial\xi)|_{\xi=L}$ is the second part of the solution of (D.12). The computation of this part proceeds exactly in the same manner as shown above; therefore,

$$\epsilon \frac{\partial G(x, \xi)}{\partial \xi}\bigg|_{\xi=L} = \frac{\epsilon \sinh \dfrac{x}{B}}{\sinh \dfrac{L}{B} - \dfrac{\beta}{B} \cosh \dfrac{L}{B}} \tag{D.19}$$

and again this checks with (10.96).

Appendix E. Symmetry of Green's Function

If $G(x, \xi)$ is the Green's function for an operator $\mathscr{L}$ in a domain $\mathscr{D}$ defined by certain boundary conditions and if $H(x, \xi)$ is the Green's function for the adjoint operator $\mathscr{M}$ in the domain defined by the adjoint boundary conditions, then

$$G(x, \xi) = H(\xi, x) \tag{E.1}$$

which is property (c) of Section 3.2.2. In particular, if $\mathscr{L}$ is self-adjoint, then $G(x, \xi)$ is a symmetric function of x and ξ. The proof is based on the properties of Section 4, as follows:

$$\mathscr{L}G(x, \xi) = \delta(x - \xi), \qquad \mathscr{M}H(x, \xi') = (x - \xi')$$

From the properties of the adjoint operator, we obtain

$$\int H(x, \xi')\, \mathscr{L}G(x, \xi)\, dx = \int \mathscr{M}H(x, \xi')\, G(x, \xi)\, dx$$

and by the use of (E.2) this equation becomes

$$\int H(x, \xi')\, \delta(x - \xi)\, dx = \int \delta(x - \xi')\, G(x, \xi)\, dx$$

or

$$H(\xi, \xi') = G(\xi', \xi)$$

This result is equivalent with (E.1).

REFERENCES

Churchill, R. V. (1958). "Operational Mathematics." McGraw-Hill, New York.

De Wiest, R. J. M. (1963a). Flow to an eccentric well in a leaky circular aquifer with varied lateral replenishment, *Geofis. Pura Appl.* **54**, 87–102.

De Wiest, R. J. M. (1963b). Replenishment of aquifers intersected by streams, *J. Hydraulics Div. Am. Soc. Civil Engrs.* **89**, 165–191.

De Wiest, R. J. M. (1967). Artificial recharge through augmented bank storage, pp. 53–68. *Intern. Assoc. Sci. Hydrology*, Publ. No. 72, *Symp.* of Haifa.

Friedman, B. (1956). "Principles and Techniques of Applied Mathematics." Wiley, New York.

Hantush, M. S., and Jacob, C. E. (1954). Plane potential flow of ground-water with linear leakage, *Trans. Am. Geophys. Union* **35**, 917–936.

Hantush, M. S., and Jacob, C. E. (1955a). Nonsteady radial flow in an infinite leaky aquifer, *Trans. Am. Geophys. Union* **36**, 95–100.

Hantush, M. S., and Jacob, C. E. (1955b). Nonsteady Green's functions for an infinite strip of leaky aquifer, *Trans. Am. Geophys. Union* **36**, 101–112.

Kellogg, O. D. (1929). "Foundations of Potential Theory." Dover, New York.

Muskat, M. (1935). A note on mixed boundary value problems in logarithmic potential theory, *Physics* **6**, 27–32.

Sommerfeld, A. (1964). "Partial Differential Equations in Physics." Academic Press, New York.

Tikhonov, A. N., and Samarskii, A. N. (1963). "Equations of Mathematical Physics." Macmillan, New York.

Watson, G. N. (1944). "A Treatise on the Theory of Bessel Functions." Macmillan, New York.

Wayland, H. (1957). "Differential Equations Applied in Science and Engineering." Van Nostrand, Princeton, New Jersey.

11

Introduction to Flow of Immiscible Liquids in Porous Media

H. J. Morel-Seytoux

Department of Civil Engineering
Colorado State University
Fort Collins, Colorado

1. Introduction

1.1. "Raison d'Être" of Chapter

There is rarely a reference to the Society of Petroleum Engineers of AIME journals in the proceedings of the relevant journals of ASCE and vice-versa. Due to the increasing amount of information published, this lack of communication is understandable. It merely emphasizes the need for works of a more comprehensive nature and it explains the popularity of books in this category, of which there is truly only one: "The Physics of Flow Through Porous Media" (Scheidegger, 1959). Taken together, Muskat's two volumes ["The Flow of Homogeneous Fluids Through Porous Media" (Muskat, 1946) and "Physical Principles of Oil Production" (Muskat, 1949)] constitute the most serious endeavor to focus the many facets of flow through porous media into a single picture.

The present volume is heavily biased toward flow through porous media from a civil engineer's or groundwater hydrologist's point of view. In light of the above comments, it would be regrettable if the student of groundwater movement, who has just mastered the early chapters, were not introduced to the closely related subject of "flow of two immiscible liquids."

The present chapter was conceived to provide such an introduction. Because of the expected audience of this book, no previous experience of two-phase flow has been assumed. The understanding of simultaneous flow of immiscible fluids in a porous medium is a prerequisite to sound operations in the practical business of production of oil and of natural gas. The petroleum industry has devoted much effort to gain a better understanding of this subject. Consequently, the major source of material used in this chapter is the petroleum literature. However, no previous encounter with petroleum terminology or the petroleum industry's problems on the part of the reader is assumed. Indeed, more is assumed of the reader from a mathematical point of view than from a physical point of view. However, even in this respect, every attempt has been made to minimize the required background for understanding. Because the reader is likely to be more interested in the techniques of solution that can be applied to his field, rather than in the solutions themselves, the techniques are stressed.

Within this introductory section, the purpose of the following paragraphs is:

1. To underline the difference in types of problems encountered in groundwater as opposed to petroleum engineering.

2. To define reservoir engineering and the basic concepts associated with it.

1.2. Groundwater and Petroleum Engineering

The hydraulic engineer is concerned about water which occurs alone in zones known as aquifers. Whether in the formation or in the well casing, a single fluid is flowing. The groundwater engineer's flow problems are classified as "single-phase" problems.

The petroleum engineer is concerned about production of oil. Oil seldom occurs alone. Even when only oil is initially present in the reservoir, (not in contact with gas or water) it has the ability to generate a second phase: gas. Gas is liberated from solution in the oil as the pressure in the oil decreases. The development of a second phase may occur either throughout the entire reservoir or in the neighborhood of the well where oil flows into the tubing and through the tubing to the surface. Thus, the petroleum engineer is confronted with a broader class of problems in fluid flow than the hydraulic engineer.

The delineation between groundwater and petroleum engineering on the basis of one phase versus several phases is an oversimplification. There are problems of concern to hydraulic engineers or agricultural or irrigation engineers in which the porous medium is not saturated. But in many instances, such as seepage through an earth dam, the size of the transition zone between the region fully air saturated and the region fully water saturated can be legitimately neglected, at least under steady-state conditions. In almost all calculation procedures a sharp free surface is assumed. In problems of salt water intrusion into a fresh aquifer, the transition zone within which salt concentration varies from its sea maximum to the uncontaminated aquifer minimum is not very wide and a sharp interface has usually been assumed.

Petroleum engineers have also often assumed sharp interfaces between fluids to render problems amenable to mathematical treatment. The simple model thus obtained has proved valuable in a variety of cases, (e.g., reservoirs in which oil and gas are thoroughly segregated). It has been used and is still in use today. However, there are cases for which the model betrays reality beyond reasonable engineering tolerance. The problem then is not to calculate the end of one fluid and the beginning of the other one, but rather to calculate how much there is of each fluid throughout the porous medium. The *concept* of *saturation* is thus basic to reservoir engineering. Saturation of a given phase is defined as *the volumetric fraction of the total pore space occupied by that phase.*

On the practical side, the depth at which the hydraulic and petroleum engineers find their fluids has a definite bearing on the operations. At the Whittier Narrows Dam treatment plant (California) there is an open-pit observation station (McMichael and McKee, 1965) where rates of infiltration can be measured. It is a different story to sample accurately a well that produces from a horizon at 21,443 ft (American Petroleum Institute, 1963) (world's deepest producer as of 1962). Stimulation of production in a well can be obtained by heating the formation in the neighborhood of the well bore. Stimulation results partly from reduction in oil viscosity. A procedure which is used to raise the oil temperature is to generate heat at the bottom of the hole by burning gas (Pantaleo and DePriester, 1963). In the oil field it has been found that this downhole gas burner procedure was today uneconomic below depths of the order of 7000 ft. Other promising ideas such as injecting hot fluids to heat the oil in place may be discarded for deep reservoirs, depending on economic conditions, because of the heat losses occuring along the thousands of feet of pipe prior to reaching the target zone.

1.3. Reservoir Engineering

Early in the production history of an oil field, oil flows by itself because of its usually high discovery pressure. Later on, as the reservoir pressure drops, it may become necessary to pump the oil out of the ground. Later on still, little recovery can be achieved merely by pumping. To maintain production, it becomes necessary to drive the oil out of the ground by displacing it with another fluid injected into the reservoir.

In the first two stages, primary recovery is said to occur. In the last stage, secondary recovery or assisted recovery occurs.

The reservoir oil may be produced by its own energy (artesian pressure). External energy, for immiscible displacement, may also be supplied by fluid injection (e.g., injection of water), or by generation of heat at the well bore from an external source (gas burner), or in the formation by combustion of a fraction of the oil in place (in situ combustion), and by injection of heat by means of a convecting fluid (steam flooding).

The reservoir engineer decides, on the basis of the relatively scant information he has, how the field should be produced best. His calculations will ultimately lead to a forecast of the future oil production for the field.

With oil production, less desirable fluids, such as gas and water, also appear at the well head. Predictions of their production are also included on the performance charts of the field because surface facilities to

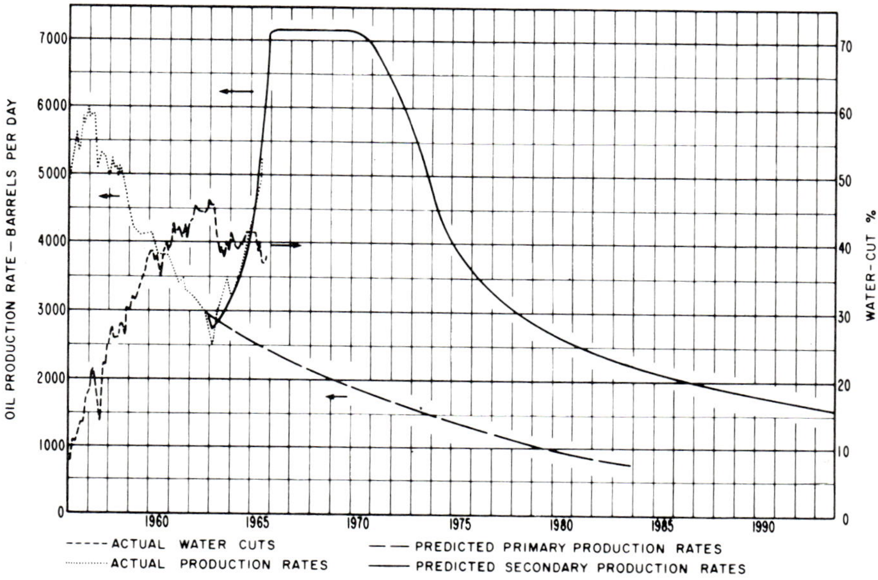

FIG. 1. North Virden Scallion Unit No. 1 production history and prediction. (Park, 1965).

handle them must also be scheduled. Figure 1 is illustrative of the past and predicted performances of an actual oil field.

1.4. Basic Definitions and Concepts

1.4.1. *Reservoir*

The oil- and gas-bearing zones form what is known as the "reservoir." (See Fig. 2.) The reservoir rock is a porous and pervious one (e.g., sand, sandstone, limestone, etc.) The fraction of the reservoir which is filled with gas is known as the *gas cap*. It need not exist at the time of discovery of the field. It may be created during the producing life of the field as gas liberated from solution proceeds updip by buoyancy. The reservoir, exclusive of the gas cap, is referred to as the *oil zone*. The oil zone and the gas cap are usually surrounded by aquifers or aquicludes.

1.4.2. *Darcy's Law for Flow of a Single Fluid*

$$\mathbf{v} = -(k/\mu) \operatorname{grad} p - (k\rho g/\mu) \operatorname{grad} z \tag{1}$$

where $\mathbf{v}$ is the (Darcy) velocity, i.e., volumetric flow-rate vector per unit bulk cross-section area normal to direction of flow, k is the intrinsic

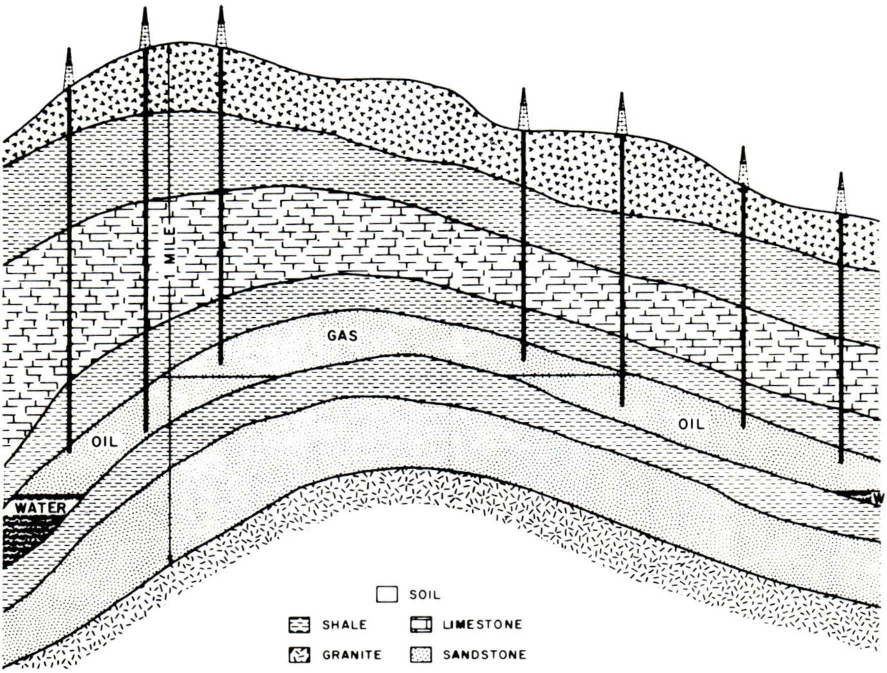

FIG. 2. Schematic cross section of an oil reservoir.

or absolute permeability, μ is the fluid viscosity, p is the pressure, ρ is the specific mass, g is the acceleration of gravity, and z is the vertical coordinate oriented upward. For fluids of constant density, e.g., incompressible liquids, Darcy's law can be written in the more condensed form:

$$\mathbf{v} = -(k/\mu)\,\mathrm{grad}(p + \rho g z) \tag{2}$$

The assumption of incompressibility is made throughout this chapter. The discussion therefore applies more accurately to liquids than to gases. The term $z + (p/\rho g)$ is known as the (hydraulic) *head*. For brevity, we shall call $p/\rho g$, p^*, which has the dimension of a length. For some petroleum applications the pressure gradient is much larger than the gravity term. Then

$$\mathbf{v} = -(k/\mu)\,\mathrm{grad}\,p \tag{3}$$

In a number of studies Darcy's law is written in the form of Eq. (3) even though Eq. (2) is meant. The symbol p in Eq. (3) is defined as the "pressure corrected to a fixed datum" and referred later as "the pressure." The misuse of the term "pressure" to describe a quantity which is not the pressure is of little consequence to the initiated. However, continual use of Eq. (3) may become misleading. In particular, it creates the habit of thinking that fluids always flow in the direction of the pressure drop. By a simple example it can be shown that fluids can flow in the direction of increasing pressure.

An apparatus (Fig. 3) is set up very similar to Darcy's experiment

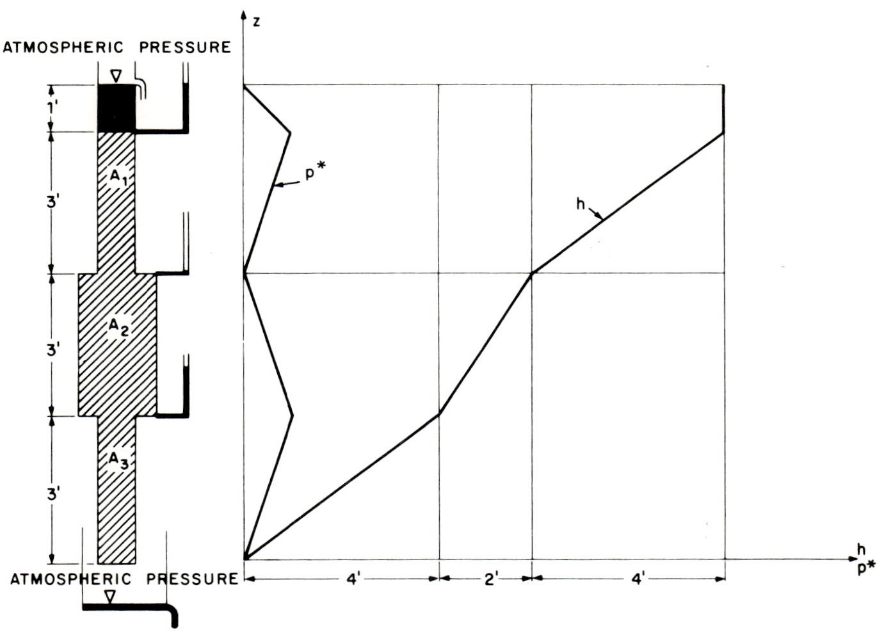

FIG. 3. Flow in the direction of increasing pressure. $A_1 = A_3 = A_2/2$.

except that in the central part, the cross-sectional area is twice what it is in the upper and lower third. The flow rate is a uniform of value Q. Because $A_1 = A_3 = A_2/2$, it follows that the flow velocity in the central part is only half of what it is in the two other sections. Consequently, the drop in head in the central section is only half of its value in the other sections. The head profile is established and, by difference, the pressure profile is established. In the central section, the fluid flows in the direction of increasing pressure. This situation will prevail whenever the slope of the head profile is less than 1. Indeed,

$$dh = dz + dp^* \qquad \text{or} \qquad dh/dz = 1 + dp^*/dz$$

If

$$dh/dz < 1, \qquad dp^*/dz < 0$$

Flow takes place in the direction of decreasing head: dh/dz is >0. A negative value of dp^*/dz indicates flow in the direction of increasing pressure.

1.4.3. *Generalized Darcy's Law*

When two phases flow together in a porous medium, we may be tempted to assumed that each one flows with a velocity proportional to its saturation. For a water–oil mixture, this would give a generalization of Darcy's law:

$$\mathbf{v}_W = -(k/\mu_W)\, S_W \,\mathrm{grad}(p_W + \rho_W gz), \qquad \mathbf{v}_O = -(k/\mu_O)\, S_O \,\mathrm{grad}(p_O + \rho_O gz)$$

which, in the limit $S_W = 1$ or 0, yields the correct velocity for single-fluid flow of either water or oil. However, experimental measurements of relative flow rates of water and oil in the flow of a mixture have not

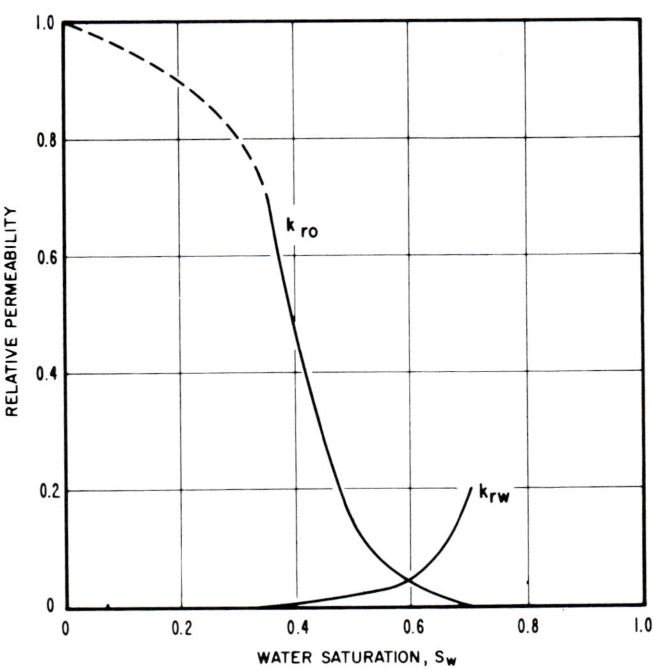

FIG. 4. Typical relative permeability of Kansas–Missouri and Wyoming heavy oil sands (Holm *et al.*, 1965).

sustained a linear relation between flow rate and saturation. Darcy's law was generalized by Muskat (1949) in the form:

$$\begin{aligned} \mathbf{v}_\mathrm{W} &= -k(k_\mathrm{rw}(S_\mathrm{W})/\mu_\mathrm{W})\,\mathrm{grad}(p_\mathrm{W} + \rho_\mathrm{W} gz) \\ \mathbf{v}_\mathrm{O} &= -k(k_\mathrm{ro}(S_\mathrm{W})/\mu_\mathrm{O})\,\mathrm{grad}(p_\mathrm{O} + \rho_\mathrm{O} gz) \end{aligned} \tag{4}$$

Naturally, Eq. (4) is not limited to water and oil. k_rw is known as the relative permeability to water. The product $k \cdot k_\mathrm{rw}$ is usually given the symbol k_w and is called the effective permeability to water (Calhoun, 1953). Figure 4 shows typical curves of relative permeability (Holm *et al.* 1965). The subscripts to pressure are necessary because the pressure in the oil and the water phase need not be the same.

In the equations of water or oil velocity, relative permeabilities are shown to depend explicitly on saturation. It must not be concluded

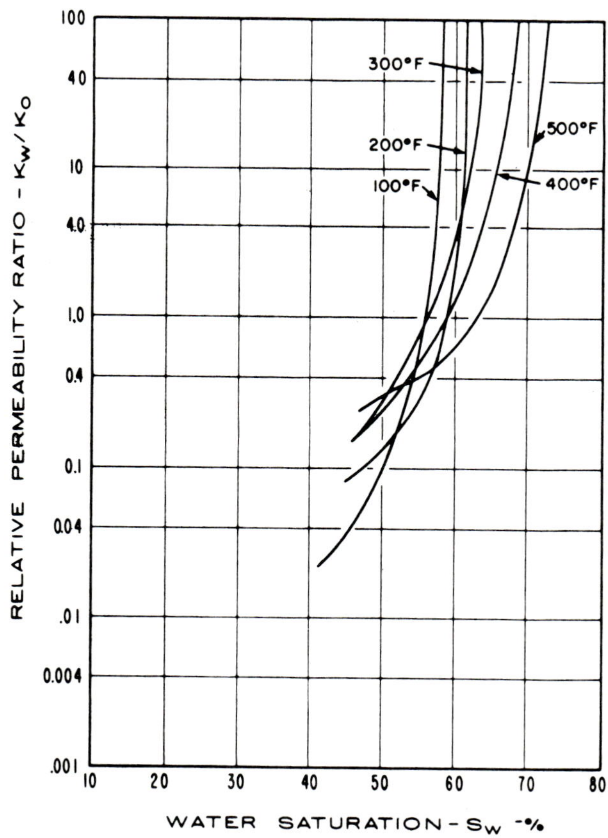

FIG. 5. Relative permeability ratio ($k_w/k_0 - S_w$) at various temperatures for core Nos. 7–15 (Edmondson, 1965).

that relative permeability is a function of saturation only. For that matter, relative permeability depends to some degree on the interfacial tension between the fluids, the preferential wettability of the rock to one or the other fluid, the type of rock, the type of flow, the flow rate, and even the temperature. Only by restricting the applicability of a given relative permeability curve to a kind of rock, a class of fluids, a type of flow, a range of flow rate, and an interval of temperature, can it be said that the relative permeability is a function of saturation only. However, saturation is the decisive variable. Relative permeabilities are used per se or as a ratio, k_{rw}/k_{ro}. The ratio is experimentally easier to measure than the individual permeabilities. Figure 5 shows the influence of temperature on the k_w/k_o curve (Edmondson, 1965). Figure 6 shows the influence of wettability on relative permeabilities (Jennings, 1957), measured by the "steady-state" method (Calhoun, 1953).

The present section introduces some basic definitions and concepts which have traditionally formed the background for studies of two-phase

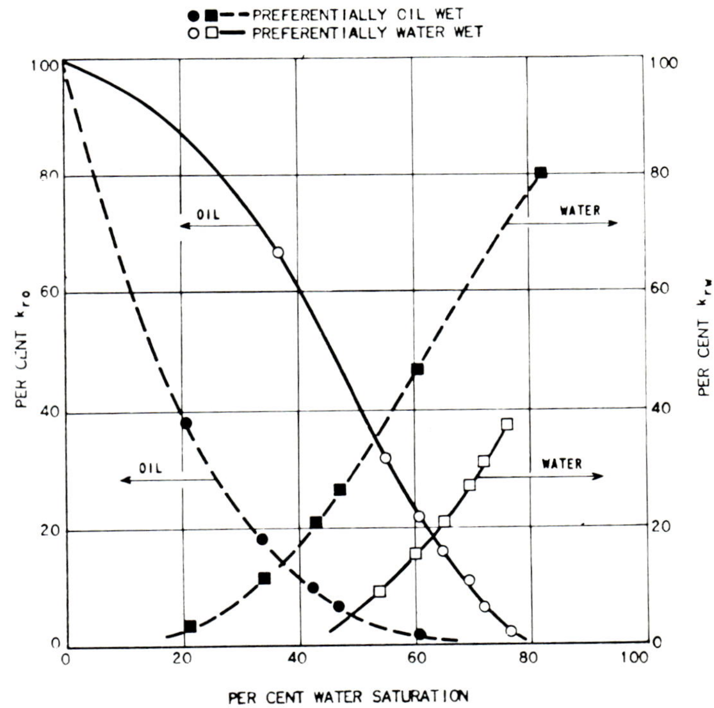

FIG. 6. Water–oil relative permeability by the steady-state method (Jennings, 1957).

flow in a porous medium. The following sections, 2 and 3, aim at providing an insight into the essential characteristics of two-phase flow. Microscopic models of the porous medium and of fluid flow within the porous medium are developed in section 2. In the absence of microscopic observations, the conception of a microscopic model usually follows an analysis of some macroscopic observations. All the conclusions that can be drawn from a particular model are then compared with facts. The model is no longer valid if a significant conflict arises. The nature of the conflict usually provides a clue toward proper medication for the ailing model. Section 2 provides several examples of this process.

Section 3 is the most important section; it focuses on the simplest conceivable two-phase flow that can be realized in an experiment. Complications from effects which can a priori be considered unessential to the physics of two-phase flow are eliminated from the experiment and from the mathematical derivations. The fundamental aspects of two-phase flow, from a macroscopic viewpoint, are thus evidenced in a simple way.

2. Microscopic Models versus Macroscopic Observations

2.1. Introduction

In order to evaluate the economic merit of different ways of producing a reservoir, the engineer must first develop a feeling for what happens in the porous medium. Unfortunately, even a small element of the reservoir is in itself a universe of complexity. Because the detailed description of fluid flow in a porous medium is highly complicated, it appears reasonable to substitute a fictitious but simpler porous medium for the actual one. Fluid flow through the model must on the mean, in some statistical sense, be equivalent to fluid flow in the actual medium. Naturally, conclusions drawn from the model must not conflict to a significant degree with actual observations. The now common operation of injecting water into the reservoir has provided over the past years much useful information on the gross behavior of simultaneous flow of water and oil in a reservoir. Prior to even conceiving a model it seems wise to review briefly the features that are, apparently, common to all waterfloods. Because of its present importance, a brief history of the practice of waterflooding in general and for a particular field is now sketched.

Once a field begins to produce at uneconomical, low rates, it is termed *depleted.* The oil industry quickly realized that huge amounts of oil were left unproduced in the so-called depleted reservoirs. Great interest then developed on how to produce the remaining oil: proven and heretofore unproducible. One possibility that came to mind was to displace the oil by flushing it with another fluid. Since the oil in a depleted field can no longer be produced by its own energy, it must be displaced. In this process, the production of one barrel of oil requires the injection of approximately one barrel of fluid (assuming there is no initial free gas). In view of the fact that annual world oil production is of the order of 10 billion barrels (*Oil and Gas Journal*, 1965), the "fluid" must be readily available in large quantities. It must also be inexpensive relative to the net market value of the recovered oil. Huge aquifers are unfortunately (from a petroleum engineer's point of view) more frequently encountered than large oil pools. Since these aquifers usually surround the oil pool, even in desert areas such as Saudi Arabia, the aquifer will provide the displacing fluid.

Though great emphasis is placed today on newer and more promising methods of secondary recovery, water flooding remains the prime technique for displacing oil. Roughly 80 % of all production from projects classified as secondary recovery is obtained by waterflooding. Waterflooding has been practiced (*API*, 1947) since 1907.

In the past, water flooding was initiated only after the field production had started to decline. The present trend is to start flooding before the pressure drops below the bubble-point pressure. Figure 7 is illustrative of the past and predicted performances of an actual oil field under natural and engineered water drives. It is apparent from this example that more water is injected than oil is produced. As large quantities of water are injected, one starts to witness water in the producer's effluent. In its tortuous path from the injector to the producer, some water no longer displaces oil but instead, displaces water from earlier injection. For every barrel of water injected, roughly (assuming no free gas) a barrel of liquid is produced, of which only a fraction is oil. The water fraction in the liquid increases with time and a point is reached when the volume of oil produced per barrel of water injected becomes uneconomic. Finding ways to produce a field at low water–oil ratio as long as possible is one of the concerns of the reservoir engineer. Because there are many factors that contribute to the efficiency of a waterflood in the field, several recognized factors will be studied. For easier understanding these factors will be covered independently.

With this background information on waterflooding in the field and observations from laboratory experiments, it will be possible to look critically at a model of the porous medium that lends itself to mathematical analysis, namely, the "bundle of tubes" model. The microscopic

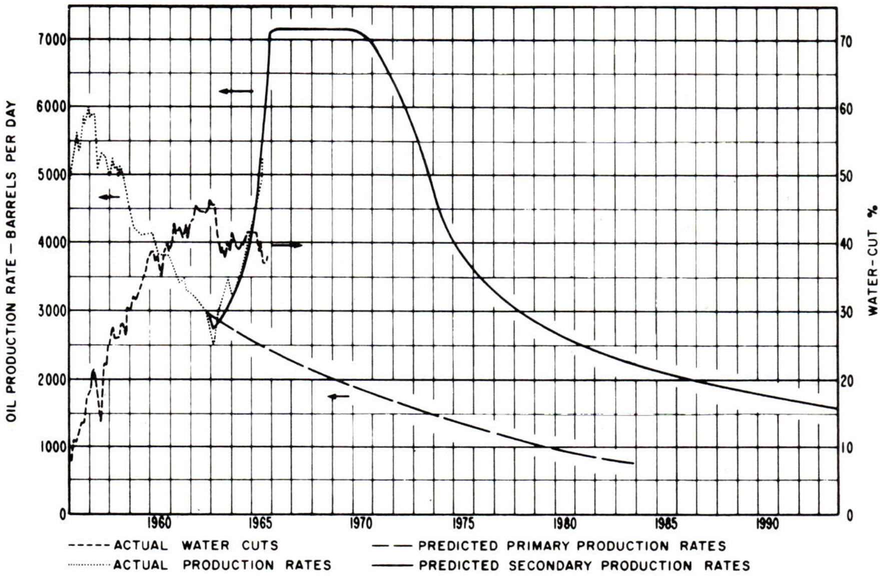

FIG. 7. North Virden Scallion Unit No. 1 production history and prediction (Park, 1965).

analysis will fail in some respects to match macroscopic observations. The introduction of the concept of capillary pressure will help to resolve the conflict. A discussion of the concept of wettability follows. The effect of capillary pressure and wettability on imbibition and oil entrapment is then considered. The final sections derive from the model theoretical relationships between relative permeability, capillary pressure, pore size distribution, and saturation.

2.2. "BUNDLE OF TUBES" MODEL

Laboratory experiments of waterflooding a rock sample that was believed homogeneous have indicated that water appeared in the receiving burette long before a volume of water equal to the volume of the pores of the rock sample was injected; or, as it is customary to say, before one pore volume was injected. As injection continued, an increasing fraction of water was produced, but significant amounts of oil were still recovered after the initial water breakthrough. In many cases, the volume of oil produced after breakthrough exceeded pre-breakthrough recovery.

The question arises: does water displace oil by dragging it along rather than by pushing it forward? Proponents of the piston-like displacement will say "No"; "the medium is not homogeneous." Microscopically, this is a fact which cannot be contested. Visualize the porous medium as a bundle of unconnected cylindric tubes, some small, some large. This highly idealized model of the porous medium is an obviously inaccurate picture of an extremely complex reality. Its merits as a model of reality are almost null. Nevertheless, it provides a convenient and simple introduction to the ulterior development of more complicated but also more applicable models. According to the formula due to Poiseuille (Schlichting, 1955) for the rate of flow under laminar conditions, the velocity with which the water–oil interface propagates (assuming the same viscosity for both fluids) is:

$$\mathbf{u} = \frac{R^2}{8\mu} \cdot \frac{\Delta p}{L} \tag{5}$$

The interface clearly moves faster in the larger tubes than in the smaller ones. The time at which water appears at the outlet face of the model (breakthrough time) varies from tube to tube. The continuing production of oil after the (first) breakthrough of water is not by itself a conclusive evidence that the displacement is not piston-like. As shown in Fig. 8, after breakthrough of water in tube 3, which has the

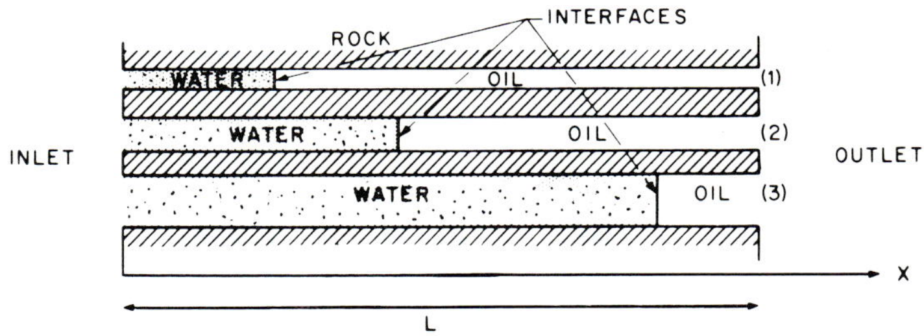

FIG. 8. "Bundle of tubes" model.

largest diameter, oil will still be produced from the smaller tubes. It is a rather simple matter to calculate the velocity with which the interface moves in any tube without assuming that oil and water have the same viscosity.

Let R_i be the radius of tube i and define $k_i = R_i^2/8$.

The velocity with which the interface moves in tube i is:

$$\mathbf{u}_i = \frac{dx_i}{dt} = \frac{k_i \, \Delta p}{L\mu_o - x_i(\mu_o - \mu_w)} \tag{6}$$

when the interface is at the abscissa x_i. For a constant pressure drop this differential equation is easily integrated

$$x_i = \frac{L\mu_o - [L^2\mu_o^2 - 2k_i \Delta p t(\mu_o - \mu_w)]^{1/2}}{\mu_o - \mu_w} \tag{7}$$

In particular, the breakthrough time, which is the value of t when $x_i = L$, is

$$t_b^i = \frac{L^2(\mu_o + \mu_w)}{2k_i \Delta p} \tag{8}$$

The larger the radius, the smaller the breakthrough time. Without pursuing the calculations any further, it is now (temporarily) well established that the bundle of parallel tubes model of the porous medium can account for oil production after breakthrough without abandoning the idea that the displacement is piston-like. It may be worth pointing out however that use of Poiseuille's formula for describing laminar viscous flow in a tube implies a velocity distribution that is parabolic. Thus, if the interface between water and oil is initially plane it will not stay so as time elapses, but will assume the shape of a paraboloid of revolution. For single-fluid flow, this shape has been well substantiated. For two-phase flow, no experimental evidence exists to indicate that even under particular conditions, the interface has such a shape.

However, a number of experiments have been performed for two-phase flow in tubes of very flat rectangular cross section. This type of tube is usually referred to as a parallel plate model, or Hele–Shaw model (Schlichting, 1955). The model may have no bearing on flow in a porous medium because the shape of the cross section of the channels in the rock are intuitively thought to be more or less circular rather than slit-like. The obvious exception to this is the case of a fractured medium.

Figure 9 shows the advance of air into glycerine as experienced in a parallel plate model (Saffman and Taylor, 1958). While nothing can be said of the shape of the interface in a cross section perpendicular to the main direction of flow, it is rather obvious that in a longitudinal cross section the displacement is not of a piston-like nature. A large amount of glycerine is left behind the leading edge of the air finger. And this may well happen on the microscopic scale, that is, within any continuous channel of pores.

This assertion is strictly speculative as it is not supported by experimental observations on a microscopic scale. To the contrary, movies of two-phase flow in a network of scratches on the surface of two glass

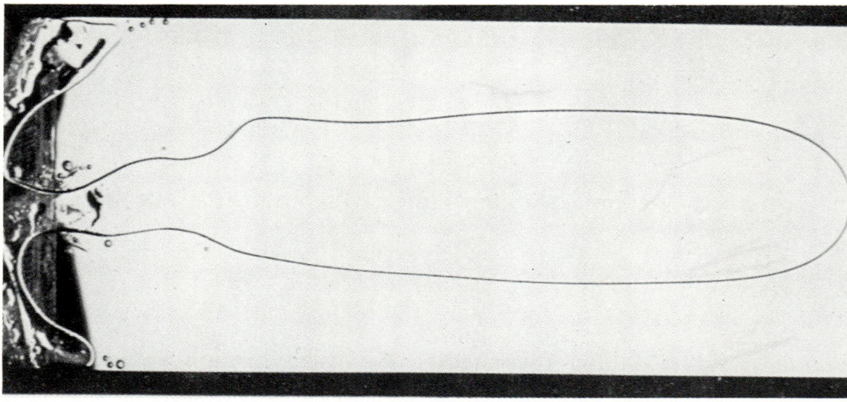

FIG. 9. Penetration of air into glycerine in a Hele–Shaw model (Saffman and Taylor, 1958).

plates in contact show a definite piston-like pattern. However, the scratches in the experiment have a rather rounded cross section shape. Consequently, the available experimental information is not altogether conclusive. Another example of fingering, that of water penetrating oil, is shown in Fig. 10. Behind the leading edge there appears to prevail an

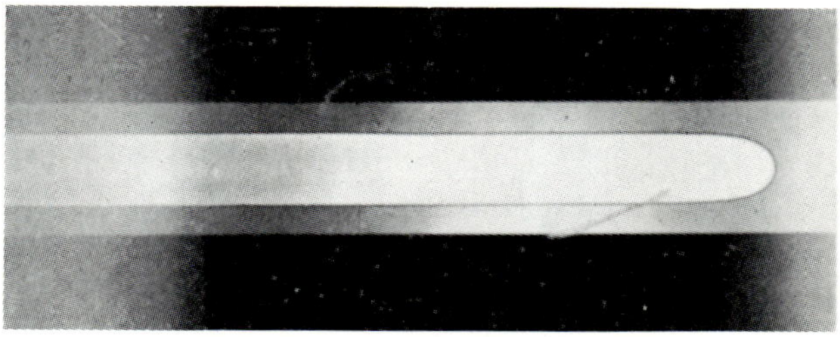

FIG. 10. Penetration of water into oil in a Hele–Shaw model. (Saffman and Taylor, 1958).

almost uniform residual oil saturation. Well-documented experimental results verify that after a laboratory core has been flushed with several pore volumes of water and only water is produced, a significant fraction of the oil remains in the sample. The amount of residual oil is calculated as the difference between the known original quantity of oil in place and that which is produced. The bundle of tubes model cannot predict a residual oil content if piston-like displacement is assumed or even if the interface has the shape of a paraboloid. However, the model can be modified to provide for entrapment of oil. The modified version of the model is shown in Fig. 11. Oil remains in the smaller capillaries but has been completely displaced from the larger ones, because flow

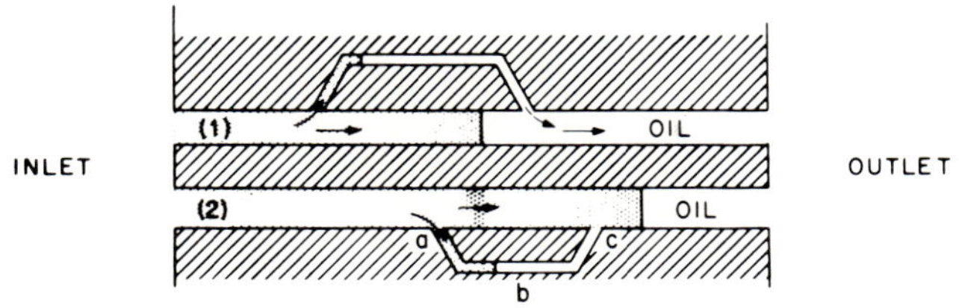

FIG. 11. "Doublet" model.

velocities are lower in the smaller capillaries. It remains to elucidate why the water and oil in the small capillary *abc* (Fig. 11) would stop moving just because the fluid in tube 2 at *c* is now water instead of oil. Prior to answering this point, we shall review the concept of capillary pressure between two immiscible fluids.

2.3. CAPILLARY PRESSURE

In the previous sections it was tacitly assumed that, given an external pressure gradient across the porous medium or the bundle of tubes, the flow was strictly conditioned by the geometric characteristics of the rock, or of the tubes, and the viscosity of the fluids. It did not seem material to consider the nature of the fluids or of the rock. No reference was made as to whether the displacing and the displaced fluids were miscible or immiscible. If the two fluids retain a distinct phase identity they are immiscible. In this case, it is possible to observe an interface at the boundary of the two phases. As a consequence of the different degrees of attraction between molecules of a different nature, such as water and oil, a tension exists at the interface, the so-called surface tension.

Let us consider the equilibrium of a spherical globule of oil in water. The whole sphere being in equilibrium, every part of it is in equilibrium. If we fictitiously partition the oil globule by a plane P (Fig. 12), the action equivalent to the effect of the lower part of the globule on the upper part is surface tension along L, the trace of the interface on P. Other forces acting on the cap are the oil pressure on the plane surface A and water pressure on the curved surface Σ. Components of these forces on the axis are $-L\sigma \sin \theta$ for the surface tension forces and $(p_o - p_w)A$

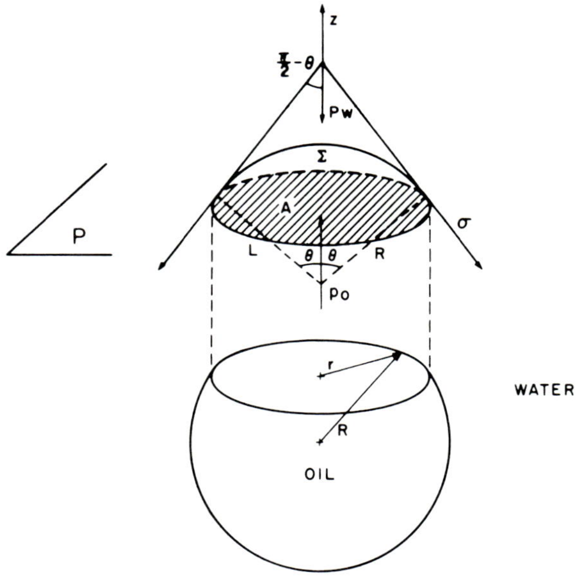

FIG. 12. Equilibrium of forces at an interface between immiscible fluids.

for the pressure forces. At equilibrium, the sum of these forces is zero from which it follows that:

$$p_\mathrm{o} - p_\mathrm{w} = \frac{L}{A}\,\sigma \sin\theta = \frac{2\pi r}{\pi r^2}\,\sigma \sin\theta = \frac{2\sigma}{r}\sin\theta = \frac{2\sigma}{R}$$

We have thus found that there exists a pressure difference across the interface of two immiscible fluids. The change in pressure is proportional to the surface tension and to the curvature. By definition, the capillary pressure is the difference between the pressure in the fluid that lies on the concave side of the interface and the pressure in the other fluid. For the case of an oil forming droplets in water:

$$p_\mathrm{c} = p_\mathrm{o} - p_\mathrm{w} = 2\sigma/R \tag{9}$$

A more general expression for capillary pressure can be obtained without imposing the spherical shape restriction. Consider a small element of interface *ABCD* (Fig. 13), more or less rectangular in shape, centered around a point *M*, and whose normal is *MN*. Let *EF* be the intersection of the interface element with a normal plane to the interface at *M*. Let *GH* be the intersection of the interface by another normal plane orthogonal to the preceding so that *EF* and *GH* are orthogonal at *M*. Then

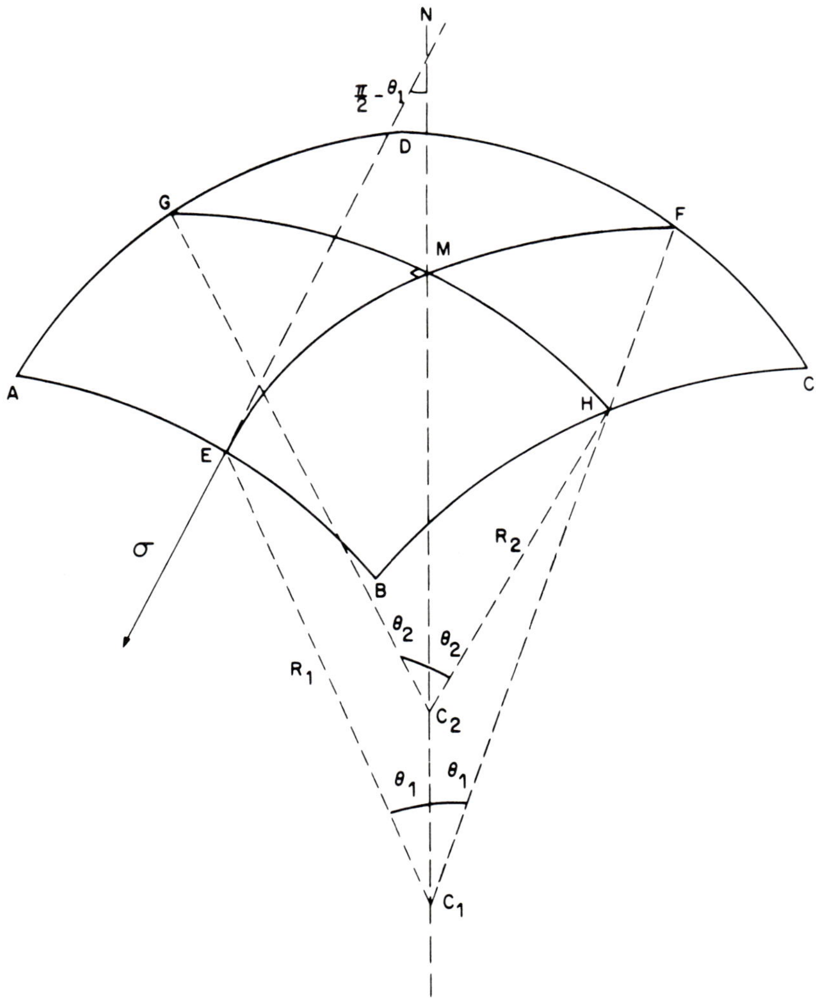

FIG. 13. Relation between capillary pressure and interface curvature.

a consideration of the equilibrium of the components of surface tension and pressure forces on the MN axis yields:

$$2\sigma EF \sin \theta_2 + 2\sigma GH \sin \theta_1 = p_c EF \cdot GH$$

As

$$EF = 2R_1\theta_1 , \qquad GH = 2R_2\theta_2$$

then

$$p_c = \sigma \left(\frac{\sin \theta_2}{R_2\theta_2} + \frac{\sin \theta_1}{R_1\theta_1} \right)$$

and in the limit for θ_1 and θ_2 tending to zero, we obtain the capillary pressure across the interface at M:

$$p_c = \sigma\left(\frac{1}{R_1} + \frac{1}{R_2}\right) \tag{10}$$

This result is alternately known as Laplace's or Plateau's equation. It is often said in the literature when Laplace's equation is introduced that R_1 and R_2 are the principal radii of curvatures of the interface. These requirements on R_1 and R_2 are an unnecessary restriction on the generality of Laplace's equation. Nowhere in the above derivations was it necessary to restrict the normal planes to be principal planes, provided they are orthogonal. Actually, the quantity $1/R_1 + 1/R_2$ is known in geometry as the mean curvature and can be shown to be an invariant characteristic of the surface at a point.

Consider a surface, a plane that is tangent to this surface at a regular point M, and the oriented normal MN to the surface. Let C be the center of curvature on MN for the intersection of the surface S and a normal plane NMT_1 (see Fig. 14). Let R_1 be the radius of curvature of

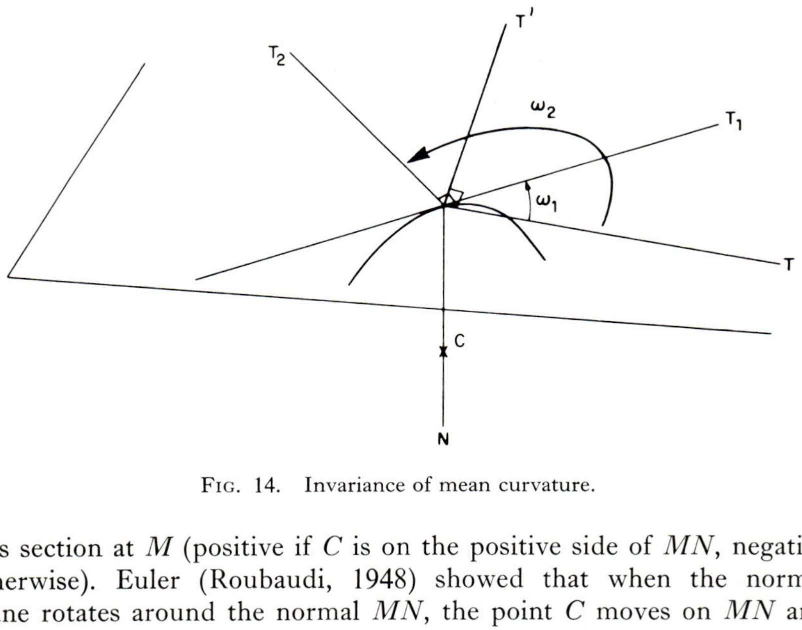

FIG. 14. Invariance of mean curvature.

this section at M (positive if C is on the positive side of MN, negative otherwise). Euler (Roubaudi, 1948) showed that when the normal plane rotates around the normal MN, the point C moves on MN and there are two normal planes MNT and MNT' for which R is a minimum or a maximum (call this minimum and maximum R and R', respectively). These two planes are called principal planes and they are orthogonal.

Euler also showed (Roubaudi, 1948) that the radius of curvature for the curve lying in the plane NMT_1 is R_1 such that

$$\frac{1}{R_1} = \frac{1}{R}\cos^2\omega_1 + \frac{1}{R'}\sin^2\omega_1$$

Consider another plane NMT_2 orthogonal to NMT_1 ; then

$$\frac{1}{R_2} = \frac{\cos^2\omega_2}{R} + \frac{\sin^2\omega_2}{R'}, \qquad \omega_2 = \omega_1 + \frac{\pi}{2}$$

Therefore,

$$\frac{1}{R_2} = \frac{\sin^2\omega_1}{R} + \frac{\cos^2\omega_1}{R'}$$

Consequently,

$$\frac{1}{R_1} + \frac{1}{R_2} = \frac{1}{R} + \frac{1}{R'} = \text{constant}$$

(independent of ω) which proves the invariance of $1/R_1 + 1/R_2$.

In summary, at the interface of two immiscible fluids there exists a pressure discontinuity known as the capillary pressure. It is proportional to the surface tension between the two fluids, and to the mean curvature

$$p_c = \sigma/R_m \tag{11}$$

where

$$\frac{1}{R_m} = \frac{1}{R_1} + \frac{1}{R_2} \tag{12}$$

and R_1 and R_2 are the radii of curvature of two orthogonal, but otherwise arbitrary, normal cross sections of the interface at a given point. In spite of its misleading name, the definition of capillary pressure in no way assumes the presence of a solid wall, as introduced by a capillary. It does assume the presence of two immiscible fluids. However, the presence of a solid wall may have a bearing on the shape of the interface and consequently, a bearing on the magnitude of the capillary pressure. We shall first introduce the concept of wettability for the ulterior purpose of evaluating the effect of the solid wall.

2.4. Wettability

It can be readily verified that a drop of liquid mercury in contact with a glass plate (and air) will not spread over the glass surface, but instead will retain a compact shape. On the other hand, a water drop will tend to stretch into a thin film. Between these two extremes, a variety of shapes can exist as shown on Fig. 15. In case (a) ($\theta > 90°$),

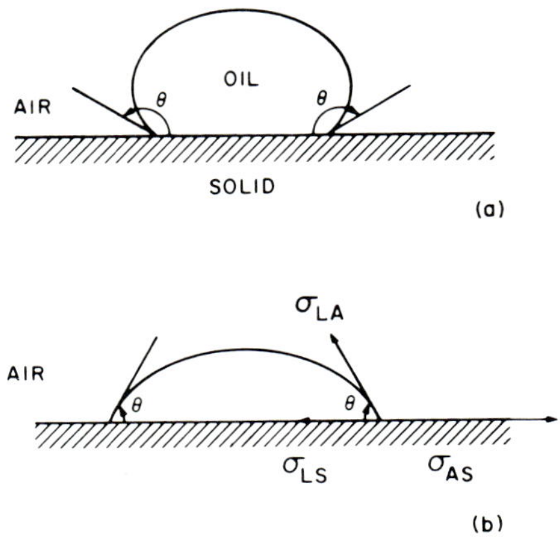

FIG. 15. Wettability.

air is said to preferentially wet the wall, whereas in case (b) ($\theta < 90°$), the liquid preferentially wets the solid.

Just as there exists a surface tension between immiscible fluids, there exists a surface tension between a fluid and a solid. Thus, σ_{AS} refers to surface tension between air and solid, σ_{LS} refers to surface tension between liquid and solid,..., etc. Expressing that there cannot exist a nonzero resultant force at the contact of three phases in equilibrium yields a relationship between the three surface tensions

$$\sigma_{AS} = \sigma_{LS} + \sigma_{LA} \cos \theta \tag{13}$$

known as Young's (or sometimes Dupre's) formula. The above relationship supposes that the three phases can coexist in equilibrium; that is,

$$\left| \frac{\sigma_{AS} - \sigma_{LS}}{\sigma_{LA}} \right| = | \cos \theta | \leqslant 1$$

So far our study of capillary phenomena has been from a static point of view. It is interesting to illustrate through the following example that preferential wettability may cause motion without an external pressure gradient. The motion of a fluid into a capillary, thereby causing displacement of the prior occupant, is called imbitition with respect to the displacing fluid.

2.5. Free Capillary Imbibition

Consider a preferentially water-wet tube of small diameter initially filled with air. The tube is brought in contact with an ample supply of water at atmospheric pressure (Fig. 16). The initial water–air interface

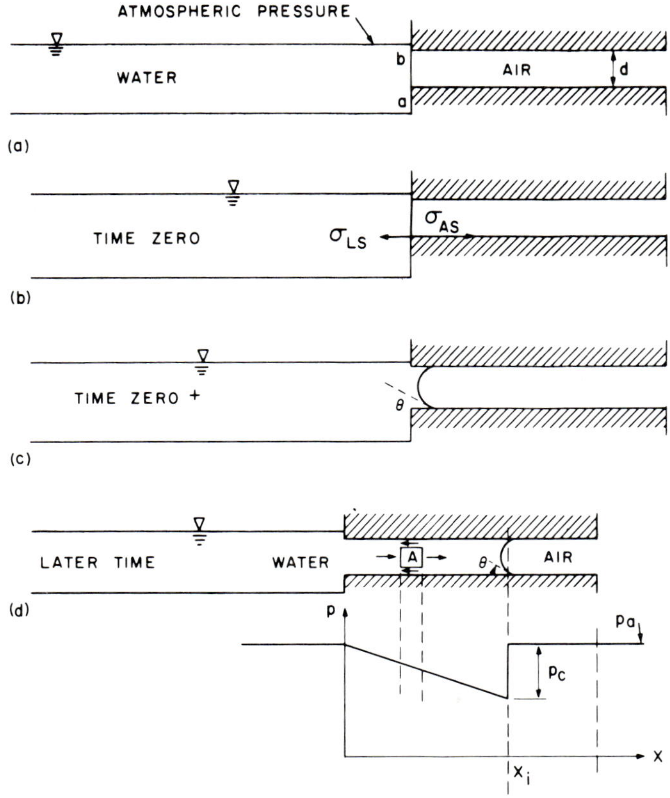

Fig. 16. Free imbibition in a capillary.

may be considered to be the plane *ab*. For an infinitesimal amount of time the force diagram will be as shown on Fig. 16(b). The interface will be in equilibrium except at the contact with the solid. There, a net impulse of magnitude proportional to $\sigma_{AS} - \sigma_{LS}$ causes the point *a* of the interface to move into the tube. Because the other points of the interface are in equilibrium, the interface curves in and it very quickly assumes the shape shown in Fig. 16(c), where θ is the equilibrium wetting angle. The interface is now in equilibrium because the surface tension force and the capillary pressure force balance one another. Because the pressure drop in the moving air phase is negligible compared to that in water, the pressure in the air just ahead of the interface is essentially atmospheric. Just behind the interface, the pressure in the water is less than atmospheric by an amount equal to the capillary pressure. At the inlet, the pressure is atmospheric. As a result there exists an unbalanced pressure gradient in the water phase which causes the water to flow and which sustains the motion initially created by unbalanced surface tension forces. Viscous shear (particle *A*, Fig. 16) will resist the driving force tending to enforce a uniform flow. Actually, flow is constantly retarded because the pressure gradient in the water phase, p_c/x_i, itself decreases with time.

In summary, motion is originated by preferential wettability and later sustained by the induced pressure gradient in the wetting phase.

2.6. Entrapment of Oil

We can now reconsider the question posed at the end of Section 2.2 of entrapment of oil (or air) in tubes when displaced by a wetting phase (water) under free imbibition, or under an external pressure gradient. First consider an experiment of the same nature as that just described. Assume that atmospheric pressure prevails at both ends of the tubes (Fig. 17). Assuming that the interface has the shape of a spherical cap, the capillary pressure at the interface in tube i is

$$p_c^i = (2\sigma/R_i)\cos\theta$$

Assuming again a negligible pressure drop in the gas phase, the pressure gradient in the water phase is

$$(2\sigma/R_i)\cdot(\cos\theta/x_i)$$

where x_i is the location of the water–air interface in tube i. Application of Poiseuille's equation yields the velocity of the interface:

$$\frac{dx_i}{dt} = \frac{R_i\sigma\cos\theta}{4\mu_w x_i} \qquad \text{and} \qquad x_i^2 = \frac{R_i\sigma t\cos\theta}{2\mu_w}$$

Under free imbibition, water flows more rapidly in the larger tube. Thus, water will reach point *a* through tube (2) first. Immediately after the water has passed section *a* in tube (2), water imbibes tube (1) due to the rock's preferential wettability to water. An interface starts to propagate into tube (1). It will come to a stop at an equilibrium position somewhere within the constriction between tube (1) and (2) where the radius is R_a such as to satisfy:

$$p_c^a = (2\sigma\cos\theta)/R_a$$

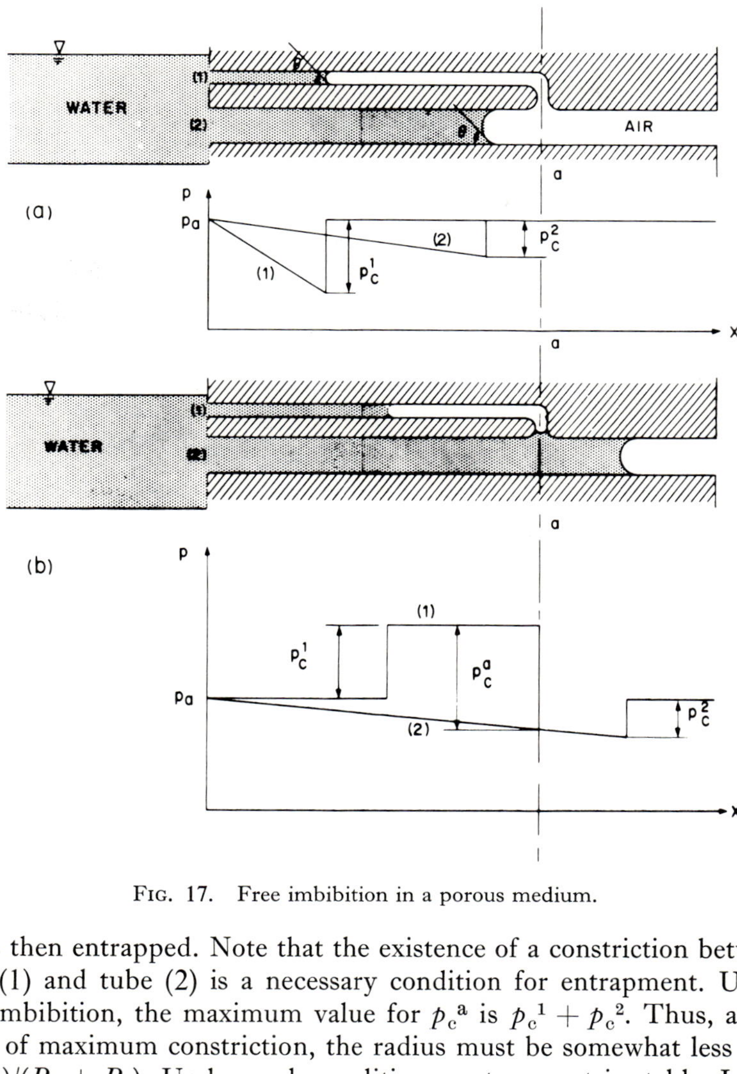

FIG. 17. Free imbibition in a porous medium.

Air is then entrapped. Note that the existence of a constriction between tube (1) and tube (2) is a necessary condition for entrapment. Under free imbibition, the maximum value for p_c^a is $p_c^1 + p_c^2$. Thus, at the point of maximum constriction, the radius must be somewhat less than $(R_1R_2)/(R_1 + R_2)$. Under such conditions, entrapment is stable. In the event of an external pressure gradient such as obtained by a column of water (Fig. 18), the constriction must be more severe to constrain air within the smaller tube. The limiting value for R_a is the one that gives

$$p_c^a = p_c^1 + p_c^2 + \Delta p$$

Thus,

$$R_a < \left(\frac{1}{R_1} + \frac{1}{R_2} + \frac{\Delta p}{2\sigma \cos \theta}\right)^{-1} \tag{14}$$

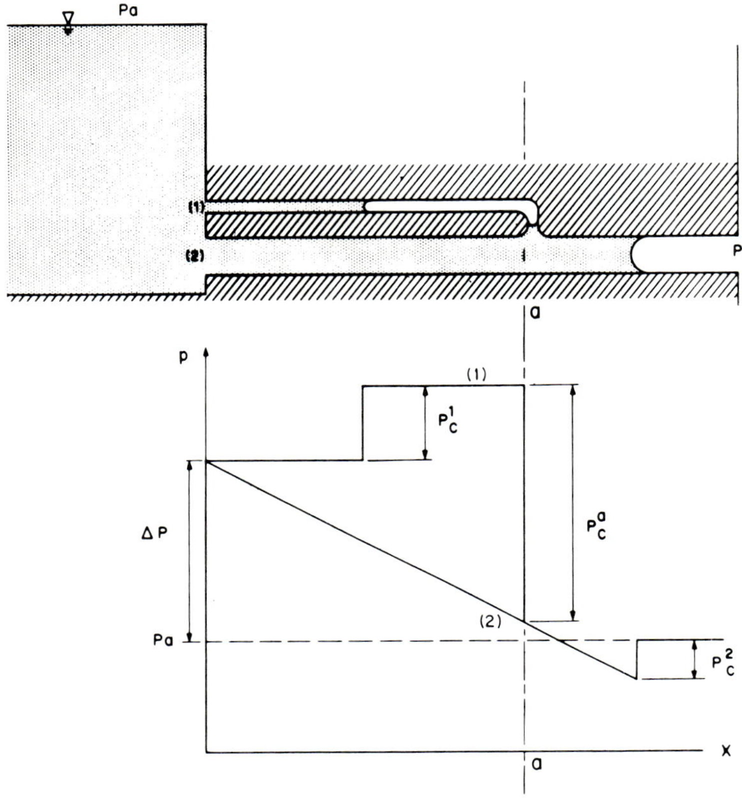

FIG. 18. Forced imbibition.

The larger the imposed pressure gradient, the less chance there is to entrap air. In the limit for an infinite pressure gradient, there would be no entrapment at all. It follows that the degree of air entrapment (or the residual air saturation, or the displacement efficiency) depends not only on the geometric characteristics of the medium, but also on the rate at which the displacement proceeds.

2.7. Capillary Pressure in a Porous Medium

Because the porous medium is composed of many interconnected capillaries of various sizes, one cannot possibly hope to describe the time evolution of water–air or water–oil interfaces in every capillary. One cannot talk of a capillary pressure, but rather of capillary pressure in this or that tube. However, under equilibrium conditions the pressure is uniform in both the water and the oil phase. It is then possible to talk about capillary pressure without ambiguity. Suppose a water-wet rock filled with water is placed in contact with oil. Pressure in the oil phase is increased and maintained constant. A certain volume of oil penetrates the core and expels a corresponding amount of water, which is measured.

The oil is retained in the porous medium by a semipermeable membrane which transmits the displaced water but not the oil. Ultimately a no-flow condition is established. Due to expulsion of a certain volume of water, the remaining water fills up only a fraction of the total pore space. The fraction of the total pore volume occupied by water is by definition called the water saturation. Pressure in the oil is increased again. When equilibrium is reached, a new and lower equilibrium water saturation prevails in the core. Ultimately, repeating the operation successively (and patiently), a curve of capillary pressure versus water saturation is obtained (Handy, 1960) (Fig. 19). Because water was

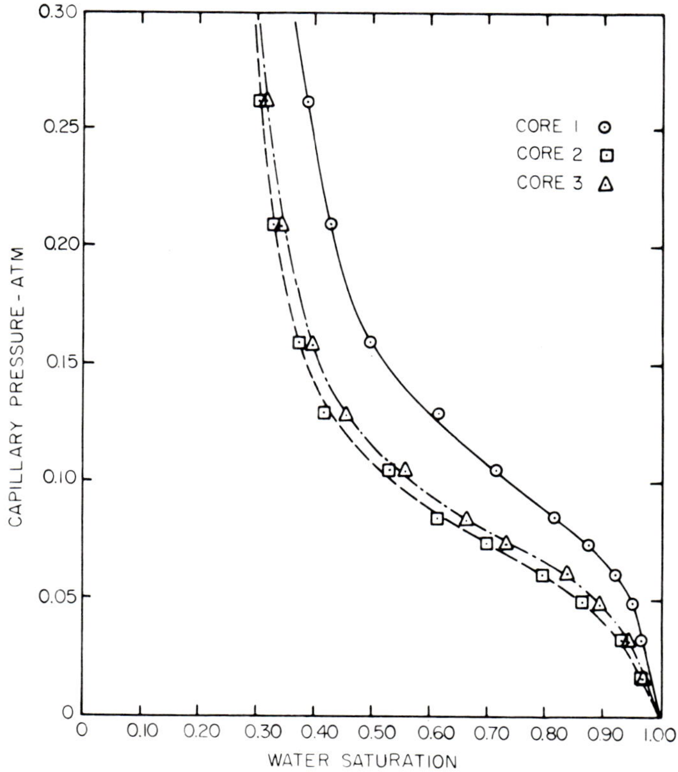

FIG. 19. Static capillary pressure as a function of saturation for sandstone cores (air–water system). (Handy, 1960.)

drained from the core in the procedure, the curve so obtained is labeled drainage curve. The term drainage is applied whenever the water saturation decreases with time. It is found experimentally that a point is reached when even a tremendous increase in capillary pressure no longer induces a saturation change. The water saturation is said to have reached its irreducible, or residual, value. The concept of irreducibility is relative to the process. For example, the residual water saturation could be reduced further by placing the core in an oven at an elevated temperature.

Once residual water saturation has been established in the core by displacement by oil, the imbibition curve can be obtained by letting the pressure drop stepwise and the water imbibe back. Imbibition is the term used if the water saturation increases with time. A different curve is obtained which implies that, given a water saturation, several equilibrium states are possible depending on previous history. The capillary pressure curve is said to display hysteresis. By the time the capillary pressure has returned to the zero value, all the oil has not been expelled. A fraction of it has been trapped: the residual oil saturation. If the cycle is now repeated the same curves are described.

2.8. Pore Size Distribution

Capillary pressure increases with decreasing water saturation. According to Laplace's equation, it is proportional to the mean curvature of the interface. Therefore, as the water saturation decreases, the water retreats into the smaller pores. These correlations make one expect that some information with respect to pore size distribution can be deduced from the capillary pressure curve.

In the following derivations, the model for the porous medium is again that of straight capillaries with circular cross sections.

Let $\delta(R)\,dR$ be the fractional number of pores with a radius between R and $R + dR$. The function $\delta(R)$ is unknown but subject to the normalization constraint that

$$\int_0^\infty \delta(R)\,dR = 1 \tag{15}$$

Let A be the cross-sectional area of the core, ϕ the porosity, and N the total number of pores. The number of pores of a radius of cross section between R and $R + dR$ is $N\,\delta(R)\,dR$. The elementary area of all these cross sections is $N\,\delta(R)\,dR \times R^2$. The total void area is the sum of the elementary area over all values of the radii.

It follows that

$$A\phi = N \int_0^\infty \pi R^2 \, \delta(R) \, dR = N\pi \bar{R}^2$$

where $\bar{R}$ is an average radius defined by the above equation.

The pore volume of the core of length L is

$$V_p = A\phi L = LN \int_0^\infty \pi R^2 \, \delta(R) \, dR \tag{16}$$

The wetting phase at equilibrium occupies all the pores of a radius less than R corresponding to

$$p_c(S) = (2\sigma \cos \theta)/R \tag{17}$$

Its saturation is

$$S = \frac{\int_0^R \pi R^2 \, \delta(R) \, dR}{\int_0^\infty \pi R^2 \, \delta(R) \, dR} \tag{18}$$

Elimination of S between Eqs. (17) and (18) yields

$$\delta(R) = \frac{2\sigma \cos \theta \bar{R}^2}{R^4} \left(\frac{dp_c}{dS} \right)^{-1} \tag{19}$$

dp_c/dS being evaluated at

$$p_c = (2\sigma \cos \theta)/R$$

The unknown value of $\bar{R}$ is obtained by the normalization condition on $\delta(R)$.

2.9. Theoretical Derivation of Relative Permeabilities

Since it turns out to be possible to derive at least a qualitative estimate of pore size distribution from experimental measurements, one can hope to derive an expression for permeability based on the model of a bundle of capillaries.

However, the same restriction applies which also existed for capillary pressure. We have seen earlier that the displacement efficiency of a waterflood would not only depend upon the geometric characteristics of the medium, the fluid and the rock properties, but also on the rate at which the displacement proceeds. The following developments are thus restricted to a steady-state condition of saturation. Water and oil are both flowing and are in capillary equilibrium. Thus water, the wetting phase, flows in the smaller capillaries and oil flows in the larger ones. The expression for saturation was given previously [Eq. (18)]. The permeability in a pipe which is not straight can be approximated as

$$k = k(R) = R^2/[8\tau(R)]$$

because the length of a path is L, the length of a core, times a "tortuosity" factor τ which varies with pore size. The expression for k is obtained by identification of the ratio $(v\mu L)/\Delta p$ as given by Darcy's law and by Poiseuille's formula. For all tubes of radius between R and $R + dR$, the flow rate is

$$dQ = \pi R^2 \cdot \frac{R^2 N\,\delta(R)\,dR}{8\tau(R)\,\mu} \cdot \frac{\Delta p}{L}$$

For all tubes of radius $<R$, the total flow rate is

$$Q = \left(\int_0^R \frac{\pi R^4 N\,\delta(R)\,dR}{8\mu\tau(R)}\right) \cdot \frac{\Delta p}{L}$$

The permeability to the wetting phase is

$$k_{\mathrm{w}} = \frac{Q_{\mathrm{w}} L \mu_{\mathrm{w}}}{A\,\Delta p} = \frac{N}{A}\int_0^R \frac{\pi R^4}{8}\,\frac{\delta(R)\,dR}{\tau(R)} \tag{20}$$

The volume of pores made up by capillaries with a radius in the interval R and $R + dR$ is

$$dV_{\mathrm{p}} = L\pi R^2 N\,\delta(R)\,dR = \phi LA\,dS \tag{21}$$

from which it follows that

$$\frac{\pi R^2 N\,\delta(R)\,dR}{A\phi} = dS \qquad \text{and} \qquad k_{\mathrm{w}} = \phi \int_0^R \frac{R^2}{8\tau(R)}\,dS \tag{22}$$

The upper limit of the integral is the largest pore size filled with water, that is

$$R = \frac{2\sigma\cos\theta}{p_{\mathrm{c}}(S)}$$

Let us approximate $\tau(R)$ as a/R^{α} because the smaller the capillary, the larger the tortuosity. Then

$$k_{\mathrm{w}} = \frac{(2\sigma\cos\theta)^{2+\alpha}}{8a}\,\phi \int_0^S \frac{dS}{[p_{\mathrm{c}}(S)]^{2+\alpha}} \tag{23}$$

The single-phase permeability being

$$k = \frac{(2\sigma \cos \theta)^{2+\alpha}}{8a} \phi \int_0^1 \frac{dS}{[p_c(S)]^{2+\alpha}} \tag{24}$$

it follows that the relative permeability to water is

$$k_{rw} = \frac{k_w}{k} = \frac{\int_0^S [dS/(p_c)^{2+\alpha}]}{\int_0^1 [dS/(p_c)^{2+\alpha}]} \tag{25}$$

It has been found that fairly good agreement with experimental data (Fatt and Dysktra, 1951) could be obtained with $\alpha = 1$. It must be emphasized that the relative permeability curves obtained experimentally were obtained under steady-state conditions. Under these conditions, it is possible to express the relative permeability to water as a function of saturation. It cannot be expected that a single relative permeability versus saturation curve will be obtained which is independent of the displacement process. Rather, several are obtained.

Relative permeability is a function of several variables among which is saturation. The functional relationship between relative permeability and the other variables has not been established satisfactorily on a quantitative basis. As a consequence, it has been customary to measure relative permeability as a function of saturation only and to indicate under which conditions the experiment was conducted (steady-state, imbibition, drainage, etc.). The results can be extrapolated only to rocks, fluids, and displacement processes that are slight variations of the laboratory conditions. With this understanding, relative permeabilities will henceforth be considered as functions of saturation only.

2.10. Measured Relative Permeabilities

Relative permeability curves have been shown in Figs. 4 and 6, and relative permeability ratios in Fig. 5. The results vary sufficiently to defeat attempts directed at obtaining a general correlation. Key parameters in the description of relative permeability to water and oil, for example, are: connate water saturation, S_{wc}, residual oil saturation, S_{or}; permeability to water at residual oil saturation, $\tilde{k}_{rw}$, and permeability to oil at connate water saturation, $\tilde{k}_{ro}$. Defining the normalized water saturation, S_w^*, as

$$S_w^* = \frac{S_w - S_{wc}}{1 - S_{or} - S_{wc}} \tag{26}$$

the relative permeabilities can be expressed as

$$k_{rw} = \tilde{k}_{rw}\kappa_w(S_w^*) \tag{27}$$

and

$$k_{ro} = \tilde{k}_{ro}\kappa_0(S_w^*) \tag{28}$$

where κ_w and κ_0 are dimensionless "shape" function of magnitude varying from 0 to 1. Based on a variety of experimental data for unconsolidated water-wet sands, a good approximation to an average shape function is:

$$\kappa_w = S_w^{*3}, \qquad \kappa_0 = (1 - S_w^*)^2 \tag{29}$$

Other types of correlation and further references can be found in Wyllie (1962).

Almost all developments in this section were of qualitative value at best. By now the engineer may have developed a feeling for two-phase flow in a porous medium. It remains to set the study of two-phase flow on a quantitative basis. This is the purpose of Section 3.

3. Two-Phase Flow in a Strictly Linear Medium

In previous sections, the subject of two-phase flow in a porous medium was defined. A traditional background of concepts was introduced. Some of the qualitative features of two-phase flow were described and to some degree explained in Section 2. The purpose of this section is to set the study of two-phase flow on a quantitative basis for a particularly simple case.

3.1. Posing the Problem

An understanding of the basic nature of two-phase flow is best obtained in the formulation and solution of the simplest conceivable flow that can actually be realized in an experiment. It is a simple task to grind a small cylindric core from a homogeneous rock sample. The core is fully saturated with oil and inserted into an impervious sleeve that prevents side flow. Water is injected at one end, the inlet face. Fluids are recovered at the outlet face in an appropriate receptacle. Injection rate is maintained constant throughout the experiment. Produced volumes of oil and water are measured at intervals while the clock is running. Ultimately, all the efforts materialize into a curve of oil recovery versus pore volume injected. The oil recovery is given in percent of original volume of oil in place, or displacement efficiency E_D. The state of the system can be described by a single variable: the number of pore volumes injected. The number of pore volumes injected, V_{pi}, at a time t since the beginning of injection is:

$$V_{pi} = Qt/V_p \tag{30}$$

where V_p is the pore volume of the core. The pore volume of the core of porosity ϕ, cross section A, and length L is: $V_p = \phi AL$.

Many experiments of waterflooding in cores have been performed. The type of curve experimentally observed (Jennings, 1966) is shown on Fig. 20 for various oil to water viscosity ratios.

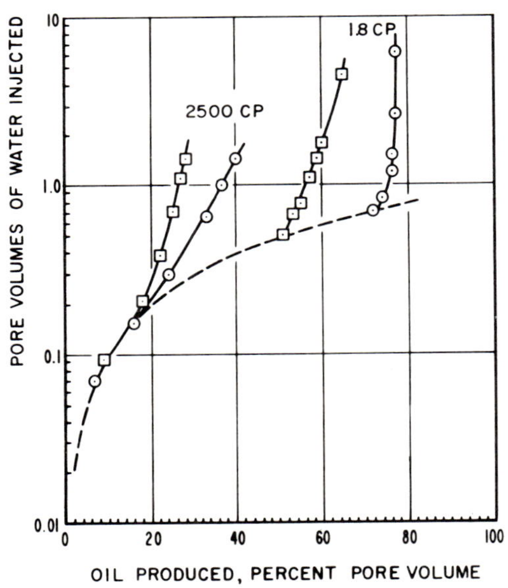

FIG. 20. Effect of wettability and viscosity on waterflood tests with alundum ⊙ water wet; ⊡ oil wet (Jennings, 1966).

In the next sections, the equations for two-phase flow are derived. The solution of the equations should match the experimental results.

3.2. Mathematical Formulation of Conservation Laws

First let us state the assumptions for the simplest case considered. The porous medium is homogeneous and isotropic. Its porosity and permeability are constants also with respect to time. The sample has a cylindric shape of uniform cross section; the liquids are incompressible; the direction of flow is horizontal; gravity effects are negligible; and, as a consequence, the saturation is uniform in any cross section. Capillary pressure is negligible so that no distinction is made between oil and water pressure. Then Darcy's law can be written for each phase:

$$v_w = \frac{Q_w}{A} = -k \frac{k_{rw}}{\mu_w} (S_w) \frac{\partial p}{\partial x} \tag{31}$$

$$v_o = \frac{Q_o}{A} = -k \frac{k_{ro}}{\mu_o} (S_w) \frac{\partial p}{\partial x} \tag{32}$$

The law of mass conservation of one phase applied to a fixed volume which contains two phases can be expressed in the following manner. If more of one phase moves in than moves out, it must of necessity accumulate within. The saturation of that phase increases. The rate of saturation increase is proportional to the rate of net influx. Given a net influx, the rate of saturation increase is inversely proportional to the pore volume. Proceeding to mathematical symbols, let us choose a volume of sample bounded by the sections at x and $x + \Delta x$. Its volume is $A\,\Delta x$ and its liquid-filled fraction is $\phi A\,\Delta x$. The volumetric water flow rate entering the volume through section x is Q_w, and that leaving through section $x + \Delta x$ is $Q_w + \Delta Q_w$. The net mass influx into the volume is $\rho_w\{Q_w - (Q_w + \Delta Q_w)\}$. It results in an equal rate of mass accumulation. Dividing by the maximum mass of water that can fill the volume yields the rate of saturation increase:

$$\rho_w \frac{\{Q_w - (Q_w + \Delta Q_w)\}}{\rho_w \phi A\,\Delta x} = \frac{\Delta S_w}{\Delta t} \qquad \text{or} \qquad \phi A \frac{\Delta S_w}{\Delta t} + \frac{\Delta Q_w}{\Delta x} = 0$$

The law of mass conservation applies regardless of volume size or time interval. It applies in particular for infinitesimal volume and time size. Proceeding to the limit $\Delta x \to 0$, $\Delta t \to 0$, one obtains:

$$\phi A \frac{\partial S_w}{\partial t} + \frac{\partial Q_w}{\partial x} = 0 \tag{33}$$

A similar equation can be derived for oil:

$$\phi A \frac{\partial S_o}{\partial t} + \frac{\partial Q_o}{\partial x} = 0 \tag{34}$$

Having assumed the presence of only two phases, what is not water in a pore must be oil and vice versa, so that symbolically the oil and water saturations are related by the condition

$$S_o + S_w = 1 \tag{35}$$

Adding Eqs. (33) and (34) and noting that $S_o + S_w$ is a constant it follows that

$$\partial(Q_o + Q_w)/\partial x = 0 \tag{36}$$

The sum of the oil and water flow rates is the total flow rate Q, so that the above condition is the expression of global mass conservation for the mixture:

$$\partial Q/\partial x = 0 \tag{37}$$

Conversely, expressing mass conservation for the mixture and one phase, automatically enforces mass conservation for the other phase. It can be verified that Eq. (34) is a consequence of Eqs. (33), (35), and (37).

By appropriate elimination of the other dependent variables, it is possible to derive an equation which involves only the saturation.

3.3. The Saturation Equation

The water flow rate is a fraction of the total flow rate, f_w.

$$Q_w = f_w Q \tag{38}$$

From Eqs. (31) and (32), the ratio Q_w/Q can be calculated. In the process the pressure term is eliminated and one obtains:

$$f_w = \frac{k_{rw}(S_w)/\mu_w}{k_{rw}(S_w)/\mu_w + k_{ro}(S_w)/\mu_o} = \left[1 + \frac{k_{ro}(S_w)}{k_{rw}(S_w)} \cdot \frac{\mu_w}{\mu_o}\right]^{-1} = f_w(S_w) \tag{39}$$

We note that f_w is a function of saturation only. A typical curve is shown in Fig. 21. Substituting this result into Eq. (33) gives

$$\phi A \frac{\partial S_w}{\partial t} + \frac{Q\,\partial f_w(S_w)}{\partial x} = 0 \tag{40}$$

The saturation S_w, as yet unknown, is a function of the two variables x and t:

$$S_w = S_w(x, t)$$

Following this trivial observation, the usual rules of calculus for differentiation of a function of a function can be applied to the term $(\partial f_w(S_w, t)/\partial x)$ with the result

$$\frac{\partial f_w(S_w)}{\partial x} = \frac{d\{f_w(S_w)\}}{dS_w}\frac{\partial S_w}{\partial x} = f_w'(S_w)\frac{\partial S_w}{\partial x}$$

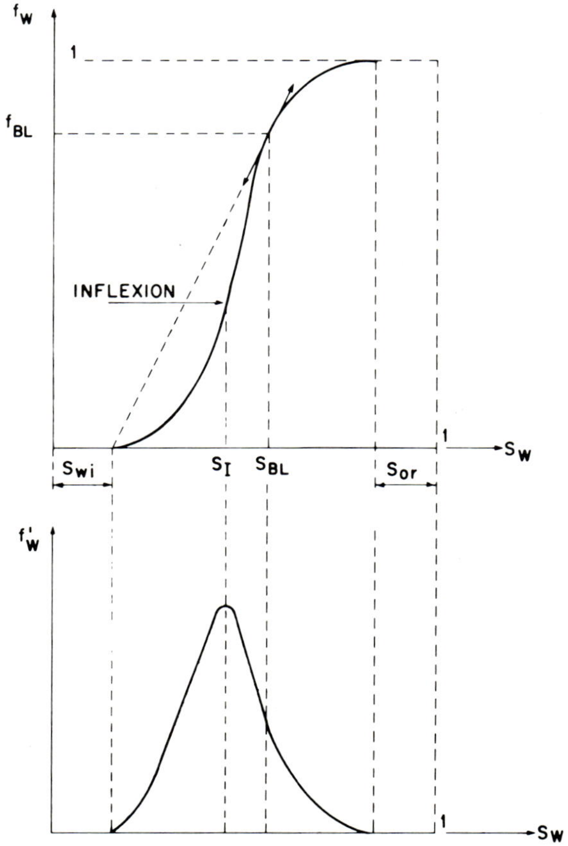

FIG. 21. Fractional flow function and derivative.

Finally, Eq. (40) becomes

$$\phi A \frac{\partial S_w}{\partial t} + Q f'_w(S_w) \frac{\partial S_w}{\partial x} = 0 \tag{41}$$

which is an equation for saturation alone. The time evolution of the saturation in the core can be obtained by solving Eq. (41) without having to solve for the pressure. This equation is known in the petroleum literature as the Buckley–Leverett equation after the authors Buckley and Leverett (1942) who first introduced it.

3.4. The Solution of the Buckley–Leverett Equation

As injection proceeds, it is expected that the saturation of water will vary with time and position along the axis of the core. The relationship between S_w, x, and t can be written in a functional form as $S_w = S_w(x, t)$ or equivalently in a differential form as:

$$dS_w = \frac{\partial S_w}{\partial x}\, dx + \frac{\partial S_w}{\partial t}\, dt \tag{42}$$

Buckley and Leverett's main and fruitful idea was to consider the history or path of a given value of saturation, that is, to find at what time t a given saturation would reach a particular location x. Thus, for any given value of S_w, a relationship can be established between x and t such that the equation $S_w = S_w(x, t) =$ constant, or equivalently, such that the differential equation

$$\frac{\partial S_w}{\partial t}\, dx + \frac{\partial S_w}{\partial t}\, dt = 0$$

is satisfied. Thus, the path of a given saturation, i.e., the relationship between x and t for a fixed S_w, is obtained by solving Eqs. (41) and (42) simultaneously when $dS_w = 0$. Summarizing, the following two equations must be solved simultaneously:

$$A\phi \frac{\partial S_w}{\partial t} + Qf'_w \frac{\partial S_w}{\partial x} = 0, \qquad \frac{\partial S_w}{\partial t}\, dt + \frac{\partial S_w}{\partial x}\, dx = 0 \tag{43}$$

The relationship between x and t is obtained in a differential form by solving the system of equations in (43).

Eliminating $\partial S_w/\partial x$ and $\partial S_w/\partial t$ from Eqs. (43) yields:

$$(dx/dt)_{S_w} = Qf'_w/A\phi \tag{44}$$

Because all characteristics of the core are assumed constant, Eq. (44) can be integrated to give the position of a particular saturation value as a function of time:

$$x_{S_w} = (Qf'_w t/A\phi) + x_0 \tag{45}$$

where x_0 is the position of the water saturation at time $t = 0$. Because f'_w and x_0 are functions of S_w, it is possible to construct the complete saturation profile at any time t by using Eq. (45) to calculate x for various values of S_w. We have assumed that initially the water saturation is uniform in the core. At time $t = 0$, all values of water saturations above S_{wi}, the initial water saturation, and below $(1 - S_{or})$, where S_{or} is the residual oil saturation, will start moving into the core at a rate proportional to $f'_w(S_w)$. Typical curves of f_w and f'_w versus S_w are illustrated in Fig. 21. As illustrated in Fig. 22, the shape of the saturation

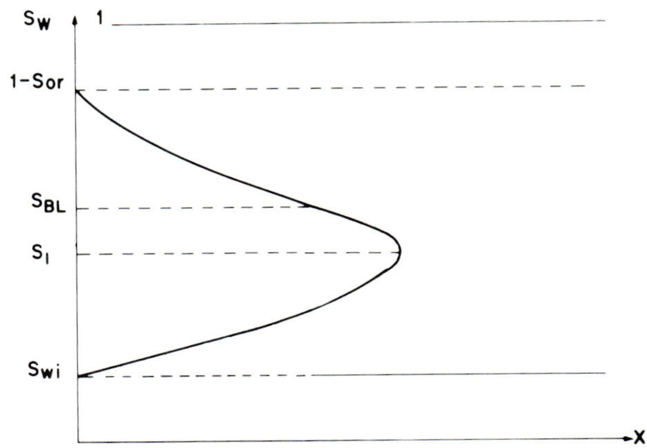

FIG. 22. Saturation profile.

profile after a period of time will be essentially the same as that of f'_w.

It is apparent from Fig. 22 that for all values of x, the saturation is not uniquely defined but has three values, which points to the fact that the straight application of Eq. (45) is not physically acceptable. It must now be pointed out that the equations describing the flow of water and oil were derived under the tacit assumption that the solution for the saturation profile is a smooth function of x and t. Therefore, the differential equation (41) does not apply in a region where the saturation profile or its slope undergoes a discontinuity or jump. Application of the law of mass conservation will yield the velocity at which a saturation front will propagate.

Consider an elemental volume (Fig. 23) bounded by two surfaces Σ^+ and Σ^- on either side of a discontinuity in saturation or front. Let A be the cross-sectional area. Conservation of mass in finite difference form yields:

$$(Q_w{}^+ - Q_w{}^-)\,\Delta t = A\phi(S_w{}^+ - S_w{}^-)\,\Delta x \tag{46}$$

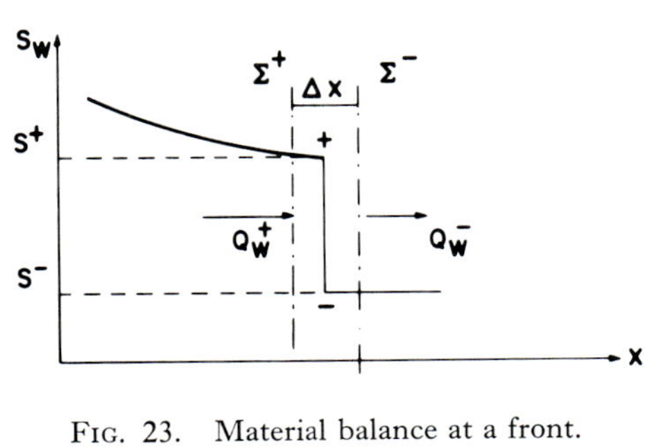

FIG. 23. Material balance at a front.

which can be rewritten equivalently as

$$\frac{\Delta x}{\Delta t} = \frac{Q_w^+ - Q_w^-}{S_w^+ - S_w^-} \cdot \frac{1}{A\phi} \tag{47}$$

where the superscripts $-$ and $+$ refer to conditions just ahead and just behind the front. In the case of a discontinuity, one can still proceed to the limit for Δx and Δt. Although when Δx and Δt tend to zero, Q^+ does not tend to Q^- nor does S^+ tend to S^-, the limit of Eq. (47) has a meaning because the limit of $(Q_w^+ - Q_w^-)/(S_w^+ - S_w^-)$ exists:

$$\frac{dx}{dt} = V_f = \frac{1}{A\phi}\frac{Q_w^+ - Q_w^-}{S_w^+ - S_w^-} = \frac{Q}{A\phi}\frac{f_w^+ - f_w^-}{S_w^+ - S_w^-} \tag{48}$$

The relation (48) gives the rate of advance of a saturation front, which can be integrated to give the position of the front as a function of time

$$x_f = \frac{Q}{A}\left\{\frac{f_w^+ - f_w^-}{S_w^+ - S_w^-}\right\} + x_{f0} \tag{49}$$

where x_f is the abscissa or location of the saturation discontinuity ΔS at time t, and x_{f0} is the abscissa of the saturation discontinuity at time $t = 0$, and $\Delta S = S_w^+ - S_w^-$.

Now we have two sets of expressions: Eqs. (44) and (45) which can be used to calculate velocity and position of a given value of saturation in a smooth profile, and Eqs. (48) and (49) which provide a means of obtaining velocity and position of a saturation jump or front.

This raises the interesting question of how to construct a judiciously blended description of saturation profile movement which eliminates the multiple values. Whenever the continuous solution is physically acceptable it will be given precedence over the discontinuous one. In other words, the proper solution is the one that is physically meaningful with the minimum amount of discontinuity. Note that introduction of a discontinuity between S_w^- and S_w^+ is tantamount to replacing the fractional flow curve between the two points by a straight line or secant (see Fig. 24), because if the discontinuity is to propagate as such, all saturations within the range S_w^-, S_w^+ must move at the same velocity.

Assuming a uniform immobile initial water saturation, four tentative solutions for the saturation profile at a given time t are illustrated in Fig. 24. The continuous profile $ab'cd'e$ is not acceptable because S_w is not a single-valued function of x. The profile constituted of front a_1b, section bb', and curve $b'cd'e$ is not acceptable for the same reason. As long as the slope of the tangent to the fractional flow curve at point b

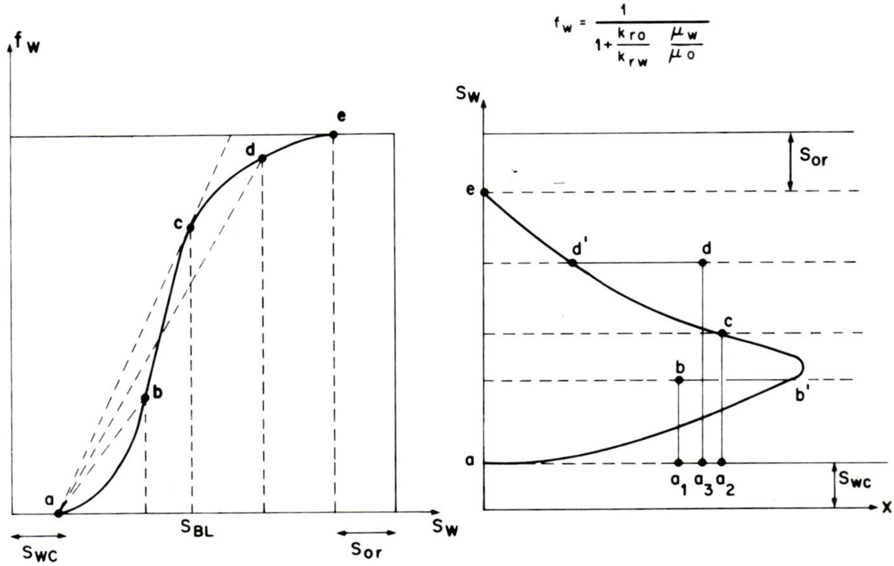

FIG. 24. Tentative saturation profiles.

is greater than the slope of the secant *ab*, the saturation profile will not be single valued. As the point *b* moves up along the fractional flow curve, this situation will prevail until *b* reaches *c*. Between *c* and *e* the solution will be single valued. Two such single-valued solutions are illustrated in Fig. 24: (1) front a_2c and curve $cd'e$; (2) front a_3d, plateau dd', and curve $d'e$. However, profile $a_2cd'e$ is the proper solution because it uses the continuous solution to the maximum extent compatible with the physical constraint that the solution must be single valued. This profile is obtained by the graphical construction due to Welge (1952), which consists of drawing a tangent to the fractional flow curve from point *a* corresponding to the initial saturation in the core S_{wc}. The saturation at the tangent point S_{BL} is the saturation established in the core immediately behind the front. The velocity at which the "Buckley–Leverett front" travels is proportional to the slope of secant *ac*. The velocity at which a saturation greater than S_{BL} travels is proportional to the slope of the tangent to the fractional flow curve.

The shape of the resulting profile, the Buckley–Leverett profile, does not change with time except for a uniform stretching in the *x* direction. The Buckley–Leverett front is created instantaneously. Its size, $S_w^+ - S_w^-$, and rate of advance do not change with time. These last conclusions hold only because of the particular choice of the initial saturation distribution, which was assumed to be uniform. Figure 25 illustrates the evolution of the saturation profile with time when the initial saturation profile is not uniform. The "normal"

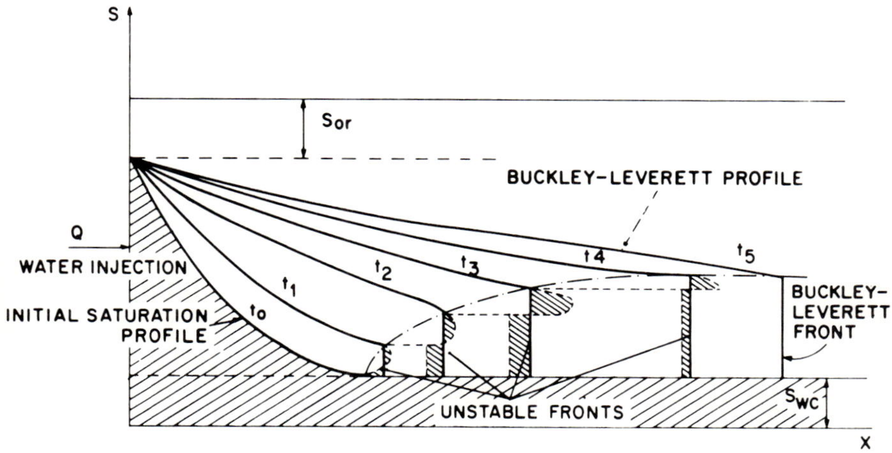

FIG. 25. Evolution of a saturation profile with time; growth of a front.

Buckley–Leverett profile occurs only after a period of time, because the larger values of saturations, which move faster than the smaller ones, start the race after the smaller values with a handicap. Thus, with increasing time, larger values of saturation continuously catch up with smaller ones and form an unstable front which grows in size until it reaches the maximum possible size and assumes a "steady-state" condition.

3.5. RECOVERY PREDICTIONS

In this and following sections, S represents water saturation and f represents the water fractional flow function. Once the velocities at which the Buckley–Leverett front and trailing saturations propagate through the core have been established, it is an easy matter to calculate the production rates or cumulative production of both fluids. Equation (44) can be rewritten:

$$\frac{dx}{L} = \frac{f'(S)Q\,dt}{\phi AL} = \frac{f'(S)Q\,dt}{V_p} = f'(S)\,dV_{pi} \tag{50}$$

a formula valid even if Q changes with time. Equation (50) can be integrated, yielding

$$x = Lf'(S)\,V_{pi} \tag{51}$$

which gives the distance swept by the saturation S for a given amount of volume of fluid injected. Multiplying Eq. (51) successively by δn and $\phi h\, \delta n$ (see Fig. 26) yields alternate expressions for (51)

$$A_s = A_p f'(S)\, V_{pi} \tag{52}$$

$$V_{ps} = V_p f'(S)\, V_{pi} \tag{53}$$

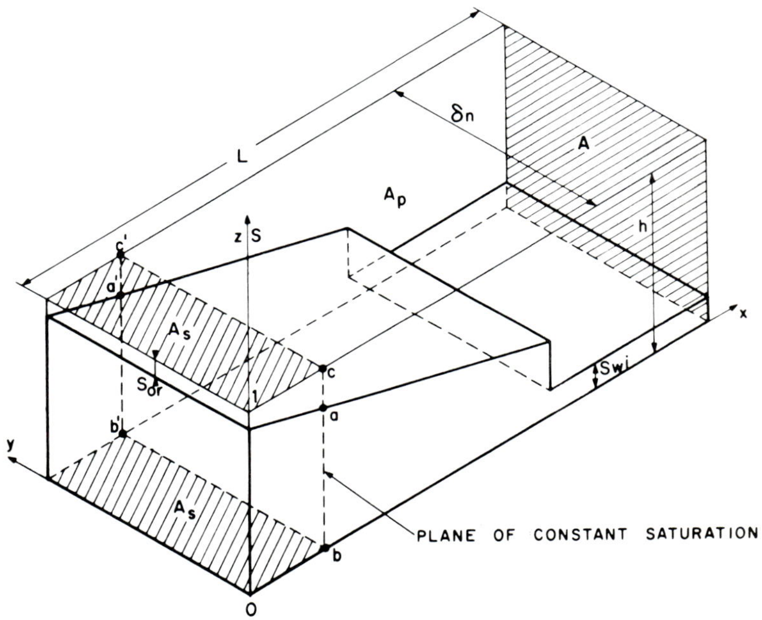

FIG. 26. Saturation profile in a linear core.

where A_s is the area swept by the saturation S, and V_{ps} is the pore volume swept by the saturation S. Prior to breakthrough of the Buckley–Leverett front at the outlet face, only oil is produced. After breakthrough, the oil fraction of the recovered incremental pore volume produced dV_{pi} during an infinitesimal time interval dt is $[1 - f(S)]\, dV_{pi}$, where S is the current prevailing saturation at the outlet face. The instantaneous water cut at that time is by definition f and the producing water–oil ratio is:

$$F_{wo} = f(S)/[1 - f(S)] \tag{54}$$

The cumulative oil recovery in units of pore volume E_R is the integral of the instantaneous oil production rate over time, and at a given stage of injection it is

$$E_R = \int_0^{V_{pi}} (1 - f)\, dV_{pi} \tag{55}$$

Integrating by parts,

$$E_R = (1 - f)\, V_{pi} + \int_0^{V_{pi}} V_{pi}\, df \tag{56}$$

For a given value of V_{pi}, there prevails at the outlet face a saturation S. From Eqs. (51), (52), or (53) it follows that this saturation must satisfy the relation

$$1 = f'(S)\, V_{pi} = (df/ds)\, V_{pi} \tag{57}$$

or

$$dS = V_{pi}\, df$$

Using Eq. (57), Eq. (56) transforms into:

$$E_R = (1 - f)\, V_{pi} + \int_{S_{wi}}^{S} dS, \qquad E_R = (1 - f)\, V_{pi} + S - S_{wi} \tag{58}$$

The parametric equations of oil recovery (or displacement efficiency) and water cut versus pore volume injected have been obtained where the parameter is S, the saturation at the outlet face.

$$\begin{aligned} V_{pi} &= 1/f'(S), \qquad f_w = f(S) \\ E_R &= S - S_{wi} + [1 - f(S)]\, V_{pi}, \qquad E_D = E_R/S_{oi} \end{aligned} \tag{59}$$

Because the volume of oil recovered is equal to the increase of the volume of water in the sand,

$$E_R = \bar{S} - S_{wi}$$

where $\bar{S}$ is the average (water) saturation in the core so that by identification with Eq. (58),

$$\bar{S} = S + (1 - f)\, V_{pi}$$

and using Eq. (57),

$$\bar{S} = S + \frac{[1 - f(S)]}{f'(S)} \tag{60}$$

A graphical solution due to Jones-Parra (1953) exists for $\bar{S}$. The equation of the tangent to the f curve at a given saturation S in the $(\tilde{S}, f)$-plane is

$$f - f(S) = f'(S)(\tilde{S} - S) \tag{61}$$

At the intersect of this tangent with the line $f = 1$, the value of the saturation $\tilde{S}$ is that obtained from Eq. (61) for $f = 1$, i.e.,

$$\tilde{S} = \frac{1 - f(S)}{f'(S)} + S \tag{62}$$

Comparison of Eqs. (60) and (62) shows that $\tilde{S} = \bar{S}$. Thus, the average saturation $\bar{S}$, when the saturation at the outlet face is S, is the value of saturation at the intersect of the tangent to the f curve at the point $(S, f(S))$ with the line $f = 1$. The graphical construction is illustrated in Fig. 27. Note that:

$$f'(S) = \tan\theta = [1 - f(S)]/(\bar{S} - S)$$

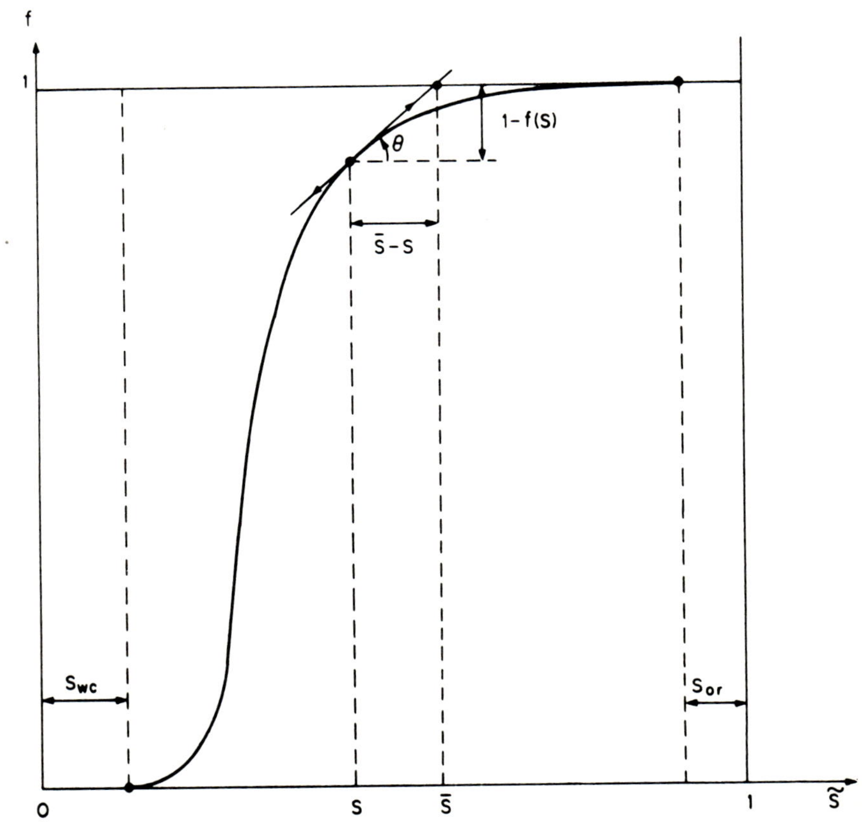

FIG. 27. Graphical determination of flood performance.

3.6. Laboratory Measurement of Relative Permeability Ratios

Having assumed the relative permeabilities to be known, the curves $f(S)$ and $f'(S)$ can be constructed. Then, by the set of Eqs. (59), the recovery performances of a core flood can be predicted. Vice versa, running a core flood and obtaining the recovery performances experimentally, the curves of relative permeability ratios can be deduced. In the laboratory, the core is initially filled with water and then flushed with oil until water is no longer produced. By difference, the initial water saturation S_{wi}, when water injection begins, is known. At a given stage of injection, V_{pi} (measured), the instantaneous water cut is f_w (measured) and the recovery is E_R (also measured). It is deduced from Eqs. (59) that the saturation S that must have existed at the outlet face of the core at that stage is

$$S = E_R + S_{wi} - (1 - f_w) V_{pi}$$

For this value of S, $f(S) = f_w$, from which it follows that k_{rw}/k_{ro} for that saturation has the value

$$(\mu_w/\mu_0)(f_w/(1 - f_w))$$

Thus, for each set of measurements, a point on the k_w/k_o curve is obtained. Actually, the slope of the $f(S)$ curve is also obtained: $1/V_{pi}$. From the flood data the k_w/k_o curve can be constructed by points. The above procedure is due to Welge and commonly called the Welge method as opposed to other procedures (the "steady-state" method, the Penn-state method) which are no longer in common use. The Welge method yields only the ratio, not the individual relative permeabilities. It would take one additional measurement to obtain this additional information. At constant water injection rate, if the pressure drop is also measured, the individual permeabilities can be obtained. A typical curve obtained by the Welge method is shown in Fig. 28.

It goes without saying that when relative permeabilities are measured in this fashion, the recovery predictions by Eqs. (59) will exactly reproduce the experimental data. That the relative permeabilities have been measured correctly is not actually established, nor is the physical validity of the Buckley–Leverett solution.

One way of verifying the validity of the Buckley–Leverett solution consists of comparing the Buckley–Leverett saturation profile with an experimentally measured one. The measurement of saturations in an experiment is a more elaborate task than the recovery of the fluids at an outlet face. Few experiments of this type have been reported. One of these is described in Section 3.7.

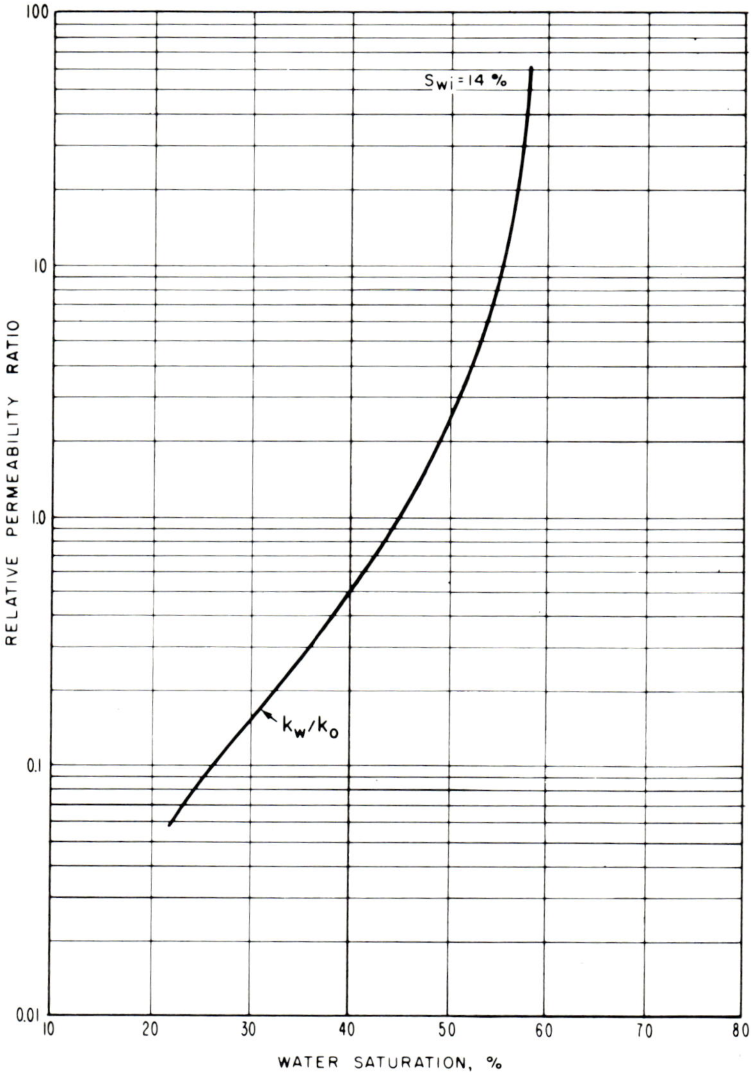

FIG. 28. Relative water–oil permeability ratio versus water saturation.

Another approach consists of including the effect of capillary pressure in the equations and of obtaining a solution. If this new solution and the Buckley–Leverett solution agree within a reasonable tolerance, the legitimacy of neglecting capillarity in the flow equations is established. This procedure is also illustrated in Section 3.7.

3.7. The Influence of Capillarity

3.7.1. *A Semiempirical Approach*

In previous sections, the influence of capillarity was neglected. It may be a valid engineering approximation, but it should be substantiated. A general expression for the fractional flow function can be derived:

$$\mathscr{F}_{\mathrm{w}} = f_{\mathrm{w}} \left\{ 1 + k \frac{k_{\mathrm{ro}} A}{Q \mu_{\mathrm{o}}} \left[\frac{\partial p_{\mathrm{c}}}{\partial x} - \Delta \rho g \sin \theta \right] \right\} \tag{63}$$

which can be rewritten

$$\mathscr{F}_{\mathrm{w}} = f_{\mathrm{w}} \left\{ 1 + k \frac{k_{\mathrm{ro}} A}{\mu_{\mathrm{o}} Q} \left[p_{\mathrm{c}}' \frac{\partial S_{\mathrm{w}}}{\partial x} - \Delta \rho g \sin \theta \right] \right\} \tag{64}$$

Without going into a formal dimensional analysis of parameters which affect $\mathscr{F}_{\mathrm{w}}$, it is apparent from Eq. (64) that the influence of capillarity will be small if the dimensionless term

$$\frac{k k_{\mathrm{ro}}}{Q \mu_{\mathrm{o}}} A p_{\mathrm{c}}'(S_{\mathrm{w}}) \frac{\partial S_{\mathrm{w}}}{\partial x}$$

is small compared to 1. It was fairly well substantiated experimentally by Leverett that capillary pressure could be correlated with rock and fluid properties in the form (Leverett, 1941):

$$p_{\mathrm{c}}(S_{\mathrm{w}}) = \frac{\sigma \phi^{\frac{1}{2}}}{k^{\frac{1}{2}}} J(S_{\mathrm{w}}) \tag{65}$$

where $J(S_{\mathrm{w}})$, the Leverett J function, is a dimensionless shape function. The effect of capillarity will be small if

$$N_{\mathrm{c}} = \sigma \frac{(k\phi)^{1/2}}{Q \mu_{\mathrm{o}}} A k_{\mathrm{ro}}(S_{\mathrm{w}}) J'(S_{\mathrm{w}}) \frac{\partial S_{\mathrm{w}}}{\partial x} \tag{66}$$

is small compared to 1. N_{c} is a dimensionless number which depends on various quantities, some of which are constants (e.g., A) or can be made constants in the displacement process (e.g., Q). Others are saturation dependent and even time dependent $\partial S_{\mathrm{w}}/\partial x$. N_{c} depends on the shape of the saturation profile since it depends on its slope $\partial S_{\mathrm{w}}/\partial x$. Therefore, it cannot be said a priori whether it is small or large. To decide whether N_{c} is small or not, one must be able to determine indirectly $\partial S_{\mathrm{w}}/\partial x$, at least in an approximate way.

Why not determine the saturation profile experimentally? If the experimentally obtained saturation profile does not deviate appreciably from the Buckley–Leverett profile, then it can be said that capillary effects were negligible. A priori it can be said that in waterflooding the oil-filled core, the process being one of imbibition, for any saturation, $\mathscr{F}_w$ is larger than F_w. Thus, the function $\mathscr{F}_w$, having a time-independent lower bound (F_w), must of necessity tend asymptotically to a limit that overlies the F_w curve. Experiments (conducted with brine and gas, displacing phase Fig. 29) showed that the saturation profile stretched with time as predicted by the Buckley–Leverett theory for the high values of saturation. However, for the low values of saturation the profiles corresponding to different times could be superimposed by translation. This implies that there exists a range of saturations which propagate at the same speed regardless of the value of saturation. From a Buckley–Leverett perspective, it further implies that the limiting shape of the $\mathscr{F}_w$ curve is straight for the low values of saturation. Ultimately, one can conclude that the time-wise asymptotic limit of the $\mathscr{F}_w$ curve is, roughly, the F_w curve as modified by the Welge construction. In mathematical terms for $S_w < S_{BL}$,

$$\lim_{t\to\infty} \mathscr{F}_w = F'(S_{BL})(S_w - S_{wc})$$

Since

$$\mathscr{F}_w = f_w \left\{1 + k \frac{k_{ro}}{\mu_o} \frac{A}{Q} \left[p_c' \frac{\partial S_w}{\partial x} - \Delta\rho g \sin\theta\right]\right\}$$

it follows that

$$\frac{\partial S_w}{\partial x} = \frac{F_w'(S_{BL})(S_w - S_{wc}) - F_w}{f_w k_{ro} J'(S_w)} \cdot \frac{Q\mu_o}{\sigma A(k\phi)^{1/2}}$$

By integration, assuming that above S_{BL} the effect of capillarity is definitely negligible, one obtains for $S_w < S_{BL}$,

$$x_{S_w}(t) = x_{BL}(t) + \frac{\sigma A(k\phi)^{1/2}}{Q\mu_o} \int_{S_{BL}}^{S_w} \frac{f_w(S)\, k_{ro}(S)\, J'(S)\, dS}{F_w'(S_{BL})(S - S_{wc}) - F_w(S)} \tag{67}$$

which describes the asymptotic shape of the saturation profile for large time for the so-called "stabilized zone." Above S_{BL}, the saturation profile is the Buckley–Leverett one.

However, the solution given by Eq. (67) is only formal because the integrand in Eq. (67) is not integrable. Expanding $F_w(S)$ about S_{BL} yields:

$$F_w(S) = F_w(S_{BL}) + F_w'(S_{BL})(S - S_{BL}) + O(S - S_{BL})^2$$

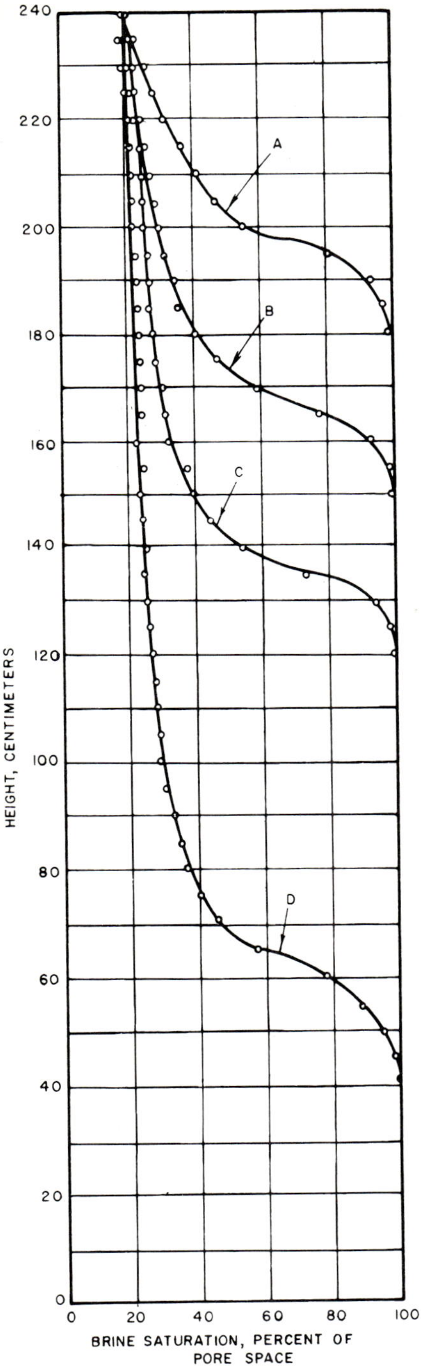

FIG. 29. Change in saturation distribution with time. Curve A = 0 hr, curve B = 24 hr, curve C = 52 hr, curve D = 120 hr. (After Terwilliger *et al.*, 1951.)

The denominator of the integrand takes the form:

$$F'_{\mathrm{w}}(S_{\mathrm{BL}})(S - S_{\mathrm{wc}}) - F_{\mathrm{w}}(S_{\mathrm{BL}}) - F'_{\mathrm{w}}(S_{\mathrm{BL}})(S - S_{\mathrm{BL}}) + O(S - S_{\mathrm{BL}})^2$$

Because

$$F_{\mathrm{w}}(S_{\mathrm{BL}}) = (S_{\mathrm{BL}} - S_{\mathrm{wc}})\, F'_{\mathrm{w}}(S_{\mathrm{BL}})$$

the expression reduces further to $-F''_{\mathrm{BL}}(S - S_{\mathrm{BL}})^2$. The integrand has a pole of order 2 which is not integrable. The approximate expression derived for $\partial S_{\mathrm{w}}/\partial x$ can only be integrated to a value of saturation somewhat smaller than S_{BL}. The integrated form of $\partial S_{\mathrm{w}}/\partial x$ can only give the shape of the stabilized zone, not its absolute position. Comparison of calculated and observed shapes is good.

As time goes on, the relative importance of the stabilized zone in contrast to the stretching one decreases. The profile in the stabilized region looks more and more like a front. The Welge method of measuring $k_{\mathrm{w}}/k_{\mathrm{o}}$ ratios is not valid unless capillary effects are indeed negligible. The actual saturation profile must not deviate appreciably from the Buckley–Leverett one. From Eq. (64) it can be inferred that this will be the case for high injection rates. Rapoport and Leas (1953) report experimentally determined scaling criteria that must be met to insure reliable results.

3.7.2. *A Perturbation Method*

First proceed to express the material balance

$$\phi A \frac{\partial S_{\mathrm{w}}}{\partial t} + \frac{\partial Q_{\mathrm{w}}}{\partial x} = 0$$

Replace Q_{w} in the above by its expression in terms of fractional flow function $\mathscr{F}_{\mathrm{w}}$

$$\mathscr{F}_{\mathrm{w}} = F_{\mathrm{w}} + \sigma \frac{(k\phi)^{1/2}}{Q\mu_{\mathrm{o}}} A k_{\mathrm{ro}} f_{\mathrm{w}} J' \frac{\partial S_{\mathrm{w}}}{\partial x}$$

For simplicity let

$$k_{r_{\mathrm{o}}}(S_{\mathrm{w}})\, f_{\mathrm{w}}(S_{\mathrm{w}})\; J'(S_{\mathrm{w}}) = E_{\mathrm{w}}(S_{\mathrm{w}})$$

The differential equation takes the form

$$\phi A \frac{\partial S_{\mathrm{w}}}{\partial t} + Q F'_{\mathrm{w}} \frac{\partial S_{\mathrm{w}}}{\partial x} + \sigma \frac{(k\phi)^{1/2}}{\mu_{\mathrm{o}}} A \frac{\partial}{\partial x}\left(E_{\mathrm{w}} \frac{\partial S_{\mathrm{w}}}{\partial x}\right) = 0$$

Dropping the subscript w, defining

$$\epsilon = \frac{\sigma (k\phi)^{1/2} A}{L Q \mu_{\mathrm{o}}}$$

where L is a characteristic length, the equation takes a dimensionless form

$$\frac{\phi AL}{Q}\frac{\partial S}{\partial t} + F'L\frac{\partial S}{\partial x} + \epsilon L^2 E'\left(\frac{\partial S}{\partial x}\right)^2 + \epsilon L^2 E\frac{\partial^2 S}{\partial x^2} = 0 \tag{68}$$

This is a second-order nonlinear partial differential equation for which no exact solution is known. If ϵ is so small as to be negligible, then neglecting the terms in ϵ we obtain the Buckley–Leverett equation. However, in the process the order of the differential equation is lowered. This creates difficulties and by extending the discussion beyond porous medium to the realm of general fluid mechanics, it can be simply illustrated by reviewing the concept of boundary layer theory. The equations for viscous fluid flow are the well-known Navier–Stokes equations, which for steady-state conditions are:

$$(\mathbf{v}\ \mathrm{grad})\ \mathbf{v} = -\mathrm{grad}\ p + \mathrm{Re}^{-1}\ \nabla^2\mathbf{v}$$

where Re is the Reynolds number. For high Reynolds number, if the last term is neglected the Navier–Stokes equations reduce to Euler equations:

$$(\mathbf{v}\ \mathrm{grad})\ \mathbf{v} = -\mathrm{grad}\ p$$

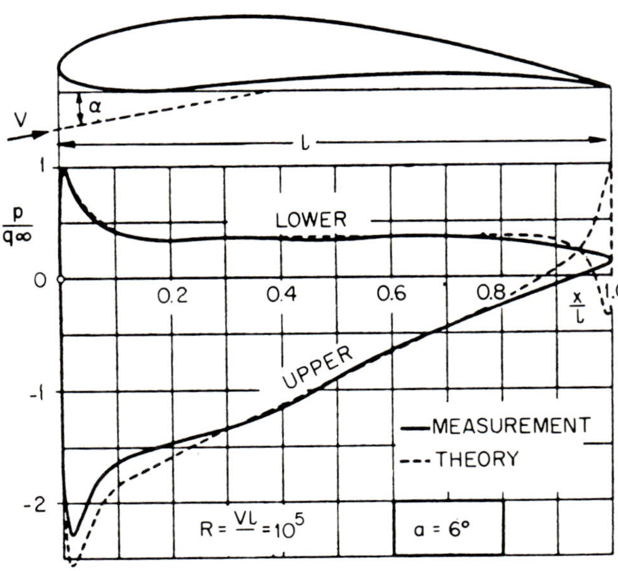

FIG. 30. Comparison of theoretical and measured pressure distribution on an airfoil (Schlichting, 1955).

In the process, the order of the differential equation has been reduced and at a solid boundary only the no-cross-flow condition can be imposed. The no-slip condition cannot be met. It turns out, however, that for many problems, the approximation is satisfactory. For example, the pressure distribution about a streamline body of revolution in a uniform stream can be predicted accurately (Fig. 30). The approximation is used to calculate the lift on an airfoil. On the other hand, it is hopelessly useless to predict the drag. To predict the drag must one solve the Navier–Stokes equation? Prandtl answered that question. To meet the no-slip condition at the wall, the order of the differential equation must be retained, but it is not necessary to retain all the higher order terms. Retain only the significant ones. Let us write the Navier–Stokes equation (steady state) in two dimensions and indicate the order of magnitude below each term (Schlichting, 1955):

$$\underset{1}{\frac{\partial u}{\partial x}} + \underset{1}{\frac{\partial v}{\partial y}} = 0 \qquad \text{(continuity equation)}$$

$$\underset{1}{u}\,\underset{1}{\frac{\partial u}{\partial x}} + \underset{\delta}{v}\,\underset{1/\delta}{\frac{\partial u}{\partial y}} = \frac{\partial p}{\partial x} + \underset{\delta^2}{\frac{1}{\text{Re}}}\left(\underset{1}{\frac{\partial^2 u}{\partial x^2}} + \underset{1/\delta^2}{\frac{\partial^2 u}{\partial y^2}}\right)$$

$$\underset{1}{u}\,\underset{\delta}{\frac{\partial v}{\partial x}} + \underset{\delta}{v}\,\underset{1}{\frac{\partial v}{\partial y}} = -\underset{1}{\frac{\partial p}{\partial y}} + \underset{\delta^2}{\frac{1}{\text{Re}}}\left(\underset{\delta}{\frac{\partial^2 v}{\partial x^2}} + \underset{1/\delta}{\frac{\partial^2 v}{\partial y^2}}\right)$$

where δ is the boundary-layer thickness [$\delta \simeq (\text{Re})^{-1/2}$]. Re is defined as UL/ν, where U is the uniform velocity away from plate, L is the characteristic length of plate, and ν is the kinematic viscosity. The procedure leads to the simplified equations:

$$\frac{\partial p}{\partial y} = 0, \qquad u\frac{\partial u}{\partial x} + v\frac{\partial u}{\partial y} = -\frac{\partial p}{\partial x} + \frac{1}{\text{Re}}\frac{\partial^2 u}{\partial y^2}$$

Within the boundary layer, a thin layer, these equations should be used. Outside the boundary layer, the Euler equations can be used. The solutions within and without the boundary layer are then matched.

Essentially, the second-order term that is retained is the one that is so large within the (thin) boundary layer that it overcomes the smallness of 1/Re. The gradient of u in the y direction is out of proportion with that in the x direction. To reestablish harmonious proportions one may try to stretch, in this instance, the y coordinate as much as necessary. This is obtained by a coordinate transformation of the form $y^* = y/\delta$. With the new system of coordinates the term $(1/\text{Re}) \cdot (\partial u^2/\partial y^2)$ becomes

$\partial^2 u/\partial y^{*2}$ which has the same order of magnitude as $u\ (\partial u/\partial x)$ and must therefore be retained. The term $(1/\mathrm{Re})(\partial^2 u/\partial x^2)$ is an order of magnitude smaller and can be neglected. The method of strained coordinates (Carrier, 1953) provides a more automatic procedure to determine which terms of the Navier–Stokes equations can be neglected than the term by term inspection of the original equations. Physically, they are equivalent as both require a prior knowledge of the order of magnitude of the boundary-layer thickness.

To solve the saturation equation with capillarity, one can use the same method. The parallel is obvious: complete saturation equation—Navier–Stokes equations; Buckley–Leverett equation—Euler equations.

Let

$$X = [x - \xi(t)]/\epsilon^\alpha L \tag{69}$$

The region of capillary effect is in the neighborhood of the Buckley–Leverett front. In contrast to the fixed boundary layer, it moves. The new system of coordinates must be relative (Lefur, 1962) to a point that more or less travels with the Buckley–Leverett front. Differentiating Eq. (69) yields

$$dX = \frac{1}{\epsilon^\alpha L}\left(dx - \frac{d\xi}{dt}\,dt\right)$$

Rules of differentiation for any function ( ) of two variables give:

$$\left[\frac{\partial(\,)}{\partial x}\right]_t = \left[\frac{d(\,)}{dx}\right]_{dt=0} = \left[\frac{\partial(\,)}{\partial X}\right]_t \left(\frac{dX}{dx}\right)_{dt=0} + \left[\frac{\partial(\,)}{\partial t}\right]_X \left(\frac{dt}{dx}\right)_{dt=0}$$

and

$$\left[\frac{\partial(\,)}{\partial t}\right]_x = \left(\frac{d(\,)}{dt}\right)_{dx=0} = \left[\frac{\partial(\,)}{\partial X}\right]_t \left(\frac{dX}{dt}\right)_{dx=0} + \left[\frac{\partial(\,)}{\partial t}\right]_X$$

$$= \left[\frac{\partial(\,)}{\partial t}\right]_X - \frac{1}{\epsilon^\alpha L}\frac{d\xi}{dt}\left[\frac{\partial(\,)}{\partial X}\right]_t$$

Use of these equations for the function S transforms Eq. (68) into the form:

$$\epsilon^{(1-\alpha)}\left[E'\left(\frac{\partial S}{\partial X}\right)^2 + E\frac{\partial^2 S}{\partial X^2}\right] + \left(F' - \frac{\phi A}{Q}\frac{d\xi}{dt}\right)\frac{\partial S}{\partial X} + \epsilon^\alpha \frac{\phi A L}{Q}\frac{\partial S}{\partial t} = 0$$

So that higher-order terms are of same order of magnitude as other terms, choose $\alpha = 1$. The equation becomes

$$\frac{\partial}{\partial X}\left(E\frac{\partial S}{\partial X}\right) + \left(F' - \frac{\phi A}{Q}\frac{d\xi}{dt}\right)\frac{\partial S}{\partial X} + \epsilon\phi\frac{AL}{Q}\frac{\partial S}{\partial t} = 0$$

The ϵ term can now be neglected because it multiplies a lower-order term. The final equation is:

$$\frac{\partial}{\partial X}\left(E\frac{\partial S}{\partial X}\right) + \left(F' - \frac{\phi A}{Q}\frac{d\xi}{dt}\right)\frac{\partial S}{\partial X} = 0 \tag{70}$$

Let $\hat{S}(x, t)$ be the Buckley–Leverett solution. We shall match the two solutions (as in the boundary layer theory) at a point $S_{\mathrm{m}}(X, t)$. At any time t, S_{m} is the solution of

$$\hat{S}(\xi(t) + \epsilon LX, t) = S(X, t), \qquad \partial\hat{S}/\partial X = \partial S/\partial X$$

Naturally ahead of the water-invaded zone the value of saturation is the initial value S_{wc}.

Equation (70) can be integrated with respect to X:

$$E(S)\frac{\partial S}{\partial X} + F(S) - \frac{\phi A}{Q}\frac{d\xi}{dt}S = C(t) \tag{71}$$

This equation holds for $S = S_{\mathrm{wc}}$. Thus,

$$C(t) = -S_{\mathrm{wc}}\frac{A\phi}{Q}\frac{d\xi}{dt}$$

and from Eq. (71), $\partial S/\partial X$ can be calculated:

$$\frac{\partial S}{\partial X} = \left[(S - S_{\mathrm{wc}})\frac{A\phi}{Q}\frac{d\xi}{dt} - F(S)\right]\Big/E(S)$$

Note that this equation is identical with Terwilliger's expression, except that $d\xi/dt$ now replaces dX_{BL}/dt. To emphasize the similarity let

$$\frac{A\phi}{Q}\frac{d\xi}{dt} = F'(S_\xi)$$

Integrating once more,

$$X = X(S, t) = \int_{S_{\mathrm{BL}}}^{S}\frac{E(S)\,dS}{F'(S_\xi)[S - S_{\mathrm{wc}}] - F(S)} + C_1(t) \tag{72}$$

or

$$x = \xi(t) + \frac{\sigma A(k\phi)^{1/2}}{Q\mu_0}\int_{S_{\mathrm{BL}}}^{S}\frac{E(S)\,dS}{F'(S_\xi)[S - S_{\mathrm{wc}}] - F(S)} + C_1(t)$$

As in boundary layer theory, there is some arbitrariness in the matching of the solutions. To correct for this, a material balance condition must also be imposed, that is, the same area is to be enclosed underneath the Buckley–Leverett profile and the Lefur profile (Fig. 31).

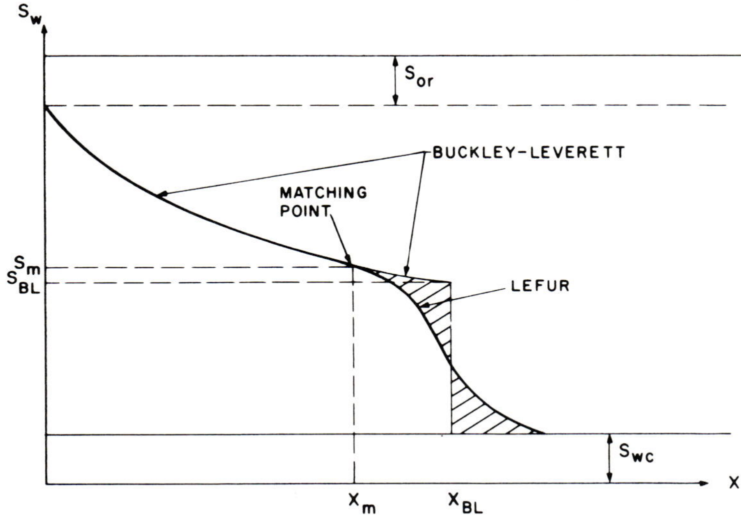

FIG. 31. Comparison of saturation profiles with and without capillarity.

(i) *Matching condition*:

$$\frac{Q}{A\phi} F'(S_{\mathrm{m}})\, t = \xi(t) + C_1(t) + \frac{\sigma A(k\phi)^{1/2}}{Q\mu_{\mathrm{o}}} \int_{S_{\mathrm{BL}}}^{S_{\mathrm{m}}} \frac{E(S)\, dS}{F'(S_\xi)(S - S_{\mathrm{wc}}) - F(S)}$$

(ii) *Smooth matching condition*:

$$\frac{QF''(S_{\mathrm{m}})\, t}{A\phi} = \left\{ \frac{E(S_{\mathrm{m}})}{(S_{\mathrm{m}} - S_{\mathrm{wc}}) F'(S_\xi) - F(S_{\mathrm{m}})} \right\} \frac{\sigma A(k\phi)^{1/2}}{Q\mu_{\mathrm{o}}}$$

This equation determines $F'(S_\xi)$ in terms of S_{m} and t. The expression for $F'(S_\xi)$ can then be substituted in Eq. (72)

$$F'(S_\xi) = \frac{F(S_{\mathrm{m}})}{S_{\mathrm{m}} - S_{\mathrm{wc}}} + \frac{\sigma A^2 k^{1/2} \phi^{3/2}}{Q^2 \mu_{\mathrm{o}}} \frac{E(S_{\mathrm{m}})}{F''(S_{\mathrm{m}})(S_{\mathrm{m}} - S_{\mathrm{wc}})} \tag{73}$$

For simplicity in writing, let

$$D(S, S_{\mathrm{m}}, t) = F'(S_\xi)[S - S_{\mathrm{wc}}] - F(S) \tag{74}$$

(iii) *Material balance condition*: From the matching point to initial conditions point, the areas must be the same underneath the two profiles (Buckley–Leverett and Lefur solutions):

$$\int_{X_m}^{X_{wc}} (S - S_{wc})\, dX = \int_{X_m}^{X_{BL}} (\hat{S} - S_{wc})\, dX$$

$$\int_{S_m}^{S_{wc}} (S - S_{wc}) \frac{E(S)\, dS}{D(S, t)} = \frac{1}{\epsilon L} \int_{X_m}^{X_{BL}} (\hat{S} - S_{wc})\, dx$$

$$= \frac{1}{\epsilon L} \frac{Qt}{A\phi} \{(S - S_{wc}) F'(S) \,|_{S_m}^{S_{BL}} - F(S_{BL}) + F(S_m)\}$$

$$= \frac{1}{\epsilon L} \frac{Qt}{A\phi} \{[(S_{BL} - S_{wc}) F'_{BL} - F_{BL}]$$

$$+ F(S_m) - (S_m - S_{wc}) F'(S_m)\}$$

Finally, an equation for S_m is obtained:

$$[F(S_m) - (S_m - S_{wc}) F'(S_m)] \frac{Qt}{A\phi} - \frac{(k\phi)^{1/2} A\sigma}{Q\mu_o} \int_{S_m}^{S_{wc}} \frac{(S - S_{wc})\, E(S)\, dS}{D(S, S_m, t)} = 0$$

This must be solved numerically. Then $S_m(t)$ is obtained. Finally, the solution takes the form:

$$x_{S_w}(t) = x_m(t) + \frac{\sigma A (k\phi)^{1/2}}{Q\mu_o} \int_{S_m}^{S_w} \frac{E(S)\, dS}{D(S, S_m, t)} \tag{75}$$

for $S < S_m$. As time increases, the matching value of saturation tends toward S_{BL}.

From Eq. (74), it can be seen that $D(S, S_m, t)$ is not zero for $S = S_{BL}$ because $S_\xi \neq S_{BL}$. From Eq. (73) it can also be verified that S_m is not a zero of $D(S, S_m, t)$. The difficulties encountered in the Terwilliger's solution have been eliminated in the Lefur' solution.

3.7.3. *The Capillary End Effect*

The influence of capillarity on the saturation distribution within the core was studied in the previous sections. If N_c [Eq. (66)] is small, capillary effects are negligible. However, we have seen that N_c cannot be small in the Buckley–Leverett front region where $\partial S_w / \partial x$ is extremely large. Similarly, N_c cannot be small in a region where $\partial p_c/\partial x$ is extremely large. This will be the case at the outlet end of the core where an abrupt change in capillary pressure takes place. Within the core a high capillary pressure prevails; outside of it, the magnitude of the pressure difference between the two phases is small as no porous medium is present. Prior to water breakthrough, only oil is flowing across the outlet face. Naturally, pressure in the oil phase is continuous across the outlet face during the primary phase of recovery. As water is about to reach the outlet face it encounters a capillary pressure barrier which at first prevents it from flowing out of the core. The water velocity is:

$$v_w = -\lambda_w \frac{\partial p_w}{\partial x} = -\lambda_w \left[\frac{\partial p_o}{\partial x} - \frac{\partial p_c}{\partial x} \right]$$

$$= -\lambda_w \frac{\partial p_o}{\partial x} + \lambda_w \frac{(p_c^- - p_c^+)}{\delta x} \tag{76}$$

where the $-$ and $+$ signs refer to conditions prevailing downstream and upstream from the outlet face.

For the low water saturations, the first ones to reach the outlet face, p_c is positive and large in magnitude. The pressure gradient in the oil phase is finite and consequently, within a small region of extent δx near the outlet face, the term $-(\lambda_w p_c^+)/\delta x$ tends to impose its sign. However, even when $\delta x \rightarrow 0$, the term $(\lambda_w p_c^+)/\delta x$ cannot make v_w negative because there is no water available to move into the core from the downstream side of the outlet face. Thus, for a small time, capillary pressure can at best prevent flow of water from the core, the time during which $\partial p_o/\partial x$ is finite and $(p_c^+ - p_c^-)/\delta x$ is very large. As a result of a decrease in water flow rate toward the end of the core, water accumulates. This in turn lessens the resistance to water outflow because p_c decreases rapidly with an increase in water saturation. Because the transition at the outlet is indeed very sharp, δx is practically zero. Even at the outlet face V_w cannot be negative. As a result, $\partial p_o/\partial x$ must be infinite at that end of the core. However, the oil flow rate at the outlet is finite. It follows that the oil saturation there must be residual.

Once water has broken through, the pressure gradient in the water phase is continuous across the outlet face. In particular, $\partial p_w/\partial x$ exists at the boundary and the oil velocity, which cannot be calculated directly from Darcy's law for that phase as $\lambda_o = 0$ and $\partial p_o/\partial x = \infty$, is $v_o = v - \lambda_w(\partial p_w/\partial x)$, where v, the total velocity, can be calculated at any cross section of the core, save at a close neighborhood of the outlet face.

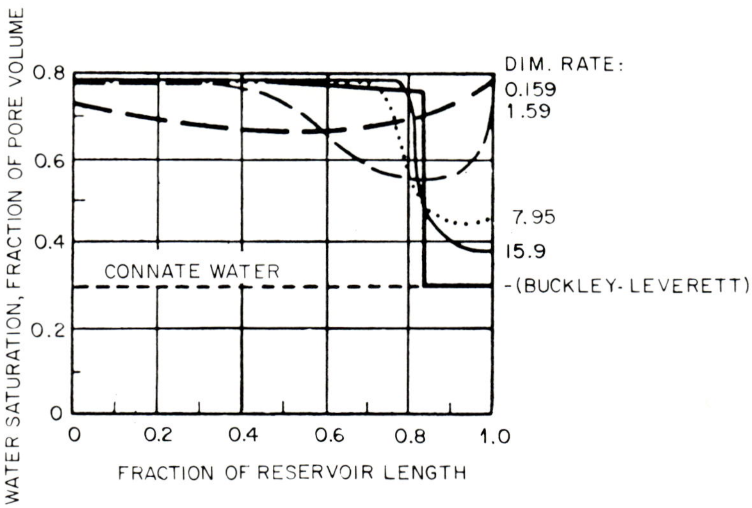

FIG. 32. Influence of capillary end effect on saturation distributions (Douglas *et al.*, 1958).

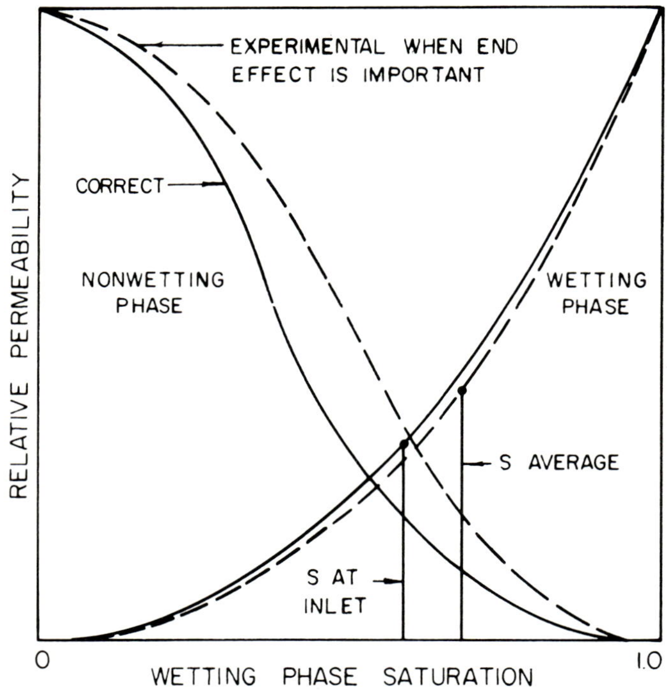

FIG. 33. Shifting of relative permeability curves when end effect is important (Hadley and Handy, 1956).

Due to the occurence of infinite pressure gradient, it is best to express the boundary conditions at the outlet face in terms of flow rate and saturation, when the saturation equation (68) is to be solved. As long as the water saturation at the outlet face has not reached its maximum, there is no water outflow: if $S_w < 1 - S_{or}$, $Q_w = 0$. As soon as water reaches the maximum saturation, water outflows and its saturation remain at the maximum value thereafter:

$$\text{if at } t, \quad S_w = 1 - S_{or}, \quad \text{then at } t' > t, \quad S_w = 1 - S_{or}$$

Douglas *et al.* (1958) used this boundary condition to solve Eq. (68) by finite-difference approximation. Their solutions for small rates of flow display the capillary influence both within the core and in the vicinity of the outlet face (Fig. 32). Hadley and Handy (1956) show the error that arises in the calculation of relative permeabilities by the steady-state method if capillary end effect is important (Fig. 33).

When proper care is exerted to minimize N_c in the laboratory experiments, the Welge method will indeed yield the proper value of the k_w/k_o ratio.

Nomenclature

f	Fractional flow function when both capillary and gravity terms are neglected
f_i, f_w	f for fluid i, for water
f^+, f^-	Fractional flow values on the two sides of a front
g	Acceleration of gravity
k	Formation permeability
k_g, k_o, k_w	Effective permeability to gas, oil, water
k_{rg}, k_{ro}, k_{rw}, k_{ri}	Relative permeability to gas, oil, water, fluid i
$\tilde{k}_{ro}$, $\tilde{k}_{rw}$	Relative permeability: to oil at connate water saturation; to water at residual oil saturation
k_i	Permeability of tube i
p	Pressure
p_c, p_c^i	Capillary pressure, in tube i
p_o, p_w	Oil pressure, Water pressure
$\mathbf{u}$, $\mathbf{u}_i$	Velocity of interface, in tube i
u, v	x and y components of velocity
v	Total flow rate per unit area or (simply) velocity
v_o, v_w	Flow rate of oil, water per unit area
x_o	Original location of saturation S
x_S	Location of saturation S at time t
A	Cross-sectional area
A_p	Area of pattern or characteristic area
A_s	Area swept by saturation S
E_D	Displacement efficiency, i.e., fraction of oil displaced over oil in place
E_R	Recovery efficiency or oil recovery in units of pore volume

F_{wo}	Instantaneous producing water–oil ratio
F, F_i, F_o, F_w	Fractional flow function when only capillary terms are neglected, for fluid i, for oil, for water
F^+, F^-	F on two sides of a front
F_{BL}	F at Buckley–Leverett saturation
$\mathscr{F}, \mathscr{F}_w$	Fractional flow function including capillary terms, for water
$J(S)$, $J_w(S_w)$	Leverett shape (J) function, for water in a water–oil system
L	Length of tube, of core, of layer, of reservoir
N	Total number of pores in a cross section
N_c	Dimensionless expression
Q	Total flow rate
Q_o, Q_w	Flow rate for oil, for water
Q^+, Q^-	Flow rates on two sides of a front
R	Radius of a capillary
$\bar{R}$	Average radius size
R_a	Radius of constriction
R_1, R_2	Radii of curvature
R, R'	Principal radii of curvature
Re	Reynolds number
R_i, R_m	Radius of tube i, of mean curvature
$S, \tilde{S}$	Saturation
S^*	Normalized saturation
S^+, S^-	Saturation on two sides of a front
S_m	Matching or minimum saturation
S_I	Saturation at an inflexion point of the fractional flow function
S_i, S_o, S_w	Saturation of fluid i, of oil, of water
S_{gi}, S_{oi}, S_{wi}	Initial gas, oil, water saturation
S_{or}	Residual oil saturation
S_{wc}	Connate water saturation
S_{BL}	Saturation behind Buckley–Leverett front
$\hat{S}(x, t)$	Buckley–Leverett solution
U	Unperturbed uniform stream velocity
V_f	Velocity of a front
V_p	Pore volume
V_{pi}	Pore volume injected
V_{ps}	Pore volume swept by saturation S
δ	Boundary-layer thickness
δn	Width of core, or layer, or streamtube
$\delta(R)$	Distribution of pores of radius between R and $R + dR$
ϵ	Smallness parameter
$\theta, \theta°$	Angle, angle in degrees
κ_o, κ_w	Relative permeability shape function
λ, λ_i	Mobility, of fluid i
μ, μ_w, μ_o	Viscosity, of water, of oil
ρ_i, ρ_o, ρ_w	Density of fluid i, of oil, of water
$\sigma, \sigma_{AS}, \sigma_{LA}, \sigma_{LS}$	Surface tension, at the air–solid, liquid–air, liquid–solid interface
τ	Tortuosity
ϕ	Porosity
Δp	Pressure drop
ΔS	Size of saturation front

Acknowledgments

The author's experience with multiphase flow in porous media was gained during his employment with Chevron Oil Field Research Company in La Habra, California.

The author wishes to express his gratitude to his colleagues, in particular, John C. Martin, Lyman L. Handy, Harley Y. Jennings, Kenneth C. G. Pritchard, James P. Brill, George R. Offen, and P. T. Woo for many useful discussions.

REFERENCES

Buckley, S. E., and Leverett, M. C. (1942). Mechanism of fluid displacement in sands, *Trans. AIME* **146**, 107–116.

Calhoun, J. C., Jr. (1953). "Fundamentals of Reservoir Engineering." Univ. of Oklahoma Press, Norman, Oklahoma.

Carrier, G. F. (1953). Boundary layer problems in applied mechanics, *Advan. Appl. Mech.* **3**, 1–18.

Douglas, J., Jr., Blair, P. M., and Wagner, R. J. (1958). Calculation of linear waterflood behavior including the effects of capillary pressure, *Petrol. Trans., AIME* **215**, 96.

Edmondson, T. A. (1965). The effect of temperature on water flooding, *J. Can. Petrol. Technol.*

Fatt, I., and Dykstra, H. (1951). Relative permeability studies, *Petrol. Trans., AIME* **192**, 249–255.

Hadley, G. F., and Handy, L. L. (1956). A theoretical and experimental study of the steady-state capillary end effect, *31st Ann. Fall Meeting*, Petrol. Branch, AIME, Los Angeles, California, October, 14–17.

Handy, L. L. (1960). Determination of effective capillary pressures for porous media from imbibition data, *Trans. AIME* **219**, 75–80.

Holm, L. W., Csaszar, A. K., and Bernard, G. G. (1965). Field test shows STP advantage in waterflood, *Oil Gas J.* March 1, 72.

Jennings, H. Y., Jr. (1957). Surface properties of natural and synthetic porous media, *Producers Monthly* **21**, 20.

Jennings, H. Y., Jr. (1966). Waterflood behavior of high viscosity crudes in preserved soft and unconsolidated cores, *J. Petrol. Technol.* Jan., 116–120.

Johnson, E. F., Bossler, D. P., and Naumann, V. O. (1959). Calculation of relative permeability from displacement experiments, *Petrol. Trans., AIME* **216**, 370.

Jones-Parra, J., and Calhoun, J. C., Jr. (1953). Computation of a linear flood by the stabilized zone method, *Petrol. Trans., AIME* **198**, 335–338.

Lefur, B. (1962). Influences de la Capillarité et de la Gravité sur le déplacement non miscible unidimensionel dans un milieu poreux, *J. Mécan.* **1**.

Leverett, M. C. (1941). Capillary behavior in porous solids, *Petrol. Trans., AIME* **142**, 152.

Long Beach Harbor Dept., Petrol. and Subsidence Control Division, Long Beach, California.

McMichael, F. C., and McKee, J. E. (1965). Research on waste water reclamation at Whittier Narrows, W. M. Keck Laboratory of Environmental Health Engineering, Calif. Inst. of Tech.

Muskat, M. (1946). "The Flow of Homogeneous Fluids Through Porous Media." Edwards, Ann Arbor, Michigan.

Muskat, M. (1949). "Physical Principles of Oil Production." McGraw-Hill, New York.

Oil Gas J. (1965). December 27, 83.

Pantaleo, A. J., and DePriester, C. L. (1963). Well stimulation by downhole gas-air burner, *J. Petrol. Technol.* December, 1297–1302.

Park, R. A. (1965). Pressure maintenance by waterflooding, North Virden Scallion Field, Manitoba, Paper presented at the 16th Ann. Tech. Meeting, *Can. Inst. Mining Met.*, Calgary, Alberta.

Petroleum facts and figures (1963). Am. Petroleum Inst., 1963.

Rapoport, L. A., and Leas, W. J. (1953). Properties of linear waterfloods, *Petrol. Trans., AIME* **198**, 139–148.

Roubaudi, C. (1948). "Traité de Géométrie Descriptive." Masson, Paris.

Saffman, P. G., and Taylor, G. I. (1958). The penetration of a fluid into a porous medium or Hele–Shaw cell containing a more viscous liquid, *Proc. Roy. Soc.* (*London*), *Ser. A* **245**, 312–329.

Sarem, A. M. (1965). Three-phase relative permeability by unsteady state method, paper SPE 1225, AIME 40th Ann. Fall Meeting, October 3–6, Denver, Colorado.

Scheidegger, A. E. (1959). "The Physics of Flow Through Porous Media," 2nd ed. Univ. of Toronto Press, Toronto.

Schlichting, H. (1955). "Boundary Layer Theory." McGraw-Hill, New York.

Secondary recovery of oil in the United States (1942). *API.*

Terwilliger, P. L., Wilsey, L. E., Hall, H. N., Bridges, P. M., and Morse, R. A. (1951). An experimental and theoretical investigation of gravity drainage performance, *Petrol. Trans., AIME* **192**, 292.

Welge, H. J. (1952). A simplified method for computing oil recovery by gas or water drive, *Trans. AIME* **195**, 91–98.

Wyllie, M. R. J. (1962). "Petroleum Production Handbook," (T. C. Frick, ed.). Vol. II, Chapt. 25. McGraw-Hill, New York.

Author Index

Numbers in italics refer to the pages on which the complete references are listed.

Subject Index

E

F

T

U

V